Communications in Computer and Information Science 793

Commenced Publication in 2007
Founding and Former Series Editors:
Alfredo Cuzzocrea, Xiaoyong Du, Orhun Kara, Ting Liu, Dominik Ślęzak,
and Xiaokang Yang

More information about this series at http://www.springer.com/series/7899

Vladimir Voevodin · Sergey Sobolev (Eds.)

Supercomputing

Third Russian Supercomputing Days, RuSCDays 2017
Moscow, Russia, September 25–26, 2017
Revised Selected Papers

 Springer

Editors
Vladimir Voevodin (iD)
Research Computing Center (RCC)
Moscow State University
Moscow
Russia

Sergey Sobolev (iD)
Moscow State University
Moscow
Russia

ISSN 1865-0929   ISSN 1865-0937 (electronic)
Communications in Computer and Information Science
ISBN 978-3-319-71254-3   ISBN 978-3-319-71255-0 (eBook)
https://doi.org/10.1007/978-3-319-71255-0

Library of Congress Control Number: 2017959610

Printed on acid-free paper

This Springer imprint is published by Springer Nature
The registered company is Springer International Publishing AG
The registered company address is: Gewerbestrasse 11, 6330 Cham, Switzerland

Preface

The Third Russian Supercomputing Days Conference (RuSCDays 2017) was held September 25–26, 2017, in Moscow, Russia. It was organized by the Supercomputing Consortium of Russian Universities and the Federal Agency for Scientific Organizations. The conference was supported by the Russian Foundation for Basic Research and our respected platinum educational partner (Intel), platinum sponsors (T-Platforms, NVIDIA, RSC), gold sponsor (Mellanox), and silver sponsors (Almaz-SP, Atos, Dell EMC, IBM, Xilinx). The conference was organized in a partnership with the ISC High-Performance conference series and the NESUS project.

The conference was born in 2015 as a union of several supercomputing events in Russia and quickly became one of the most notable Russian supercomputing meetings. The conference caters to the interests of a wide range of representatives from science, industry, business, education, government, and students – anyone connected to the development or the use of supercomputing technologies. The conference topics cover all aspects of supercomputing technologies: software and hardware design, solving large tasks, application of supercomputing technologies in industry, exaflops computing issues, supercomputing co-design technologies, supercomputing education, and others.

All papers submitted to the conference were reviewed by three referees in the first review round. The papers were evaluated according to the quality of relevance to the conference topics, scientific contribution, presentation, approbation, and related works description. After notification of the conditional acceptance of a paper, the second review round was arranged. It was aimed at the final polishing of papers and also at evaluating the authors work after the referees' comments. After the conference, the final selection was made, and the 42 best works were carefully selected to be included in this volume.

The proceedings editors would like to thank all conference committee members, especially the Organizing and Program Committee members as well as other referees and reviewers for their contributions. We also thank Springer for producing these high-quality proceedings of RuSCDays 2017.

October 2017

Vladimir Voevodin
Sergey Sobolev

Organization

The Third Russian Supercomputing Days Conference (RuSCDays 2017) was organized by the Supercomputing Consortium of Russian Universities and the Federal Agency for Scientific Organizations, Russia. The conference organization coordinator was Moscow State University Research Computing Center.

Steering Committee

V.A. Sadovnichiy (Chair)	Moscow State University, Russia
V.B. Betelin (Co-chair)	Russian Academy of Sciences, Moscow, Russia
A.V. Tikhonravov (Co-chair)	Moscow State University, Russia
J. Dongarra (Co-chair)	University of Tennessee, Knoxville, USA
A.I. Borovkov	Peter the Great Saint-Petersburg Polytechnic University, Russia
Vl.V. Voevodin	Moscow State University, Russia
V.P. Gergel	Lobachevsky State University of Nizhni Novgorod, Russia
G.S. Elizarov	NII Kvant, Moscow, Russia
V.V. Elagin	Hewlett Packard Enterprise, Moscow, Russia
A.K. Kim	MCST, Moscow, Russia
E.V. Kudryashova	Northern (Arctic) Federal University, Arkhangelsk, Russia
N.S. Mester	Intel, Moscow, Russia
E.I. Moiseev	Moscow State University, Russia
A.A. Moskovskiy	RSC Group, Moscow, Russia
V.Yu. Opanasenko	T-Platforms, Moscow, Russia
G.I. Savin	Joint Supercomputer Center, Russian Academy of Sciences, Moscow, Russia
V.A. Soyfer	Samara University, Russia
L.B. Sokolinskiy	South Ural State University, Chelyabinsk, Russia
I.A. Sokolov	Russian Academy of Sciences, Moscow, Russia
R.G. Strongin	Lobachevsky State University of Nizhni Novgorod, Russia
A.N. Tomilin	Institute for System Programming of the Russian Academy of Sciences, Moscow, Russia
A.R. Khokhlov	Moscow State University, Russia
B.N. Chetverushkin	Keldysh Institutes of Applied Mathematics, Russian Academy of Sciences, Moscow, Russia
E.V. Chuprunov	Lobachevsky State University of Nizhni Novgorod, Russia
A.L. Shestakov	South Ural State University, Chelyabinsk, Russia

Program Committee

Vl.V. Voevodin (Chair)	Moscow State University, Russia
R.M. Shagaliev (Co-chair)	Russian Federal Nuclear Center, Sarov, Russia
M.V. Yakobovskiy (Co-chair)	Keldysh Institutes of Applied Mathematics, Russian Academy of Sciences, Moscow, Russia
T. Sterling (Co-chair)	Indiana University, Bloomington, USA
A.I. Avetisyan	Institute for System Programming of the Russian Academy of Sciences, Moscow, Russia
D. Bader	Georgia Institute of Technology, Atlanta, USA
P. Balaji	Argonne National Laboratory, USA
M.R. Biktimirov	Russian Academy of Sciences, Moscow, Russia
A.V. Bukhanovskiy	ITMO University, Saint Petersburg, Russia
J. Carretero	University Carlos III of Madrid, Spain
V.E. Velikhov	National Research Center Kurchatov Institute, Moscow, Russia
V.Yu. Volkonskiy	MCST, Moscow, Russia
V.M. Volokhov	Institute of Problems of Chemical Physics of Russian Academy of Sciences, Chernogolovka, Russia
R.K. Gazizov	Ufa State Aviation Technical University, Russia
B.M. Glinskiy	Institute of Computational Mathematics and Mathematical Geophysics, Siberian Branch of Russian Academy of Sciences, Novosibirsk, Russia
V.M. Goloviznin	Moscow State University, Russia
V.A. Ilyin	National Research Center Kurchatov Institute, Moscow, Russia
I.A. Kalyaev	NII MVS, South Federal University, Taganrog, Russia
H. Kobayashi	Tohoku University, Japan
V.V. Korenkov	Joint Institute for Nuclear Research, Dubna, Russia
V.A. Kryukov	Keldysh Institutes of Applied Mathematics, Russian Academy of Sciences, Moscow, Russia
J. Kunkel	University of Hamburg, Germany
J. Labarta	Barcelona Supercomputing Center, Spain
A. Lastovetsky	University College Dublin, Ireland
M.P. Lobachev	Krylov State Research Centre, Saint Petersburg, Russia
Y. Lu	National University of Defense Technology, Changsha, Hunan, China
T. Ludwig	German Climate Computing Center, Hamburg, Germany
V.N. Lykosov	Institute of Numerical Mathematics, Russian Academy of Sciences, Moscow, Russia
M. Michalewicz	University of Warsaw, Poland
L. Mirtaheri	Kharazmi University, Tehran, Iran

A.V. Nemukhin	Moscow State University, Russia
G.V. Osipov	Lobachevsky State University of Nizhni Novgorod, Russia
A.V. Semyanov	Lobachevsky State University of Nizhni Novgorod, Russia
Ya.D. Sergeev	Lobachevsky State University of Nizhni Novgorod, Russia
H. Sithole	Centre for High Performance Computing, Cape Town, South Africa
A.V. Smirnov	Moscow State University, Russia
R.G. Strongin	Lobachevsky State University of Nizhni Novgorod, Russia
H. Takizawa	Tohoku University, Japan
M. Taufer	University of Delaware, Newark, USA
H. Torsten	ETH Zurich, Switzerland
V.E. Turlapov	Lobachevsky State University of Nizhni Novgorod, Russia
E.E. Tyrtyshnikov	Institute of Numerical Mathematics, Russian Academy of Sciences, Moscow, Russia
V.A. Fursov	Samara University, Russia
L.E. Khaymina	Northern (Arctic) Federal University, Arkhangelsk, Russia
B.M. Shabanov	Joint Supercomputer Center, Russian Academy of Sciences, Moscow, Russia
N.N. Shabrov	Peter the Great Saint-Petersburg Polytechnic University, Russia
L.N. Shchur	Higher School of Economics, Moscow, Russia
R. Wyrzykowski	Czestochowa University of Technology, Poland
M. Yokokawa	Kobe University, Japan

Industrial Committee

A.A. Aksenov (Co-chair)	Tesis, Moscow, Russia
V.E. Velikhov (Co-chair)	National Research Center Kurchatov Institute, Moscow, Russia
V.Yu. Opanasenko (Co-chair)	T-Platforms, Moscow, Russia
Yu.Ya. Boldyrev	Peter the Great Saint-Petersburg Polytechnic University, Russia
M.A. Bolshukhin	Afrikantov Experimental Design Bureau for Mechanical Engineering, Nizhny Novgorod, Russia
R.K. Gazizov	Ufa State Aviation Technical University, Russia
M.P. Lobachev	Krylov State Research Centre, Saint Petersburg, Russia
V.Ya. Modorskiy	Perm National Research Polytechnic University, Russia

A.P. Skibin	Gidropress, Podolsk, Russia
S. Stoyanov	T-Services, Moscow, Russia
N.N. Shabrov	Peter the Great Saint-Petersburg Polytechnic University, Russia
A.B. Shmelev	RSC Group, Moscow, Russia
S.V. Strizhak	Hewlett-Packard, Moscow, Russia

Educational Committee

V.P. Gergel (Co-chair)	Lobachevsky State University of Nizhni Novgorod, Russia
Vl.V. Voevodin (Co-chair)	Moscow State University, Russia
L.B. Sokolinskiy (Co-chair)	South Ural State University, Chelyabinsk, Russia
Yu.Ya. Boldyrev	Peter the Great Saint-Petersburg Polytechnic University, Russia
A.V. Bukhanovskiy	ITMO University, Saint Petersburg, Russia
R.K. Gazizov	Ufa State Aviation Technical University, Russia
S.A. Ivanov	Hewlett-Packard, Moscow, Russia
I.B. Meerov	Lobachevsky State University of Nizhni Novgorod, Russia
V.Ya. Modorskiy	Perm National Research Polytechnic University, Russia
I.O. Odintsov	RSC Group, Saint Petersburg, Russia
N.N. Popova	Moscow State University, Russia
O.A. Yufryakova	Northern (Arctic) Federal University, Arkhangelsk, Russia

Organizing Committee

Vl.V. Voevodin (Chair)	Moscow State University, Russia
V.P. Gergel (Co-chair)	Lobachevsky State University of Nizhni Novgorod, Russia
V.Yu. Opanasenko (Co-chair)	T-Platforms, Moscow, Russia
B.M. Shabanov (Co-chair)	Joint Supercomputer Center, Russian Academy of Sciences, Moscow, Russia
S.I. Sobolev (Scientific Secretary)	Moscow State University, Russia
A.A. Aksenov	Tesis, Moscow, Russia
A.P. Antonova	Moscow State University, Russia
K.A. Barkalov	Lobachevsky State University of Nizhni Novgorod, Russia
M.R. Biktimirov	Russian Academy of Sciences, Moscow, Russia
O.A. Gorbachev	RSC Group, Moscow, Russia
V.A. Grishagin	Lobachevsky State University of Nizhni Novgorod, Russia
V.V. Korenkov	Joint Institute for Nuclear Research, Dubna, Russia

I.B. Meerov	Lobachevsky State University of Nizhni Novgorod, Russia
I.M. Nikolskiy	Moscow State University, Russia
N.N. Popova	Moscow State University, Russia
N.M. Rudenko	Moscow State University, Russia
L.B. Sokolinskiy	South Ural State University, Chelyabinsk, Russia
V.M. Stepanenko	Moscow State University, Russia
A.V. Tikhonravov	Moscow State University, Russia
A.Yu. Chernyavskiy	Moscow State University, Russia
M.V. Yakobovskiy	Keldysh Institutes of Applied Mathematics, Russian Academy of Sciences, Moscow, Russia

Russian
Supercomputing
Days 2017

Contents

High Performance Architectures, Tools and Technologies

Parallel Algorithms

Parallel Numerical Methods Course for Future Scientists and Engineers

Iosif Meyerov, Sergey Bastrakov, Konstantin Barkalov, Alexander Sysoyev,
and Victor Gergel[✉]

Lobachevsky State University of Nizhni Novgorod, Nizhni Novgorod, Russia
{iosif.meyerov,sergey.bastrakov,konstantin.barkalov,
alexander.sysoyev,victor.gergel}@itmm.unn.ru

Abstract. The rise of computational science has facilitated rapid progress in many areas of science and technology over the last decade. There is a growing demand in computational scientists and engineers capable of efficient collaboration in interdisciplinary groups. Training such specialists includes courses on numerical analysis and parallel computing. In this paper we present a new Master's course Parallel Numerical Methods which bridges the gap between theoretical aspects of numerical methods and issues of implementation for modern multicore and manycore systems. The course aims to guide students through the complete process of solving computational problems, from a problem statement to developing parallel software and analyzing results of computational experiments. An important feature is that many of practical classes are based on research done at the HPC Center of the University of Nizhni Novgorod and therefore illustrate issues, which students may encounter in their research and future career.

Keywords: Education in computational science · Numerical analysis · Parallel computing · Master's program

1 Introduction

The importance and relevance of modern methods of computational science can hardly be overestimated. The progress in development of computer systems and applications for solving scientific and technical problems confronts more and more new ambitious challenges to scientists and engineers. In many fields there is a demand for non-ordinary solutions which allow replacing natural experiments with computational ones, therefore essentially shortening the way from an innovative idea to its technological implementation. Among such fields are computer-aided design, computational physics, computational biomedicine and others. These areas can greatly benefit from collaboration of experts in different areas: researchers in natural and social sciences, theoretical and applied, mathematicians, and software engineers. However, efficient collaboration in such multidisciplinary groups is not always easy, as different professional communities tend to have specific traditions, methods and terminology.

V. Voevodin and S. Sobolev (Eds.): RuSCDays 2017, CCIS 793, pp. 3–13, 2017.
https://doi.org/10.1007/978-3-319-71255-0_1

A way to approach this issue is to train specialists oriented towards multidisciplinary collaboration as a part of Master's programs. The institute of IT, mathematics and mechanics at the Lobachevsky State University of Nizhni Novgorod (UNN) has created a Master's program in computational science, which includes a wide range of topics concerning numerical simulation, applied mathematics, computational mathematics, and computer science. Many students on this program are members of multidisciplinary groups carrying out research projects at the UNN HPC center [1]. By the time of graduation these students have some real-world experience in computational science, which can be valuable for their career.

This paper describes a core course in our Master's program in computational science, Parallel Numerical Methods. To date, a considerable amount of educational and methodical literature on numerical methods is available, for example, the latest editions of the classical textbooks [2–4]. In the literature on numerical methods, the issues of development, application and theoretical substantiation of algorithms for numerical solution of various classes of mathematical problems are considered. Courses on theoretical aspects of numerical methods have been developed for decades with lots of excellent courses and materials available. Parallel programming, performance analysis and optimization are much more rapidly developing areas. Evolution of hardware, tools and technologies constantly creates new challenges and requires development and modernization of course materials. There are respectable textbooks on key technologies for parallel programming, for example, [5, 6]. Some books consider optimization of applications from various areas for modern architectures [7–9]. Our Parallel Numerical Methods course aims to guide students through the complete process of solving computational problems, from a problem statement to developing parallel software and analyzing results of computational experiments. The course forms skills in studying a problem at hand and its mathematical model, choosing appropriate numerical methods, developing a parallel algorithm and its implementation for multicore and manycore systems, performing computational experiments and analyzing results in terms of accuracy and performance. An important feature of the course is that most problems considered are based on the experience from research projects done at the UNN HPC Center. These examples illustrate the common issues, which students are likely to encounter in their future career.

This paper is organized as follows. Section 2 contains a short overview of courses on numerical analysis and numerical methods. Section 3 presents the main ideas and principles of our Parallel Numerical Methods course. Course structure is described in Sect. 4 with examples of lectures and practical classes given in Sect. 5. Section 6 is devoted to assessment of student performance. Section 7 concludes the paper.

2   Related Work

Courses on numerical analysis and numerical methods, for example [10–14] are delivered in many universities worldwide. The textbooks with several editions released, including [15, 16], form a methodical basis for such courses. In general, these are mostly classical courses on numerical methods with the main focus on theoretical material: theorems on approximation, stability and convergence.

There are also courses which cover the classical topics of numerical methods and are directed particularly towards the issues of implementation for modern computational systems. A notable example is the Introduction to Numerical Methods course at MIT [10]. The course begins with considering the issues of performance, software optimization, and floating-point arithmetic. Then, the basic numerical algorithms of linear algebra (solving the eigenvalue problem, direct and iterative methods for solving systems of linear equations) are considered. There are several advanced courses concerning parallel aspects of numerical algorithms, most notably in linear algebra [17, 18]. Linear algebra problems are rather intuitive, and, therefore, very suitable to demonstrate the basics of parallel computing. Other numerical methods are typically part of courses on scientific computing, for example [19–21].

This paper presents the Parallel Numerical Methods course developed at the UNN HPC Center based on 15 years' experience of research in computational science. The course covers numerical methods and issues of parallel implementation for a wide range of problems: dense and sparse linear algebra, direct and iterative solvers, finite-difference schemes for ordinary and partial differential equations, Monte Carlo methods.

3   Course Description

The Parallel Numerical Methods course described in this paper is a core course of the Master's program in computational science at the Lobachevsky State University of Nizhni Novgorod. The goals of the course are mastery of numerical algorithms and considering the issues of implementation, performance and scalability on modern hardware. The course covers parallel aspects of the classical topics of numerical methods, including dense and sparse linear algebra, ordinary and partial differential equations, Monte Carlo methods.

Course prerequisites include fundamentals of linear algebra, mathematical analysis, numerical methods, and parallel programming. This set of skills is rather typical for graduates of Bachelor's programs in applied mathematics and computer science, such as [22]. Since some students with a solid mathematical background may not be familiar with parallel programming, for example, Bachelor's in mathematics, our curriculum offers an optional parallel programming course in the same semester, which completely covers demands of the Parallel Numerical Methods course.

The course is based on the following main principles:

1. *A wide range of topics*: the course covers basic topics of numerical methods, widely used for scientific and engineering computing in various areas.
2. *Balance between numerical analysis and computer science*: the course combines mathematically strict presentation of material with proper attention to efficient implementation for parallel hardware.
3. *Integrity*: the course demonstrates the whole chain of stages required to solve a computational problem (problem statement, mathematical model, serial algorithm,

parallel algorithms, implementation and parallelization, computational experiment and analysis).

4. *Real-world experience*: demonstrating approaches used by research groups to solve state-of-the-art problems of computational science.

5. *Assessment based on applications*: assessment of student performance is done based mostly on ability to solve a problem going through all stages, from problem statement to computational experiment and analysis.

6. *Flexibility*: the course is designed in such a way that modules/practical classes are to a large degree independent.

Based on the above mentioned principles and course prerequisite we have decided to give basic mathematical statements and theorems in the lectures without proofs, making references to textbooks on numerical methods. Most lectures combine theoretical descriptions of methods with approaches to parallel implementation and demonstrations of performance results. Each practical class is a detailed study and development of a parallel implementation for a computational problem, using tools part of Intel Parallel Studio (C++ Compiler, Cilk Plus, TBB, MKL, Amplifier). The course contains a large number of case studies demonstrating applications from computational physics, computational finance, computational biology, and other areas.

4   Course Outline

Below we give a list of basic modules of the course with brief descriptions.

1. *Elements of computer arithmetic*. The topic of this module is representation of floating point numbers in computer memory [23]. The problems of computational error accumulation and methods for its reduction and control are discussed. Typical examples, where error accumulation may result in incorrect computation results, are presented.

2. *Direct methods for solving systems of linear equations*. This module is devoted to direct methods of solving systems of linear algebraic equations: Gaussian elimination, Cholesky decomposition, Thomas and reduction methods. The classical methods are presented and estimates of complexity given. We demonstrate insufficient efficiency of naïve implementations of these methods on modern computational architectures. The idea of block data processing is highlighted consistently. The problems of sparse algebra are considered here as well. A brief review of the data structures for storing sparse matrices is given, typical problems arising when performing the basic operations with sparse matrices are considered. Comparison of the matrix-vector and matrix-matrix multiplication algorithms for the cases of dense and sparse matrices is given. Cholesky decomposition is considered as an example of a more complex computational algorithm for sparse matrices. The issue of increasing amount of nonzero elements after factorization is demonstrated, several algorithms of matrix reordering to reduce the fill-in of the resulting matrix (minimum degree and nested dissection methods) are presented.

3. *Iterative methods for solving systems of linear equations*. This module considers iterative methods for solving the systems of linear equations, from the basic methods (simple iteration, Jacobi, Seidel, upper relaxation methods) to Krylov-type methods (generalized minimal residual, conjugated and biconjugated gradient methods). We discuss approaches to parallelization, give theoretical and experimental estimates of speed-up. The module also covers some methods of preconditioning: the basic methods (Jacobi method, Gauss-Seidel method) and the methods based on the incomplete LU-decomposition (ILU(0) and ILU(p) factorization).

4. *Methods for solving ordinary differential equations*. This module concerns the basic methods for solving ODEs: Runge-Kutta methods and Adams methods. The parallel variants of the methods for solving systems of ODEs are considered. For Runge-Kutta methods, the pipelining scheme of solving systems of ODEs with a sparse right-hand part is given. Solving a system of ODEs arising from simulating a neural system is considered as an illustrative example.

5. *Methods for solving differential equations in partial derivatives*. The module encompasses the issues of parallel solving differential equations in partial derivatives. Typical equations in partial derivatives (of hyperbolic, parabolic, and elliptical types) are considered. The finite differences method is delivered to the students as a method of reduction of differential equations to algebraic ones, leading to solving the difference equations. The explicit and implicit schemes of solving parabolic and hyperbolic equations and issues of parallel implementation are considered. The advantages and drawbacks of each approach are discussed. The pentadiagonal system of linear equations arising while solving 2D Poisson equation is discussed separately. The wave scheme of data processing in parallel solving of this system by iterative methods is presented.

6. *Monte Carlo methods*. This module introduces general concepts of the Monte Carlo methods. It describes issues of utilizing pseudo-random number generators in parallel programs, ways of reducing variance and presents applications for multidimensional integration, computational physics, and computational finance.

5 Conducting the Classes

The lecture part of the course concerns construction and analysis of efficient parallel algorithms from various topics of numerical methods. The presentation is accompanied by the results of the computational experiments and analysis. For example, a lecture on Cholesky factorization of a dense matrix is organized as follows. We start with the definition of Cholesky factorization and describe applications for solving systems of linear equations with a symmetric positive-definite matrix. Then we show how a naive algorithm can be constructed based on the definition and estimate its complexity. Approaches to parallelization are considered and scaling efficiency obtained for our implementation is demonstrated. We proceed to estimating cache efficiency of the naive algorithm and introducing the idea of blocking to increase cache reuse. Serial and parallel block Cholesky factorization algorithms are presented along with performance and scaling efficiency of our implementation compared to the naive algorithm. Analysis of the results concludes the lecture.

Another section of the course covers sparse linear algebra algorithms. One of the lectures considers sparse direct solvers. We discuss advantages and disadvantages of direct methods, give a general computational scheme, review main approaches to parallelization of sparse matrix factorization, and introduce widely used software. The demonstration is done using the open source sparse matrix reordering library PMORSy [24] developed at the UNN HPC Center. The example of workload distribution during a sparse matrix reordering is shown below (Fig. 1).

Fig. 1. Task mapping for a test matrix on 16 threads. Logical tasks are nodes of the graph, dependencies between them are edges. Descendant nodes correspond to the tasks generated after the parent task is completed. Same colored nodes are processed by the same thread [24].

Each practical class is a study of a selected computational problem. A problem description includes a problem statement, brief information on the research area, numerical method, possible approaches of parallelization, analysis of correctness, performance and scaling efficiency, and possible ways to improve it. The class is conducted either in form of a demonstration and analysis done by a teacher or in form of students gradually developing and analyzing their implementation following the description.

Let us describe several practical classes, which are part of the course. One group of classes is based on research done by a group of mathematicians and computer scientists on computational finance. A feature of this area is that problems are often seemingly simple; however the models and methods used are rather complicated and rely on statistics, differential equations and mathematical optimization. Nevertheless, all formalisms used have a clear financial interpretation, which makes it easier to introduce financial terms while keeping the material mathematically strict. Some of the methods used in computational finance can also be applied for other areas. Below we describe two concrete examples of this group.

The first example is performance optimization of Black-Scholes pricing. The problem is to calculate the fair prices for a set of European options [25], which is a fairly simple problem of financial mathematics. In this case, the result can be calculated analytically. From the programming point of view, this is a trivial problem (just to apply a formula for input data); however, it demonstrates that the computational time can vary by an order of magnitude even in such a simple program depending on programming and optimization skills and techniques. First, we introduce a model and basic concepts of a financial market and some intuitive descriptions of the option pricing problem briefly. We create a basic implementation, analyze its performance and improve it in a step-by-step fashion: eliminate unnecessary type casts, carry out invariants, perform mathematical transforms that replace heavy math routines with the lighter ones, vectorize and parallelize, perform warm-up to reduce overhead on thread creation, try reducing precision of floating-point operations, utilize streaming stores. The effects of these optimization techniques are demonstrated on both CPU and Intel Xeon Phi. The main methodological direction of this work is to teach pragmatics of using mathematical routines (choosing efficient mathematical library, controlled reduction of precision if justified), vectorization by compiler directives and optimization for manycore architectures. The detailed description of this work is published in [25].

The second example on computational finance is performance optimization of Monte Carlo option pricing (Fig. 2). We consider the case where the fair prices cannot be computed analytically. A widely used method is Monte Carlo simulation, which is relatively easy to implement and has a huge degree of parallelism. We cover topics of correct pseudo-random number generation in parallel applications and demonstrate typical errors in this area. Efficiency of low-discrepancy sequences and approaches to parallel implementation are shown. We demonstrate methods to check accuracy of a Monte Carlo simulation. The main value for students is to learn how to correctly use pseudo-random number generators in parallel programs.

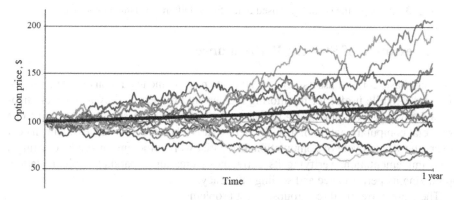

Fig. 2. Evolution of option price in time. Several Monte Carlo trajectories and the average are shown.

Another group of practical classes is devoted to computational physics. One example is based on a research project in plasma physics done by a large group of theoretical and experimental physicists, mathematicians and software developers from the UNN HPC Center, Institute of Applied Physics of Russian Academy of Sciences and Chalmers University of Technology. The example concerns solving Maxwell's equations in 3D space using the Finite Difference Time Domain method, a cell of the grid used is given at Fig. 3. We discuss choosing data layout, vectorization, scaling efficiency on Intel Xeon Phi. Another example is Monte Carlo simulation of brain sensing by optical diffuse spectroscopy based on a joint research by the UNN HPC Center and Institute of Applied Physics. We show problem statement and demonstrate results of a straightforward implementation of Monte Carlo simulation. Using a profiler, we show an approach to change data structures in order to improve memory efficiency and load balancing on Xeon Phi. The methodical value of these two examples is to demonstrate a pragmatic choice of data structures and approaches to load balancing on many-core architectures.

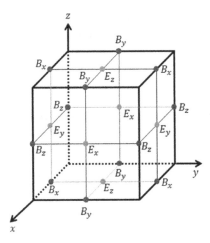

Fig. 3. A cell of the spatial grid used in the Finite Difference Time Domain method.

6   Assessment of Student Performance

As a very basic form of assessment, all students pass online testing on every module of the course. While useful for monitoring the current progress, it only focuses on theoretical knowledge, not practical skills. Thus, the main form of assessment is solving one or several computational problems. We believe it is a better form of assessment since it covers the whole cycle of computational scientist work: studying methods, creating a serial implementation, verifying its correctness, creating a parallel implementation, optimizing its performance and scaling efficiency.

There are currently three groups of test problems:

1. Block algorithms for dense linear algebra problems (e.g. block LU and Cholesky factorization).
2. Iterative solvers for sparse systems of linear equations (e.g. conjugate gradient method).

3. Solvers for ODEs and PDEs (e.g. finite-difference schemes).

Each student is randomly assigned one problem from each group to provide a good coverage of the course material.

For each problem we provide sets of parameters to be used for testing and requirements for performance and scaling efficiency. An implementation is accepted once it passes all tests in terms of correctness, performance and scaling efficiency.

On a technical side, we use an automated checking system based on the open source edge software (https://ejudge.ru/). Students upload source code files and a make file via a web interface. The system builds the submitted code, runs it on all test items, and checks correctness and performance. The specific way of checking the correctness depends on the type of a problem. For example, for the direct methods for solving systems of linear equations, the checking is performed by substitution (with a tolerance depending on a norm of the matrix); for the methods for solving differential equations, the accordance of behavior of the error when increasing the grid dimensionality to the theoretical properties of the methods is checked. The efficiency of parallel implementations is checked by means of comparing the speed-up relative to the sequential version with the threshold value dependent on the problem.

Let us present two examples of problem statements.

Example 1. *The conjugate gradient method for sparse systems of linear equations.*

Problem statement: Implement the conjugate gradient method for solving a sparse symmetric system of linear equations in form $Ax = b$. Choose an exact solution x^* and use Ax^* as the right-hand side. Use the norm of the difference between the consecutive approximations as the stop condition. Output the norm of the residual on the last step of the method.

Input format: a sparse symmetric matrix in .mtx format (from the University of Florida Sparse Matrix Collection).

Output format: a single number that is the norm of the residual on the last step of the method.

Verification: checking that the norm of the residual does not exceed 1% of the norm of the input matrix.

Limitations of the problem size: no more than 100 000 000 non-zero elements in the input matrix.

Requirements on scalability: the scaling efficiency is not less than 50%.

Example 2. *The Crank-Nicolson method for solving the 1D heat equation.*

Problem statement: Implement the Crank-Nicolson method for solving the 1D dynamic heat equation with the Dirichlet boundary conditions. Choose a non-trivial function as the exact solution and construct the right-hand side, initial and boundary conditions accordingly. Use the cyclic reduction method for solving the resulting system of linear equations with a tridiagonal matrix.

Input format: the number of grid nodes on space and time axes.

Output format: the maximum difference between the exact and numerical solutions on the grid.

Verification: steadily increasing the number of grid nodes checking that the error is proportional so the sum of squares of space and time steps.

Limitations of the problem size: the total number of grid nodes is no more than 1 000 000.

Requirements to scalability: the scaling efficiency is not less than 40%.

7   Conclusion

This paper describes the Parallel Numerical Methods course for Master's program in computational science at the Lobachevsky State University of Nizhni Novgorod. The main goal of the course is to bridge the gap between theoretical aspects of numerical methods and issues of implementation for modern multicore and manycore systems. This is an important chain in training specialists capable of working in multidisciplinary scientific and engineering groups. The course relies on basic knowledge of numerical methods and parallel programming obtained during Bachelor's programs and concentrates of parallelization and efficiency.

The course has a flexible modular structure. Each module is devoted to a key area of numerical methods. Most lectures demonstrate a whole cycle from a mathematical model to results of computational experiments in terms of accuracy and efficiency. Most practical classes are devoted to solving computational problems in different areas. An important feature is that many of practical classes are based on research done at the UNN HPC Center and therefore illustrate issues, which students may encounter in their research and future career. Assessment of student performance is mostly done based on solving test computational problems. By means of an automated system, we control accuracy of submitted solutions as well as performance and scaling efficiency.

Course materials are currently available in Russian on the website http://www.hpcc.unn.ru/?doc=491. These materials have been used for several training programs for teachers and researchers. Over 500 students have been trained since 2012. There is an ongoing process of extending the materials and translating them to English; a preliminary English version of materials for some modules is available at http://hpc-education.unn.ru/en/trainings/collection-of-courses. Another direction of future work is creating a course for one of the widely used e-learning systems.

References

1. Bastrakov, S., Meyerov, I., Gergel, V., Gonoskov, A., Gorshkov, A., Efimenko, E., et al.: High performance computing in biomedical applications. Procedia Comput. Sci. **18**, 10–19 (2013)
2. Stoer, J., Bulirsch, R.: Introduction to Numerical Analysis, vol. 12. Springer, Heidelberg (2013)
3. Mathews, J.H., Fink, K.D.: Numerical Methods Using MATLAB, vol. 31. Prentice hall, Upper Saddle River (1999)
4. Hamming, R.: Numerical Methods for Scientists and Engineers. Courier Corporation (2012)
5. Andrews, G.R.: Foundations of Parallel and Distributed Programming. Addison-Wesley Longman Publishing Co. Inc., Boston (1999)

6. Prasad, S.K., Gupta, A., Rosenberg, A.L., Sussman, A., Weems, C.C.: Topics in Parallel and Distributed Computing: Introducing Concurrency in Undergraduate Courses, 1st edn. Morgan Kaufmann, San Francisco (2015)
7. Jeffers, J., Reinders, J. (Eds.): High Performance Parallelism Pearls: Multicore and Many-core Programming Approaches, 1st edn. (2014)
8. Jeffers, J., Reinders, J., Sodani, A. (Eds.): Intel® Xeon Phi™ Processor High Performance Programming, Knights Landing Edition (2016)
9. Hwu, W.-M.W. (Ed.): GPU Computing Gems Jade Edition. Morgan Kaufmann (2011)
10. Introduction to numerical methods (2010). Accessed Jan 2017. MIT Open Courseware: http://ocw.mit.edu/courses/mathematics/18-335j-introduction-to-numerical-methods-fall-2010
11. Introduction to numerical analysis (2004). http://ocw.mit.edu/courses/mathematics/18-330-introduction-to-numerical-analysis-spring-2004
12. CME206 – Introduction to Numerical Methods for Engineering (2016). http://scpd.stanford.edu/search/publicCourseSearchDetails.do?method=load&courseId=11683
13. Math 128A: Numerical Analysis (2014). Accessed Jan 2017. http://persson.berkeley.edu/128A
14. Course MAT321 Numerical Methods (2014). Accessed Jan 2017. https://www.math.princeton.edu/undergraduate/course/MAT321
15. Burden, R., Faires, J.: Numerical Analysis, 9th edn. Brooks-Cole, Boston (2010)
16. Kincaid, D., Cheney, E.: Numerical Mathematics and Computing, 7th edn. Brooks-Cole, Boston (2012)
17. Demmel, J.: Matrix Computations/ Numerical Linear Algebra (2016). https://people.eecs.berkeley.edu/~demmel/ma221_Spr16
18. Saad, Y.: Computational Aspects of Matrix Theory, Sparse Matrix Computations (2015). http://www-users.cs.umn.edu/~saad/teaching.html
19. Dongarra, J.: Scientific Computing for Engineers: Spring 2012 (2012). http://www.netlib.org/utk/people/JackDongarra/WEB-PAGES/SPRING-2012/cs594-2012.htm
20. Heath, M.: Parallel numerical algorithms (2013). http://www.mat.unimi.it/users/pavarino/heath_2013
21. Edelman, A.: Numerical Computing with julia (2016). http://courses.csail.mit.edu/18.337/2016/calendar.html
22. Gergel, V., Liniov, A., Meyerov, I., Sysoyev, A.: NSF/IEEE-TCPP curriculum implementation at University of Nizhni Novgorod. In: Proceedings of Fourth NSF/TCPP Workshop on Parallel and Distributed Computing Education, pp. 1079–1084 (2014)
23. Muller, J.M., Brisebarre, N., De Dinechin, F.: Handbook of Floating-Point Arithmetic. Springer (2009)
24. Pirova, A., Meyerov, I., Kozinov, E., Lebedev, S.: PMORSy: parallel sparse matrix ordering software for fill-in minimization. Optim. Method. Softw. **32**, 274–289 (2016)
25. Meyerov, I., Sysoyev, A., Astafiev, N., Burylov, I.: Performance optimization of Black-Scholes pricing. In: Jeffers, J., Reinders, J. (Eds.) High Performance Parallelism Pearls: Multicore and Many-core Programming Approaches, pp. 319–340 (2014)

GPU Acceleration of Dense Matrix and Block Operations for Lanczos Method for Systems over Large Prime Finite Field

Nikolai Zamarashkin$^{(\boxtimes)}$ and Dmitry Zheltkov

INM RAS, Gubkina 8, Moscow, Russia
`nikolai.zamarashkin@gmail.com, dmitry.zheltkov@gmail.com`
`http://www.inm.ras.ru`

Abstract. GPU based acceleration of computations with dense matrices and blocks over large prime finite field are studied. Particular attention is paid to the following algorithms:

- multiplication of rectangular $N \times K$ blocks with $N \gg K$;
- multiplication of $N \times K$ blocks by square $K \times K$ matrices;
- LU-decomposition of matrices.

Several approaches for optimal use of GPU resources are proposed.

Efficiency analysis of implemented algorithms is provided for prime finite field with number of elements about 2^{512}, 2^{768}, 2^{1024} and GPUs of different computational performance and architecture generations. Numerical experiments prove efficiency of proposed solutions.

From numerical results it follows that GPU usage allows to accelerate block operations and to expand area of almost linear parallel scalability of Lanczos method implementation by INM RAS. Moreover, a sparse system of size about 2 millions, with 82 average nonzero elements per row, over field with about 2^{512} elements, on 128 nodes of Lomonosov supercomputer will be solved 2 times faster in case of GPUs used.

Keywords: GPGPU · RSA · Large prime finite field · Block Lanczos method

1 Introduction

This research is the result of analysis made for implementation of the improved block Lanczos method for the linear systems over large prime finite field (see, [6]).

Until recently, large data exchanges were considered as the main reason for poor scalability of block Lanczos method implementations on powerful computing systems [3–5,7]. Moreover, in case of low speed communication network the acceleration noticeably deviated from linear for the number of nodes about 100.

The basic idea of [6] was the efficient way of data storage. Thanks to this, the data exchange is significantly reduced, and is *perfectly scalable for block size K*. As a result, in the improved implementation the time for data exchange turned out to be less than the time for operations with dense matrices and blocks, which

V. Voevodin and S. Sobolev (Eds.): RuSCDays 2017, CCIS 793, pp. 14–26, 2017.
https://doi.org/10.1007/978-3-319-71255-0_2

does not depend on the block size K. As a matter of fact, the larger the block size K the greater the difference.

Thus, it is the computations with dense matrices and blocks that determine the limit of linear scalability for the improved implementation of block Lanczos method. Namely, while the time for symmetrized sparse matrix by the block multiplication significantly exceeds the time for operations with dense matrices and blocks, the parallel properties of the method are almost perfect. In practice, this is valid up to a few hundreds or a thousand nodes.

In order to spread the ideal scalability even further (for example, up to 2^{13} nodes), we have to speed up the computation with blocks.

As modern computing nodes are multi-core systems, they are very advantageous for operations with dense matrices and blocks. The use of multicore accelerates these operations proportionally to the number of cores. However, the number of cores per node is usually limited (typical values about $8-16$). But for really hard problems this is not enough. The systems with much larger number of cores are needed. The most common example of such system are the graphic accelerators (GPUs).

This paper explores the possibility of the using GPUs for computations with dense matrices and blocks with elements from large prime fields. The examples are considered for the fields with the number of elements of order 2^{512}, 2^{768}, and 2^{1024}.

The choice of algorithms for efficient implementation is constricted due to restrictions on access to GPU resources and dependence of time on presence of branches in the program. For these reasons, we prefer simple algorithms with a regular structure. However, if possible we use the Winograd's idea to reduce the number of multiplications twice. This is important, since in large prime fields the complexity of multiplying greatly exceeds the complexity of addition and subtraction.

Our main purpose is to clarify the possibility of *significant acceleration (more than 10 times)* of calculation dense matrix by block product.

The applicability of this research is not just limited to calculations in the block Lanczos method. The same improvements can be useful for Thome type algorithm implementations [1,2].

2   GPU Acceleration of Operations with Dense Matrices and Blocks

2.1   Algorithms

The algorithms with simple structure are preferred for GPU accelerators since the limited resource management, and dependence of algorithm running time on the presence of branches in the program.

Let us assume that elements in the large prime field can be specified using 512, 768, or 1024 bits. Further we show that the field size (the number of elements in the field) can significantly affect the implementation efficiency.

We are interested in two particular cases:

1. implementation of block-by-block multiplication $X^T Y$ for $N \times K$ blocks X, Y;
2. implementation of block-by-matrix multiplication XU for $N \times K$ block X, and $K \times K$ matrix U;

We decided on "naive algorithm", as well as on an algorithm by Winograd, which reduces the number of multiplications twice.

In addition, we use Winograd idea for efficient LU decomposition, which is used in block Lanczos method for $K \times K$ matrix inversion. However, the total time for operations with $K \times K$ matrices is still significantly smaller than the time for operations with $N \times K$ blocks [6, 7]. So the improvement of LU decomposition via Winograd method is considered only as a theoretical result.

The Winograd method is based on the elementary equality for the elements of matrix $C = AB$. Assuming the number of columns and rows of A to be $2m$, we write

$$
\begin{aligned}
c_{ij} &= \sum_{k=1}^{2m} a_{ik} b_{kj} \\
&= \sum_{k=1}^{m} \left(a_{i,2k-1} + b_{2k,j} \right) \left(a_{i,2k} + b_{2k-1,j} \right) \\
&\quad - \sum_{k=1}^{m} a_{i,2k-1} a_{i,2k} - \sum_{k=1}^{m} b_{2k-1,j} b_{2k,j}.
\end{aligned}
\tag{1}
$$

The last two sums have low complexity and can be pre-calculated in advance. The main calculation corresponds to the sum (1). It is easy to see that the number of multiplications in this sum is 2 times less than the one in "naive algorithm".

While Winograd method for matrix product multiplication is well known, the similar technique for Gaussian elimination is not. Since a complete description of an algorithm would be unnecessarily cumbersome, we only give its main idea.

Consider a *strictly regular* matrix A of order N in the block form

$$
A = \begin{bmatrix} A_{11} & A_{12} \\ A_{21} & A_{22} \end{bmatrix},
\tag{2}
$$

with 2×2 block A_{11}. The first two steps of elimination can be written as

$$
A \to A - \begin{bmatrix} A_{11} \\ A_{21} \end{bmatrix} A_{11}^{-1} \begin{bmatrix} A_{11} & A_{12} \end{bmatrix} = \begin{bmatrix} 0 & 0 \\ 0 & A_{22}^{(1)} \end{bmatrix},
\tag{3}
$$

where the submatrix $A_{22}^{(1)}$ of A is used as a starting point for the next steps of Gaussian eliminations. Thus calculation of $A_{22}^{(1)}$ determines the complexity of the whole algorithm.

Indeed, let us transform (3) by removing A_{11}^{-1} from it. For this, we represent A_{11} using the strict regularity of A and $\mathcal{O}(1)$ multiplications as

$$
A_{11} = L_{11} U_{11},
\tag{4}
$$

with lower triangular 2×2 matrix L_{11} and upper-triangular 2×2 matrix U_{11}. Then

$$A \to A - \begin{bmatrix} L_{11} \\ A_{21}U_{11}^{-1} \end{bmatrix} \begin{bmatrix} U_{11} \ L_{11}^{-1}A_{12} \end{bmatrix} = A - \begin{bmatrix} L_{11} \\ \hat{A}_{21} \end{bmatrix} \begin{bmatrix} U_{11} \ \hat{A}_{12} \end{bmatrix}, \qquad (5)$$

Note that $\mathcal{O}(N)$ multiplications are enough to find $\hat{A}_{21}$ and $\hat{A}_{12}$.

Now we will show that it is possible to calculate $A_{22}^{(1)} = A_{22} - \hat{A}_{21}\hat{A}_{12}$ with just $(N-2)^2 + \mathcal{O}(N-2)$ multiplications. In order to do this consider a matrix product

$$C = \begin{bmatrix} A_1 \ A_2 \end{bmatrix} \begin{bmatrix} B_1 \\ B_2 \end{bmatrix}, \qquad (6)$$

where A_1, A_2 are columns, and B_1, B_2 are rows of order $N-2$ (take into account that the sizes of $\hat{A}_{21}$ and $\hat{A}_{12}$ are equal to $(n-2) \times 2$ and $2 \times (N-2)$, respectively). Then using Winograd technique we can write

$$C_j^i = (a_{i1} + b_{j2})(a_{i2} + b_{j1}) - a_{i1}a_{i2} - b_{j1}b_{j2}, \qquad (7)$$

where a_{i1}, a_{i2}, b_{j1}, and b_{j2} are components of the vectors A_1, A_2, B_1 and B_2, respectively. The statement about the number of multiplications for two steps of Gaussian eliminations directly follows from (7), and the general result follows from the induction on the matrix size.

2.2 Algoritm Mapping on GPU Architecture

"Naive Algorithm" for Matrix Multiplication. The organization of calculations is similar to the one proposed in [9]. Consider multiplication of $N \times M$ matrix A and $M \times K$ matrix B with the elements in a large prime field.

Suppose that the memory size of GPU is sufficient to store the matrices A, B and the resulting matrix $C = AB$. Let the matrix A be represented in the following row-block form

$$A = \begin{bmatrix} A^1 \\ A^2 \\ \dots \\ A^{\frac{M}{t}} \end{bmatrix}, \qquad (8)$$

and B in column-block form

$$B = \begin{bmatrix} B_1 \ B_2 \ \cdots \ B_{\frac{K}{t}} \end{bmatrix}, \qquad (9)$$

with parameter t denoting the block size (number of rows/columns).

Each executable block relates to calculation of $C_j^i = A^i B_j$. Since the submatrices C_j^i do not intersect, the operating results for different blocks are independent. The total number of executable blocks in the algorithm is $N_b = \frac{MK}{t^2}$.

Now let's turn to the threads inside the executable blocks. Each thread calculates a product of a row of A^i by a column of B_j, which corresponds to one

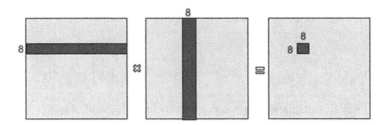

Fig. 1. Calculation in one executable block

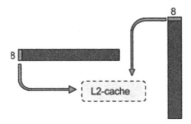

Fig. 2. "Naive" matrix multiplication algorithm: data loading for one executable block.

element in the submatrix C_j^i. Thus, the number of threads is equal to t^2 (Figs. 1 and 2).

As follows from the above, the number of executable blocks and the number of threads in blocks depend on the value t, namely with increasing t the number of blocks decreases, while the number of threads increases. We propose the following heuristic principle to obtain the optimal value of t:

the more blocks, the better.

Without going into details, we note that a large number of blocks has the following advantages:

1. More multiprocessors on GPU are filled (and more uniformly);
2. The scheduler can more effectively "hide" the time for data and instructions swapping.

But this rule is applicable only for reasonable value of t, as there are several GPU architecture limitations on number of threads per block and number of blocks for multiprocessor [10]. Formally, the maximum number of blocks is obtained with $t = 1$. But there are four objections to this choice.

First, the number of blocks is limited by the computational grid size of the particular GPU. This restriction, however, is not too strong. For example, it can be avoided by considering multiplication of smaller submatrices, such as parts of the rows of A, and parts of the columns of B.

Second limitation is the linear dependence of the number of downloads from global memory from t. Actually, due to high complexity of the calculations with long numbers the loading time does not have a decisive influence.

Third, maximal number of blocks per multiprocessor is limited, so with small number of threads per block total number of threads per multiprocessors would be less than maximal available (occupancy will be low). In this case multiprocessor would worse hide data and instruction fetch.

Fourth reason is provided by the condition that the number of threads in the block must be a multiple of 32 (the number of threads in a warp). Choosing $t = 1$, we use only one of the 32 threads that are allocated anyway. This is absolutely ineffective. Therefore, we must chose t such that t^2 is a multiple of 32 and the optimal t is $t = 8$ (the minimal t such that t^2 is a multiple of 32).

Note, that with $t = 8$ occupancy of multiprocessor is not limited by maximal number of blocks for the most modern GPU: each block uses 64 threads, maximal number of blocks is equal to 32. So, in this case block number limitation allows to use 2048 threads per multiprocessor and that is exactly limitation of thread number per multiprocessor. For older architectures such t limits occupancy by maximal number of blocks per multiprocessors, but for this architectures real limiter for occupancy would be number of used registers.

Consider executable block algorithm.

Algorithm 1. Multiplication of $N \times 8$ blocks. "Naive approach"

1. *Two 1×8 vectors are loaded from the device's memory in the shared memory (could be considered as equivalent of shared L2 cache of multicore CPU) of streaming multiprocessor (SM): one vector corresponds to the column of the row-block, and another is the row of the column-block;*
2. *Each of the 64 threads loads two numbers (elements of a large prime field) into the registers (Cache L1) of its stream processor (SP);*
3. *Each thread executes a product of its own numbers and sums it with the current value of the result;*
4. *Montgomery conversion is performed once at the end of all calculations; the necessary constants are loaded from the constant memory.*

Despite the simplicity of the Algorithm 1, the very possibility of its execution on a GPU is nontrivial. Let us consider the necessary resources for its execution. Each thread is associated with:

1. $2W + 1$ 32-bit registers for storing the result, where W is the number of 32-bit words necessary for storing elements of a prime field. For example, for a field with 512 bits per element, $W = 16$; for 768-bit field $W = 24$, and for 1024-bit field $W = 32$.
2. $2W$ registers for storing the inputs (i.e. elements of the corresponding row and column).
   Note that without loss of performance, we can store only one of the input numbers on registers, and load the second one word by word as needed. Therefore, only $W + 1$ registers are needed to store the inputs.

Thus, even by the most primitive calculations $3W + 2$ registers per an executable block are needed, that is:

512-bit field: not less than 50 registers;
768-bit field: not less than 74 registers;
1024-bit field: not less than 98 registers.

These elementary estimates show that GPUs with 63 registers per thread have a very limited applicability resource. The latter, of course, does not mean that it is impossible to organize calculations on such GPUs. One can certainly get an implementation for any large field by arranging calculations involving additional work with memory. But the aim of our research is to obtain *the maximum acceleration*. Therefore, we are primarily interested in situations without unnecessary obstacles to the most rapid implementation.

Remark 1. *We are interested in two types of block operations for the block Lanczos method: block-by-block multiplications in form $X^T Y$, and block-by-matrix multiplications in form XU (with $K \times K$ matrix U, and $N \times K$ blocks X and Y). Note that in applications the parameter N, is usually very large, but the block size K can be insignificant (for example, about 8). In this case, there are certain difficulties in choosing t. This is especially characteristic for $X^T Y$ calculation. Indeed, by the above scheme, for $t = 8$ we get only one executable block. And reducing t would result in inefficient use of threads.*
We can partially solve the problem by considering X and Y in a form

$$X = \begin{bmatrix} X_1 \\ X_2 \\ \dots \\ X_l \end{bmatrix}, \quad Y = \begin{bmatrix} Y_1 \\ Y_2 \\ \dots \\ Y_l \end{bmatrix}, \tag{10}$$

with the same number of rows in each block X_i, Y_j. Since in this case

$$X^T Y = \sum_{j=1}^{l} X_j^T Y_j, \tag{11}$$

one can consider computations of the form $X_j^T Y_j$ as executable blocks.
However, this solution is not perfect. The results of calculations for individual threads are not independent. In addition, it becomes necessary to synchronize the calculations. Both factors negatively affect the efficiency of computing $X^T Y$. We emphasize, that the problem arises only for small K and we are mostly interested in situations with K large. In this case the problem is not so critical.

Finally, due to the large number of registers in use, the number of threads on SM will be noticeably less than the maximum possible. For older architectures only 20 registers could be used to achieve full occupancy, for new – about 30. However, since the number of downloads is smaller than the number of calculations, only the instructions loading is worse compensated, which leads to uncritical decrease in performance.

For 512 bit numbers achieved occupancy is good enough — performance profiler show that instruction and data fetch are successfully hided and more than 90 percent of time is spent by arithmetic operations. Nevertheless, in case of fields with more than 512 bits per element for more optimal use of GPU resources, it is necessary to further consider the algorithms that use fewer registers, that is, multiplying long numbers in several stages.

Matrix Multiplication with Winograd Approach. The organization of matrix multiplication with Winograd approach is similar to the "naive algorithm" (see Sect. 2.2) (Fig. 3).

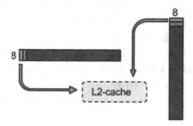

Fig. 3. Matrix multiplication algorithm with Winograd approach: data loading for one executable block

Analogously to the "naive algorithm" each executive block performs the calculation of 8×8 submatrix $C_j^i = A^i B_j$, and each thread calculates one element of C_j^i. The difference is that for Winograd approach a two columns of A^i and two rows of B_j are loaded to L2 and L1 Cache. This is necessary for the following

$$\text{elementary calculation} = (a_{i,2k-1} + b_{2k,j})(a_{i,2k} + b_{2k-1,j}). \tag{12}$$

Below we describe the main ideas of the algorithm for GPU.

Algorithm 2. Multiplication of $N \times 8$ blocks with Winograd approach

1. *Two 2×8 blocks are loaded from the device's memory in the shared memory (could be considered as equivalent of shared L2 cache of multicore CPU) of streaming multiprocessor (SM) : one block corresponds to the column of the row-block, and another is the row of the column-block;*
2. *Each of the 64 threads loads 4 numbers (elements of a large prime field) into the registers (Cache L1);*
3. *Each thread executes (12) for its own 4 numbers and sums it with the current value of the result;*
4. *Montgomery conversion is performed once at the end of all calculations; the necessary constants are loaded from the constant memory.*

Let us consider the necessary resources for the Algorithm 2 execution. Each executable thread is associated with:

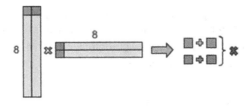

Fig. 4. Calculations in matrix multiplication algorithm with Winograd approach.

1. $2W + 1$ 32-bit registers for storing the result. Recall that W is the number 32-bit words required per element of the large field.
2. $4W$ registers for storing the inputs.
   Analogously to the "naive algorithm", only $2W + 2$ registers are enough to store the inputs without loss of performance. Moreover we can reduce the number of input data registers to $W + 3$ by a slight increase of the operations number (not more than $3W$ extra additions for one multiplication of numbers) (Fig. 4).

However, the most estimate gives $4W + 2$ registers per an executable block, that is:

512-bit field: not less than 67 registers;
768-bit field: not less than 99 registers;
1024-bit field: not less than 131 registers.

Note that Winograd algorithm requires a larger number of registers, so the need of economical algorithms for it is more critical.

2.3 Important Realization Details

An important feature of GPU is the instruction *madc* in the pseudo-assembler (CUDA PTX). This instruction multiplies two numbers with obtaining the first or the last word of the result, and adding it to the third number, taking into account the carry flag. Also it can change the carry flag in case of overflow. This makes it easy to implement the arithmetic with numbers from large prime finite fields [8].

Note that for architectures of the second and third generations, the instruction for 32-bit numbers is translated into assembler instruction which is slower than 32-bit instructions with floating-point numbers (2 – 3 times for the second generation, and 6 times for the third one). For the newer architectures, it is translated into a set of 16-bit instructions that are executed with the same speed as 32-bit floating-point instructions. Thus, in general, the instruction is executed 4 times slower than the instructions with a floating point numbers. And the peak performance of calculations with long numbers is 2 – 6 times lower than the one for floating-point numbers.

However, the performance of CPUs for this task is also far from peak:

- due to the lack of integer instruction for simultaneous multiplication and addition;
- due to the lack of a vector instructions for addition with a carry flag, and for multiplication with obtaining the major word.

As a result, for modern CPUs performance for long arithmetic is 8 - 32 times below the peak performance for 32-bit floating point numbers. Thus, theoretically, in case of such tasks GPU should be so many times faster than CPU, as GPU single-precision peak performance is higher than CPU one. Thus, theoretically, in case of such tasks the gain of GPU performance to CPU performance coincides to the proportion of the peak performances for single-precision tasks.

An important task for the GPU programming is to get rid of branches. Since all threads within a group (warp) must perform the same instruction, branching (with threads executing different branches) is converted to a sequential code, where each thread executes all branches. This leads to more registers and slower execution.

For matrix multiplication over a large prime field, such branching occur only at the stage of reduction, and can significantly affect the performance only for small block sizes. These branching compare two long numbers, and subtract if the first one is greater. However, it is easy enough to get rid of it. To do this, we subtract the second number from the first one, and then add the second number multiplied by the carry flag occurred in the subtraction.

2.4   Numerical Experiments

We compare results for CPUs and GPUs of different generations on the following problems: $2^{21} \times K$ block by $K \times K$ matrix multiplication (with $K = 8, 16$), and multiplication of square matrices of order 1024.

We use implementations of Winograd approach and Strassen method for CPU (Strassen only for square matrix multiplication), and "naive" implementation and Winograd method for GPU.

The experiments were performed on the following devices (note, that due to frequency boost technologies peak performance of the newest hardware is given approximately):

- 4-core CPU Intel Core i5-4440, 3.1 GHz, power consumption 84W.
  It is CPU of 4th generation of Intel Core microarchitecture. From that generation (and till the latest available at the moment) CPU core could execute per clock 2 fused multiply-add vector instruction with 256-bit vector.
  For single precision floats each such instruction performs 16 floating point operation (8 multiplications and 8 additions). Thus, single core executes up to 32 floating point operations per clock, 4 cores — 128 flop per clock.
  As considered CPU has 4 cores and its clock frequency is 3.1 GHz, its theoretical single precision peak performance is 396.8 Gflops.
- Nvidia Tesla C2070, power consumption 250W, compute capability 2.0, peak single precision performance — 1.03Tflops.

- Nvidia Tesla K40, power consumption 235W, compute capability 3.5, peak single precision performance — 4.2Tflops.
- Nvidia GeForce GTX 1050, power consumption 75W, compute capability 6.1, peak single precision performance — 2 Tflops.

Results of multithread experiments on CPU are presented in Table 1.

The results of operations on GPU are given in Table 2 ($2^{21} \times 8$ block by 8×8 matrix multiplication), in Table 3 ($2^{21} \times 16$ block by 16×16 matrix), and in Table 4 (for matrices of order 1024).

The considered large prime fields required 512 bits, 768 bits, and 1024 bits per element.

Table 1. Time for matrix multiplications on CPU (sec.)

Matrix size	$2^{21} \times 8$	$2^{21} \times 16$	1024×1024	1024×1024, Strassen
512 bits	3.98	13.41	20.92	12.62
768 bits	7.24	24.28	39.53	23.57
1024 bits	12.6	54.97	69	40.9

Table 2. Time for $2^{21} \times 8$ block by 8×8 matrix multiplications on GPU (sec.)

GPU	C2070	K40	GTX1050
Naive algorithm, 512 bits	0.35	0.28	0.41
Winograd approach, 512 bits	0.26	0.19	0.28
Naive algorithm, 768 bits	0.85	0.58	1.15
Winograd approach, 768 bits	0.8	0.63	1
Naive algorithm, 1024 bits	2.04	1.06	2.83
Winograd approach, 1024 bits	1.53	1.07	2.12

Table 3. Time for $2^{21} \times 16$ block by 16×16 matrix multiplications on GPU (sec.)

GPU	C2070	K40	GTX1050
Naive algorithm, 512 bits	1.31	0.89	1.56
Winograd approach, 512 bits	0.89	0.6	0.95
Naive algorithm, 768 bits	3.1	2	3.88
Winograd approach, 768 bits	2.86	2.07	3.5
Naive algorithm, 1024 bits	7.59	3.91	10.82
Winograd approach, 1024 bits	5.47	3.57	6.8

Table 4. Time for 1024×1024 matrix multiplications on GPU (sec.)

GPU	C2070	K40	GTX1050
Naive algorithm, 512 bits	2.38	1.57	2.53
Winograd approach, 512 bits	1.48	0.91	1.42
Naive algorithm, 768 bits	5.75	3.67	6.49
Winograd approach, 768 bits	5.38	3.31	5.52
Naive algorithm, 1024 bits	13.55	7.16	18.11
Winograd approach, 1024 bits	9.37	5.77	9.74

It is noticeable that, due to the use of a larger number of registers, the Winograd approach usually accelerates the computation much less than twice, especially in case of 768 bit numbers. In addition, the advantage of the GPU over the CPU becomes smaller with increasing the field sizes. This proves the necessity of algorithms for long numbers that require smaller number of registers (with multiplying via several stages).

Nevertheless, all the GPUs significantly outperform CPUs for this problem both in terms of computing speed and performance per watt of power. Note that the algorithm used on CPU is quite efficient, and although the CPU is not the most modern, but has almost the same performance on this task as the most modern Intel CPUs (especially at the same clock frequency).

Now consider the results on the Tesla C2070 adapter for 512 bit numbers. This accelerator is similar to the Tesla X2070, which is used on "Lomonosov". Matrix multiplication with Winograd method is 15 times faster than on CPU. Note, than CPU used in our experiments is even slightly faster on such task than 2 Intel Xeon X5570 4-core CPUs (which "Lomonosov" node contains). It has a slightly higher clock frequency and the execution of 64-bit instructions ADC and MUL (which dominate in the algorithms) requires 2 and 3 times less cycles, respectively.

Thus, the matrix multiplication on GPU of supercomputer "Lomonosov" will be no less than 15 times faster than the one on its CPU. This means that in case of the same time spent on the above operations, the block size in the algorithm can be increased in 15 times, and the time spent for data exchanges will be reduced approximately in 15 times too.

For a linear system of order about 2 million, with 82 nonzero elements, over a large simple field of size 512 bits, on 128 nodes of "Lomonosov", the time for data exchanges was about 55%. Thus, the implementation of matrix operations on the GPU reduce calculation time in approximately 2 times.

3   Conclusion

The possibility of using GPU for a significant acceleration of computations with dense matrices and blocks with elements from the large prime fields is experimentally substantiated. The implementation of "naive algorithm" and the algorithm using the Winograd approach for multiplication number reduction are described. Numerical simulations were made for various graphics accelerators architecture and performance. It is shown that for the prime fields with more than 2^{512} elements, in order to obtain the greatest possible acceleration, the multi-stage algorithm should be implemented for the multiplications of long numbers.

Acknowledgments. The work was supported by the Russian Science Foundation, grant 14-11-00806.

References

1. Kleinjung, T., Aoki, K., Franke, J., Lenstra, A.K., Thomé, E., Bos, J.W., Gaudry, P., Kruppa, A., Montgomery, P.L., Osvik, D.A., te Riele, H., Timofeev, A., Zimmermann, P.: Factorization of a 768-Bit RSA modulus. In: Rabin, T. (ed.) CRYPTO 2010. LNCS, vol. 6223, pp. 333–350. Springer, Heidelberg (2010). https://doi.org/10.1007/978-3-642-14623-7_18
2. Thome, E., et al.: Factorization of RSA-704 with CADO-NFS. Preprint, pp. 1–4 (2012)
3. Dorofeev, A.Ya.: Vychislenie logarifmov v konechnom prostom pole metodom lineinogo resheta. [Computation of logarithms over finite prime fields using number sieving]. Trudy po diskretnoi matematike, vol. 5. pp. 29–50 (2002)
4. Dorofeev, A.Y.: Solving systems of linear equations arising in the computation of logarithms in a finite prime field. Math. Aspects Crypt. **3**(1), 551 (2012). Russian
5. Popovyan, I.A., Nestrenko, Y.V., Grechnikov, E.A.: Vychislitelno slozhnye zadachi teorii chisel. Uchebnoe posobie [Computationally hard problems of number theory. Study guide] Publishing of the Lomonosov Moscow State University (2012)
6. Zamarashkin, N., Zheltkov, D.: Block Lanczos–Montgomery method with reduced data exchanges. In: Voevodin, V., Sobolev, S. (eds.) RuSCDays 2016. CCIS, vol. 687, pp. 15–26. Springer, Cham (2016). https://doi.org/10.1007/978-3-319-55669-7_2
7. Zamarashkin, N.L.: Algoritmy dlya razrezhennykh sistem lineinykh uravneniy v GF(2). Uchebnoe posobie [Algorithms for systems of linear equations over GF(2). Study guide]. Publishing of the Lomonosov Moscow State University (2013)
8. Efficient basic linear algebra operations for solution of large sparse linear systems over finite fields. Russian Supercomputing Days (2016)
9. Nath, R., Tomov, S., Dongarra, J.: An improved MAGMA GEMM for Fermi graphics processing units. Int. J. High Perform. Comput. Appl. **24**(4), 511–515 (2010)
10. Nvidia Corporation, CUDA C. Programming guide. http://docs.nvidia.com/cuda/cuda-c-programming-guide

Means for Fast Performance of the Distributed Associative Operations in Supercomputers

Gennady Stetsyura[✉]

Institute of Control Sciences of Russian Academy of Sciences, Moscow, Russia
gstetsura@mail.ru

Abstract. This study proposes architectural solutions and operations for the rapid implementation of distributed associative operations in supercomputers. The operations are carried out by means of interactions between supercomputer devices using wireless optical links. Some operations result in improved distributed versions of the local operations of associative memory devices and associative processors. The operations for distributed fast digital calculations are also included. The operations for analog-digital counting are proposed for quick counting the number of records in distributed big data. The structure of the connections between devices can be completely changed in comparable time to the execution time of the processor command.

Keywords: Wireless optical network · Retroreflector · Dynamical reconfiguration · Distributed synchronization · Barrier synchronization · Distributed computing · Fault tolerance

1 Introduction

The distributed associative operations (DAO) refer to the operations of distributed search and data processing while analyzing data from many records included in supercomputer devices (objects). The associative operations (AO) are similar in function to DAO but act with the records located within the same device. The associative (or Content Addressable) memory devices (AM) and the associative parallel processors (APP) were created to quickly perform AO operations.

The AM device performs a parallel search in the base of records, which have keys equal to those in AO. The search also retrieves the records of keys-number values in a given interval, with a maximum value and so on. The AM counts the number of records found and resolves conflicts when multiple records meet the search criteria. The APP simultaneously separates the array of records into clusters, performs a limited set of logical and arithmetic operations.

The history of the use of associative operations in computers has many stages. A large number of studies on AM and APP were done in the 1960s of the last century.

By the early 1970s, these studies had led to the creation of several large computers, focused on the implementation of associative operations. Reviews of these areas of work are contained in several books [1–3]. However, ever-increasing demands on the processing of large amounts of data led to the fact that AO, as a rule, is now carried out

© Springer International Publishing AG 2017
V. Voevodin and S. Sobolev (Eds.): RuSCDays 2017, CCIS 793, pp. 27–39, 2017.
https://doi.org/10.1007/978-3-319-71255-0_3

by programming. The programmed associative operations are more powerful than the operations in *AM* or *APP* but are much slower. Such operations are used in many algorithms, such as when using associative rules [4, 5].

The hardware implementation of the *DAO* in the supercomputers (*SC*) could accelerate the implementation of these algorithms, but this requires the active cooperation of distributed objects, the exchange of short messages, and distributed quick computations. Standard communications between *SC* devices are not effective at such actions, since they were designed for long message exchanges and do not support calculations directly in communication media. In the article, this disadvantage is eliminated, and objects quickly perform distributed associative operations.

There are opportunities for new types of relations between the SC objects used, as considered by the author at the conference "Supercomputer Days in Russia in 2016" [6, 7] and in [8]. These opportunities allow each facility to operate as a standalone device that performs the *AO* on its local data, but the *DAO* and distributed computations are carried out by means of communications between the objects.

2   Structure of the SC That Supports Distributed Associative Operations

The structure of the optical connections between the objects [6–8] will be used to perform the *DAO* with additions that the *DAO* require (Fig. 1).

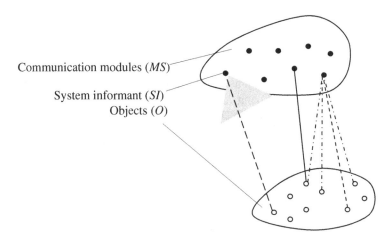

Communication modules (*MS*)

System informant (*SI*)

Objects (*O*)

Fig. 1. The structure of the SC connections

The structure contains two sets of nodes - the objects (*O*) and the communication modules (*MSs*). The wireless signals are transmitted between the communication modules and the objects. One *MS*—the systems informant (*SI*)—has special properties.

Any object can send signals of three types—f_1, f_2, f_3—to a selected *MS* or simultaneously to groups of *MSs* (including *MSs* not associated with objects). The signals f_1 and f_3 have arbitrary lengths. The objects transmit the messages by means of the f_2 signal.

As the *MS* receives signals from f_2 objects, it simultaneously modulates all entering into *MS* f_1 signals without delaying the f_2 signals. This results in the f_1 signals informing the objects about the f_2 signals coming into the *MS*. The f_3 signals prohibit the *MS* to return the f_1 signals to the objects. If the object is associated with the *MS* for receiving signals, it sends the f_1 signal continuously to the MS (solid line in Fig. 1). The receiver and its *MS* is the single device consisting of two spaced components.

The source sends f_2 signals into the *MS* modules of the receivers. The *MS* module receives f_2 signals from the source (dashed line), modulates them with continuous f_1 signals coming from all sources (dash-dot line), and returns them to the receiver and to all sources of f_1 signals.

The object receiver acts like the source transmitting f_2 signals for the sources of the f_1 signals. However, it only sends the signals to its *MS*, which the other sources watch.

Technically, all operations of the objects in interaction in the system are carried out by the network controller part of the object. The module (*SI*) is different from the *MS*: in obtaining f2 signal, the module creates the non-directional f_{si} signal, which is specific only for *SI*, and sends it for all network objects.

Section 4 of the article requires the next addition in the *SI*. The photodetector (*SI*) will summarize the energy of the f_2 signal from the objects (The VCSEL sources of the objects may have 30 ppm/°C stability [9]). Its analog output is connected to an analog-digital converter (ADC) that digitizes the analog signal from the photodetector and returns it to all objects.

The following characteristics of network resources are used below [6–8]:

1. The fast synchronization of the objects—message sources are obtained. If the group of sources receives the synchronization start signal from the receiver, they send the signals or messages to the receiver so that they arrive at the *MS* receiver simultaneously or sequentially, without pause time between sending sources.

This principle of fast synchronization is as follows: Let the source O_i know the delivery times of the signal T_{ij} to an arbitrary communication modulus MS_j.

For synchronization, the object O_i sends a signal to the MS_j with a delay of $^*T_i$. Relative to the time of arrival from the MS_j, the clock signal $^*T_i = T_{max} - T_{ij}$, where $T_{max} \geq max\ T_{ij}$. Then, the signals of all objects acting in the same way will go to MS_j simultaneously, with the same delay of T_{max}.

If the objects transmit messages at the same time, the same-named bits of the messages will be combined and represented as a single message.

2. If there are conflicts in the message access to the *MS*, they will be quickly detected and eliminated by conflict resolution algorithms using rapid synchronization. There are several ways to resolve the conflict, one of which is discussed in Sect. 3.1 below.
3. There is the quick barrier synchronization operation. The barrier synchronization is widely used in computers. Its main purpose is to allow interacting computers (or programs) to determine the total time of completion of the task with minimum latency in order to ensure that the results obtained are correct. Usually, this is a lengthy operation, but a quick way is given in [6–8].

Let us consider its variant. Let all the sources of the interacting group complete the work and then transmit the messages—the results of their work to the receivers waiting for the message. The sources need different amounts of time to complete the work. After the completion of work by all sources, they must transmit messages to the receivers as a single message without time delays between the individual messages.

To synchronize in a group of sources, one member of the group is allocated. Its module MS^* is known to all sources and receivers that monitor MS^* by sending it a f_1 signal. (A free module that is not associated with the object can be taken as MS^*.)

When preparing the message, the sources transmit a continuous f_3 signal to the MS^*, which prohibits the return of the f_1 signals. Having prepared the message, the source removes the prohibiting signal. After all sources are ready, the MS^* will start to return the f_1 signal, which will be a clock signal for the objects. After receiving the signal, the objects will transmit messages synchronously (using fast synchronization) to the MS^*, and all receivers will receive it as a single message.

4. Simultaneous arrival of the messages from the objects in MS allow bitwise logical addition and multiplication, as well as finding the max and min values in times that do not depend on the number of participants in the operation. The object connections in the chain allow the logical operations and the arithmetic operations of addition, subtraction and multiplication without delay the calculations [6–8].

It is useful to consider the organization of the associative memory for comparison with the organization of the system in Fig. 1. Let us turn to Fig. 2. The associative memory contains the memory of the cells with the records (1), with the separate logical unit connected to each of its cells. The aggregate of these units is the distributed control unit (2) cells. The units from (2) store the result of the search task impacted on each cell in (1). On (1) and (2), a search query (3) is received from the computer containing the AM, and each unit from (2) stores the result of the request for this record, allocating the record corresponding to the request. Next, the unit state is used to perform associative operations (Sect. 1). The AM has the central control unit (4), which interacts with the computer by exchanging control signals (5). It also receives search results (6) outputted from (1) and (2) to the computer and acts on (1) and (2) together with the signals (3).

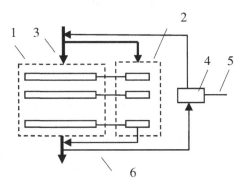

Fig. 2. The structure of the associative device

Thus, the AM is an orderly structure of simple devices, where many records are analyzed simultaneously. The structure of the *APP* is close to the structure of the *AM*, but since the *APP* performs more complex operations, its control unit (2) is more complicated. In the early version of the *APP* [10], the array of records is simultaneously divided into clusters directly in the associative memory of *APP*.

The flexibility of the structure of AM, APP and the structure in Fig. 1 are significantly different. In AM and APP the structure is fixed—all units of devices and their connections are unchanged. The structure in Fig. 1 changes cardinally during the execution of the processor command. In this case, the functions of the devices also change—the search initiator becomes its executor, a group of interacting initiators of the search is quickly created, and so on. As will be shown below, distributed associative devices that perform *DAO* obtain results with significantly expanded capabilities compared to those available in *AM* and *APP*.

3   Distributed Associative Units and Their Actions

In the article, the distributed associative unit (*DAU*) is a collection of objects using wireless optical communications and performs both the *AO*, like the devices *AM* and *APP*, and a number of additional operations, which may be implemented only in the structure of Sect. 2.

3.1   The Simple Search in *DAU*

Let the objects be combined in accordance with Fig. 1 and have the records, among which the operation *DAO* is making the distributed associative search. This search depends on a particular implementation of the object. It runs in the *AM* of the object, in its operative storage device, or directly in the small *AM* in the network controller of the object.

One of the objects O_i is the search initiator and sends the *SI* a command of the simple search for implementation in the *DAU* devices. This command contains the search key (Q). In the simple search the value of the key should have an exact match in the records. Objects get this information, conduct a local search in their memory devices (or in the *AM*), and prepare a response for their initiators.

An answer or multiple answers are placed in the network controller of the object for use outside the object. Depending on the tasks involved in the operation, the object sends the reply message to the SI or into the module MS of the initiator.

A conflict will occur in the transfer if the objects are transmitting messages simultaneously, and it must be resolved. We offer the method of using the binary scale to resolve this conflict, which is slightly modified compared to [6–8]. The object does not have information about the conflict, and it sends their messages in *SI*, starting with its name (address). If there is a conflict, the address will be distorted, and the object receives the distorted message from the *SI*.

The object perceives such distortion as a synchrosignal for synchronous transmission of these messages. The physical addresses in these messages, assigned to N objects, are divided into n groups with m objects in the group. Each object knows its affiliation to the group and its serial number between the m objects of the group.

We introduce the scale of A—the binary string of n positions, each of which is one of n groups. We will be writing the value one into the position of the string that corresponds to the value of the digit if the object has the answer for *DAO*.

The objects synchronously send their messages—the scales A into the module *SI* with superposition of the bits in these scales A. The combined scale A is returned from *SI* to the objects.

Only the objects, which record the digit one in the scale, send the new scale B consisting of m bits, where each bit is allocated to one of the objects of the group. This is similar to scale A. The scale B comes to *SI*, and the combined scale returns to the objects.

The objects, which record the digit one in the scale B, create the new scale C. The scale C has slots where the objects point to the number of ready answers. After returning the scale to the objects, they consistently convey their messages without pauses. For small N, the types of scales can be reduced and may even have only one scale C. The conflict is resolved.

For many years, the *AM* used the paraphase presentation of binary digits with the active zero signal [6–8] for searching. Each bit is encoded by a pair of binary digits: 10 for 1, 01 for 0 and 00 for the mask M. The mask in Q coincides with any value of the corresponding bit in the records.

The arrival on the object of several responses with the paraphase pair of the bits 11 (U) indicates the difference in the responses. The paraphase encoding is also used in the *DAO* for speeding up the distributed logical and arithmetic operations [6–8].

3.2   The *DAO* That Have Keys with Numeric Values

Consider a search of records that have keys with numerical values in the *DAU*. This search includes searching in the records that have keys with maximum (minimum) values, in the records that have keys with values closest to the question, and in the records that have a key with a value in the given interval.

The p-ary positional system is used with digits that represent the strings of the bits [6, 11]. We will include digit one in the position of the string, corresponding to the value of the digit. For example, the number 36, with p equal to 10, is 000000100_000100000; for the paraphase bit encoding, it is 010101010101100101_010101100101010101.

The unary representation of the digits greatly speeds up the work when the *DAO* is executed directly on the network, such as when calculating the maximum or minimum value of the number of numbers sent to the network by objects. We will use the result from [6–8].

To determine the maximum or minimum, a group of the objects (performers of the *DAO*) synchronously sends the messages to the *MS* module of the initiator of the *DAO*, and the bits in the same position of the record of the digit are combined together. At first, each source-participant of the *DAO* sends a message with the high-order digit of the compared numbers to the *MS* module. The *MS* module returns the messages received as a result of superimposing the bits of the messages from the sources, and if the source sending the digit to the module detects the presence of a larger number, then it stops attempting to transmit its number. This operation continues for all other digits of the compared numbers. As a result, the maximum value of the numbers sent by the objects will be detected simultaneously for all sources. By inverting the representations of the signals one and zero, the minimum value will be found in a similar way.

The result of the operation is created by the *MS* module without involvement from the object computing facilities, and the execution time of the operation does not depend on the number of the objects participating in the operation. The result is sent in parallel to all objects. Similarly, the value closest to the specified value is found.

Now let us go to the search for the numbers in a given interval. The source sets the search interval and directs it to the distributed objects. It is assumed that the object or its network controller has an *AM* device, and it is required to select the search request form that is directly perceived by the *AM* without the use of the processor. We use a slightly modified solution from [11, 12].

For example, let us say that it is a requirement to find the records with values of the parameter U in the interval $137 \leq U \leq 628$ in the *AM* of objects with the searching rule "bitwise AND $\neq 0$". We perform the searches with the specification of the intervals U: 13 ($y \geq 7$); 1 ($y \geq 4$) z; ($2 \leq y \leq 5$)$z\dot{z}$; 62 ($y \leq 8$); 6 ($y \leq 1$) z, where y is the value of the digit and z is any value of the digit. For example, the record $2 \leq y \leq 5$ in the paraphase code has the form 000000001010101000. Digits 2–5 are selected. The searching rule "bitwise AND $= 0$" selects digit "0".

It is easy to check that these searches select all records with values in the specified interval.

Let the digit "0" is represented by an additional position containing «1», then, for example, the searching "bitwise AND $\neq 0$" with the form 00000010101010101010 selects digits 0÷6 simultaneously.

Now let the *DAO* source require delivery of the numbers from the objects in order to carry out the next steps of the *DAO* on their basis. Objects simultaneously send the p-bit strings of the code of the highest digits (or the intervals of digits, given by the chain "1") of numbers corresponding to the requirements of the request to the *DAO* source. The source decides which bits in the string are stored to refine the search, sends the next refined query, etc. The additional controls for the search steps appear if analog-to-digital computation is used (Sect. 4): for each bit one in the string of the digit, the number of records that generate this bit is calculated.

3.3   Distributed Associative Parallel Processor (*DAPP*)

Let us consider the implementation of two *DAO* operations in the *DAPP*, close to the operations in the *APP*. It is the separation of the set of records into clusters and the ordering of records in clusters. The difference from the *APP* arises from the distribution of the records between the objects.

- **Selecting the clusters of objects and records.** Let one of the above *DAO*s be performed in the *DAPP*, and the objects put the records—the results of the local search—in the *AM* or in the registers of the network controllers. Let there be additional bits in the records of the analyzed array, and the *DAO* specifies the additional bit and requires all objects with the correct answer to write the value equal to one into this bit for all the records found. Thus, a cluster of records distributed among objects will be allocated, and it can be accessed by its name.

- **Sorting the records in the clusters.** The required order of the objects and the records in them will be created by repeating the scheme for eliminating conflicts with the scales *A*, *B*, and *C*.

The *DAO* states the keys for selecting a group of objects which contain the allocated record clusters. The group objects attempt to send messages to the initiator of the search, and a conflict arises. It is eliminated using the scales *A*, *B* and *C*, after which the objects transmit special messages containing their physical addresses. The distributed cluster is ordered by these addresses, and the source of the DAO can change this order.

Then, the *DAO* conducts an analysis of the data, already taking into account the order of the location of records in the objects. Communication facilities allow the distributed parts of the array of records to quickly form and be ordered into a single array, ensuring interaction with it as a single entity. Such actions are easily supplemented by distributed computations with distributed records, which are performed directly on the network in accordance with [6–8].

This ordering of the records stored in different objects speeds up the analysis of logical, spatial and temporal relationships between the records. The same task was typical for the *APP*.

For example, one of the first developments of the *APP* was intended for grammatical analysis of texts in the information-logical computer [13] with a variable structure. Specialized devices were developed for this computer—the *AM* and *APP* used in this article [10–12].

4   Distributed Analog-Digital Operations

4.1   Distributed Analog-Digital Counting and Summation

Many tasks require counting the number of objects and records corresponding to the condition specified in the *DAO*. Such counts are often iterative. At the beginning of the process, it is enough to have inaccurate but quickly obtained results, and only at the last steps may an exact calculation be required.

If an exact solution is required, ordering records in sub-sets is used (Sect. 3.3). After the completion of work with the scales *A*, *B* and *C*, the total number of messages sent by the objects is determined, and the quantities of records found by the object are summarized in each message.

The operation of distributed summation from [6–8], in which the objects are connected in a chain, is also applicable. If the records in the object satisfy the condition in the *DAO*, then the object adds their number to the number in the message passing through the chain of objects. The operation is performed without delaying the message to perform the summation.

For a quick approximate calculation, the analog-to-digital method of interaction via *SI* (or *MS*) is proposed below, for which it was required in Sect. 2 to modify *SI* in comparison with [6–8]. The objects perform the following steps for the approximate count.

Step 1

The initiator of the *DAO* chooses the objects for the counting and conditions of the counting. As objects are programmable, complex conditions are admissible to require a sequence of searches within the object and interaction with other objects within the same *DAO*.

The time of execution of such operations is unknown, so in general, the operation should be performed in barrier synchronization mode. For its conduct, the initiator of the *DAO* sends the participants of the operation the name of the communication module MS_{br} and the indicator of the moment when the *DAO* be completed.

Step 2

Each *DAO* executor sends the command in the MS_{br} that prohibits the MS_{br} from returning the f_1 signals to the objects that sent them in MS_{br}. After that, the object conducts the analysis of records specified by the initiator, and after completing it, removes the prohibition of MS_{br} from returning f_1 signals. All *DAO* performers watch the MS_{br}, and returning the f_1 signal to them from MS_{br} is the start of the count in step 3.

Step 3

- **The general provisions for step 3.** The source of the DAO sends information about the search condition and the keys K in the records to the objects. Each object containing the required records must send to the source of the *DAO* a message consisting of a string of references. Each reference corresponds to one of the keys K.

  The reference contains the number N_b, which fixes the number of records found by the object with such a key. The number of digits in N_b is given by the source. Each digit is represented by a scale S - a string of binary digits in an amount equal to p – which is the base of the chosen number system (Sect. 3.2). The reference has a binary digit R_o, where the object puts "one", if it finds the corresponding key K in the records.

- **The object's actions in step 3.** Each object sends a message to *SI*. In the references of the messages, the object sends the f_2 signals to R_o and to the strings S into the bits represented the digits of the numbers.

  The strings are transmitted synchronously, and their same name bits must coincide in time when they enter the *SI*. As a result, the *SI* photodetector will receive a signal from each bit of the string with the total energy sent by all objects.

  The signal will be digitized, and the result is sent simultaneously to the initiator of the search and to all objects. Having received digital values, objects determine the total number of records found and the number of objects that have them. The computation is completed.

  If the source only needs the number of objects that meet the request, then the messages are limited to the R_o bit. The counting time does not depend on the number of objects participating in the *DAO*, and consists of a double time interval that includes the time required for the signal to pass between the *SI* and the object most distant from it and the time required to convert the analog signal to a digit.

Another solution is possible. The *DAO* selects certain modules of *MS**, to which the objects will send signals now instead of to *SI*. The module does not create a digital message as *SI does*, but reduces the transparency of the light filter for the f_1 signal in proportion to the energy of all incoming f_2 signals. Each object has a photodetector and an analog-to-digital converter. The object forms the digital value as it does *SI*.

To reduce the number of O^* objects that perform analog-to-digital conversion, we will create a small number of such objects, and we will provide the object O^* in dynamics to different initiators of the *DAO*.

After performing the analog-to-digital conversion, the O^* object will send the result to *MS**, and the result will receive the *DAO* initiator and other objects that watch *MS**.

If the range of energy levels of the total signal arriving at the photodetector *SI* and O^* exceeds the linear region the photodetector, two methods may be applied to reduce the energy of the signals arriving at the photodetector.

The first method: the ADC reduces the throughput of the light filter-modulator receiving the f_2 signals and supplements the message sent to the objects with information about the small precision of the sample.

The second method is the logical method: the request initiator selects groups of objects that must simultaneously send signals to *SI* and O^*. To do this, the initiator details the request in the *DAO*, reducing the number of the sources of the messages, and/or indicates the area of physical addresses of objects that are allowed to send messages.

If different groups of objects are allocated different O^* values, then this reduces the total energy of the signals arriving at O^*.

Additional information is provided using paraphase binary signals. Let us give an example. The group of records, in which the search is performed, usually, has an unknown size. However, to assess the significance of the data found, their share in the total search volume must be known. The paraphase code in the position R_0 will be applied. The number of analyzed records may be determined by a count of active signals one and zero in the R_0.

The counting in large numbers of records is required by many applied algorithms, for example, algorithms that work with associative rules [4, 5], and algorithms using a naive Bayesian classifier [14]. The counting operations of keys in various combinations in large sets of records take considerable time in such algorithms.

Our solutions turn the distributed operations of counting records into the summation of the energy of signals within a single message, which is created simultaneously by all objects.

It should be noted that an operation analogous to the summation of the number of records allows the summation of any numbers to be simultaneously transmitted by objects to the communication module. In *MS*, the energy of the signals in the scales representing the digits is summed. From these sums, the objects get the sum of numbers.

Such summation is also performed during the simultaneous sending of messages by objects.

4.2 Associative Operations as a Means of Managing *SC* Objects

The *DAO* is a quick tool for monitoring the state of *SC* objects, but it is also useful for managing their behavior. Let us turn to Fig. 3.

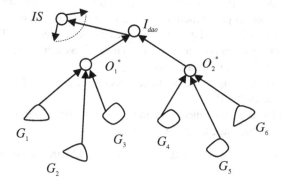

Fig. 3. Managing the state of objects

Let the object I_{dao}, initiator of the *DAO*, send a group program (a sequence of the *DAO* commands) via the *SI* module that collects information about the state of objects. The analog-digital calculations and barrier synchronization is used.

Let I_{dao} initially receive a number that exceeds the resolution of its analog-to-digital converter. Upon discovering this, I_{dao} refines the requests (Sect. 4.1) by dividing the set of responding objects into groups $G_1, G_2 ..., G_6$. The groups G_1, G_2 and G_3 are assigned an object O_1^* to convert the analog-digit, and the object O_2^* is assigned to groups G_4, G_5 and G_6. The objects O_1^* and O_2^* forward the answers to the I_{dao} after they receive it.

Now I_{dao} can proceed to the management of the objects, sending the new *DAO* to the group G_i via the *SI* to all objects.

In Fig. 3, the object I_{dao} is somehow allocated in advance. However, unlike *AM* and *APP*, the objects participating in the *DAO* are active and each takes into account its state and the state of the objects it observes. They can each receive the rights of the initiator of the *DAO*. Therefore, within the framework of Fig. 3, there is a place for a common control center of the system, but there is an additional control capability. Any object is allowed to promptly intervene in the behavior of the system, leaving slow work for the center to manage access to the shared resources.

5 Conclusions

The considered organization of the *DAO* with the use of wireless optical connections makes it possible to obtain the following main results.

1. This study proposes the extension of local operations of associative memory and associative processors up to the distributed associative operations that operate between the groups of interacting *SC* objects.

2. Data exchange with distributed computations requires the same amount of time as data exchange without computation.
3. Analog-to-digital approximate calculations are developed that accelerate the distributed summation of numbers, each of which is located in a separate device of the *SC*. The time of the summation does not depend on the number of participants in the addition operation and is performed during the time of simultaneous message transmission by devices with overlapping messages in time (Sect. 4). The options to increase the accuracy of calculations are shown. Such a calculation method greatly speeds up the analysis of large data sets.
4. Proposed DAOs do not require sophisticated technical aids in addition to the technical aids proposed in [6–8].
5. The structure of the device connections and their functions are quickly changed by sending a broadcast message to the *SC* devices.

New *SC* functions are obtained through the use of optical wireless connections, using retroreflectors.

References

1. Krazmer, L.P.: Associative memory devices. Energy, Leningrad (1967). (in Russian)
2. Kohonen, T.: Content-Addressable Memories. Springer, Berlin (1980)
3. Foster, C.: Content Addressable Parallel Processors. Van Nostrand, New York (1976)
4. Agrawal, R., Shafer, J.C.: Parallel mining of association rules. IEEE Trans. Knowl. Data Eng. **8**(6), 962–969 (1996)
5. Vasoya, A., Koli, N.: Mining of association rules on large database using distributed and parallel computing. Procedia Comput. Sci. **79**, 221–230 (2016)
6. Stetsyura, G.G.: Addition for supercomputer functionality. In: Reports of the International Supercomputing Conference "Russian Supercomputing Days 2016", Moscow, MSU, 314–324 (2016). (in Russian). http://russianscdays.org/files/pdf16/314.pdf
7. Stetsyura, G.: Addition for supercomputer functionality. In: Voevodin, V., Sobolev, S. (eds.) RuSCDays 2016. CCIS, vol. 687, pp. 251–263. Springer, Cham (2016). https://doi.org/10.1007/978-3-319-55669-7_20
8. Stetsyura, G.G.: The computer network with distributed fast changing of structure, and with data processing during transmission. Control Sci. (1), 47–56 (2017). (in Russian). http://pu.mtas.ru/archive/Stetsyura_117.pdf
9. Downing, J., Babićb, D., Hibbs-Brennerc, M.: An ultra-stable VCSEL light source. In: SPIE Photonics West 2013 and 2015 Conferences on VCSELs. USL Technologies, 22 p. (2015). http://www.ultrastablelight.com/papers/SPIEPapers.pdf (2012)
10. Stetsyura, G.G.: Decisive device for translation tasks. In: Reports at the Conference on Information Processing, Machine Translation and Automatic Reading of Text. Acad. Sciences of the USSR. Institute of Scientific. Information, no. 8, 9 p. (1961). (in Russian)
11. Stetsyura, G.G.: A new principle of building memory devices. Rep. Acad. Sci. USSR **132**(6), 1291–1294 (1960). (in Russian)
12. Stetsyura, G.G.: Automatic search device with feedback. Reports at the Conference on Information Processing, Machine Translation and Automatic Reading of Text. Acad. Sciences of the USSR. Institute of Scientific. Information, no. 4, 10 p. (1961). (in Russian)

13. Kleinerman, G.I., Kosarev, A.A., Nisnevich, L.B., Stetsyura, G.G.: On the question of constructing universal digital machines with many solving devices. ib, no. 9, 5 p. (1961). (in Russian)
14. Naive Bayes classifier (Wikipedia). https://en.wikipedia.org/wiki/Naive_Bayes_classifier

Scalability Evaluation of NSLP Algorithm for Solving Non-Stationary Linear Programming Problems on Cluster Computing Systems

Irina Sokolinskaya and Leonid B. Sokolinsky$^{(\boxtimes)}$

South Ural State University, 76 Lenin prospekt, Chelyabinsk 454080, Russia
{Irina.Sokolinskaya,Leonid.Sokolinsky}@susu.ru

Abstract. The paper is devoted to a scalability study of the NSLP algorithm for solving non-stationary high-dimension linear programming problem on the cluster computing systems. The analysis is based on the BSF model of parallel computations. The BSF model is a new parallel computation model designed on the basis of BSP and SPMD models. The brief descriptions of the NSLP algorithm and the BSF model are given. The NSLP algorithm implementation in the form of a BSF program is considered. On the basis of the BSF cost metric, the upper bound of the NSLP algorithm scalability is derived and its parallel efficiency is estimated. NSLP algorithm implementation using BSF skeleton is described. A comparison of scalability estimations obtained analytically and experimentally is provided.

Keywords: Non-stationary linear programming problem · Large-scale linear programming · NSLP algorithm · BSF parallel computation model · Cost metric · Scalability bound · Parallel efficiency estimation

1 Introduction

The Big Data phenomenon has spawned the large-scale linear programming (LP) problems [1]. Such problems arise in the following areas: scheduling, logistics, advertising, retail, e-commerce [2], quantum physics [3], asset-liability management [4], algorithmic trading [5–8] and others. The similar LP problems include up to tens of millions of constraints and up to hundreds of millions of decision variables. In many cases, especially in mathematical economy, these LP problems are nonstationary (dynamic). It means that input data (matrix A, vectors b and c) is evolving with time, and the period of data change is within the range of hundredths of a second.

The reported study has been partially supported by the RFBR according to research project No. 17-07-00352-a, by the Government of the Russian Federation according to Act 211 (contract No. 02.A03.21.0011.) and by the Ministry of Education and Science of the Russian Federation (government order 2.7905.2017/8.9).

V. Voevodin and S. Sobolev (Eds.): RuSCDays 2017, CCIS 793, pp. 40–53, 2017.
https://doi.org/10.1007/978-3-319-71255-0_4

Until now, one of the most popular methods solving LP problems is the class of algorithms proposed and designed by Dantzig on the base of the simplex method [9]. The simplex method has proved to be effective in solving a large class of LP problems. However, in certain cases the simplex method has to move across all the vertices of the polytope, which corresponds to an exponential time complexity [10]. Karmarkar in [11] proposed a method for linear programming called "Interior point method" which runs in polynomial time and is also very efficient in practice.

The simplex method and the method of interior points remain today the main methods for solving the LP problem. However, these methods may prove ineffective in the case of large scale LP problems with rapidly evolving input data. To overcome the problem of non-stationarity of input data, the authors proposed in [12] the scalable algorithm *NSLP* (*Non-Stationary Linear Programming*) for solving large-scale non-stationary LP problems on cluster computing systems. It includes two phases: *Quest* and *Targeting*. The *Quest* phase calculates a solution of the system of inequalities defining the constraint system of the linear programming problem under condition of the dynamic changes of input data. The point of pseudo-projection on n-polytope M is taken as a solution. Polytope M is the set of feasible solutions of the LP problem. The pseudo-projection is an extension of the projection, which uses Fejer (relaxation) iterative process [13–16]. A distinctive feature of the Fejer process is its "self-guided" capability: the Fejer process automatically corrects its motion path according to the polytope position changes during the calculation of the pseudo-projection. The *Quest* phase was investigated in [12], where the convergence theorem was proved for the case when the polytope is translated with a fixed vector in the each unit of time. In the paper [17], the authors demonstrated that Intel Xeon Phi multi-core processors can be efficiently used for calculating the pseudo-projections.

The *Targeting* phase forms a special system of points having the shape of the n-dimensional axisymmetric cross. The cross moves in the n-dimensional space in such a way that the solution of the LP problem permanently was in the ε-vicinity of the central point of the cross. The Targeting phase can be effectively implemented as a parallel program for a clustered computing system by using the "master-workers" framework [18–20]. In this paper, we discuss a parallel implementation of the NSLP algorithm using the BSF computational model presented in [21]. On the base of the described BSF-implementation, a quantitative scalability analysis of the NSLP algorithm is performed.

The rest of the paper is organized as follows. Section 2 gives a formal statement of a LP problem and presents the brief description of the NSLP algorithm. Section 3 provides an outline of the BSF computational model and presents corresponding cost metrics. Section 4 describes a BSF-implementation of the NSLP algorithm, calculates the upper bound of scalability and evaluates the parallel efficiency depending on the percentage of initial data being changed dynamically. Section 5 describes an implementation of the NSLP algorithm based on the BSF skeleton in C language and compares the results obtained analytically and experimentally. Section 6 summarizes the results obtained and proposes the directions for future research.

2   NSLP Algorithm

Let we be given a non-stationary LP problem in the vector space $\mathbb{R}^n$:

$$\max\left\{\langle c_t, x\rangle \,|\, A_t x \le b_t,\ x \ge 0\right\}, \tag{1}$$

where the matrix A_t has m rows. The non-stationarity of the problem means that the values of the elements of the matrix A_t and the vectors b_t, c_t depend on the time $t \in \mathbb{R}_{\ge 0}$. We assume that the value of $t = 0$ corresponds to the initial instant of time:

$$A_0 = A, b_0 = b, c_0 = c \tag{2}$$

Let M_t be a polytope defined by the constraints of the non-stationary LP problem (1). Such a polytope is always convex. The *Quest* phase calculates a point z belonging to the polytope M_t. This phase is described in detail in [12]. The *Quest* Phase is followed by the *Targeting* phase. At the *Targeting* phase, a n-dimensional axisymmetric cross is formed. The *n-dimensional axisymmetric cross* is a finite set $G = \{g_0, \ldots, g_{P-1}\} \subset \mathbb{R}^n$ having the cardinality equals $P+1$, where P is a multiple of $n \ge 2$. Among points of the cross, the point g_0 called the *central point* is single out. The initial coordinates of the central point are assigned the coordinates of the point z calculated in the *Quest* phase. The set $G\backslash\{g_0\}$ is divided into n disjoint subsets C_i $(i = 0, \ldots, n-1)$ called the *cohorts*:

$$G\backslash\{g_0\} = \bigcup_{i=0}^{n-1} C_i.$$

Each i-th cohort $(i = 0, \ldots, n-1)$ consists of

$$K = P/n \tag{3}$$

points lying on the straight line, which is parallel to the i-th coordinate axis and passing through the central point g_0. By itself, the central point does not belong to any cohort. The distance between any two neighbor points of the set $G \cup \{g_0\}$ is equal to the constant s. It can be changed during computing. An example of the two-dimensional cross is shown in Fig. 1. The number of points in one dimension excluding the central point is equal to K. The symmetry of the cross supposes that K takes only even values greater than or equal to 2. Using Eq. (3), we obtain the following equation giving the total number of points in the cross:

$$P + 1 = nK + 1 \tag{4}$$

Since K can take only even values greater than or equal to 2 and $n \ge 2$, from Eq. (4), it follows that P can also take only even values and $P \ge 4$. In Fig. 1, we have $n = 2$, $K = 6$, $P = 12$.

Each point of the cross G is uniquely identified by a *marker* being a pair of integers numbers (χ, η) such that $0 \le \chi < n$, $|\eta| \le K/2$. Informally, χ specifies the number of the cohort, and η specifies the sequence number of the point in the

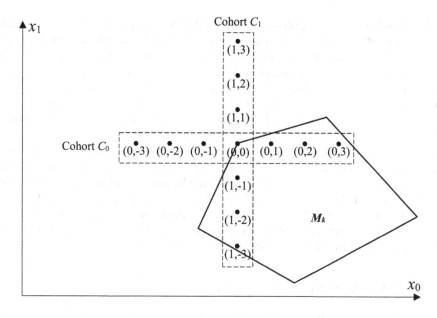

Fig. 1. Two-dimensional cross G: $K = 6$, $P = 12$

cohort C_χ, being counted out of the central point. The corresponding marking of points for the two-dimensional case is given in Fig. 1. The coordinates of the point $x_{(\chi,\eta)}$ having the marker (χ, η) can be reconstructed as follows:

$$x_{(\chi,\eta)} = g_0 + (0, \ldots, 0, \underbrace{\eta \cdot s}_{\chi}, 0, \ldots, 0) \tag{5}$$

The vector being added to g_0 in the right part of the Eq. (5) has a single non-zero coordinate in the position χ. This coordinate equals $\eta \cdot s$, where s is the distance between neighbor points in a cohort.

The *Targeting* phase includes the following steps.

1. Build the n-dimensional axisymmetric cross G that has K points in each cohort, the distance between neighbor points equaling s, and the center at point $g_0 = z_k$, where z_k is obtained in the *Quest* phase.
2. Calculate $G' = G \cap M_k$.
3. Calculate $C'_\chi = C_\chi \cap G'$ for $\chi = 0, \ldots, n - 1$.
4. Calculate $Q = \bigcup\limits_{\chi=0}^{n-1} \{\arg\max \{\langle c_k, g \rangle \mid g \in C'_\chi, C'_\chi \neq \emptyset\}\}$.
5. If $g_0 \in M_k$ and $\langle c_k, g_0 \rangle \geq \max\limits_{q \in Q} \langle c_k, q \rangle$, then $k := k + 1$, and go to step 2.
6. $g_0 := \frac{\sum\limits_{q \in Q} q}{|Q|}$.
7. $k := k + 1$.
8. Go to step 2.

Thus, in the Targeting phase, the steps 2–7 form a perpetual loop in which the approximate solution of the non-stationary LP problem is permanently recalculated. From the non-formal point of view, in the step 2, we determine which points of the cross G are belonged to the polytope M_k. In the step 3, points that do not belong to the polytope are dropped out of each cohort. In the step 4, the point with the maximum value of the objective function is chosen among the residuary points of each cohort. In the step 5, we check if the value of the objective function at the central point of the cross is greater than all the maximums found in the step 4. If this condition is true then the cross does not shift, the time counter t is incremented by one unit and the next iteration is started. If this condition is false then we go to step 6 where the new center point is calculated as the centroid of the set of points obtained in the step 4. In the step 7, the time counter t is incremented by one unit. In the step 8, we go to the new iteration. In such a way, the center g_0 of the cross G permanently performs the role of an approximate solution of the non-stationary problem (1).

3   BSF Computational Model

We use the BSF parallel computation model proposed in [21] to evaluate the upper bound of the scalability of the NSLP algorithm in the Targeting phase. The BSF (Bulk Synchronous Farm) model was proposed to multiprocessor systems with distributed memory. A *BSF-computer* consists of a collection of homogeneous computing nodes with private memory connected by a communication network that delivers messages among the nodes. Among all the computing nodes, one node called the *master-node* is single out. The rest of the nodes are the *slave-nodes*. The BSF-computer must include at least one master-node and one slave-node. Thus, if P is the number of slave-nodes then $P \geq 1$.

BSF-computer utilizes the *SPMD* programming model [22] according to which all nodes executes the same program but process different data. A BSF-program consists of sequences of macro-steps and global barrier synchronizations performed by the master and all the slaves. Each macro-step is divided into two sections: master section and slave section. A master section includes instructions performed by the master only. A slave section includes instructions performed by the slaves only. The sequential order of the master section and the slave section within the macro-step is not important. All the slave nodes act on the same data array, but the base address of the data assigned to the slave-node for processing is determined by the logical number of this node. The BSF-program includes the following sequential sections (see Fig. 2):

- initialization;
- iterative process;
- finalization.

Initialization is a macro-step, during which the master and slaves read or generate input data. The initialization is followed by a barrier synchronization. The *iterative process* repeatedly performs its body until the exit condition checked

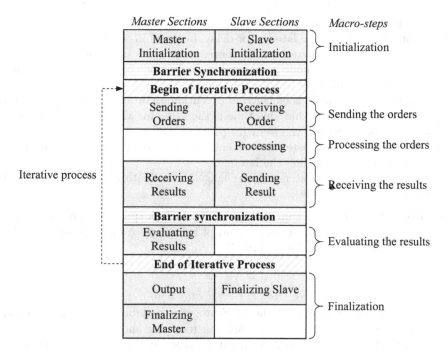

Fig. 2. BSF-program structure

by the master becomes true. In the *finalization macro-step*, the master outputs the results and ends the program.

Body of the iterative process includes the following macro-steps:

(1) sending the orders (from master to slaves);
(2) processing the orders (slaves);
(3) receiving the results (from slaves to master);
(4) evaluating the results (master).

In the first macro-step, the master sends the same orders to all the slaves. Then, the slaves execute the received orders (the master is idle at that time). All the slaves execute the same program code but act on the different data with the base address depending on the slave-node number.

It means that all slaves spend the same time for calculating. During processing the order, there are no data transfers between nodes. In the third step, all slaves send the results to the master. After that, global barrier synchronization is performed. During the fourth step, the master evaluates received results. The slaves are idle at that time. After result evaluations, the master checks the exit condition. If the exit condition is true then iterative process is finished, otherwise the iterative process is continued.

The BSF model provides an analytical estimation of the scalability of a BSF-program. The main parameters of the model are [21]:

P: the number of slave-nodes;

L: an upper bound on the latency, or delay, incurred in communicating a message containing one byte from its source node to its target node;

t_s: time that the master-node is engaged in sending one order to one slave-node excluding the latency;

t_v: time that a slave-node is engaged in execution an order within one iteration (BSF-model assumes that this time is the same for all the slave-nodes and it is a constant within the iterative process;

t_r: total time that the master-node is engaged in receiving the results from all the slave-nodes excluding the latency;

t_p: total time that the master-node is engaged in evaluating the results received from all the slave-nodes.

Lets denote $t_w = P \cdot t_v$ – summarized time which is spent by slave-nodes for order executions. Then, the upper bound of a BSF-program scalability can be estimated by the following inequality [21]:

$$P \le \sqrt{\frac{t_w}{2L + t_s}}. \tag{6}$$

Note that the upper bound of the BSF-program scalability does not depend on the time, which the master is engaged in receiving and evaluating the slave results. The speedup of BSF-program can be calculated by the following equation [21]:

$$a = \frac{P(2L + t_s + t_r + t_p + t_w)}{P^2(2L + t_s) + P(t_r + t_p) + t_w}. \tag{7}$$

One more important property of a parallel program is the parallel efficiency. The parallel efficiency of a BSF-program can be calculated by the following approximate equation [21]:

$$e \approx \frac{1}{1 + (P^2(2L + t_s) + P(t_r + t_p))/t_w}. \tag{8}$$

4   BSF-implementation of NSLP Algorithm

In this section, we demonstrate how the algorithm presented in Sect. 2 can be implemented in the BSF-program form. Based on this implementation, we calculate the time complexity of one iteration and give an analytical estimation of the scalability upper bound of the NSLP algorithm in the Targeting phase. For calculating, we use the synthetic scalable linear programming problem of the dimension called *Model-n* [17]. This LP problem has the matrix A of the size $n \times 2(n + 1)$. We assume $n > 10^4$.

The NSLP algorithm can be implemented in the BSF-program form by the following way. In the Initialization macro-step, the master and all the slaves read (generate) and store in the local memory all the initial data of the nonstationary LP problem (1); the master executes the Quest phase and finds a point z belonging to the polytope M_t. Then, the iterative process begins. In each iteration, the following steps are performed:

(1) sending the orders from master to slaves;
(2) processing the orders by slaves;
(3) sending the results from slaves to master;
(4) evaluating the results by master.

Table 1. Structure of message "Order for slaves"

No.	Attribute ID	Attribute semantic	Overhead
1	θ	New central point coordinates of the n-dimensional cross	t_θ
2	α	New values of matrix A entries	t_α
3	β	New values of column b elements	t_β
4	γ	New values of objective function coefficients	t_γ

The *order* includes the information given in Table 1. Suppose that the fraction of the changed elements of matrix A, column b and objective function c coefficients equals to $\delta(n)$, where $\forall n \, (0 \leq \delta(n) \leq 1)$. In that case, the time t_s that the master-node is engaged in sending one order to one slave-node (excluding the latency) can be approximated according to Table 1 as follows:

$$
\begin{aligned}
t_s &= t_\theta + t_\alpha + t_\beta + t_\gamma = O(n) + O(\delta(n) \cdot n \cdot 2(n+1)) + O(\delta(n) \cdot 2(n+1)) + O(\delta(n)n) \\
&= O(n) + O(\delta(n) \cdot n(n+1)) + O(\delta(n)(n+1)) + O(\delta(n)n) \\
&< O(n+1) + O(\delta(n) \cdot (n+1)^2) + O(\delta(n)(n+1)) + O(\delta(n)(n+1)) \\
&= O(n+1) + O(\delta(n) \cdot (n+1)^2) + O(\delta(n)(n+1)) \\
&< O(\delta(n) \cdot (n+1)^2) + O(n+1).
\end{aligned}
$$

Hence,

$$
t_s < O(\delta(n) \cdot (n+1)^2) + O(n+1). \tag{9}
$$

The smallest unit of parallelization in the BSF-implementation of the NSLP algorithm is the cohort. The number of cohorts equals to the space dimension n. Thus, the number P of slave-nodes should be less than or equal to the space dimension n. We shall assume $n \gg P$. A slave-node sequentially process all the cohorts assigned to it. In the current cohort, the coordinates of every point x are calculated using Eq. (5). The time complexity of this operation is $O(n)$. Then the point x is checked to be belonged to the polytope M_t. To do this, it is sufficient for the slave to verify the truth of the condition $A_t x = b_t$. Since A_t is of size $n \times 2(n+1)$, the time complexity of this operation is $O(n^2 + n)$. The number of points in a cohort excluding the central one is equal to the constant K. According to the Eq. (4), the total number of points in the cross excluding the central one is equal to nK. Hence, the time complexity of the calculations performed for all points of the cross in steps 2–3 can be estimated as $O(n^3 + n^2)$. After this, the slaves partially (for their cohorts only) execute the step 4 of Targeting phase (see Sect. 2). The total time complexity of these operations is $O(n^2)$. Thus, the total time complexity of all the calculations performed by slaves has the following estimation:

$$t_w = O(n^3 + n^2) + O(n^2) + O(n) \leq O(n^3 + n^2 + n). \tag{10}$$

As a result, each slave sends to the master a summarized vector of points which belong to the polytope and have the maximum value of the objective function in the corresponding cohort. Thus, the total time complexity of transferring the results from the slaves to the master is

$$t_r = O(PKn) \approx O(n). \tag{11}$$

Having received the results from the slaves, the master sums them up to complete the step 4 of the algorithm. The time complexity of these calculations will have the following estimation:

$$t_{\text{step 4}} \approx O(n^2). \tag{12}$$

Because of the non-stationarity of the LP problem, the condition in step 5 will be rarely true. Hence, we may assume that the next step after the step 4 will be the step 6 in most cases. Since the number of cohorts equals to n, the total time complexity of the step 6 of the Targeting phase is

$$t_{\text{step 6}} = O(n^2). \tag{13}$$

Thus, the total time complexity of processing the results obtained by the master from the slaves is

$$t_p \approx t_{\text{step 4}} + t_{\text{step 6}} = O(n^2). \tag{14}$$

Substituting the values from (10) and (9) into Eq. (6), we obtain the following estimation for the upper bound of the NSLP algorithm scalability:

$$P_{NSLP} \leq \sqrt{\frac{O(n^3 + n^2 + n)}{2L + O(\delta(n) \cdot (n+1)^2) + O(n+1)}}. \tag{15}$$

Suppose all the input data of the problem are changed at each iteration. It corresponds to $\delta(n) = 1$. In this case, inequality (15) is converted to the following form

$$P_{NSLP} \leq \sqrt{\frac{O(n^3 + n^2 + n)}{2L + O((n+1)^2) + O(n+1)}} \approx O(\sqrt{n}). \tag{16}$$

It means that the upper bound of the BSF-program scalability increases proportionally to the square root of the problem dimension. Hence, the NSLP algorithm implementation in the form of a BSF-program has limited scalability in this case.

Now suppose that the fraction of the changed problem input data at each iteration is

$$\delta(n) = \frac{1}{2(n+1)}. \tag{17}$$

It corresponds to a situation where the matrix has only one changed row, the column has only one changed element, and the objective function has no more than one changed coefficient. In this case, we obtain the following estimation

$$P_{NSLP} \leq \sqrt{\frac{O(n^3 + n^2 + n)}{2L + O(n+1) + O(n+1)}} \approx O(\sqrt{n^2}) = O(n) \qquad (18)$$

substituting the value of $\delta(n)$ from the Eq. (17) into the Eq. (15). It means that the upper bound of the BSF-program scalability increases proportionally to the problem dimension. Hence, the NSLP algorithm implementation in the form of a BSF-program is scalable well in this case.

We can also estimate the BSF-implementation parallel efficiency of the NSLP algorithm using approximate Eq. (8):

$$e = \frac{1}{1 + \frac{P^2 \cdot (2L + t_s) + P \cdot (t_r + t_p)}{t_w}}$$
$$= \frac{1}{1 + \frac{P^2 \cdot (2L + O(\delta(n) \cdot (n+1)^2) + O(n+1)) + P \cdot (O(n) + O(n^2))}{O(n^3 + n^2)}}. \qquad (19)$$

For $\delta(n) = 1$, $n \to \infty$ and $P \to \infty$, we get from (19) the following estimation

$$e = \frac{1}{1 + \frac{P^2(2L + O((n+1)^2) + O(n+1)) + P(O(n) + O(n^2))}{O(n^3 + n^2)}}$$
$$\approx \frac{1}{1 + \frac{P^2(O(n^2) + O(n)) + P \cdot (O(n^2) + O(n))}{O(n^3 + n^2)}} \qquad (20)$$
$$= \frac{1}{1 + (P^2 + P)\frac{O(n^2) + O(n)}{O(n^3 + n^2)}} \approx \frac{1}{1 + \frac{P^2 + P}{O(n)}} \approx \frac{1}{1 + P^2/O(n)}.$$

In such a way, we obtain

$$e \approx \frac{1}{1 + P^2/O(n)}. \qquad (21)$$

Hence for $\delta(n) = 1$, the high parallel efficiency is achieved when $n \gg P^2$. For $\delta(n) = \frac{1}{2(n+1)}$ we get from (19) the following estimation

$$e \approx \frac{1}{1 + P^2/O(n^2) + P/O(n)}. \qquad (22)$$

Hence for $\delta(n) = \frac{1}{2(n+1)}$, the high parallel efficiency is achieved when $n \gg P$.

5   Numerical Experiments

The implementation of the Qwest phase was described and evaluated by us in the paper [17]. In the present work, we have done the implementation of the Targeting phase in C language using the BSF skeleton. The source code of this program is freely available on Github, at https://github.com/leonid-sokolinsky/BSF-NSLP. We investigated the speedup and parallel efficiency of this BSF-program on the supercomputer "Tornado SUSU" [23] using the synthetic scalable linear

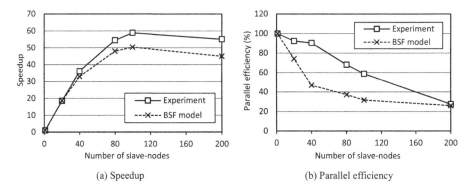

(a) Speedup

(b) Parallel efficiency

Fig. 3. Experiments for $n = 400$.

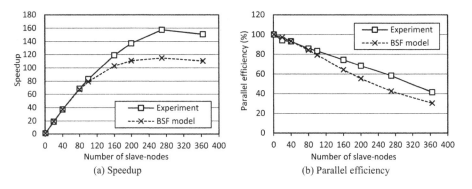

(a) Speedup

(b) Parallel efficiency

Fig. 4. Experiments for $n = 800$.

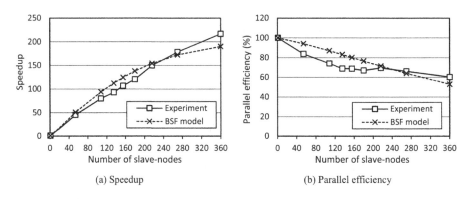

(a) Speedup

(b) Parallel efficiency

Fig. 5. Experiments for $n = 1080$.

programming problem *Model n* [17] mentioned in the Sect. 4. The calculations were performed for the dimensions $400, 800$ and 1080. At the same time, we plotted the curves of speedup and parallel efficiency for these dimensions using Eqs. (7) and (8). We assumed that $\delta(n) = 1/2(n+1)$. The results are presented in Figs. 3, 4 and 5. In all cases, the analytical estimations were very close to experimental ones. Moreover, the performed experiments show that the upper bound of the BSF-program scalability increases proportionally to the problem dimension. It was analytically predicted using the Eq. (18) in Sect. 4.

6 Conclusion

In this paper, the scalability and parallel efficiency of the NSLP algorithm used to solve large-scale non-stationary linear programming problems on cluster computing systems were investigated. To do this, we used the BSF (Bulk Synchronous Farm) parallel computation model based on the "master-slave" paradigm. The BSF-implementation of the NSLP algorithm is described. A scalability upper bound of the BSF-implementation of the NSLP algorithm is obtained. This estimation tells us the following. If all the input data of the problem are changed at each iteration then the upper bound of the BSF-program scalability increases proportionally to the square root of the problem dimension. In this case, the NSLP algorithm implementation in the form of a BSF-program has limited scalability. If each inequality of the constraint system has no more than one coefficient changed during an iteration then the upper bound of the BSF-program scalability increases proportionally to the problem dimension. In this case, the NSLP algorithm implementation in the form of a BSF-program is scalable well. The equations for estimating the parallel efficiency of the BSF-implementation of the NSLP algorithm are also deduced. These equations allow us to conclude the following. If during the iteration all the problem input data are dynamically changed then for the high parallel efficiency it is necessary that the problem dimension is much greater than the square of the number of slaves: $n \gg P^2$. However, if in each inequality of the constraint system no more than one coefficient changes during each iteration then for a high parallel efficiency it is necessary that the problem dimension be much greater than the number of slaves: $n \gg P$. The numerical experiments with a synthetic problem showed that the BSF model accurately predicts the upper bound of the scalability of the program that implements the Targeting phase using the BSF skeleton.

As future research directions, we intend to do the following:

(1) implement the Qwest phase in C language using the BSF skeleton and MPI-library;
(2) carry out numerical experiments on a cluster computing system using synthetic and real LP problems;
(3) compare the scalability boundaries of the Qwest phase obtained experimentally and analytically.

References

1. Chung, W.: Applying large-scale linear programming in business analytics. In: Proceedings of the 2015 IEEE International Conference on Industrial Engineering and Engineering Management (IEEM), pp. 1860–1864. IEEE (2015)
2. Tipi, H.: Solving super-size problems with optimization. Presentation at the meeting of the 2010 INFORMS Conference on O.R. Practice, Orlando, Florida, April 2010. http://nymetro.chapter.informs.org/prac_cor_pubs/06-10%20Horia%20Tipi%20SolvingLargeScaleXpress.pdf. Accessed 7 May 2017
3. Gondzio, J., et al.: Solving large-scale optimization problems related to Bell's Theorem. J. Comput. Appl. Math. **263**, 392–404 (2014)
4. Sodhi, M.S.: LP modeling for asset-liability management: a survey of choices and simplifications. Oper. Res. **53**(2), 181–196 (2005)
5. Brogaard, J., Hendershott, T., Riordan, R.: High-frequency trading and price discovery. Rev. Finan. Stud. **27**, 2267–2306 (2014)
6. Budish, E., Cramton, P., Shim, J.: The high-frequency trading arms race: frequent batch auctions as a market design response. Q. J. Econ. **130**, 1547–1621 (2015)
7. Goldstein, M.A. and Kwan, A., Philip, R.: High-frequency trading strategies. https://ssrn.com/abstract=2973019
8. Hendershott, T., Jones, C.M., Menkveld, A.J.: Does algorithmic trading improve liquidity? J. Finan. **66**, 1–33 (2011)
9. Dantzig, G.B.: Linear Programming and Extensions. Princeton University Press, Princeton (1998)
10. Klee, V., Minty, G.J.: How good is the simplex algorithm? In: Inequalities III (Proceedings of the Third Symposium on Inequalities Held at the University of California, Los Angeles, California, 1–9 September 1969, dedicated to the memory of Theodore S. Motzkin), pp. 159–175. Academic Press, New York (1972)
11. Karmarkar, N.: A new polynomial-time algorithm for linear programming. Combinatorica **4**, 373–395 (1984)
12. Sokolinskaya, I., Sokolinsky, L.B.: On the Solution of Linear Programming Problems in the Age of Big Data. In: Sokolinsky, L., Zymbler, M. (eds.) PCT 2017. CCIS, Vol. 753. pp. 86–100. Springer, Cham (2017). https://doi.org/10.1007/978-3-319-67035-5_7
13. Agmon, S.: The relaxation method for linear inequalities. Can. J. Math. **6**, 382–392 (1954)
14. Motzkin, T.S., Schoenberg, I.J.: The relaxation method for linear inequalities. J. Can. Math. **6**, 393–404 (1954)
15. Eremin, I.I.: Fejerovskie metody dlya zadach linejnoj i vypukloj optimizatsii [Fejer's Methods for Problems of Convex and Linear Optimization]. Publishing of the South Ural State University, Chelyabinsk (2009)
16. González-Gutiérrez, E., Hernández Rebollar, L., Todorov, M.I.: Relaxation methods for solving linear inequality systems: converging results. TOP **20**, 426–436 (2012)
17. Sokolinskaya, I., Sokolinsky, L.: Revised pursuit algorithm for solving nonstationary linear programming problems on modern computing clusters with manycore accelerators. In: Voevodin, V., Sobolev, S. (eds.) RuSCDays 2016. CCIS, vol. 687, pp. 212–223. Springer, Cham (2016). https://doi.org/10.1007/978-3-319-55669-7_17
18. Sahni, S., Vairaktarakis, G.: The master-slave paradigm in parallel computer and industrial settings. J. Glob. Optim. **9**, 357–377 (1996)

19. Silva, L.M., Buyya, R.: Parallel programming models and paradigms. High Perform. Cluster Comput. Architect. Syst. **2**, 4–27 (1999)
20. Leung, J.Y.-T., Zhao, H.: Scheduling problems in master-slave model. Ann. Oper. Res. **159**, 215–231 (2008)
21. Sokolinsky, L.B.: Analytical estimation of scalability of iterative numerical algorithms on distributed memory multiprocessors. http://arxiv.org/abs/1710.10490
22. Darema, F., George, D.A., Norton, V.A., Pfister, G.F.: A single-program-multiple-data computational model for EPEX/FORTRAN. Parallel Comput. **7**, 11–24 (1988)
23. Kostenetskiy, P.S., Safonov, A.Y.: SUSU supercomputer resources. In: Proceedings of the 10th Annual International Scientific Conference on Parallel Computing Technologies (PCT 2016). CEUR Workshop Proceedings, vol. 1576, pp. 561–573 (2016)

Dynamic Optimization of Linear Solver Parameters in Mathematical Modelling of Unsteady Processes

Dmitry Bagaev[1,2(✉)], Igor Konshin[1,3], and Kirill Nikitin[1]

[1] Institute of Numerical Mathematics of the Russian Academy of Sciences, Moscow 119333, Russia
bvdmitri@gmail.com, igor.konshin@gmail.com, nikitin.kira@gmail.com
[2] Lomonosov Moscow State University, Moscow 119991, Russia
[3] Dorodnicyn Computing Centre, FRC CSC RAS, Moscow 119333, Russia

Abstract. The optimization of linear solver parameters in unsteady multiphase groundflow modelling is considered. Two strategies of dynamic parameters setting for the linear solver are proposed when the linear systems properties are modified during simulation in the INMOST framework. It is shown that the considered algorithms for dynamic selection of linear solver parameters provide a more efficient solution than any prescribed set of parameters. The results of numerical experiments on the INM RAS cluster are presented.

Keywords: Parallel linear solver · Mathematical modelling · Unsteady process · Automatic performance tuning · Black-Oil Simulator

1 Introduction

The problem of software performance tuning is of great importance for efficient usage of the modern supercomputer facilities. It is very important for numerical modelling applications exploiting such a software.

One of the most famous examples of automatic software tuning is the ATLAS package [1] which carry out the performance optimization for several BLAS functions during the installation of the package.

Another important and very popular idea of software performance tuning is the usage of data mining techniques. For example, Self-Adapting Numerical Software (SANS) [2] and Self-Adapting Large-scale Solver Architecture (SALSA) [3] perform the analysis of the input data to select the linear solver from the set of available ones. The machine learning techniques is used for the same goal as well [4].

The genetic algorithms are used in [5] in a software system called Intelligent Performance Assistant (IPA) to improve the performance of ExxonMobil's proprietary reservoir simulator, EMpower[TM].

© Springer International Publishing AG 2017
V. Voevodin and S. Sobolev (Eds.): RuSCDays 2017, CCIS 793, pp. 54–66, 2017.
https://doi.org/10.1007/978-3-319-71255-0_5

In the present paper we would like to return 'back to basics' of linear algebra and to knowledge on the mathematical properties of the preconditioned iterative algorithms considered. For this reason we consider the solution of unsteady problems that comes from multiphase black-oil reservoir simulation. The main difficulty of selecting the optimal parameters of the linear solvers at each simulation time step is the modification of the stiffness matrix properties. If the linear solver is already selected prior to the unsteady problem solution, then one has at least a possibility to select the input set of linear solver parameters. The main idea is to construct the procedure of automatic and dynamic selection of parameters that are close to the optimal ones, i.e. provide the minimum of the solution time. In the present paper we propose two different algorithms for this approach.

For our numerical experiments on dynamic parameters tuning we have exploited an INMOST software platform [6]. Besides the ability to operate with the distributed meshes of general form, this platform includes a convenient interface for solving large sparse linear systems. It allows user to forget about specific implementations of each particular linear solver and to focus only on parameters optimizations. INMOST provides a large variety of different linear solvers, some of them are implemented inside the platform, the others can be enabled as external libraries, such as PETSc [7] or Trilinos [8].

We consider the INMOST linear solver BIILU2 for our numerical experiments. This solver is the combination of the second order incomplete triangular factorization ILU2(τ) and the incomplete inverse LU factorization BIILU(q) (as a replacement of additive Schwarz preconditioning AS(q)) [9,10]. Here, τ is the factorization threshold and q is the number of overlap levels for blocks corresponding to each processor. The use of ILU2(τ) factorization is chosen due to it's robust and efficient preconditioning in comparison with the conventional structural incomplete factorization ILU(k) or the conventional incomplete threshold factorization ILU(τ).

As an example of simulation we use the multi-phase flow model based on the fully implicit time discretization and the nonlinear monotone two-point approximation for the Darcy fluxes in Jacobian matrix [11].

2 Algorithm's Description

2.1 The Choice of Appropriate Optimization Algorithm

The function to be optimized can be defined as

$$T_k = G(A_k, b_k, p, \varepsilon) \equiv F(A_k, b_k, p) \pm \varepsilon, \tag{1}$$

where A_k is the linear system matrix on kth time step, b_k is the right-hand side vector on kth time step, p is the parameter (or parameters) of some liner solver to be optimized, T_k is the return value of function which is equal to the time needed for solving the linear system $A_k x = b_k$.

A lot of difficulties associated with the real unsteady processes should be taken into account while choosing algorithms for parameter optimization of linear solvers:

1. Function G can behave differently from run to run, as we solve the problem in parallel mode using the MPI library. The time of messages delivery between processors is nondeterministic, so the function value may vary on some small but essential unknown value ε which is impossible to predict. Therefore the target function may have several local minima and maxima.
2. During unsteady process both A_k and b_k will be modified with each time step and as a result optimal parameters p will be changed as well. The optimization algorithm should be able to find these parameters (or close to it) regardless their modification in time.
3. The value of function G can be calculated only once for given A_k, b_k and for the selected parameters p. There is no reason to solve linear system $A_k x = b_k$ again even with more optimal parameters.
4. As A_k may vary with the simulation time, the minimum value of function F may increase. This is why it is really hard to use the previous values of $F(A_k, b_k, p)$.
5. The algorithm should not be computationally expensive and time spent on the parameters optimization should not affect the total time of solving the unsteady problem.

To deal with the above difficulties we should also use a number of assumptions on the function F:

1. For given A_k and b_k the function is continuous by parameters p and has the form close to a paraboloid, and since $F > 0$ the global minima exists and finite.
2. In a real simulation matrices A_k may differ, however they have about the same structure and properties, and as a result we expect that the optimal parameters based on the kth time step are moved in its small neighborhood and within this area the values of the T_k are roughly equal.
3. We also assume that at some time step k' the minimal solution time T_k is not increasing and depends only on parameters p.

Based on the above issues and assumptions we have proposed to use two optimization algorithms.

2.2   Very Fast Simulated Re-annealing

Annealing (simulated annealing, SA) is a probabilistic technique for approximating the global optimum of a given function. At each step, the SA heuristic considers some neighbouring state s' of the current state s, and probabilistically decides between moving the system to state s' or staying in state s. These probabilities ultimately lead the system to move to states of lower energy. Typically this step is repeated until the system reaches a state that is good enough for the application, or until a given computation budget has been exhausted.

The probability of making the transition from the current state s to a candidate new state s' is specified by an *acceptance probability function* $h(e, e', T)$, that depends on the energies $e = E(s)$ and $e' = E(s')$ of the two states, and on a global time-varying parameter T called the *temperature*. States with a smaller energy are better than those with a greater energy. The probability function P must be positive even when e' is greater than e. This feature prevents the method from becoming stuck at a local minimum that is worse than the global one.

When T tends to zero, the probability $h(e, e', T)$ must tend to zero if $e' > e$ and to a positive value otherwise. For sufficiently small values of T, the system will then increasingly favor moves that go "downhill" (i.e., to lower energy values), and avoid those that go "uphill". With $T = 0$ the procedure reduces to the *greedy algorithm*, which makes only the downhill transitions [12].

The method of simulated annealing consists of three functional relationships:

g – probability density of state-space of D parameters $x = \{x^i, i = 1, D\}$;
h – probability density for acceptance of new cost-function given the just previous value;
$T(k)$ – schedule of annealing temperature T in annealing time steps k, i.e. of changing volatility or fluctuations of the two previous probability densities.

The acceptance probability is based on the chances of obtaining a new state s' relative to a previous state s,

$$h = \frac{\exp(-e'/T)}{\exp(-e'/T) + \exp(-e/T)} \approx \frac{1}{1 + \exp(\Delta E/T)}, \tag{2}$$

where ΔE represents the *energy* difference between the present and previous values of the cost-function appropriate to the physical problem, i.e. $\Delta E = e' - e$ (see [13]).

The algorithm itself can be described by the following steps:

1. Select a random state s. The energy values of the system is set to $E(s)$.
2. On kth step:
   (a) Compare the energy of the system $E(s)$ in the state s with the global minimum. If it is smaller then change the global minimum value.
   (b) Generate a new state s' and calculate $E(s')$.
   (c) Generate a random number α uniformly distributed over $[0, 1]$. If $\alpha < h(\Delta E, T(k))$ then set s' as the current state and go to the next iteration $k + 1$. Otherwise repeat the previous step until a suitable state s' will be found.

In the present paper we are using the "very fast annealing scheme" produced by Ingber [13]. In this scheme different parameters may have different finite ranges, fixed by physical considerations, and different annealing-time-dependent sensitivities, measured by the curvature of the cost-function at local minima.

Consider parameters x_k^i in ith dimension generated by an annealing step k with the following range

$$x_k^i \in [A_i, B_i] \tag{3}$$

calculated with the random variable ξ_i:

$$x_{k+1}^i = x_k^i + \xi_i(B_i - A_i), \qquad \xi_i \in [-1, 1]. \tag{4}$$

The above formula can be applied several times until $x_{k+1}^i \in [A_i, B_i]$.

Generating function defined as

$$g_T(\xi) = \prod_{i=1}^{n} \frac{1}{2(|\xi_i| + T_i)\ln(1 + 1/T_i)} \equiv \prod_{i=1}^{n} g_{(i;T)}(\xi_i), \qquad \xi_i \in [-1, 1],$$

$$\xi_i = \text{sgn}\left(\alpha_i - \frac{1}{2}\right) T_i((1 + 1/T_i)^{|2\alpha_i - 1|} - 1), \tag{5}$$

where α_i are random numbers, uniformly distributed over segment $[0, 1]$.

Annealing schedule will be defined as

$$T_i(k) = T_{(i;0)} \exp(-c_i k^{1/D}), \qquad c_i > 0. \tag{6}$$

It is proven [13], that the very fast annealing algorithm are one of the most effective method of random search of optimal solutions for a wide class of problems.

2.3 Alternating Parameters Probe Based Tuning

Another idea for constructing the algorithm for dynamic parameters tuning for unsteady problem is the attempt to stay at a local minimum probing a nearby area. If the current parameters set is near to the minimum or the minimum is moving not too fast then the algorithm may track the minimum.

The algorithm (1U) for unsteady problem can be formulated as follows:

Specify initial values for τ, q, and probe direction dir from $\{\delta_{\tau+}, \delta_{q+}, \delta_{\tau-}, \delta_{q-}\}$
while simulation stopping criterion **do**
    Make time step
    Solve linear system
    if new minimum found **then**
        Update minimum set (τ, q)
    end if
    if dir= $\delta_{\tau+}$ **then**
        ind$(\tau) + +$
    else if dir= δ_{q+} **then**
        ind$(q) + +$
    else if dir= $\delta_{\tau-}$ **then**
        ind$(\tau) - -$

else if dir= δ_{q-} **then**
    ind(q) $--$
else
    Stay with no change of (τ, q)
end if
end while

2.4   Linear Brute-Force Searching

Linear brute-force search is the simplest algorithm, which can find the global minimum of the given function $F(\overline{x})$, where $\overline{x} = (x_1, x_2, \ldots x_n)$ on an arbitrary grid D.

Linear brute-force search algorithm implies optimizing each variable x_i independently and was implemented in the following way:

- The set of runs for different values of τ from $\tau_{\min}$ to $\tau_{\max}$ for a fixed value of overlap size parameter $q = 3$ was performed, and a quasi-optimal value of τ^* was found.
- The set of runs for different values of q from $q_{\min}$ to $q_{\max}$ for a fixed value of quasi-optimal τ^* was performed, and a quasi-optimal pair of parameters (τ^*, q^*) was found.

This method are very computationally expensive and therefore can't be recommended for solving real problems. However it can be used to find the almost precise global minimum on quite dense grids and enable us to verify the other parameter tuning approaches.

3   Numerical Experiments

3.1   INM Cluster Configuration

All numerical experiments was performed on INM RAS cluster. The configuration of the cluster computational nodes, used for numerical experiments [14]:

- Compute Node Arbyte Alkazar+ R2Q50;
- 16 cores (two 8-core processors Intel Xeon E5-2665@2.40 GHz);
- 64 Gb RAM;
- SUSE Linux Enterprise Server 11 SP1 (x86_64).

3.2   Dependance on Parameters for a Sample Problem

As a sample linear system we have used the system (called below N14) obtained from the INM RAS Black-Oil Simulator for Scholars (BOSS) for the well-known SPE-10 problem [15]. The size of the model mesh is $60 \times 220 \times 85$ cells ($1.122 \cdot 10^6$ cells). The top 35 layers of the model is a Tarbert formation, and is a representation of a prograding near shore environment, while the bottom 50

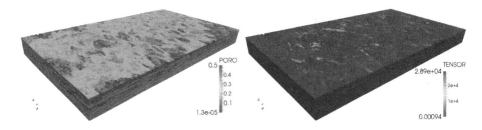

Fig. 1. The porosity and permeability distributions for SPE-10 problem

layers represents Upper Ness which is fluvial. The coefficients of the media are very contrast. The porosity varies from $1.3 \cdot 10^{-5}$ to 0.5 (see Fig. 1, left) and the permeability varies from 10^{-3} to $3 \cdot 10^{4}$ (see Fig. 1, right). The model has 5 vertical wells completed throughout formation. The central well is an injector and the other 4 wells in the corners are producers.

The dimension of the obtained linear system N14 is 3 896 013 unknowns. The dependences of solution time T (in seconds) on parameters τ and q is demonstrated in Fig. 2 for 16 cores and it is quite smooth with the minimum pronounced.

Figure 3 shows the 2D surface of the solution time T in variables τ and q for the same problem N14 solved on 16 cores. The obtained surface is of paraboloid type.

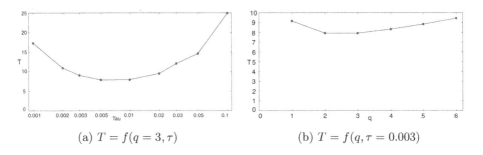

(a) $T = f(q = 3, \tau)$ (b) $T = f(q, \tau = 0.003)$

Fig. 2. Total solution time T in s. for N14 depending on τ and q for $p = 16$

3.3 Dynamic Function Simulation

We consider the following two-parameter function for the research purposes:

$$f(\tau, q) = \left(\frac{16}{25} (\lg(\tau/\tau_0))^2 + 1 \right) \left(\frac{1}{25} \left(\frac{17.5(q - q_0)}{7.5 + q - q_0} \right)^2 + 1 \right), \tag{7}$$

$$\tau_0 = 0.003, \quad q_0 = 3. \tag{8}$$

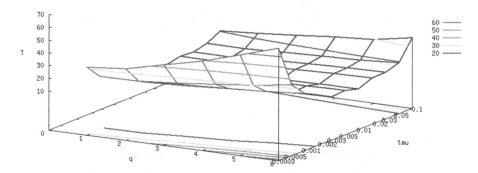

Fig. 3. Total solution time T in s. for N14 in variables τ and q for $p = 16$

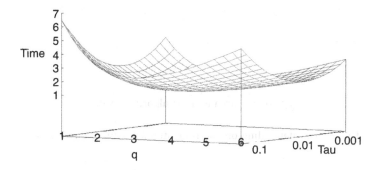

Fig. 4. Two-parameter function (7) and (8)

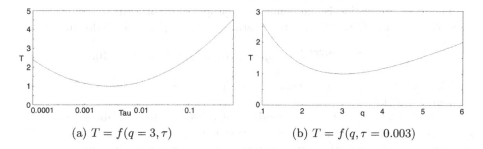

(a) $T = f(q = 3, \tau)$ (b) $T = f(q, \tau = 0.003)$

Fig. 5. Cross-sections for $q = q_{\mathrm{opt}} = 3$ and $\tau = \tau_{\mathrm{opt}} = 0 : 003$

This function can be used as a solution time T measured in seconds instead of that for real black-oil simulation process and demonstrates more strong dependence on $\lg(\tau)$ as well as more weak one on overlap parameter q. Figure 4 demonstrates the respective paraboloid for the above function, which is qualitatively similar to the paraboloid on Fig. 3. The minimum of this two-parameter function is in $(\tau = 0.003, q = 3)$ in accordance with (8). Figures 5a and b show the cross-sections for $q = q_{\mathrm{opt}} = 3$ and $\tau = \tau_{\mathrm{opt}} = 0.003$, respectively. With this simple

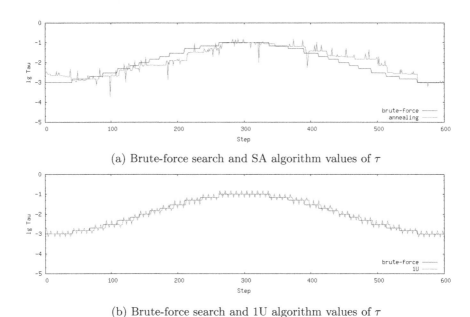

(a) Brute-force search and SA algorithm values of τ

(b) Brute-force search and 1U algorithm values of τ

Fig. 6. τ_{opt} depending on the time step k for function (7), (9)

function we can easily examine the proposed parameter tuning approaches as well as provide the complete repeatability of our numerical experiments.

The most interesting is the behavior of proposed algorithms in the unsteady case. We can modify the above steady state function (7) in the following way:

$$\tau_0 = 10^{-2-\cos(2\pi t/t_0)}, \quad q_0 = 2 + \cos(2\pi t/t_0), \quad t_0 = 100 \tag{9}$$

where we have the local optimal values $\lg \tau \in [-3; -1]$ and $q \in [1; 3]$ for time moment $t \in [0; t_0]$.

Figures 6a and b plot τ_{opt} depending on the time step for above unsteady-state function (7), (9). This figures show that proposed algorithms SA and 1U are able to track the optimal parameters even if they change in time.

3.4   Unsteady Black-Oil Simulation

We consider the two-phase flow model of the INM RAS BOSS simulator for the real unsteady problem. We simulate 6000 days of the quarter five spot problem with one injector and one producer wells. The initial water saturation is equal to residual saturation which results in rather sharp front. The starting time step is 0.0001 days, which increases to 25 days later in the simulation. An incremental time step leads to an increase in the complexity of linear systems, so does the water breakthrough which results in higher flow velocities. In our simulation the water breakthrough occurs at about time 1400 or at about 65th time steps.

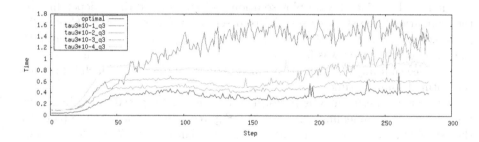

Fig. 7. Unsteady black-oil simulation times with fixed parameters and the dynamic optimal ones depending on the simulation time step k

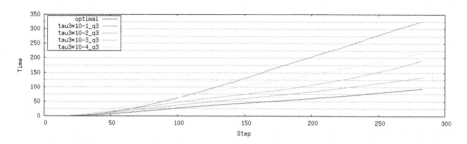

Fig. 8. Unsteady black-oil simulation cumulative times with the fixed parameters and the dynamic optimal ones depending on the simulation time step k

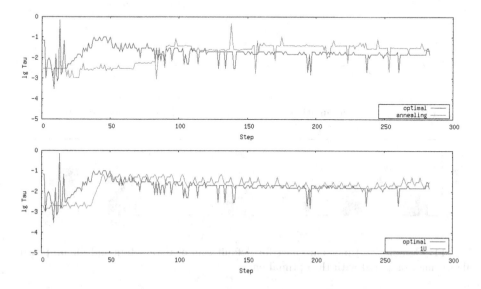

Fig. 9. Optimizing τ for black-oil simulator

First, in Figs. 7 and 8 we plot solution and cumulative times depending on the simulation time step for several fixed sets of parameters: ($\tau = 0.3, q = 2$), ($\tau = 0.03, q = 3$), ($\tau = 0.003, q = 3$), ($\tau = 0.0003, q = 3$) and compare it with the optimal one, which was found using the linear brute-force search algorithm. One can see that any fixed set of parameters produce the result which is far from the optimal solution time.

The same experiment was performed for the two proposed algorithms very fast simulated re-annealing (SA) and 1U. Figures 9 and 10 present the plots for the estimated value τ_{opt}, local and cumulative solution time T and T_Σ, respectively, depending on the simulation time step k. One can observe that the results of the proposed algorithms are very close to that for the optimal set of parameters. The cumulative solution time for all the proposed algorithms is less than that for any observed fixed set of parameters (τ, q) (Fig. 11).

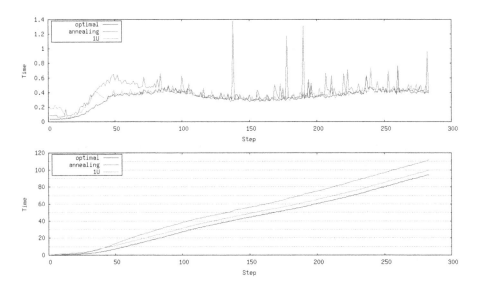

Fig. 10. Local and cumulative times depending on the simulation time step k

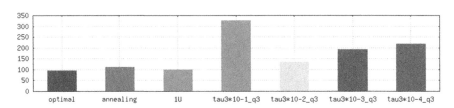

Fig. 11. Cumulative times bar chart for default sets of parameters and for proposed algorithms compared with the optimal one

4 Conclusion

The proposed linear solver parameters tuning algorithms were implemented in the INMOST framework as an optional toolkit named Trace and Tuning Software Platform (TTSP). In conclusion of the present paper, we formulate the most important issues of the progress in this area:

- The influence of the linear solver parameters on the real black-oil simulation performance was examined;
- The set of optimization algorithms for linear solver parameters tuning were proposed;
- The TTSP toolkit for parallel linear solver parameters tuning was developed and verified for the INM RAS black-oil reservoir simulator for scholars (BOSS);
- It was shown that proposed algorithms essentially increase the performance of the real unsteady black-oil simulation in comparison with even the best fixed set of parameters.

Acknowledgements. This work has been supported in part by RFBR grant 17-01-00886, Russian Federation President Grant MK-2951.2017.1, and ExxonMobil Upstream Research Company.

References

1. Automatically Tuned Linear Algebra Software (ATLAS). http://math-atlas. sourceforge.net/. Accessed 15 Apr 2017
2. Eijkhout, V., Fuentes, E., Eidson, T., Dongarra, J.: The component structure of a self-adapting numerical software system. Int. J. Parallel Program. **33**(2), 14–18 (2005)
3. Self-Adapting Large-scale Solver Architecture (SALSA). http://icl.cs.utk.edu/ salsa/index.html. Accessed 15 Apr 2017
4. Bhowmick, S., Eijkhout, V., Freund, Y., Fuentes, E., Keyes, D.: Application of machine learning to the selection of sparse linear solvers. Int. J. High Perform. Comput. Appl. 1–24 (2006)
5. Mishev, I.D., Beckner, B.L., Terekhov, S.A., Fedorova, N.: Linear solver performance optimization in reservoir simulation studies. In: Society of Petroleum Engineers. SPE Reservoir Simulation Symposium, pp. 1–9. The Woodlands, Texas (2009)
6. INMOST - a toolkit for distributed mathematical modelling. http://www.inmost. org. Accessed 15 Apr 2017
7. PETSc - Portable, Extensible Toolkit for Scientific computation. https://www. mcs.anl.gov/petsc/. Accessed 15 Apr 2017
8. The Trilinos Project. https://trilinos.org. Accessed 15 Apr 2017
9. Kaporin, I.E., Konshin, I.N.: Parallel solution of large sparse SPD linear systems based on overlapping domain decomposition. In: Malyshkin, V. (ed.) PaCT 1999. LNCS, vol. 1662, pp. 436–446. Springer, Heidelberg (1999). https://doi.org/10. 1007/3-540-48387-X_45

10. Kaporin, I.E., Konshin, I.N.: A parallel block overlap preconditioning with inexact submatrix inversion for linear elasticity problems. Numer. Linear Algebra Appl. **9**(2), 141–162 (2002)

11. Nikitin, K.D., Terekhov, K.M., Vassilevski, Y.V.: A monotone nonlinear finite volume method for diusion equations and multiphase OWS. Comp. Geosci. **18**(3), 311–324 (2014)

12. Simulated annealing - Wikipedia, The Free Encyclopedia. https://en.wikipedia.org/wiki/Simulated_annealing. Accessed 15 Apr 2017

13. Ingber, L.: Very fast simulated re-annealing. Math. Comput. Model. **12**, 967–973 (1989)

14. INM RAS cluster. http://cluster2.inm.ras.ru. Accessed 15 Apr 2017. (in Russian)

15. SPE Comparative Solution Project. http://www.spe.org/web/csp. Accessed 15 Apr 2017

Optimization of Numerical Algorithms for Solving Inverse Problems of Ultrasonic Tomography on a Supercomputer

Sergey Romanov[⊠]

Lomonosov Moscow State University, Moscow, Russia
romanov60@gmail.com

Abstract. The paper is dedicated to optimizing numerical algorithms to solve wave tomography problems by using supercomputers. The problem is formulated as a non-linear coefficient inverse problem for the wave equation. Due to the huge amount of computations required, solving such problems is impossible without the use of high-performance supercomputers. Gradient iterative methods are employed to solve the problem. The gradient of the residual functional is calculated from the solutions of the direct and the "conjugate" wave-propagation problems with transparent boundary conditions. Two formulations of the transparency condition are compared. We show that fourth-order finite-difference schemes allow us to reduce the size of the grid by a factor of 1.5–2 in each coordinate compared to second-order schemes. This makes it possible to significantly reduce the amount of computations and memory required, which is especially important for 3D problems of wave tomography. The primary application of the method is medical ultrasonic tomography.

Keywords: Ultrasound · Coefficient inverse problems · Supercomputer · Wave tomography · Finite-difference schemes

1 Introduction

Currently, intensive works are being carried out to develop new tomographic devices that use wave radiation sources. The most promising technology is ultrasonic tomography. The most important applications of ultrasonic tomography are in medical research, primarily the differential diagnosis of breast cancer, which is one of the most pressing issues of medical diagnostics. Wave tomography technology can also be used in many other applications, such as seismic studies, non-destructive testing, and medical ultrasonic imaging [1–3].

One of the problems in the development of ultrasonic tomography is associated with the nonlinearity of inverse problems of wave tomography. These inverse problems are formulated as coefficient inverse problems for the wave equation [4, 5]. The developments of ultrasonic tomography devices are currently at the stages of modelling and prototypes [6–8]. These works employ simplified mathematical models. The most promising approach is to develop methods for solving inverse problems of wave tomography under models that account for both wave diffraction effects

© Springer International Publishing AG 2017
V. Voevodin and S. Sobolev (Eds.): RuSCDays 2017, CCIS 793, pp. 67–79, 2017.
https://doi.org/10.1007/978-3-319-71255-0_6

(diffraction, refraction, multiple scattering) and absorption. The derivation of the gradient of the residual functional of the coefficient inverse problem, as obtained in [9–14], was the breakthrough result in this field.

The approximate gradient-based methods for solving inverse problems of ultrasonic tomography have been developed in [15–19]. The developed algorithms are designed for supercomputers. These algorithms implement iterative gradient methods to minimize the residual functional between the wave field measured experimentally and the numerically simulated wave field. To calculate the gradient of the residual functional at each iteration of the method, it is necessary to solve the "direct" problem of simulating the wave propagation process in an inhomogeneous medium in forward time and the "conjugate" problem in reverse time. The efficiencies of the developed numerical methods were evaluated by benchmarking numerous model problems on the "Lomonosov" supercomputer. The developed methods allow effective parallelization. The numerical algorithms practically linearly scale with the number of processors in CPU- and GPU-based systems.

The aim of this study is to optimize the developed numerical algorithms. The first way to optimize the algorithms is to use a finite-difference scheme. The numerical methods for solving the inverse problem of wave tomography that have been implemented in the previous works are based on the finite-difference time-domain (FDTD) method that provides a second-order approximation of the wave equation. The FDTD method has been chosen because it has a very large potential for relatively simple parallelization of computations. Because of the large amount of computations, highly parallel computing is required. Supercomputer technologies drastically reduce the computation time required to solve inverse problems. However, one of the issues is the accuracy of the calculations. To solve the ill-posed inverse problems of wave tomography, very high accuracy is required. For second-order FDTD schemes, this leads to the need to use very large grids. With the increase of the sounding frequencies, the volume of data becomes unacceptably large. This is especially true for 3D problems, where high-performance GPU processors are required, and GPUs have a limited memory capacity. Additionally, there are numerical error accumulation issues associated with large grids.

For second-order FDTD schemes, it is necessary to use at least n = 1000 grid points in each coordinate for the numerical error to be no more than a few percent. In the 3D version of the method, this results in a 1000^3-point grid. Even when solving such problems on powerful GPU clusters, such an amount of data does not fit into the internal memory of the GPU devices. Another problem is the dependence of the number of operations in the gradient iterative algorithms on the number of grid points, which is of the form $O(n^4)$. A fourth power of n means that whereas it takes one hour to solve a 3D problem for $n = 400$ on a GPU cluster, it would take approximately 5 h to solve a problem for $n = 600$. Therefore, one of the ways to optimize the algorithms is to use higher-order approximations. As will be shown, using a fourth-order approximation scheme results in a decrease in the number of grid points n by a factor of 1.5–2, which significantly reduces the computation time.

The second important issue is the problem of boundary conditions. This problem arises because we have to solve direct and inverse problems in a bounded domain.

As a result, reflection of waves occurs at the boundary of the computational domain. In this paper, two methods of implementing a "transparent" boundary are considered.

The third issue is that the inverse problem of wave tomography typically has incomplete input data — the sources and detectors are not located on all sides of the studied object. In ultrasonic mammography applications, the data incompleteness results from the fact that we cannot place sources and detectors at the patient's chest-wall side. Thus, it is an incomplete-data tomography.

2   Formulation of the Inverse Problem of Ultrasonic Tomography with Incomplete Data and Solution Methods

Let us consider the "direct" problem of computing the acoustic pressure $u(r,t)$ for the time $(0; T)$ in the region $\Omega \subset \mathbf{R}^N$ ($N = 2, 3$), bounded by the surface $\partial\Omega$ (Fig. 1), with a point source at the point r_0:

$$c(r)u_{tt}(r,t) - \Delta u(r,t) = \delta(r - r_0)\, g(t). \tag{1}$$

Let us assume that $u(r,t)$ satisfies the zero initial and boundary conditions

$$u(r, t = 0) = u_t(r, t = 0) = 0,\ \partial_n u(r,t)|_{\partial\Omega} = 0. \tag{2}$$

Here, $c^{-0.5}(r) = v(r)$ is the sound speed in the medium, $r \in \mathbf{R}^N$; Δ is the Laplace operator with respect to r. The pulse generated by the source is described by the function $g(t)$; $\partial_n u(r,t)|_{\partial\Omega}$ is the derivative along the normal to the boundary $\partial\Omega$. It is assumed that the inhomogeneities of the medium are sound-speed variations and are localized within the studied object G. Outside of the object, $v(r) = v_0$ is constant and v_0 is known. The acoustic pressure is measured at the boundary of the domain R, G $\subset$ R. The sources insonify the studied object from different directions. We assume that the sources and the region R are located far enough from the boundary $\partial\Omega$ such that the conditions (2) are satisfied.

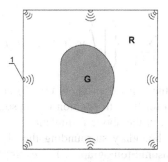

Fig. 1. The scheme of the 2D experiment.

Figure 1 illustrates the arrangement of sources and detectors in the two-dimensional inverse problem of wave tomography. The number 1 denotes the positions of the sources of ultrasound waves, and the measurements are taken at the boundary ∂R. The studied object G is located inside the domain R, which is filled with a homogeneous medium with a known sound speed v_0.

The inverse problem consists of determining the sound speed $c(r)$ from the experimental data $U(s, t)$ measured at the boundary ∂R of the domain R during the time $(0, T)$ with different positions r_0 of the source. In the formulation with incomplete data, the acoustic pressure $U(s, t)$ is not measured on the whole boundary ∂R. The inverse problem with incomplete data can be formulated as a problem of minimizing the residual functional

$$\Phi(u(c)) = \frac{1}{2} \sum_{j=1}^{M} \int_0^T \int_{\partial R} E^2(s, t) ds dt, \tag{3}$$

$$\text{where } E(s, t) = \begin{cases} u(\mathbf{s}, t) - U(\mathbf{s}, t), & \text{for } \mathbf{s} \in \partial R, \text{ where } U(\mathbf{s}, t) \text{ is known} \\ 0, & \text{otherwise} \end{cases}. \tag{4}$$

Here, $u^j(c)$ is the solution of the problem (1)–(2) for some $c(r)$; the index $j = 1,\ldots,$ M denotes the position of the source. The inverse problem is formulated as the problem of finding a function $\bar{c}(r)$ that minimizes the residual functional $\min_{c(r)} \Phi(u(c)) = \Phi(u(\bar{c}))$.

Let us consider another problem, which we call "conjugate" to the "direct" problem (1)–(2):

$$c(r)w_{tt}(r, t) - \Delta w(r, t) = E(r.t)|_{r \in \partial R}, \tag{5}$$

$$w(r, t = T) = w_t(r, t = T) = 0, \ \partial_n w(r, t)|_{\partial \Omega} = 0, \tag{6}$$

where $E(r,t)$ from (4) is derived from the measured data $U(s,t)$ and the solution u of the direct problem (1)–(2). Then, as shown in [9, 11, 14], the gradient of the functional (3) has the form

$$\Phi'(u(c), dc) = \sum_{j=1}^{M} \int_{\Omega} \left\{ \left[\int_0^T w_t^j(r, t) u_t^j(r, t) dt \right] dc(r) \right\} dr, \tag{7}$$

where u^j is the solution of the "direct" problem (1)–(2) and w^j is the solution of the "conjugate" problem (5)–(6) for the j-th position of the source.

In contrast to [9, 11, 14], in the above formulation, the experimental data may be absent on some part of the boundary surrounding the object. Such incomplete data problems are typical in ultrasonic tomography. For example, in ultrasonic mammography the data cannot be measured at the chest-wall side. Nevertheless, the expression for the gradient (7) is mathematically exact. The formulations of the "direct" and "conjugate" problems considered in this paper also differ from those used in previous works [9, 11, 14].

3 Numerical Algorithms for Solving Inverse Problems of Ultrasonic Tomography

3.1 Finite-Difference Approximations of the Wave Equation

To solve the coefficient inverse problem for the wave equation, we used a finite-difference time-domain method (FDTD). In this formulation, solving the differential wave equation reduces to solving finite-difference equations. Let us present the discretization scheme of the problem in the two-dimensional case. On the computational domain defined by the spatial coordinates (x, y) and the time t, we introduce a uniform discrete grid with a space step of h and a time step of τ. To approximate the second-order partial derivatives in Eq. (1), we use second-order finite differences. We obtain the following explicit finite-difference scheme for Eq. (1) for the region that does not contain any sources:

$$u_{ij}^{k+1} = \frac{1}{c_{ij}}\tau^2 \Delta u_{ij}^k + 2u_{ij}^k - u_{ij}^{k-1}, \tag{8}$$

where $\Delta u_{ij}^k = \frac{u_{i+1j}^k - 2u_{ij}^k + u_{i-1j}^k}{h^2} + \frac{u_{ij+1}^k - 2u_{ij}^k + u_{ij-1}^k}{h^2}$ is the discrete Laplacian, u_{ij}^k are the values of $u(r,t)$ at the point (i, j) at the time step k, and c_{ij} are the values of $c(r)$ at the point (i, j). The parameters h and τ are connected by the Courant stability condition $c^{-0.5}\tau < h/\sqrt{2}$ for a 2D problem. The "conjugate" problem (5)–(6) is computed using a similar FDTD scheme.

This explicit 2nd-order FDTD scheme for the wave equation is the simplest and is quite effective for the numerical simulation of wave propagation on a supercomputer. Nevertheless, when the ultrasound pulse propagates distances much larger than the wavelength, the errors of the finite-difference approximation accumulate, which leads to dispersion of the wave. One of the ways to overcome the numerical dispersion is to increase the number of grid points. Model calculations showed that for typical problems of ultrasonic mammography, 25–30 grid points per wavelength are required to obtain sufficient precision. This means that the computational grid size should be approximately $n = 1000$ points in each spatial coordinate and in time. In the 2D case, such grid sizes do not pose a problem for modern supercomputers. However, in the 3D case, the amount of computation grows as n^4 and the required memory capacity grows as n^3. A large grid size in the 3D case requires a very large number of computing nodes and faces memory size limitations on the GPU processors.

One possible approach to resolving this issue in ultrasonic tomography problems is to increase the accuracy of the finite-difference approximation, which would reduce the size of the grid, the amount of computations and the required memory capacity while maintaining the accuracy of the calculations. In this paper, the use of fourth-order FDTD schemes is considered. Model calculations were performed to compare the performances of second- and fourth-order FDTD schemes for the wave tomography problem (Fig. 2).

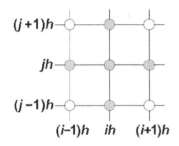

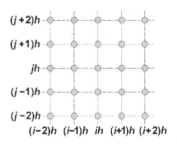

Fig. 2. The stencil of the second-order 2D FDTD scheme.

Fig. 3. The stencil of the fourth-order 2D FDTD scheme.

Following the work [20], we construct a 2D FDTD scheme that provides fourth-order accuracy with respect to the spatial coordinates. This FDTD scheme is shown in Fig. 3. It has the following general form:

$$
\begin{aligned}
u_{ij}^{k+1} + u_{ij}^{k-1} = {} & \lambda^2 a \left(u_{ij+1}^{k} + u_{ij-1}^{k} + u_{i+1j}^{k} + u_{i-1j}^{k} \right) \\
& + \lambda^2 b \left(u_{i+1j+1}^{k} + u_{i+1j-1}^{k} + u_{i-1j+1}^{k} + u_{i-1j-1}^{k} \right) + \lambda^2 c \left(u_{i+2j}^{k} + u_{i-2j}^{k} + u_{ij+2}^{k} + u_{ij-2}^{k} \right) \\
& + \lambda^2 d \left(u_{i+2j+1}^{k} + u_{i+2j-1}^{k} + u_{i-2j+1}^{k} + u_{i-2j-1}^{k} + u_{i+1j+2}^{k} + u_{i+1j-2}^{k} + u_{i-1j+2}^{k} + u_{i-1j-2}^{k} \right) \\
& + \lambda^2 e \left(u_{i+2j+2}^{k} + u_{i+2j-2}^{k} + u_{i-2j+2}^{k} + u_{i-2j-2}^{k} \right) + \lambda^2 f u_{ij}^{k}.
\end{aligned}
$$

$$(9)$$

For this scheme to approximate the wave equation to the fourth order, the parameters must satisfy the following relations:

$$
a = 14d + 32e + 4/3, \quad b = -8d - 16e,
$$
$$
c = -2d - 2e - 1/12, \quad f = 2/\lambda^2 - 24d - 60e - 5.
$$

For the scheme to be direction-independent up to the sixth order of accuracy, an additional condition $d/2 + 2e = -1/60$ must be satisfied. The parameter $\lambda = -v\tau/h$ is determined from the Courant stability condition. The choice of the parameters d and e specifies various variants of the scheme. In this paper, for simplicity, we assume that $d = 0$; therefore, $e = -1/120$. The variant with $d = 0$ and $e = 0$ reduces the accuracy of the scheme but also reduces the computation time because the diagonal elements are excluded from the calculations.

Figure 5 shows the results of the numerical simulations for the 2D case. The propagation of a short pulse in a homogeneous medium was computed using the 2nd-order scheme (8) and the 4th-order scheme (9). The cross-sections of the pulse generated by the source according to formula (1) has the waveform shown in Fig. 4. Figure 5 shows the cross-sections of the wave function $u(r,t)$ perpendicular to the wave front at the same time step for the 2nd-order scheme (solid line) and the 4th-order scheme (dashed line). The X-axis shows the grid-point number. The computational

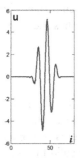

Fig. 4. Waveform of the sounding pulse.

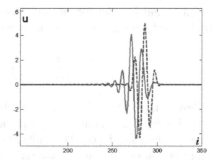

Fig. 5. Cross-sections of the propagating waves for the 2nd-order (solid line) and 4th-order (dashed line) FDTD schemes.

domain size is 200×200 mm, and the number of grid points is 350×350. The central wavelength of the pulse is 7 mm, so there are approximately 12 grid points per wavelength.

The time moment in Fig. 5 is chosen so that the wave propagation distance reaches approximately 200 mm. It is evident that, for the 4th-order scheme, the distortion of the pulse is insignificant and that, for the 2nd-order scheme, the grid step turned out to be too large, which resulted in the distorted waveform and appearance of a "tail". If we use a grid that is two times finer in the 2nd-order scheme, we can obtain a waveform similar to that in Fig. 5 for the 4th-order scheme. Thus, the use of the 4th-order scheme makes it possible to reduce the grid size by a factor of 1.5–2 in each coordinate compared to the 2nd-order scheme for the problem of wave tomography, given parameters that are typical for medical imaging.

3.2 Transparency Conditions for the Boundary of the Computational Domain

When solving the "direct" (1)–(2) and "conjugate" (5)–(6) problems numerically, the boundary conditions must be applied at the boundary $\partial\Omega$. The boundary is assumed to be located far enough from the domain R such that during the time T the waves from the sources do not reach the boundary. In this case, the zero boundary conditions (2) are automatically satisfied. When carrying out the calculations, we can either choose a sufficiently large computational domain Ω or assume that $\Omega = R$, which has a much smaller volume, and apply the non-reflecting ("transparent") boundary conditions.

In this paper, the numerical simulations are implemented with approximate non-reflecting boundary conditions. There are various options for the "transparency" conditions [21–25]. The first option considered in this study is to create a border zone with a width of M grid points. Within this zone, an absorbing term $a(\mathbf{r})u_t\ (\mathbf{r},t)$ is added to the left-hand side of the wave Eq. (1). The absorption coefficient quadratically increases for the points closer to the boundary ∂R.

The second option is to apply non-reflecting boundary conditions (NRBC) at the boundary of the computational domain. The exact NRBC formulation is non-local and is quite difficult to calculate. The first-order approximation of the NRBC has the form

$v \, \partial_n u|_{\partial R} = -\partial_t u|_{\partial R}$ and is exact for incident waves propagating perpendicular to the boundary. In this study, we use a second-order approximation, which has the form

$$\frac{\partial^2 u}{\partial x \partial t} - \frac{1}{v}\frac{\partial^2 u}{\partial t^2} + \frac{v}{2}\frac{\partial^2 u}{\partial y^2} = 0. \tag{10}$$

Figure 6 shows the results of the 2D numerical simulations of a reflected pulse that has a width of 10 mm (25 grid points) and amplitude of 1. Figure 6a shows the results of the first method (absorbing layer). The width M of the absorbing layer is 50 points in this case. Figure 6b shows the results of the second method (10). The solid line shows the cross-section of the incident wave propagating perpendicular to the boundary, and the dashed line shows that of the wave propagating at a 45-degree angle.

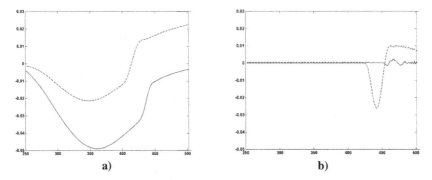

Fig. 6. The reflected waves for the 90-degree angle of incidence (solid line) and the 45-degree angle (dashed line): (a) using an absorbing layer, (b) using a 2nd-order approximate NRBC.

The X-axis shows the grid-point number, and the boundary is located at the right edge of the plots. The maximum amplitude of the reflections in Fig. 6a is 5%, and in Fig. 6b, it is 3%. In the second case, the reflections for the 90-degree angle of incidence are practically absent. In the first case, the reflected signal is almost 10 times wider than the original.

As follows from the numerical simulations, both methods allow us to approximate the boundary transparency condition with good accuracy, including steep angles of incidence. This made it possible to significantly reduce the size of the computational domain compared to the size at which the wave does not reach the boundary in time T. Even if a supercomputer is available, the reduction of the grid size is very important, especially in the 3D case, where the number of operations grows as n^4 and the data volume grows as n^3.

3.3   The Iterative Process of Solving the Inverse Problem

The following iterative process was used to solve the inverse problem numerically. As the initial approximation, we use the value $c^{(0)} = c_0 = const$, which corresponds to the speed of sound in pure water, $v_0 = 1500$ m·s^{-1}. At each iteration (m), the following actions are performed:

1. The "direct" problem (1)–(2) is solved for the current approximation $c^{(m)}$. The propagation of the wave $u^{(m)}(r,t)$ is computed using formula (8) or (9). The values of $u(r,t)$ at each detector are computed.
2. The residual $\Phi^{(m)} = \Phi(u^{(m)}(r))$ is computed using formula (3).
3. The "conjugate" problem (5)–(6) is solved for $w^{(m)}(r,t)$ The gradient $\Phi'_C(u^{(m)}(r))$ is computed using formula (7) for all sources.
4. The current approximation is updated: $c^{(m+1)} = c^{(m)} + \lambda^{(m)}\Phi'_C(u^{(m)}(r))$. The process returns to step 1.

The iteration process is stopped if the residual becomes smaller than some predetermined value, which corresponds to the a priori known precision of the measured data. The step of the gradient descent $\lambda^{(m)}$ is chosen based on a priori considerations. Determining the step more precisely requires performing additional iterations and would increase the computation time by a factor of 2 or more. If the residual $\Phi^{(m)}$ at the current iteration becomes larger that $\Phi^{(m-1)}$, the step $\lambda^{(m)}$ is reduced by a factor of 1.5.

4 Numerical Simulations of Ultrasonic Tomography for the Second- and Fourth-Order FDTD Schemes

The numerical simulations for the 2D ultrasonic tomography problem were performed according to the scheme shown in Fig. 1. First, the direct problem of wave propagation through the simulated test object was solved using the 4th-order FDTD scheme (9). The wave field at the perimeter of the square (Fig. 1) was recorded and used as simulated measurement data to solve the inverse problem. The inverse problem was solved using both the 4th- and 2nd-order FDTD schemes. The approximate non-reflecting boundary condition (10) was applied.

The central wavelength of the pulse was 7 mm, the sound-speed range in the test object — 1430–1600 m·s^{-1}, the sound speed in the environment — 1500 m·s^{-1}, the size of the computational domain — 200 × 200 mm, and the size of the FDTD grid — 350 × 350 points. In the numerical simulations, we used eight sources that were located in the middle of each side of the square and in the corners of the square, as shown in Fig. 1. The detectors were located at the sides of the square with a pitch of 0.6 mm.

Figure 7a shows an image of the simulated test object, Fig. 7b shows the image reconstructed using the 4th-order FDTD scheme, and Fig. 7c shows the image reconstructed using the 2nd-order FDTD scheme.

Comparing Fig. 7b and c, we can see that the numerical dispersion shown in Fig. 5 becomes significant for the 2nd-order FDTD scheme and significantly deteriorates the image quality, thus producing numerous artefacts. Using the 4th-order scheme allows the reconstruction of not only the shapes of irregularities but also the sound-speed function with high precision. Even small inclusions of size of 2–3 mm are reconstructed, and the precision of the sound speed reconstruction is 10 m·s^{-1} or better.

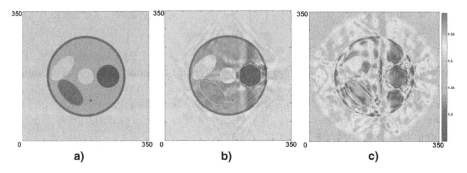

Fig. 7. (a) Simulated test object, (b) the image reconstructed using the 4th-order FDTD scheme, (c) the image reconstructed using the 2nd-order FDTD scheme.

The computing time for the 2D problem using eight computing cores of the "Lomonosov-1" supercomputer was approximately 2 h for 400 iterations of the gradient method. Figure 7c shows the result obtained after 150 iterations; then, the process stopped because the residual functional did not decrease any further.

The developed program for solving ultrasonic tomography problems is realized in the C++ language, designed for operation on high-performance cluster computer systems under the control of one of the Linux OS clones. For interprocessor exchange, the MPI interface was selected. Computations were carried out on the "Lomonosov-1" supercomputer of the Lomonosov Moscow State University Supercomputer Center on CPU Intel Xeon X5570 2.93 GHz processors, 1.5 GB of memory per core, 8 x86 cores per node, Infiniband [26].

When carrying out computation in the problem under consideration on the CPU of the cluster system, it is natural to have a two-level parallelization based on the number of sources at the first level and then decomposition of the calculation area at the second level. This approach was implemented when performing computations with the 2nd-order FDTD scheme for 7-point stencil and showed high efficiency and scalability up to several tens of thousands of computing cores [14]. Moreover, scalability by sources is practically linear, since calculations for different sources are practically independent. When parallelizing the decomposition of the calculation area, it is necessary to perform data exchanges between neighboring regions, so the decomposition into too small areas is impractical. In the present work, in computations with the 4th-order FDTD scheme for 27-point stencil, parallelization by sources was performed, which showed linear scalability. Parallelization by the technique of domain decomposition is supposed to be implemented in subsequent works.

As follows from the results of the work, using the 4th-order FDTD scheme allows us to reduce the size n of the computational grid by a factor of 1.5–2 compared to the 2nd-order scheme while maintaining the accuracy of the calculations. This fact is very important for inverse problems of wave tomography, especially in the 3D case, because the computation time increases as n^4. Although the increase of the stencil size causes approximately a two-fold increase of the computation time per grid point and a two-fold increase in inter-processor communications if the domain is divided among multiple processors, the number of grid points decreases by at least a factor of 3.

Moreover, for GPU processors, the memory requirement is a very important factor. For efficient GPU computing, all of the data used in the FDTD scheme must reside in the on-board memory of the GPU device. The volume of these data grows as n^3 in the 3D case. The memory capacity limits the problem size to $\sim 500^3$ points per computing node. This number of points is insufficient for precise calculations using a 2nd-order finite-difference scheme.

5 Conclusions

This paper is concerned with the optimization of numerical algorithms for solving the inverse problem of wave tomography using supercomputers. To reduce the computational grid dimensions, required memory capacity and computation time and to improve the accuracy of the calculations, the use of higher-order finite-difference schemes is proposed. It is shown that the fourth-order FDTD scheme allows us to decrease the grid size by a factor of 1.5–2 in each dimension compared to a second-order scheme while preserving the accuracy of the calculations. This approach significantly reduces both the computation time and the required memory capacity.

An important issue for numerical methods is the problem of boundary conditions for solving direct and inverse problems in a bounded domain. The article discusses two approximate methods that implement boundary "transparency". It is shown that both methods allow precise calculations and have significantly smaller computational complexities than exact non-reflecting boundary conditions.

The inverse problem is formulated as the incomplete-data tomography problem, where the sources and the detectors cannot be located on all sides of the examined object. To solve this important problem mathematically strictly, a method of calculating the gradient of the residual functional is proposed, which includes solving special "direct" and "conjugate" problems.

The proposed optimization scheme of the numerical algorithms is relevant due to the very large amount of computations required to solve the problems of wave tomography. The method is easy to implement on supercomputers.

Acknowledgements. This research was supported by Russian Science Foundation (project No. 17-11-01065). The study was carried out at the Lomonosov Moscow State University.

References

1. Bazulin, A.E., Bazulin, E.G., Vopilkin, A.K., Kokolev, S.A., Romashkin, S.V., Tikhonov, D.S.: Application of 3D coherent processing in ultrasonic testing. Russ. J. Nondestr. Test. **50**(2), 92–108 (2014)
2. Ruvio, G., Solimene, R., D'Alterio, A., Ammann, M.J., Pierri, R.: RF breast cancer detection employing a noncharacterized vivaldi antenna and a MUSIC-inspired algorithm. Int. J. RF Microwave Comp. Aid Eng. **23**, 598–609 (2012)
3. Tran-Duc, T., Linh-Trung, N., Do, M.N.: Modified distorted born iterative method for ultrasound tomography by random sampling. In: International Symposium on Communications and Information Technologies (ISCIT), pp. 1065–1068, Gold Coast, QLD (2012)

4. Goncharskii, A.V., Romanov, S.Y.: On a three-dimensional diagnostics problem in the wave approximation. Comput. Math. Math. Phys. **40**(9), 1308–1311 (2000)
5. Goncharskii, A.V., Ovchinnikov, S.L., Romanov, S.Y.: On the one problem of wave diagnostic. Moscow Univ. Comput. Math. Cybern. **34**(1), 1–7 (2010)
6. Duric, N., Littrup, P., Li, C., Roy, O., Schmidt, S., Janer, R., Cheng, X., Goll, J., Rama, O., Bey-Knight, L., Greenway, W.: Breast ultrasound tomography: bridging the gap to clinical practice. In: Proceedings of SPIE, Medical Imaging: Ultrasonic Imaging, Tomography, and Therapy, vol. 8320, p. 83200O (2012)
7. Gemmeke, H., Berger, L., Birk, M., Gobel, G., Menshikov, A., Tcherniakhovski, D., Zapf, M., Ruiter, N.V.: Hardware setup for the next generation of 3D ultrasound computer tomography. In: IEEE Nuclear Science Symposuim and Medical Imaging Conference, pp. 2449–2454 (2010)
8. Wiskin, J., Borup, D., Andre, M., Johnson, S., Greenleaf, J., Parisky, Y., Klock, J.: Three-dimensional nonlinear inverse scattering: Quantitative transmission algorithms, refraction corrected reflection, scanner design, and clinical results. J. Acoust. Soc. Am. **133**, 3229 (2013)
9. Natterer, F.: Possibilities and limitations of time domain wave equation imaging. Contemp. Math. Providence Am. Math. Soc. **559**, 151–162 (2011)
10. Natterer, F.: Sonic imaging. In: Handbook of Mathematical Methods in Imaging, pp. 1253–1278. Springer, New York (2015)
11. Beilina, L., Klibanov, M.V., Kokurin, M.Y.: Adaptivity with relaxation for ill-posed problems and global convergence for a coefficient inverse problem. J. Math. Sci. **167**, 279–325 (2010)
12. Beilina, L., Klibanov, M.V.: Approximate Global Convergence and Adaptivity for Coefficient Inverse Problems. Springer, New York (2012)
13. Goncharskii, A.V., Romanov, S.Y.: Two approaches to the solution of coefficient inverse problems for wave equations. Comput. Math. Math. Phys. **52**, 245–251 (2012)
14. Goncharsky, A.V., Romanov, S.Y.: Supercomputer technologies in inverse problems of ultrasound tomography. Inverse Probl. **29**(7), 075004 (2013)
15. Goncharsky, A.V., Romanov, S.Y.: Inverse problems of ultrasound tomography in models with attenuation. Phys. Med. Biol. **59**(8), 1979–2004 (2014)
16. Goncharsky, A.V., Romanov, S.Y., Seryozhnikov, S.Y.: Inverse problems of 3D ultrasonic tomography with complete and incomplete range data. Wave Motion **51**(3), 389–404 (2014)
17. Goncharsky, A.V., Romanov, S.Y., Seryozhnikov, S.Y.: A computer simulation study of soft tissue characterization using low-frequency ultrasonic tomography. Ultrasonics **67**, 136–150 (2016)
18. Goncharskii, A.V., Romanov, S.Y., Seryozhnikov, S.Y.: Low-frequency three-dimensional ultrasonic tomography. Dokl. Phys. **61**(5), 211–214 (2016)
19. Goncharsky, A.V., Romanov, S.Y.: Iterative methods for solving coefficient inverse problems of wave tomography in models with attenuation. Inverse Prob. **33**(2), 025003 (2017)
20. Bilbao, S.: Numerical Sound Synthesis: Finite Difference Schemes and Simulation in Musical Acoustics. Wiley, Chichester (2009)
21. Manolis, G.D., Beskos, D.E.: Boundary Element Methods in Elastodynamics. Unwin Hyman, London (1988)
22. Givoli, D., Keller, J.B.: Non-reflecting boundary conditions for elastic waves. Wave Motion **12**(3), 261–279 (1990)
23. Clayton, R., Engquist, B.: Absorbing boundary conditions for acoustic and elastic wave equations. Bull. Seismol. Soc. Am. **67**(6), 1529–1540 (1977)

24. Engquist, B., Majda, A.: Absorbing boundary conditions for the numerical simulation of waves. Math. Comput. **31**, 629 (1977)
25. Alpert, B., Greengard, L., Hagstrom, T.: Boundary conditions for the time-dependent wave equation. J. Comput. Phys. **180**, 270–296 (2002)
26. Sadovnichy, V., Tikhonravov, A., Voevodin, V., Opanasenko, V.: "Lomonosov": Super-computing at Moscow State University. In: Vetter, Contemporary High Performance Computing: from Petascale Toward Exascale, pp. 283–307. Chapman and Hall/CRC, Boca Raton (2013)

The Comparison of Large-Scale Graph Processing Algorithms Implementation Methods for Intel KNL and NVIDIA GPU

Ilya Afanasyev[(✉)] and Vladimir Voevodin

Lomonosov Moscow State University, Moscow, Russia
afanasiev_ilya@icloud.com

Abstract. The paper describes implementation approaches to large-graph processing on two modern high-performance computational platforms: NVIDIA GPU and Intel KNL. The described approach is based on a deep a priori analysis of algorithm properties that helps to choose implementation method correctly. To demonstrate the proposed approach, shortest paths and strongly connected components computation problems have been solved for sparse graphs. The results include detailed description of the whole algorithm's development cycle: from algorithm information structure research and selection of efficient implementation methods, suitable for the particular platforms, to specific optimizations for each of the architectures. Based on the joint analysis of algorithm properties and architecture features, a performance tuning, including graph storage format optimizations, efficient usage of the memory hierarchy and vectorization is performed. The developed implementations demonstrate high performance and good scalability of the proposed solutions. In addition, a lot of attention was paid to profiling implemented algorithms with NVIDIA Visual Profiler and Intel® VTune ™ Amplifier utilities. This allows current paper to present a fair comparison, demonstrating advantages and disadvantages of each platform for large-scale graph processing.

Keywords: Graph algorithms · GPU · KNL · CUDA · Vectorization · SSSP · SCC · Large-scale graph processing

1 Introduction

The interest to large-scale graph processing is growing rapidly, since graphs successfully emulate real-world objects and connections between them. In many areas, people need to identify some patterns and rules from object relationships that results into processing large amounts of data. The examples of such objects and relationships are: analysis of social, semantic and Internet networks, infrastructural problems solution (analysis of transport and energy networks), biology (analysis of the network of protein-protein interactions), health-care (epidemic spreading analysis), social-economic modelling. All those problems have one common property: their model graphs have a very large size, so a parallel approach is required to perform computations in reasonable amount of time.

V. Voevodin and S. Sobolev (Eds.): RuSCDays 2017, CCIS 793, pp. 80–94, 2017.
https://doi.org/10.1007/978-3-319-71255-0_7

The question, which parallel computational platforms are able to process graphs more efficiently, is still open. Graphic accelerators and coprocessors perform really well for solving traditional problems, such as linear algebra computations, image processing or solving molecular dynamics problems, since they provide high performance and energy efficient computational power together with high throughput memory. The most well-known and widely used families of coprocessors are NVidia GPU and Intel Xeon Phi. Recent important trend is that vendors try to combine central processors and coprocessor functions, which results into modern KNL Xeon PHI architecture.

2   Target Architectures

2.1   NVidia GPU

Modern NVidia GPUs belong to three architectures: Kepler, Maxwell and Pascal. Currently, Kepler is the most common and widely used architecture in HPC. Tesla K40 accelerator, which has been used for all testing in current paper, has 2880 cores with clock signal rate of 745 MHz. This GPU provides peak performance up to 4.29 TFLOPs on single precision computations and 12 GB device memory with 288 GB/s bandwidth. The PCI-express 3.0 bus with 32 Gbps bandwidth is used to maintain connection between host and device. Memory hierarchy also includes L1 (64 KB), and L2 (1.5 MB) caches. Device computational model is very important: thread is a single computational unit; 32 threads are grouped into a warp, which works using SIMD approach. The warp performance is also very affected by memory access data pattern (coalesced memory access) and conditional operations presence.

During the tests the corresponding host was equipped with Intel(R) Xeon(R) CPU E5-2697 v3 @ 2.60 GHz processor. For compilation NVCC v6.5.12 from CUDA Toolkit 6.5 has been used with –O3 –m64 –gencode arch = compute_35,code = sm_35 options.

2.2   Intel KNL

The newest architecture of Intel Xeon Phi coprocessors is Knights Landing (KNL). Each processor has 64-72 cores (in current paper a 68-core accelerator is used) with a clock signal rate of 1.3-1.5 GHz. Processor provides a peak performance up to 6 TFLOPs on single precision computations. It also has two memory levels: high-bandwidth MCDRAM memory with a capacity up to 16 GB and bandwidth up to 400 GB/s, and DDR4 memory with a capacity up to 384 GB and bandwidth up to 90 GB/s. Cores are grouped in Tile-s (pair of cores), each having a common 1 MB size L2 cache. Another important feature of Intel KNL is the support of vector instructions AVX-512, containing gather and scatter operations, which are necessary for graph processing. For compilation ICC 17.0.0 has been used with –O3 –m64 options.

3   State of Art

Algorithms for solving shortest paths problem for CPU and GPU are described following papers: [1–3] This approach can be applied for KNL architecture too, which is demonstrated in the current paper. Sequential (Tarjan) algorithm for solving strongly connected components problem is presented in [4]. Another algorithm (Forward-Backward), which has a much larger parallelism potential, but also have a larger computational complexity, is presented in the papers [5, 6]. CUDA implementation of this algorithm is also researched in papers [6, 8].

4   Research Methodology

The current paper uses the following structure to describe both graph problems. First, an accurate mathematical problem definition is formulated, to prevent any ambiguity. After that, a review of most important possible algorithms is presented together with target architecture features. Based on the results of this survey, well-suited algorithms for all architectures are selected.

After that, first implementation of all selected algorithm is developed, followed by a series of iterative optimizations and profiling. It is extremely important to analyze the final implementation perfomance, and how well the implementations use target hardware features. As a result, conclusions about advantages and disadvantages of both architectures for solving a specific graph problem can be presented.

5   Shortest Paths Problem

5.1   Mathematical Description

A directed graph $G = (V, E)$ with vertices $V = (v_1, v_2, \ldots, v_n)$ and edges $E = (e_1, e_2, \ldots, e_m)$ is given. Each edge $e \in E$ has a weight value $w(e)$. The path is defined as edges sequence $\pi_{u,v} = (e_1, \ldots, e_k)$, beginning in vertex u and ending in vertex v, so that each edge follows another one. A path length can be defined as $w(\pi_{u,v}) = \sum_{i=1}^{k} w(e_i)$. A path $\pi_{u,v}^*$ with minimal possible length between vertices u and v is called a shortest path: $d(u, v) = w(\pi_{u,v}^*) = \min w(\pi_{u,v})$.

Depending on a vertices pair choice, between which a search is performed, the shortest paths problem can be formulated in three different ways:

- SSSP (single source shortest paths) — shortest paths from a single selected source vertex are computed.
- APSP (all pairs shortest paths) — shortest paths between all pairs of graph vertices are computed.
- SPSP (some pairs shortest paths) — shortest paths between some pre-selected pairs of vertices are computed.

In the current paper the SSSP problem will be researched, since it is the simplest and most basic between these problems: for example, APSP problem for large-scale graphs can be solved by repeated calls of SSSP operation for each source vertex, since traditional algorithms, such as Floyd-Warshall, can not be applied because high memory requirements.

5.2 Algorithm Descriptions

SSSP problem can be solved with two traditional algorithms: Dijkstra and Bellman-Ford.

- **Dijkstra's algorithm** is designed to solve the problem in graphs with edges, having only non-negative weights. A variation of the algorithm, implemented with a Fibonacci heap has the most efficient time complexity $O(|E| + |V|\log|V|)$. The algorithm's computational core includes sequential traversals of vertices beginning from the source vertex; during each traversal algorithm while puts adjacent vertices to the stack (or heap), so they can be processed later. As a result, the global vertices traversal in the algorithm can be performed only sequentially, while local adjacent vertices traversal can be executed in parallel as described in [10], but it's usually provides not enough parallelism for significant GPU utilization.
- **Bellman-Ford algorithm** is designed to solve the problem in graphs, including those which have edges with negative weights. The computational core of the algorithm consists of a few iterations, each of which requires traverses of all graph edges. Computations continue until there are no changes in the distance array. The algorithm has a sequential complexity equal to $O(p|E|)$, where p is the maximum possible length of the shortest path from the source vertex to any other. As a consequence, the worst-case complexity is equal to $O(|V||E|)$. However, for many real-world graphs, the algorithm is terminates in a much smaller amount of steps. Moreover, the algorithm has a significant parallel potential: its parallel complexity is equal to $O\left(p\dfrac{|E|}{N}\right)$, where N is the number of processors being used.

5.3 Algorithm Selection for Target Architectures

Before the implementation, one needs to select the algorithms, most suitable for all target architectures. Both KNL and NVidia GPU have a large number of cores with a relatively low clock rate. If Dijkstra's algorithm, which is strictly sequential, is used for computations, all those cores will be idle, and, in addition, it will be very difficult to handle a stack or queue complex data structures on cores with a low clock rate. At the same time, Bellman-Ford algorithm does not require a processing of complex data structures; moreover, on any iteration this algorithm performs a parallel traversal of all graph edges. It will be shown later, that those properties will compensate algorithm's greater arithmetical complexity. In addition, it is possible to develop Bellman-Ford algorithm modification, which allows to process graphs with a size larger than the amount of available memory. This property is very important advantage for architectures with a limited amount of available memory, such as GPUs.

Before implementing the chosen algorithm, it is important to determine the storage data format for input graphs. For Bellman-Ford algorithm, the most suitable format is an edges list, where each edge is stored as a triple {vertex-start, vertex-end, edge's weight}; all edges are stored in a single array in any order.

5.4   GPU Implementation

5.4.1   Basic Version

CUDA-kernel, implementing the basic version of Bellman-Ford algorithm for the GPU is presented in listing 1:

Listing 1: Bellman-Ford algorithm's CUDA-kernel

```
1  register const long long idx = blockIdx.x * blockDim.x + threadIdx.x;
2
3  if (idx < _edges_count) // for all graph edges do
4  {
5      register int src_id = _src_ids[idx];
6      register int dst_id = _dst_ids[idx];
7      register _TEdgeWeight weight = _weights[idx];
8
9      register _TEdgeWeight src_distance = __ldg(&_distances_for_read[src_id]);
10     register _TEdgeWeight dst_distance = __ldg(&_distances_for_read[dst_id]);
11
12     if (dst_distance > src_distance + weight)
13     {
14         _distances_for_write[dst_id] = src_distance + weight;
15         _modif[0] = _iter + 1;
16     }
17 }
```

The presented non-optimized kernel fully corresponds to the classical version of Bellman-Ford algorithm. The kernel is executed on number of threads, equal to the graph edges count. Each thread gets its corresponding edge's data (5) (6) (7), then reads current distance data of source and destination vertices of corresponding edge (9) (10). If those values minimize current distance to the destination vertex, the array of distances (12), (14) is updated. In Fig. 1 the results of profiling (obtained with NVidia Visual Profiler) of the kernel are presented, clearly demonstrating kernel's disadvantages.

A first important observation is that this kernel is memory-bound, since every two arithmetic operations are followed by 6 memory access operations. Second, there is an indirect addressing during (9), (10), (12) and (14) memory accesses, where the optimal memory access pattern for the GPU is violated (coalesced memory access). In addition, values src_id and dst_id may point into completely different memory locations, what prevents efficient using of GPU caches.

The results of profiling clearly demonstrate the main reason of basic kernel's low performance — inefficient usage of GPU memory bandwidth ("device memory total" metric, due to non-coalesced memory accesses), as well as weak usage of L1 and L2 caches (due to weak locality of data accesses). These problems can be avoided with

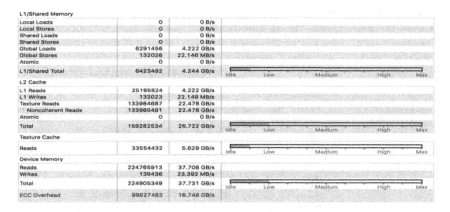

Fig. 1. Analysis of memory bandwidth usage for the basic kernel implementation.

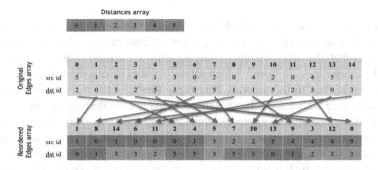

Fig. 2. Graph edges reordering example for graph with 5 vertices and 16 edges. (Color figure online)

graph storage format optimization, which allows changing memory access pattern, making it more suitable for GPU architecture.

5.4.2 Graph Storage Format Optimization

In the current section, the main optimization (graph edges reordering) is described. It allows improving memory access pattern, to achieve higher performance, since the data with indirect memory accesses will be placed more locally and stored in the caches. Modern K40 GPUs of Kepler architecture are equipped with 64 KB of L1 cache and 1.5 MB of L2 cache. The distances array has 1 MB size for a graph with 2^{18} vertices, 2 MB for a graph with 2^{19} vertices; so, even for medium-scaled graphs, the whole distance array can't be fully placed in caches.

That is why edges rearrangement strategy is used to make sure that the data from distances arrays remains in cache memory as long as possible. The reordering process is illustrated in Fig. 3. A similar reordering approach is described in [7].

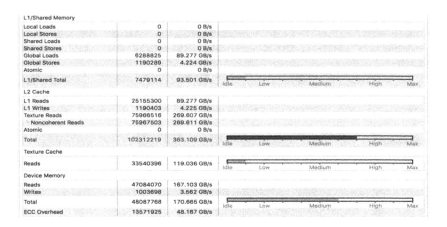

Fig. 3. Analysis of memory bandwidth of basic kernel with optimized graph format storage

An array of distances is divided into segments (red, green and blue colors – in Fig. 2), which size is equal to the size of the lowest level cache - L2 for GPU (size 2 on Fig. 2). After that, the edges are placed into the array in the following way: in the beginning of the array edges are stored, which source vertices belong the first segment of the distances array, then to the second, then to the third. Edges with the same segment number are sorted with the similar strategy, applied to their destination vertices.

Due to described sorting approach, threads from the same warp will be accessing data within one or two segments; this result into a smaller number of different memory cells accessed by a single warp. Without this optimization, 32 different memory cells could be accessed, which would lead to a 32-times slowdown. The profiling report of optimized kernel is presented on Fig. 3. It demonstrates almost 5 times increase in used memory bandwidth (device memory reads/ total). In addition, for the threads from a single block, distance array data will be stored in L2 cache for a much longer time, which can be also observed on presented profiling report: L2 cache used bandwidth is 15x times better now.

5.4.3   GPU Implementation Results

The performance comparison between basic and fully optimized kernel versions (with graph storage format optimization) is demonstrated on Fig. 4. Another important GPU-program characteristic is the percentage of program execution time, spent for data transfers between host and device. That is why fully optimized version is represented with two curves: with and without time spent for memory copy operation. The performance is measured with TEPs metrics — number of traversed edges per second (the edge is considered "traversed" when the data about it's source and destination vertices is requested). RMAT graphs with average connection count equal to 32 and vertices count from 2^{15} to 2^{24} are used as input data.

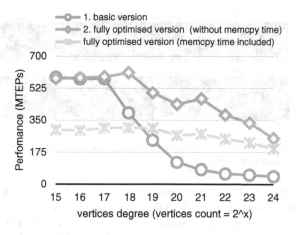

Fig. 4. Performance comparison of GPU Bellman-Ford algorithms versions, RMAT graphs with 2^{15} — 2^{24} vertices.

Figure 4 demonstrates two important implementation properties. First, non-optimized and optimized versions have similar behavior on graphs with size less than 2^{18}, since on smaller sizes graphs corresponding to the distances array can be fully placed into L2 cache of GPU. Second, the presented results demonstrate, that data transfers between host and device indeed require a significant amount of time. However, in other shortest paths problem variations (APSP, SPSP), data transfers will be less important, since more computations will be performed after coping data into device memory.

5.5   KNL Implementation

First of all, it is important to decide which parallel technology should be used for KNL implementations [9]. The two most widely-used technologies are openMP and Intel TBB. Experimental results confirm that openMP technology is more suitable for graph algorithms implementations, since it requires fewer overheads for threads creation and synchronizations.

Simple openMP implementation is universal, since it can be compiled and executed on both classic CPUs and on Intel Xeon co-processors. However, even for the simplest version it is important to take into an account some KNL features, discussed below.

First, the threads creation must be performed only once in the beginning of the algorithm. Second, the number of synchronizations between threads should be minimized, since those synchronizations are extremely expensive with a larger number of parallel threads (60-70 for KNL). Last, it is important to select thread scheduler correctly between static, guided and dynamic thread scheduling policies. VTune Amplifier analysis on Fig. 5, demonstrates the crucial difference between static and guided modes.

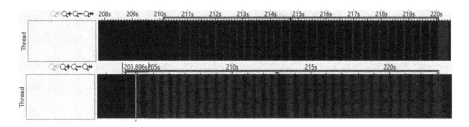

Fig. 5. Threads occupation analysis for static (top) and guided (bottom) modes, red color shows threads stall time. (Color figure online)

In addition, Intel KNL has two types of memory: DDR4 and MCDRAM. The simplest usage of high-performance MCDRAM memory is possible with the following command: *numactl -m 1./program_name*, where 1 is the MCDRAM memory node number. Also, hbwmalloc library can be used to allocate MCDRAM memory region inside the program; it allows allocating only certain arrays in high-bandwidth memory, if program memory requirements are larger than MCDRAM memory size. It can be very useful for large-scale graph processing, where only the distances array can be stored in MCDRAM memory, while edges arrays can be stored in usual DDR4 memory with larger size (Fig. 6).

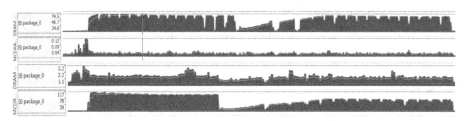

Fig. 6. Memory throughput usage analysis for different types of launches: program launched on MCDRAM node (bottom) and DDR4 node (top)

As second optimization, a similar reordering of graph edges (discussed in Sect. 5.4.2) was performed. Segment size was chosen equal to KNL L2 cache size, devided on two (since L2 cache is shared by 2 cores in Tile).

The last important optimization was vectorization. An important feature of vectorization is the possibility to manually load distance data into the cache using *_mm512_prefetch_i32extgather_ps* instructions. As a result, vectorization allowed to achieve in average 1.5 times acceleration, which can be observed on Fig. 7.

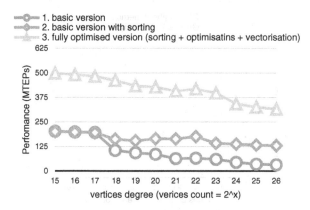

Fig. 7. Performance comparison of KNL Bellman-Ford algorithm implementations for KNL, RMAT graphs with 2^{15} — 2^{26} vertices.

5.6 GPU and KNL Implementations Comparison

Current section demonstrates general comparison of the two architectures in the context of solving shortest paths problem. The two GPU implementations are presented: with and without memory copies from host to device and back. For Intel KNL, the most optimized version with vectorization is presented. All graphs used for testing have RMAT and SSCA-2 structure and average connections count equal to 32.

Results from Fig. 8 demonstrate, that, first, KNL is able to process graphs with larger size. GPU is limited with 12 GB device memory, which can only contain graphs with 2^{24} vertices and 2^{29} edges. KNL processors can be equipped with up to 384 GB memory, which is able to contain graphs with up to 2^{29} vertices and 2^{34} edges.

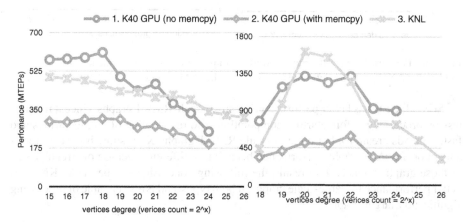

Fig. 8. Performance of Bellman-Ford algorithm implementations for different architectures. RMAT graphs with 2^{15} — 2^{26} vertices (left), SCCA-2 graphs with 2^{18} — 2^{26} vertices (right).

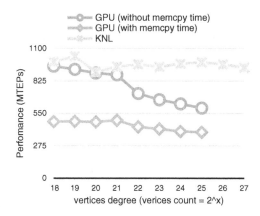

Fig. 9. Breadth-first search performance comparison for different architectures. RMAT graphs with 2^{18} — 2^{25} vertices.

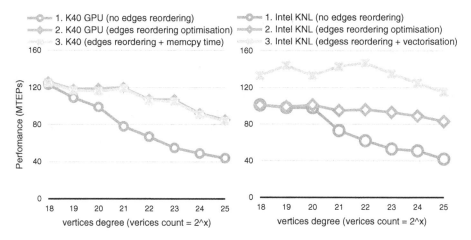

Fig. 10. Forward-Backward-Trim algorithm implementations performance for different architectures: NVidia GPU (left) and Intel KNL (right). RMAT graphs with 2^{18} — 2^{25} vertices.

Second, GPU has better performance on small-scale RMAT graphs, since it requires less preprocessing before starting computations (no reallocation of aligned arrays and faster threads creation), but on large-scale RMAT graphs KNL show higher performance. For SSCA-2 graphs performance behavior is different, because of irregular size of those graphs cliques. As a result, the following conclusion can be made: KNL has better performance for large-scale graphs of both types, and is also capable of processing significantly larger graphs.

6   Strongly Connected Components

6.1   Mathematical Description

A directed graph $G = (V, E)$ with vertices $V = (v_1, v_2, \ldots, v_n)$ and edges $E = (e_1, e_2, \ldots, e_m)$ is given. Edges may not have any data assigned (so graphs without edges weights are discussed in the current section). A strongly connected component (SCC) of a directed graph G is a strongly connected subgraph, which is maximal within the following property: no additional vertices from G can be included in the subgraph without breaking its property of being strongly connected.

6.2   Algorithm Descriptions

Strongly connected components can be found with one of the following algorithms.

- **Tarjan's algorithm** is based on a single depth first search (DFS) and uses $O(|E|)$ operations. Due to the fact that the algorithm is based on the DFS, only a sequential implementation is possible.
- **The DCSC algorithm** (Divide and Conquer Strong Components), or FB (Forward-backward) is based on BFS and requires $O(|V| * \log(|V|))$ operations. This algorithm is initially designed for parallel implementations: at each step it finds a single strongly connected component and allocates up to three subgraph, each of which may contain other strongly connected components, and, as a result, can be processed in parallel.
- **Variations of the DCSC algorithm**, such as Coloring and FB with step-trim. These modified versions of the DCSC algorithm are described in detail in papers [5, 6].

6.3   Algorithm Selection for Target Architectures

For obvious reasons, Tarjan's algorithm is not suitable for solving SCC problem on parallel architectures, since it is based on a depth first search, as well as complex data structures (stack and queue) processing, which can not be implemented efficiently on GPUs.

A large number of papers, such as [6], have already investigated different variation of DCSC algorithms, which can be more or less effective for different types of graphs; paper [6] concludes that the forward-backward-trim algorithm is the most efficient way to process RMAT graphs; the same was also proved during the current research.

The Forward-Backward-Trim algorithm is designed in the following way: in the first step, the removal of the strongly connected components of size 1 is performed. After that, on each step the algorithm finds one nontrivial strongly connected component and allocates up to three subgraphs, each of which contains other components, and, more important, can be processed in parallel. This step heavily relies on breadth-first search to find all vertices, which can be reached from the selected pivot, and all vertices, from which the current pivot vertex can be reached. Thus, this algorithm has two levels of parallelism: "BFS level" and "parallel subgraphs handling level", which appears to be a big advantage for target parallel architectures.

6.4 GPU Implementation

A forward-Backward-Trim algorithm is based on three important stages — a trim step, a pivot selection and BFS in selected subgraphs. At the trim step the number of edges, adjacent to each vertex (equal to number of incoming and out-coming edges), is calculated, with a removal of vertices, which incoming or outgoing degrees are equal to zero. Since the graph is stored in edges list format, these values can be computed using new atomic operations, added in Kepler architecture. Random pivot selection can be implemented with a simple kernel, based on random nature of thread execution. Breadth-first search can be performed by the algorithm, similar to Bellman-Ford shortest paths computations. The downward is that it has a higher computational complexity, compared to the traditional BFS algorithm (using queues), but for RMAT graphs the efficiency of proposed approach has already been demonstrated.

Since all steps can be implemented for a graph, which is stored in edges list format, this format is selected again for graph storage. Since FB algorithm will be performing BFS and trim steps both in original and transposed graphs, it is even more important to sort graph edges using approach, already discussed in Sect. 5.4.2. Without edges reordering, sub step performance (such as BFS) in transposed graph will be much lower, compared to the performance in original graph.

There is another way how this problem can be avoided — with a pre-processing transpose of the input graph before SCC operation (as proposed in [8]), but edges reordering is proved to be much more efficient for two reasons. First, edges reordering can be performed much faster on parallel architectures (such as KNL), since it can be based on parallel sorting algorithms, and does not require complex data structures (like maps or dictionaries) to be supported. Second, proposed reordering is universal for many different operations. For example, this reordering can be used to improve performance for shortest paths, breadth-first search, bridges and transitive closure computational problems. As a result, input graphs can be optimized right after generation, and stored in reordered intermediate representation to allow more efficient graph processing in the

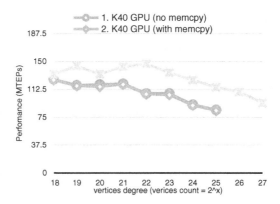

Fig. 11. Forward-Backward-Trim algorithm implementations performance for different architectures. RMAT graphs with 2^{18} — 2^{27} vertices.

future. Figure 11 in next section demonstrates computational time difference between two approaches: when input graph is optimized and not.

It is important to notice the percentage of time, required for trim and BFS steps. Later it will be shown, that these values greatly differ for both architectures. For GPU architecture and RMAT graphs this ratio is approximately equal to 6:10; the trim step requires slightly less time, since atomic operations implementation in Kepler architecture is very effective.

6.5 KNL Implementation

First of all, it is important to study algorithms, implementing sub steps performance separately for all steps: trim, BFS and pivot selection. Trim step on KNL is executed in average 1.1-1.2x times slower, compared to GPU, since the openMP atomic operations appear to be less efficient compared to GPU ones. The breadth-first search, in contrary, can be implemented much more efficiently on KNL, using vectorization and similar to Bellman-Ford approach. Figure 9 demonstrate BFS-only step performance for single graph traversal; for RMAT graphs trim/BFS ratio is almost 1:1 on KNL.

As shown in Fig. 9, the Intel KNL BFS implementation has significantly better performance on large-scale graphs, compared to GPU implementation (Fig. 10).

6.6 GPU and KNL Implementations Comparison

Similar to previously discussed shortest paths problem, SCC algorithm implementation for Intel KNL is also capable of processing larger graphs (up to 2^{27}) vertices. Since strongly connected components problem doesn't require edges weights stored, this value is bigger, compared to shortest paths one (2^{26}). Inetl KNL also demonstrates slightly better execution time, since its BFS implementation for RMAT graphs demonstrates better performance.

7 Conclusion

In the current paper, an implementation comparison of two important graph-processing problems on modern high-performance architectures (NVidia GPU and Intel KNL) has been discussed in details. Algorithms have been selected for both architectures, based on algorithm properties and target architecture features. As a result of many optimizations, high-performance and scalable parallel implementations have been created; moreover, the implementations have been examined in details using profiling utilities and theoretical research, which granted the ability to find potential bottlenecks and significantly improve final performance.

The best performance was achieved by Intel KNL processor for both investigated problems. Moreover, it was shown that Intel KNL is capable of processing much larger graphs with up to 134 million vertices and 42 billion edges. On K40 GPU, the maximum processed graph consisted from 33 million vertices and 10 billion edges. It is important, that Kepler architecture accelerators are currently outdated, while new GPUs from Pascal generation can achieve higher performance.

The results were obtained in the Lomonosov Moscow State University with the financial support of the Russian Science Foundation (agreement N 14-11-00190).

References

1. Harish, P., Narayanan, P.J.: Accelerating large graph algorithms on the GPU using CUDA. Center for Visual Information Technology, International Institute of Information Technology Hyderabad, India
2. Katz1, G.J., Kider, J.: All-Pairs-Shortest-Paths for Large Graphs on the GPU. University of Pennsylvania
3. Ortega-Arranz, H., Torres, Y., Llanos, D.R., Gonzalez-Escribano, A.: A new GPU-based approach to the shortest path problem, Dept. Informática, Universidad de Valladolid, Spain
4. Tarjan, R.E., Vishkin, U.: An efficient parallel biconnectivity algorithm. SIAM J. Comput. **14**(4), 862–874 (1985)
5. Fleischer, Lisa K., Hendrickson, B., Pınar, A.: On identifying strongly connected components in parallel. In: Rolim, J. (ed.) IPDPS 2000. LNCS, vol. 1800, pp. 505–511. Springer, Heidelberg (2000). https://doi.org/10.1007/3-540-45591-4_68
6. Barnat, J., Bauch, P., Brim, L., Ceska, M.: Computing strongly connected components in parallel on CUDA. In: Proceedings of the 2011 IEEE International Parallel & Distributed Processing Symposium, IPDPS 2011 (2011)
7. Kolganov, A.: Evaluating GPU performance on data-intense problems (translated from Russian), http://agora.guru.ru/abrau2014/pdf/079.pdf
8. Barnat, J., Bauch, P.: Computing strongly connected components in parallel on CUDA. Faculty of Informatics, Masaryk University, Botanická 68a, 60200 Brno, Czech Republic
9. Florian, R.: Choosing the right threading framework (2013), https://software.intel.com/en-us/articles/choosing-the-right-threading-framework
10. Pore, A.: Parallel implementation of Dijkstra's algorithm using MPI library on a cluster, http://www.cse.buffalo.edu/faculty/miller/Courses/CSE633/Pore-Spring-2014-CSE633.pdf

Two Approaches to Speeding Up Dynamics Simulation for a Low Dimension Mechanical System

Stepan Orlov$^{(\boxtimes)}$, Alexey Kuzin, and Nikolay Shabrov

Computer Technologies in Engineering Department,
Peter the Great St. Petersburg Polytechnic University,
St. Petersburg, Russian Federation
{majorsteve,kuzin_aleksei}@mail.ru, shabrov@rwwws.ru

Abstract. A dynamical model of continuously variable transmission (CVT) is considered. The model is described by ordinary differential equations (ODE) of motion with about 1800 generalized coordinates, and the same number of generalized speeds. Despite the low dimension of the model, the times of numerical simulations of global dynamics are high due to the properties of the system, namely its stiffness. This work presents our activities aimed on the reduction of simulation time. Two approaches are covered. The first one is to parallelize the code computing ODE right-hand side using OpenMP. The other one is to find or develop a faster numerical integration method. The paper presents results of performance tests of the parallelized algorithm on various computer systems and describes scalability problems related to peculiarities of the NUMA architecture. The second approach is illustrated by the results of application of several explicit and implicit numerical methods.

Keywords: Dynamics simulation · Initial value problem · Numerical integration · Parallel algorithm

1 Introduction

In this paper we consider numerical simulations of global dynamics for a model of continuously variable transmission (CVT).

Mathematical model of CVT has been obtained in the framework of Lagrangian mechanics and contains about 1800 generalized coordinates, plus the same number of generalized speeds, so the total problem dimension is about 3600. To numerically simulate dynamics, one has to solve an initial value problem for ordinary differential equations (ODE).

Taking into account today's sizes of numerical problems solved in the fields of structural mechanics, computational fluid dynamics, and others, we have to state that our problem has a low dimension. Nevertheless, simulation running times are high: sequential code takes several hours of CPU time to simulate

© Springer International Publishing AG 2017
V. Voevodin and S. Sobolev (Eds.): RuSCDays 2017, CCIS 793, pp. 95–107, 2017.
https://doi.org/10.1007/978-3-319-71255-0_8

one second of real time. Speeding up simulations is important since it enables user to apply new analysis types, such as optimization, based on global dynamics simulations. Therefore, a significant speedup factor is highly desired for practical applications. To achieve this goal, we use two approaches.

First of all, it is possible to parallelize the code implementing the numerical simulation, aiming on modern multi-core or hybrid hardware architectures. However, the scalability of parallelization is very limited due to low dimension of the problem and the heterogeneity of the model.

The paper is organized as follows. Section 2 presents an overview of the model. Section 3 discusses current results of OpenMP-based parallelization of the code. Section 4 illustrates the behavior of various numerical integration methods applied to the problem of CVT dynamics. Section 5 provides a summary of the results obtained and outlines future work.

2   CVT Model Overview

The model of CVT includes two elastic shafts, the input and the output one, on nonlinearly elastic supports. There are two pulleys on each shaft, one motionless and one moving (Fig. 1). The pulleys have toroidal (almost conical) contact surfaces. There is a chain consisting of *rocker pins* and *plates* (Fig. 2). Each pin has two halves that roll over each other during chain motion. End surfaces of pin halves are in contact with the pulleys. The application of clamping forces to pulleys leads to certain chain configuration such that pins are at certain contact radius at each pulley set; the gear ratio can be changed by shifting the moving pulleys along the shafts. The torque is transmitted due to the friction forces at pin-pulley contact points. Mathematical models of CVT parts for global dynamics simulation have to be as simple as possible, while being able to correspond to the reality good enough and represent stressed and deformed state in individual elements, such as pins and links. Therefore, CVT shafts, rocker pins, and plates are modeled as elastic bodies. To describe the state of CVT chain, there are 21 generalized coordinates per link: 10 for each pin half (at each end of pin half axis, there are three coordinates determining its position and two angles determining its small slope; rotation of a pin half about the axis is determined by the positions of the neighboring pins) and one to determine the position of pack of plates along pin axis (see Fig. 3). Those coordinates fully determine the deformed state of each pin half and each plate in our model of CVT chain.

There are many contact interactions in the CVT model: first of all, we have pin-pulley contact; there are two more types of contact, namely the interactions between pin halves and between a pins and plates (Fig. 4).

Special attention should be paid to the contact interaction between pins and pulleys because the torque is transmitted solely due to the friction forces at pin-pulley contact points. The model of contact interaction is physically very simple: for each end surface of a pin half, the interaction is localized at one point; in that point, normal reaction force $\mathbf{N}$ and tangential friction force $\mathbf{R}$ are applied to pin half, and the opposite forces are applied to the pulley (Fig. 6).

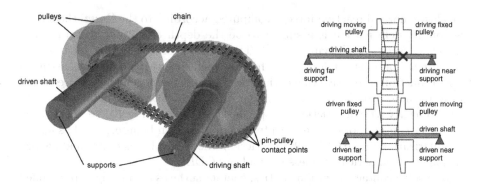

Fig. 1. CVT model general view

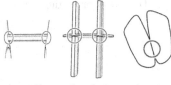

Fig. 2. CVT chain

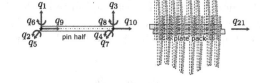

Fig. 3. CVT chain generalized coordinates

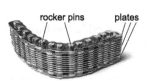

pin-pulley   pin-plate   pin-pin

Fig. 4. Types of contact interaction in CVT

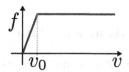

Fig. 5. Friction law for pin-pulley contact

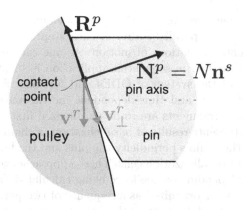

Fig. 6. Pin-pulley contact forces

The elastic normal reaction force is computed according to the Hertz' theory [1]; the contact deformation is assumed to be the depth of mutual penetration of contact surfaces, which remain rigid. The friction force is proportional to the normal reaction magnitude and the friction coefficient f. The latter is assumed to be a function of relative tangential speed v, and the dependency corresponds to the Coulomb friction at speeds higher than v_0 (a constant parameter), and to the linearly viscous friction at speeds less than v_0 (Fig. 5), so there is no sticking at contact point. The accepted friction law can be interpreted as a kind of regularization of the Coulomb dry friction; the value of v_0 is quite small, which is a source of numerical stiffness of resulting ODEs.

To resolve contact point kinematics, contact surfaces are locally approximaed with quadratic functions, which allows to determine the right position of contact point on pin half end surface. This is important because the contact point positions ultimately determine the deformed state of pins and plates.

The model of CVT is described in more detail in [2].

Differential equations of motion for the CVT model are obtained in the framework of Lagrangian mechanics, so they have the following form:

$$\frac{d}{dt}\frac{\partial L}{\partial \dot{q}_k} - \frac{\partial L}{\partial q_k} = \tilde{Q}_k, \quad k = 1, \ldots, n, \tag{1}$$

where t is the time, q_k are generalized coordinates, L is the Lagrangian, $\tilde{Q}_k$ are non-potential applied generalized forces, and n is the number of degrees of freedom.

3   Parallelization with OpenMP

Initially CVT simulation application was rather complex sequential code written in C++ that is why OpenMP was treated as preferable technology of parallelization. OpenMP's important advantage is its relative simplicity when it is applied to existing sequential code. Of course, it does not exclude abilities of the code restyling if necessary.

Due to the problem pecularities the most obvious way of introduction of parallel computing is the parallelization of each step of numerical integration procedure of differential equations of motion. As one can see in Fig. 7 most of the time of integration step in sequential application is spent on calculation of the right-hand side of the system of ODEs and foremost for the chain forces calculation, contact forces and time-dependent inertia matrix evaluation and factorization. So these fragments are to be parallelized first.

Current work represents results of parallelization of chain and contact forces evaluation. Due to the chain's periodicity the pins and the links of the chain are natural candidates to parallelization cells. These approaches to the cell definition are both used now, depending on the force being calculated. Therefore the chain forces calculation block is organized as a sequence of two parallel loops (**pragma omp for**) with static scheduling across the links and pins respectively, combined in common parallel section (**pragma omp parallel**). For example, the link-based loop contains evaluation of the forces of the link plates deformation and elastic and damping forces of pin halves deformation. At the other hand, the pin-based loop contains calculation of pin halves interaction of the same pin, such as pin halves contact and friction forces. It also contains evaluation of damping forces in joints.

As it has been mentioned above, the problem has a low dimension so it is possible to use thread-local buffers for the vector of generalized forces of the chain. Due to usage of thread-local buffers OpenMP threads require few synchronization and the procedure of chain force calculation can be scaled well. But this approach has drawback too. The buffers should be initialized with zeros before each forces calculation takes place and also the results should be gathered into common forces vector after it. In the worst case these steps can not be scaled at all, because the amount of arithmetical operations per thread does not depend on the thread count. Of course, it is worth taking into account that the count of non-zero elements in each thread-local buffer decreases when the thread count increases and, namely, has an order of N/n, where N is the length of the vector of the forces of the chain and n is the count of threads.

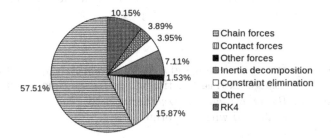

Fig. 7. CPU time consumption in CVT simulation. Sequential code

Therefore, the sequence of chain force calculation at each simulation time point consists of the next steps: thread-local buffers initialization, chain force calculation, gathering of the thread-local results into common force vector. All the steps are being executed in parallel inside **pragma omp parallel** block.

The results of simulation of the same CVT model are presented in Figs. 8 and 9. The simulation is performed on *Tesla* computer of Computer Technologies in Engineering dept. (CTM). It consists of 2 NUMA nodes and its hardware and software parameters are presented in Table 1. CPU affinity was managed with environment variable **GOMP_CPU_AFFINITY** so that when $n \leq 6$ only one NUMA node is in use. And only when $n > 6$ the cores of the second node are used. Therefore, the influence of non-uniformity of memory access in NUMA architecture becomes more explicit in this case.

The values of Y axis of Fig. 8 is time and the X axis shows number of threads. The data of the curves in the Fig. 8, refered as "CPU time" is calculated in the following way. Let us denote as $T_{i,t}^{(init)}$, $T_{i,t}^{(calc)}$ and $T_{i,t}^{(g)}$ durations of buffer

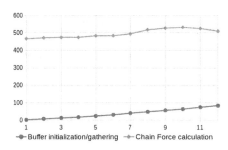
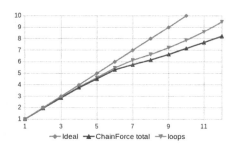

Fig. 8. CPU time

Fig. 9. Relative speedup, chain forces, Tesla

Table 1. Hardware parameters and OS/GCC versions of computers used in simulations

	Tesla	Tornado
Cores per socket	6	14
NUMA nodes	2	2
CPUs	Intel Xeon X5660 2.80 GHz	Intel Xeon E5-2697 v3 2.60 GHz
Linux	Ubuntu 12.04.05 LTS	CentOS Linux release 7.0.1406 (Core)
GCC version	4.6.3	4.8.2

initialization, chain force calculation and gathering respectively, measured in i-th thread with help of `omp_get_wtime` function at simulation time moment t. Then the data points of *Buffer initialization/gathering* and *Chain Force calculation* curves in Fig. 8 are evaluated with the formulas:

$$T_1 = \sum_t \sum_{i=1}^n \left(T_{i,t}^{(init)} + T_{i,t}^{(g)} \right), \qquad T_2 = \sum_t \sum_{i=1}^n T_{i,t}^{(calc)} \qquad (2)$$

and represent overall amount of time spent in all threads for buffers initialization/gathering and forces calculation respectively over all simulation steps. This chart demonstrates scalability of the code: in ideal case both curves should be straight horizontal lines, which means that there are no extra CPU time consumption when number of threads grows. One can see that the time of initialization/gathering grows faster than the time of the forces calculation and it will degrade efficiency of the code when the number of threads becomes large. But contribution of initialization/gathering is relatively small in the interval of thread numbers considered from 1 to 12.

The values shown in Fig. 8 do not take into account time that has been spent on parallel section creation/closing (`pragma omp parallel` block creation). This fraction can be significant, as it is shown in Fig. 9. The speeding up of chain forces parallel calculation with respect to single-thread case is presented there. The Y axis contains ratios of calculation time at 1 core to calculation time at n cores and X axis is the numbers of cores. Both curves refer to the same simulation but use time evaluated in the different way. The time used in curve $ChainForceTotal$

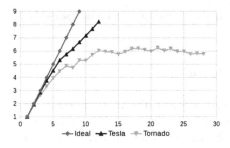

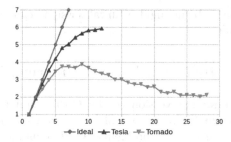

Fig. 10. Relative speedup, chain forces

Fig. 11. Relative speedup, contact forces

evaluation takes into account time consumed on `omp parallel` section. This time has been measured as a difference of `omp_get_wtime` calls after and before `omp parallel` section, therefore it shows real time of calculation. The time used in curve *loops* evaluation is the sum of curves 1 and 2 from Fig. 8 divided by corresponding number of threads and characterizes calculation time with `omp parallel` excluded. One can see that the difference between these curves becomes especially significant when both NUMA nodes are involved $(n > 6)$.

This CVT configuration has been also simulated on another machine described in the second column of the Table 1. This is two nodes of computational cluster "Polytechnic RSK Tornado" of Supercomputer Center "Polytechnic" of SPbPU. Further it is referred as *Tornado*. Unlike the simulation on *Tesla* this computation does not use explicit thread binding with `GOMP_CPU_AFFINITY` variable, so the system assigns threads to cores implicitly.

Figure 10 represents chart analogous to Fig. 9: relative speed up in dependency on the number of threads. The curve *Tesla* is the same as *ChainForceTotal* in Fig. 9, i.e., relative speedup of chain forces calculation on *Tesla*. *Tornado* curve is analogous result obtained at *Tornado*. One can see that in the second case the scalability is much worse and there is no speed up since the level of 11–12 threads is reached. This dependency on architecture of hardware used is subject to future investigation.

The calculation of contact forces between chain and pulleys takes place in a separate parallel block and because it is being performed faster relatively to chain forces calculation, the contribution of parallel section creation/closing in this case is more significant. Contact forces calculations on both machines are presented in Fig. 11. Meanings of the curves are the same as in Fig. 10 with respective replacement of "chain forces" with "contact forces". One can see the loss of performance on *Tornado* with the number of threads growing. Code of contact forces calculation does not contain explicit synchronization structures, therefore the genesis of such slowing down is not obvious and requires additional investigation. Measuring of time of contact forces calculation without taking into account parallel section creation/closing shows much better scalability so the problem might be in rather significant contribution of `pragma omp parallel` code.

4   Investigation of Numerical Methods

Production version of the CVT simulation code has always been using the Runge–Kutta numerical integration scheme of fourth order (RK4) [3] to solve the initial value problem of CVT dynamics. It is known that the RK4 scheme, as well as other explicit numerical integration schemes, have a step size limitation due to the stability requirement: in general, for a linear system the value $h\lambda$, where h is the step size and λ is an eigenvalue of ODE right-hand side Jacobian, must belong to the stability region, which for an explicit scheme is always a bounded area in the complex plane; for nonlinear ODEs, it is usually the same.

The step size used for CVT numerical simulations with the RK4 scheme has to be quite small, between 10^{-8} and 10^{-7} due to the above mentioned stability limitation. As a consequence, CPU time required for a simulation is high. The analysis of ODE system Jacobian has shown that without friction, maximum eigenvalue magnitude is about 10^6 and corresponds to pin axial vibrations; due to the friction, there are also real negative eigenvalues up to -10^8. Therefore, the ODE system can be considered mildly stiff.

While the actual goal of the entire investigation is to decrease CPU time of simulations, in this section we try to achieve a different goal: find a method that can be applied with significantly larger step sizes than those currently in use. Once such a method is found, its performance has to be further optimized.

Sections below illustrate our attempts to apply different numerical methods to CVT dynamics simulation; we cover explicit methods (Sect. 4.2), semi-implicit methods (Sect. 4.3), and one completely implicit method (Sect. 4.4).

4.1   Numerical Experiment Setup

For each numerical integration scheme, two tests have been done. In the first test, the dependency of step local error on the step size is investigated. The error is computed simply by comparison with the "exact" solution obtained with a very small step size of $2 \cdot 10^{-9}$ using the RK4 scheme. The test can be used, in particular, to verify scheme order of accuracy. In the second test, a dynamics simulation is performed during 0.005 second of real time; a sample history curve is obtained (namely, the axial force in a pin half entering a pulley set) as an evident indicator of the acceptability of numerical results.

To illustrate the impact of nonsmoothness of friction law on the accuracy of numerical results, we also included the results of testing for smooth friction law $f = f_0 \arctan \frac{v}{v_0 f_0}$, where f_0 is the saturation value of friction coefficient.

Sections below present the results of first test for the original and smoothed friction law, and the results of second test for both friction laws and for selected step sizes.

4.2   Explicit Methods

Explicit methods covered in this subsection are the following classical ones.

– Three embedded Dormand–Prince schemes with step size control [3]. The step
   size control was disabled in test simulation. An embedded scheme provides
   two solutions of different orders of accuracy at each step, which can be used
   to control the step size; those orders are encoded in the name of the scheme.
   The three schemes are DOPRI45 (orders 4, 5), DOPRI56 (orders 5, 6), and
   DOPRI78 (orders 7, 8).
– Gragg–Bulirsch–Stoer method (GBS) with smoothing step [3]. It is an extrapolation method with the symmetric Gragg's scheme used as the reference
   scheme. We tried this method with a fixed number of extrapolation stages
   (2, 4, 6) and the harmonic step size sequence (the schemes are referred to as
   GBS2, GBS4, GBS6 below).
– Extrapolated explicit Euler scheme, with 2 extrapolation stages and the harmonic step size sequence (referred to as Euler-x2 below).

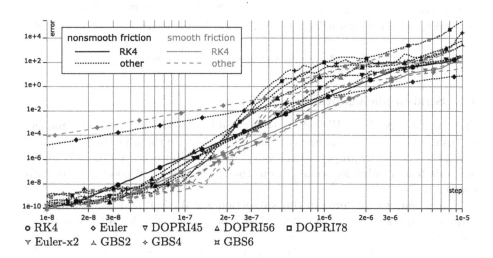

Fig. 12. Step local error for explicit methods

The step local error test (Fig. 12) shows that the local error is generally less
for smooth friction law; further data processing also indicates that the local error
behaves according to scheme order only in a limited step size ranges, different
for different schemes; some schemes (DOPRI45, DOPRI78, GBS4, GBS6) do not
show the expected behavior at all, although they do in tests with simple ODEs.

The dynamics test (Fig. 13) confirms that all explicit schemes considered
have severe step size limitation that is about 10^{-7} for nonsmooth friction law
and schemes GBS2, DOPRI56, and is less for other schemes; for smooth friction
law, the limit is higher yet it is less than $5 \cdot 10^{-7}$. We can also conclude that low
order schemes (2–4) are preferrable in model with non-smooth friction low; in
model with smooth friction law, higher order schemes may be preferrable.

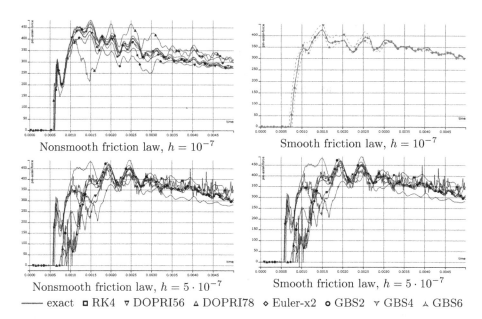

Nonsmooth friction law, $h = 10^{-7}$

Smooth friction law, $h = 10^{-7}$

Nonsmooth friction law, $h = 5 \cdot 10^{-7}$

Smooth friction law, $h = 5 \cdot 10^{-7}$

—— exact ▫ RK4 ▽ DOPRI56 ▵ DOPRI78 ◇ Euler-x2 ○ GBS2 ▽ GBS4 ▵ GBS6

Fig. 13. Sample curves for explicit methods

4.3   Semi-implicit Methods

There was a hope that a W-method [4] is capable of producing acceptable numerical solution at steps much greater than 10^{-7}, because those methods generally have better stability properties than explicit ones. However, in our case all W-methods tested failed for some reason, though they worked good in tests with simple ODEs.

The schemes considered in this subsection are W24 [4] and the W1 method extrapolated according to the Richardson's procedure [3]. The W1 scheme is as follows:

$$\mathbf{x}_{k+1} = \mathbf{x}_k + h\mathbf{f}(t_k, \mathbf{x}_k) + hd\mathbf{A}(\mathbf{x}_{k+1} - \mathbf{x}_k), \tag{3}$$

where $\mathbf{x}$ is the numerical solution vector, the subscript k denotes the step number, h is the step size, t is the time, $\mathbf{f}$ is the ODE right-hand side vector, d is a parameter (usually between 0 and 1), and $\mathbf{A}$ is the matrix approximating the ODE system Jacobian $D\mathbf{f}/D\mathbf{x}$. W-methods are attractive compared to Rosenbrock methods [5] due to the ability to keep $\mathbf{A}$ constant during many time steps, thus eliminating the necessity to compute it and factorize the matrix $\mathbf{I} - hd\mathbf{A}$ at each time step.

Figure 14 shows that all schemes tested have much greater local step error than explicit schemes. The expected order of schemes is observed at step sizes less than 10^{-7}; the higher the order, the less the range in which scheme order is obeyed.

Figure 15 shows that all W-method schemes produce inacceptable solution even at step 10^{-7}. We have to conclude that they didn't work in our case.

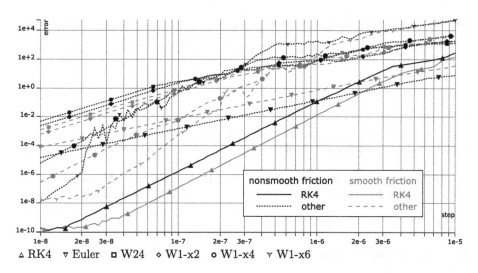

Fig. 14. Step local error for semi-implicit W-methods

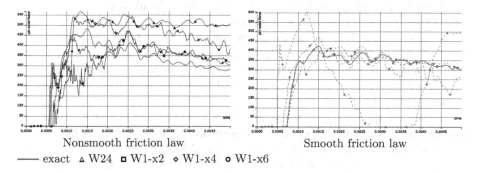

Nonsmooth friction law   Smooth friction law

—— exact ▵ W24 ▫ W1-x2 ◇ W1-x4 ○ W1-x6

Fig. 15. Sample curves for W-methods, $h = 10^{-7}$

4.4 Trapezoidal Rule Method

Among many implicit methods, we chose the trapezoidal rule (2-nd order scheme):

$$\mathbf{x}_{k+1} = \mathbf{x}_k + \frac{h}{2}[\mathbf{f}(t_k, \mathbf{x}_k) + \mathbf{f}(t_k + h, \mathbf{x}_{k+1})], \qquad (4)$$

Figure 16 shows the step local error for the trapezoidal rule. Notice that it is less than for any other scheme tested at steps greater than $4 \cdot 10^{-7}$.

Sample curve shown in Fig. 17 is obtained at step size $2 \cdot 10^{-6}$ and practically coincides with the exact solution. It is possible to use larger step sizes, but only with step size control because the Newton's method used at a time step may fail to converge.

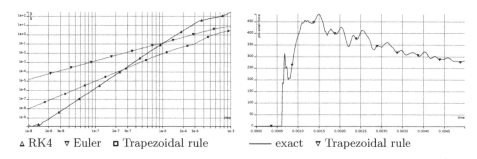

△ RK4   ▽ Euler   ⊡ Trapezoidal rule   —— exact   ▽ Trapezoidal rule

Fig. 16. Step local error for trapezoidal rule, nonsmooth friction law

Fig. 17. Sample curves for trapezoidal rule, $h = 2 \cdot 10^{-6}$, nonsmooth friction law

5   Conclusions and Future Work

The paper considers two approaches for speeding up the numerical integration of about 3600 ODEs of CVT dynamics. The ODE right-hand side is quite numerically expensive, and the ODE system is mildly stiff.

The first approach is to parallelize the computation of ODE right-hand side. Usage of pins and chain links as cells of parallelization allows to calculate forces in the chain in natural way. But scalability of present implementation strongly depends on parameters of machine used and may be rather poor. The cause of it is the goal of future investigation.

The second approach is to find a numerical method faster than RK4 currently used in the production version of CVT software. The investigation has shown that traditional explicit numerical integration schemes and W-methods don't work in our case. Implicit methods give good results; however, to make those methods run faster than RK4, additional efforts are required: for example, ODE right-hand side Jacobian could be computed much faster but it requires tedious programming (the idea is to combine the approach presented in [6] with the decomposition of the ODE right-hand side into a sum and providing faster code for the Jacobian of contact forces).

Future plans include performance improvements for implicit schemes. In addition, we are planning to test so called stabilized explicit Runge–Kutta methods because they have not been covered in this research, while seem to be quite apropriate for the case of Jacobian eigenvalues that we really have.

References

1. Johnson, K.: Contact Mechanics. Cambridge University Press, Cambridge (1987)
2. Shabrov, N., Ispolov, Y., Orlov, S.: Simulations of continuously variable transmission dynamics. ZAMM **94**(11), 917–922 (2014)
3. Hairer, E., Nørsett, S.P., Wanner, G.: Solving Ordinary Differential Equations I, 2nd Revised. Edn. Nonstiff Problems. Springer, New York (1993)

4. Steihaug, T., Wolfbrandt, A.: An attempt to avoid exact Jacobian and nonlinear equations in the numerical solution of stiff differential equations. Math. Comp. **33**, 521–534 (1979)
5. Rosenbrock, H.H.: Some general implicit processes for the numerical solution of differential equations. Comput. J. **5**, 329–330 (1963)
6. Ypma, T.J.: Efficient estimation of sparse Jacobian matrices by differences. J. Comput. Appl. Math. **18**(1), 17–28 (1987)

Solving Time-Consuming Global Optimization Problems with Globalizer Software System

Alexander Sysoyev$^{(\boxtimes)}$, Konstantin Barkalov, Vladislav Sovrasov,
Ilya Lebedev, and Victor Gergel

Lobachevsky State University of Nizhny Novgorod, Nizhny Novgorod, Russia
{alexander.sysoyev,konstantin.barkalov,ilya.lebedev,
victor.gergel}@itmm.unn.ru, sovrasov.vlad@gmail.com

Abstract. In this paper, we describe the Globalizer software system for solving global optimization problems. The system implements an approach to solving the global optimization problems using the block multistage scheme of the dimension reduction, which combines the use of Peano curve type evolvents and the multistage reduction scheme. The scheme allows an efficient parallelization of the computations and increasing the number of processors employed in the parallel solving of the global optimization problems many times.

Keywords: Multidimensional multiextremal optimization · Global search algorithms · Parallel computations · Dimension reduction · Block multistage dimension reduction scheme

1 Introduction

The development of optimization methods that use high-performance computing systems to solve time-consuming global optimization problems is an area receiving extensive attention. The theoretical results obtained provide efficient solutions to many applied global optimization problems in various fields of scientific and technological applications.

At the same time, the practical software implementation of these algorithms for multiextremal optimization is quite limited. Among the software for the global optimization, one can select the following systems:

- LGO (Lipschitz Global Optimization) [1] is designed to solve global optimization problems for which the criteria and constraints satisfy the Lipschitz condition. The system is a commercial product based on diagonal extensions of one-dimensional multiextremal optimization algorithms.
- GlobSol [2] is oriented towards solving global optimization problems as well as systems of nonlinear equations. The system includes interval methods based on the branch and bound method. There are some extensions of the system for parallel computations, and it is available to use for free.
- LINDO [3] is features by a wide spectrum of problem solving methods that can be used for these include linear, integer, stochastic, nonlinear, and global optimization problems. The ability to interact with the Microsoft Excel software environment is a

© Springer International Publishing AG 2017
V. Voevodin and S. Sobolev (Eds.): RuSCDays 2017, CCIS 793, pp. 108–120, 2017.
https://doi.org/10.1007/978-3-319-71255-0_9

key feature of the system. The system is widely used in practical applications and is available to use for free.

- IOSO (Indirect Optimization on the basis of Self-Organization) [4] is oriented toward solving of a wide class of the extremal problems including global optimization problems. The system is widely used to solve applied problems in various fields. There are versions of the system for parallel computational systems. The system is a commercial product, but is available for trial use.

- MATLAB Global Optimization Toolkit [5], includes a wide spectrum of methods for solving the global optimization problems, including multistart methods, global pattern search, simulated annealing methods, etc. The library is compatible to the TOMLAB system [6], which is an additional extension the widely-used MATLAB. It is also worth noting that similar libraries for solving global optimization problems are available for MathCAD, Mathematica, and Maple systems as well.

- BARON (Branch-And-Reduce Optimization Navigator) [7], is designed to solve continuous integer programming and global optimization problems using the branch and bound method. BARON is included in the GAMS (General Algebraic Modeling System) system used widely [8].

- Global Optimization Library in R [9] is a large collection of optimization methods implemented in the R language. Among these methods, there are stochastic and deterministic global optimization algorithms, the branch and bound method, etc.

The list provided above is certainly not exhaustive – additional information on software systems for a wider spectrum of optimization problems can be obtained, for example, in [10–12], etc. Nevertheless, even from such a short list the following conclusions can be drawn (see also [13]).

- The collection of available global optimization software systems for practical use is insufficient.

- The availability of numerous methods through these systems allows complex optimization problems to be solved in a number of cases, however, it requires a rather high level of user knowledge and understanding in the field of global optimization.

- The use of the parallel computing to increase the efficiency in solving complex time-consuming problems is limited, therefore, the computational potential of modern supercomputer systems is very poorly utilized.

In this paper, a novel Globalizer software system is considered. The development of the system was conducted based on the information-statistical theory of multiextremal optimization aimed at developing efficient parallel algorithms for global search – see, for example, [14–16]. The advantage of the Globalizer is that the system is designed to solve time-consuming multiextremal optimization problems. In order to obtain global optimized solutions within a reasonable time and cost, the system efficiently uses modern high-performance computer systems.

The paper is further structured as follows. In Sect. 2, the general statement of the multidimensional global optimization problem is considered. In Sect. 3, the Globalizer software system is presented and its architecture is described. In Sect. 4, the approaches to solving the multidimensional global optimization problem based on the information-statistical theory of multiextremal optimization is given. In Sect. 5, the results of applied problem solving with the Globalizer system are described. Finally, Sect. 6 presents the conclusion.

2 Statement of Multidimensional Global Optimization Problem

In this paper, the core class of optimization problems which can be solved using the Globalizer is examined. This involves multidimensional global optimization problems without constraints, which can be defined in the following way:

$$\varphi(y) \rightarrow \inf, y \in D \subset R^N, \tag{1}$$

$$D = \{y \in R^N : a_i \leq y_i \leq b_i, 1 \leq i \leq N\}, \tag{2}$$

i.e., a problem of finding the globally optimal values of the objective (minimized) function $\varphi(y)$ in a domain D defined by the coordinate bounds (2) on the choice of feasible points $y = (y_1, y_2, \ldots, y_N)$.

If y^* is an exact solution of problem (1) – (2), the numerical solution of the problem is reduced to building an estimate y^0 of the exact solution matching to some notion of nearness to a point (for example, $\|y^* - y^0\| \leq \varepsilon$ where $\varepsilon > 0$ is a predefined accuracy) based on a finite number k of computations of the optimized function values.

Regarding to the class of problems considered, the fulfillment of the following important conditions is supposed:

1. The optimized function $\varphi(y)$ can be defined by some algorithm for the computation of its values at the points of the domain D.
2. The computation of the function value at every point is a computation-costly operation.
3. Function $\varphi(y)$ satisfy the Lipschitz condition:

$$|\varphi(y_1) - \varphi(y_2)| \leq L\|y_1 - y_2\|, \text{where } y_1, y_2 \in D, 0 < L < \infty, \tag{3}$$

that corresponds to a limited variation of the function value at limited variation of the argument.

The multiextremal optimization problems i.e. the problems, which the objective function $\varphi(y)$ has several local extrema in the feasible domain D in, are the subjects of consideration in the present paper. The dimensionality affects the difficulty of solving such problems considerably. For multiextremal problems so called "curse of dimensionality" consisting in an exponential increase of the computational costs with increasing dimensionality takes place.

3 Globalizer Architecture

The Globalizer considered in this paper expands the family of global optimization software systems successively developed by the authors during the past several years. One of the first developments was the SYMOP multiextremal optimization system [17], which has been successfully applied for solving many optimization problems. A special place is occupied by the ExaMin system [18], which was developed and used extensively to investigate the application of novel parallel algorithms to solve global optimization problems using high-performance multiprocessor computing systems.

The program architecture of Globalizer system is presented in Fig. 1.

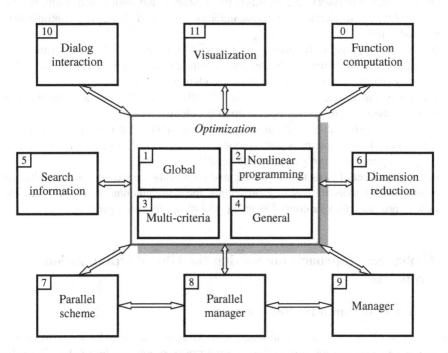

Fig. 1. Program architecture of Globalizer system (Blocks 1-2, 5-7 have been implemented; Blocks 3-4 and 8-11 are under development)

The structural components of the systems are:

- Block 0 is an external block. It consists of the procedures for computing the function values (criteria and constraints) for the optimization problem being solved.
- Blocks 1-4 form the optimization subsystem and solve the global optimization problems (Block 1), nonlinear programming (Block 2), multicriterial optimization (Block 3), and general decision making problems (Block 4). It is worth noting the successive scheme of interaction between these components – the decision making problems are solved using the multicriterial optimization block, which, in turn, uses the nonlinear programming block, etc.

- Block 5 is a subsystem for accumulating and processing the search information; this is one of the main subsystems – the amount of search information for time-consuming optimization problems may appear to be quite large on the one hand, but, on the other hand, the efficiency of the global optimization methods depends to a great extent on how completely all of the available search data is utilized.
- Block 6 contains the dimensional reduction procedures based on the Peano evolvents; this block also provides interaction between the optimization blocks and the initial multidimensional optimization problem.
- Block 7 organizes the choice of parallel computation schemes in the Globalizer system subject to the computing system architecture employed (the numbers of cores in the processors, the availability of shared and distributed memory, the availability of accelerators for computations, etc.) and the global optimization methods applied.
- Block 8 is responsible for managing the parallel processes when performing the global search (determining the optimal configuration of parallel processes, distributing the processes between computing elements, etc.).
- Block 9 is a management subsystem, which fully controls the whole computational process when solving global optimization problems.
- Block 10 is responsible for organizing the dialog interaction with users for stating the optimization problem, adjusting system parameters (if necessary), and visualizing and presenting the global search results.
- Block 11 is a set of tools for visualizing and presenting the global search results; the availability of tools for visually presenting the computational results enables the user to provide efficient control over the global optimization process.

4   Globalizer Approach for Solving the Global Optimization Problems

4.1   Methods of Dimension Reduction

Globalizer implements a block multistage scheme of dimension reduction [18], which reduces the solving of initial multidimensional optimization problem (1) – (2) to the solving of a sequence of «nested» problems of less dimensionality.

Thus, initial vector y is represented as a vector of the «aggregated» macro-variables

$$y = (y_1, y_2, \ldots, y_N) = (u_1, u_2, \ldots, u_M) \qquad (4)$$

where the i-th macro-variable u_i is a vector of the dimensionality N_i from the components of vector y taken sequentially i.e.

$$
\begin{aligned}
u_1 &= (y_1, y_2, \ldots, y_{N_1}), \\
u_2 &= (y_{N_1+1}, y_{N_1+2}, \ldots, y_{N_1+N_2}), \ldots \\
u_i &= (y_{p+1}, \ldots, y_{p+N_i}) \text{ where } p = \sum_{k=1}^{i-1} N_k, \ldots
\end{aligned}
\qquad (5)
$$

at that, $\sum_{k=1}^{M} N_k = N$.

Using the macro-variables, the main relation of the well-known multistage scheme can be rewritten in the form

$$\min_{y \in D} \varphi(y) = \min_{u_1 \in D_1} \min_{u_2 \in D_2} \ldots \min_{u_M \in D_M} \varphi(y), \tag{6}$$

where the subdomains D_i, $1 \leq i \leq M$, are the projections of the initial search domain D onto the subspaces corresponding to the macro-variables u_i, $1 \leq i \leq M$.

The fact, that the nested subproblems

$$\varphi_i(u_1, \ldots, u_i) = \min_{u_{i+1} \in D_{i+1}} \varphi_{i+1}(u_1, \ldots, u_i, u_{i+1}), 1 \leq i \leq M, \tag{7}$$

are the multidimensional ones in the block multistage scheme is the principal difference from the initial scheme. Thus, this approach can be combined with the reduction of the domain D (for example, with the evolvent based on Peano curve) for the possibility to use the efficient methods of solving the one-dimensional problems of the multiextremal optimization [19].

The Peano curve $y(x)$ lets map the interval of the real axis [0,1] onto the domain D uniquely:

$$\{y \in D \subset R^N\} = \{y(x) : 0 \leq x \leq 1\}. \tag{8}$$

The evolvent is the approximation to the Peano curve with the accuracy of the order 2^{-m} where m is the density of the evolvent.

Application the mappings of this kind allows reducing multidimensional problem (1) – (2) to a one-dimensional one

$$\varphi(y^*) = \varphi(y(x^*)) = min\{\varphi(y(x)) : x \in [0, 1]\}. \tag{9}$$

4.2   Method for Solving the Reduced Global Optimization Problems

The information-statistical theory of global search formulated in [14, 16] has served as a basis for the development of a large number of efficient multiextremal optimization methods – see, for example, [20–23], [24–27], etc. Within the framework of information-statistical theory, a general approach to parallelization computations when solving global optimization problems has been proposed – the parallelism of computations is provided by means of simultaneously computing the values of the minimized function $\varphi(y)$ at several different points within the search domain D – see, for example, [15, 16]. This approach provides parallelization for the most costly part of computations in the global search process.

Let us consider the general computation scheme of Parallel Multidimensional Algorithm of Global Search that is implemented in Globalizer.

Let us introduce a simpler notation for the problem being solved

$$f(x) = \varphi(y(x)) : x \in [0, 1]. \tag{10}$$

Let us assume $k > 1$ iterations of the methods to be completed (the point of the first trial x^1 can be an arbitrary point of the interval $[a; b]$ – for example, the middle of the interval). Then, at the $(k + 1)$-th iteration, the next trial point is selected according to the following rules.

Rule 1. To renumber the points of the preceding trials $x^1, \ldots, x^n$ (including the boundary points of the interval $[a; b]$) by the lover indices in the order of increasing values of the coordinates,

$$0 = x_0 < x_1 < \ldots < x_i < \ldots < x_k < x_{k+1} = 1 \tag{11}$$

The function values $z_i = \varphi(x_i)$ have been calculated in all points $x_i, i = 1, ..k$. In the points $x_0 = 0$ and $x_{k+1} = 1$ the function values has not been computed (these points are used for convenience of further explanation).

Rule 2. To compute the values:

$$\mu = \max_{1 \le i \le k} \frac{|z_i - z_{i-1}|}{\Delta_i}, M = \begin{cases} r\mu, & \mu > 0, \\ 1, & \mu = 0, \end{cases} \tag{12}$$

where $r > 1$ is the *reliability* parameter of the method, $\Delta_i = x_i - x_{i-1}$.

Rule 3. To compute the characteristics for all intervals $(x_{i-1}; x_i), 1 < i < k+1$, according to the formulae:

$$R(1) = 2\Delta_1 - 4\frac{z_1}{M}; \quad R(k+1) = 2\Delta_{k+1} - 4\frac{z_k}{M};$$

$$R(i) = \Delta_i + \frac{(z_i - z_{i-1})^2}{M^2 \Delta_i} - 2\frac{z_i + z_{i-1}}{M}, 1 < i < k+1. \tag{13}$$

Rule 4. To arrange the characteristics of the intervals obtained according to (13) in decreasing order

$$R(t_1) \ge R(t_2) \ge \ldots \ge R(t_k) \ge R(t_{k+1}) \tag{14}$$

and to select p intervals with the highest values of characteristics (p is the number of processors/cores used for the parallel computations).

Rule 5. To execute new trials at the points

$$x_{k+j} = \begin{cases} \frac{x_{t_j} + x_{t_j-1}}{2}, t_j \in \{1, k+1\}, \\ \frac{x_t + x_{t_j-1}}{2} - \text{sign}(z_{t_j} - z_{t_j-1})\frac{1}{2r}\left[\frac{|z_{t_j} - z_{t_j-1}|}{M}\right]^N, 1 < t_j < k+1. \end{cases} \tag{15}$$

4.3   Implementation of Parallel Algorithm of Global Optimization

Let us consider a parallel implementation of the block multistage dimension reduction scheme described in Subsection 4.1.

For the description of the parallelism in the multistage scheme, let us introduce a vector of parallelization degrees

$$\pi = (\pi_1, \pi_2, \ldots, \pi_M), \tag{16}$$

where $\pi_i, 1 \leq i \leq M$, is the number of the subproblems of the $(i+1)$-th nesting level being solved in parallel, arising as a result of execution of the parallel iterations at the i-th level. For the macro-variable u_i, the number π_i means the number of parallel trials in the course of minimization of the function $\varphi_M(u_1, \ldots, u_M) = \varphi(y_1, \ldots, y_N)$ with respect to u_i at fixed values of $u_1, u_2, \ldots, u_{i-1}$, i.e. the number of the values of the objective function $\varphi(y)$ computed in parallel.

In the general case, the quantities $\pi_i, 1 \leq i \leq M$ can depend on various parameters and can vary in the course of optimization, but we will limit ourselves to the case when all components of the vector π are constant.

Thus, a tree of MPI-processes is built in the course of solving the problem. At every nesting level (every level of the tree) PMAGS is used. Let us remind that the parallelization is implemented by selection not a single point for the next trial (as in the serial version) but p points, which are placed into p intervals with the highest characteristics. Therefore, if p processors are available, p trials can be executed in these points in parallel. At that, the solving of the problem at the i-th level of the tree generates the subproblems for the $(i+1)$-th level. This approach corresponds to such a method of organization of the parallel computations as a «master-slave» scheme.

When launching the software, the user specifies:

- A number of levels of subdivision of the initial problem (in other words, the number of levels in the tree of processes) M;
- A number of variables (dimensions) at each level ($\sum_{k=1}^{M} N_k = N$ where N is the dimensionality of the problem);
- A number of the MPI-processes and the distribution of these ones among the levels ($\pi = (\pi_1, \pi_2, \ldots, \pi_M)$).

Let us consider an example:

$$N = 10, M = 3, N_1 = 3, N_2 = 4, N_3 = 3, \pi = (2, 3, 0).$$

Therefore, we have 9 MPI-processes, which are arranged into a tree (Fig. 2: at every function φ_i varied parameters are shown only, the fixed values are not shown in the figure). According to N_1, N_2, N_3 we have the following macro-variables: $u_1 = (y_1, y_2, y_3), u_2 = (y_4, y_5, y_6, y_7), u_3 = (y_8, y_9, y_{10})$. Each node solves a problem from relation (10). The root (level #0) solves the problem with respect to the first N_1 variables of the initial N-dimensional problem. The iteration generates a problem of the next level at any point. The nodes of level #1 solve the problems with respect to N_2 variables with the fixed values of the first N_1 variables, etc.

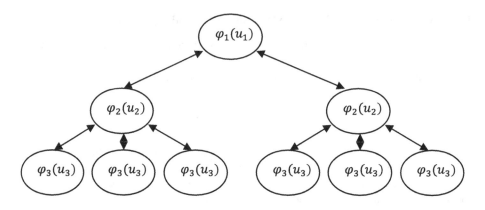

Fig. 2. Scheme of organization of parallel computations

5 Numerical Results

5.1 Test Problems Solving

The computational experiments were conducted using the Lobachevsky supercomputer at the State University of Nizhny Novgorod (http://hpc-education.unn.ru/en/ resources). The problems generated by the GKLS-generator [21] were selected for the test problems.

The results of the numerical experiments with Globalizer on an Intel Xeon Phi are provided in Table 1. The computations were performed using the Simple and Hard function classes with the dimensions equal to 4 and 5.

Table 1. Average number of iterations

		p	N = 4		N = 5	
			Simple	*Hard*	*Simple*	*Hard*
I	**Serial computations** *Average number of iterations*	1	11953	25263	15920	>148342(4)
II	**Parallel computations on CPU** *Speedup*	2	2.51	2.26	1.19	1.36
		4	5.04	4.23	3.06	2.86
		8	8.58	8.79	4.22	6.56
III	**Parallel computations on Xeon Phi** *Speedup*	60	8.13	7.32	9.87	6.55
		120	16.33	15.82	15.15	17.31
		240	33.07	27.79	38.80	59.31

In the first series of experiments, serial computations using MAGS were executed. The average number of iterations performed by the method for solving a series of problems for each of these classes is shown in row I. The symbol ">" reflects the situation where not all problems of a given class were solved by a given method. It means that the algorithm was stopped once the maximum allowable number of iterations K_{max} was achieved. In this case, the K_{max} value was used for calculating the average number of iterations corresponding to the lower estimate of this average value. The number of unsolved problems is specified in brackets.

In the second series of experiments, parallel computations were executed on a CPU. The relative "speedup" in iterations achieved is shown in row II; the speedup of parallel computations was measured in relation to the serial computations (p = 1).

The final series of experiments was executed using a Xeon Phi. The results of these computations are shown in row III; in this case, the speedup factor is calculated in relation to the PMAGS results on a CPU using eight cores (p = 8).

5.2   The Problem of Optimal Vibration Isolation for the Multi-Degree-of-Freedom System

Consider the vibration isolation problem for a multidegree-of-freedom system consisting of a base and elastic body to be isolated modeled by two material points connected each other by elastic and damping elements [28]. This mechanical system is described by the equations

$$
\begin{aligned}
&\ddot{\xi}_1 = -\beta\left(\dot{\xi}_1 - \dot{\xi}_2\right) - \xi_1 + \xi_2 + u + v, \\
&\ddot{\xi}_2 = -\beta\left(\dot{\xi}_2 - \dot{\xi}_1\right) - \xi_2 + \xi_1 + v, \\
&\xi_1(0) = \xi_2(0) = 0, \dot{\xi}_1(0) = \dot{\xi}_2(0) = 0.
\end{aligned}
\tag{17}
$$

where ξ_1 and ξ_2 are coordinates of the material points, v is the base acceleration up to sign (the external excitation), u is the control force, β is a positive damping parameter. Rewrite the Eq. (26) in the standard form

$$
\begin{aligned}
&\dot{x}_1 = x_3, \\
&\dot{x}_2 = x_4, \\
&\dot{x}_3 = -x_1 + x_2 - \beta x_3 + \beta x_4 + v + u, \\
&\dot{x}_4 = x_1 - x_2 + \beta x_3 - \beta x_4 + v, \\
&x_1(0) = x_2(0) = x_3(0) = x_4(0) = 0.
\end{aligned}
\tag{18}
$$

This model can describe the typical situations of vibration isolation for devices, apparatuses and humans located on moving vehicles.

Choose two criteria for this system to characterize the process of vibration isolation

$$
J_1(u) = \sup_{v \in L_2} \frac{\sup_{t \geq 0}|x_1(t)|}{\|v\|_2}, \quad J_2(u) = \sup_{v \in L_2} \frac{\sup_{t \geq 0}|x_2(t) - x_1(t)|}{\|v\|_2}.
\tag{19}
$$

The first criterion characterizes the maximal displacement of the body to be isolated with respect to the base, while the second one the maximal deformation of the elastic body. Consider two-objective control problem for state-feedback case. The Pareto optimal front computed by Globalizer is presented on Fig. 3.

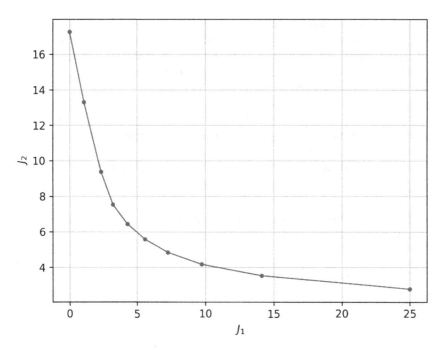

Fig. 3. Pareto optimal front for the vibration isolation problem

6   Conclusion

In this paper, the Globalizer global optimization software system was presented for implementing a general scheme for the parallel solution of globally optimized decision making. The work is devoted to the investigation of the possibility to speedup the process of searching the global optimum when solving the multidimensional multi-extremal optimization problems using the approach based on the application of the parallel block multistage scheme of the dimension reduction.

The architecture of Globalizer system has been considered. The usage of Globalizer has been demonstrated by solving the applied problem of control theory.

Acknowledgements. This research was supported by the Russian Science Foundation, project No 16-11-10150 "Novel efficient methods and software tools for the time consuming decision making problems with using supercomputers of superior performance".

References

1. Pintér, J.D.: Global Optimization in Action (Continuous and Lipschitz Optimization: Algorithms, Implementations and Applications). Kluwer Academic Publishers, Dordrecht (1996)
2. Kearfott, R.B.: Globsol user guide. Optim. Methods Softw. **24**, 687–708 (2009)

3. Lin, Y., Schrage, L.: The global solver in the LINDO API. Optim. Methods Softw. **24**, 657–668 (2009)
4. Egorov, I.N., Kretinin, G.V., Leshchenko, I.A., Kuptzov, S.V.: IOSO optimization toolkit - novel software to create better design. In: 9th AIAA/ISSMO Symposium on Multidisciplinary Analysis and Optimization, Atlanta, Georgia (2002)
5. Venkataraman, P.: Applied optimization with MATLAB programming. Wiley, Hoboken (2009)
6. Holmström, K., Edvall, M.M.: The TOMLAB optimization environment. In: Modeling Languages in Mathematical Optimization, vol 88, pp. 369–376. Springer, Boston (2004). https://doi.org/10.1007/978-1-4613-0215-5_19
7. Sahinidis, N.V.: BARON: a general purpose global optimization software package. J. Glob. Optim. **8**(2), 201–205 (1996)
8. Bussieck, M.R., Meeraus, A.: General algebraic modeling system (GAMS). In: Kallrath, J. (ed.) Modeling Languages in Mathematical Optimization. Applied Optimization, vol 88, pp. 137–157. Springer, Boston (2004). https://doi.org/10.1007/978-1-4613-0215-5_8
9. Mullen, K.M.: Continuous global optimization in R. J. Stat. Softw. **60**(6) (2014)
10. Mongeau, M., Karsenty, H., Rouzé, V., Hiriart-Urruty, J.B.: Comparison of public-domain software for black box global optimization. Optim. Methods Softw. **13**(3), 203–226 (2000)
11. Pintér, J.D.: Software development for global optimization. In: Pardalos, P.M., Coleman, T. F. (eds.), Lectures on Global Optimization, Fields Institute Communications, vol. 55, pp. 183–204 (2009)
12. Rios, L.M., Sahinidis, N.V.: Derivative-free optimization: a review of algorithms and comparison of software implementations. J. Global Optim. **56**(3), 1247–1293 (2013)
13. Liberti, L.: Writing global optimization software. In: Global Optimization: From Theory to Implementation, vol. 84, pp. 211–262. Springer, Boston (2006). https://doi.org/10.1007/0-387-30528-9_8
14. Strongin, R.G.: Numerical Methods in Multiextremal Problems (Information-Statistical Algorithms). Nauka (In Russian), Moscow (1978)
15. Strongin, R.G., Gergel, V.P., Grishagin, V.A., Barkalov, K.A.: Parallel Computations for Global Optimization Problems. Moscow State University (In Russian), Moscow (2013)
16. Strongin, R.G., Sergeyev, Ya.D.: Global Optimization with Non-convex Constraints: Sequential and Parallel Algorithms. Kluwer Academic Publishers, Dordrecht (2000)
17. Gergel, V.P.: A software system for multiextremal optimization. Eur. J. Oper. Res. **65**(3), 305–313 (1993)
18. Barkalov, K.A., Gergel, V.P.: Multilevel scheme of dimensionality reduction for parallel global search algorithms. In: Proceedings of the 1st International Conference on Engineering and Applied Sciences Optimization, pp. 2111–2124 (2014)
19. Sergeyev, Ya.D., Strongin, R.G., Lera, D.: Introduction to Global Optimization Exploiting Space-Filling Curves. Springer, New York (2013)
20. Barkalov, K., Gergel, V.: Parallel global optimization on GPU. J. Global Optim. **66**(1), 3–20 (2016)
21. Barkalov, K., Gergel, V., Lebedev, I.: Use of Xeon Phi Coprocessor for Solving Global Optimization Problems. In: Malyshkin, V. (ed.) PaCT 2015. LNCS, vol. 9251, pp. 307–318. Springer, Cham (2015). https://doi.org/10.1007/978-3-319-21909-7_31
22. Gergel, V.P., Grishagin, V.A., Gergel, A.V.: Adaptive nested optimization scheme for multidimensional global search. J. Global Optim. **66**(1), 1–17 (2015)
23. Gergel, V., Lebedev, I.: Heterogeneous parallel computations for solving global optimization problems. Procedia Comput. Sci. **66**, 53–62 (2015)
24. Sergeyev, Y.D.: An information global optimization algorithm with local tuning. SIAM J. Optim. **5**(4), 858–870 (1995)

25. Sergeyev, Y.D.: Multidimensional global optimization using the first derivatives. Comput. Math. Math. Phys. **39**(5), 743–752 (1999)
26. Sergeyev, Y.D., Grishagin, V.A.: A parallel method for finding the global minimum of univariate functions. J. Optim. Theor. Appl. **80**(3), 513–536 (1994)
27. Sergeyev, Y.D., Grishagin, V.A.: Parallel asynchronous global search and the nested optimization scheme. J. Comput. Anal. Appl. **3**(2), 123–145 (2001)
28. Balandin, D.V., Kogan, M.M.: Pareto-optimal generalized H2-control and vibration isolation problems. Autom. Remote Control (2017). (in press)

An Approach for Parallel Solving the Multicriterial Optimization Problems with Non-convex Constraints

Victor Gergel$^{(\boxtimes)}$ and Evgeny Kozinov

Lobachevsky State University of Nizhni Novgorod, Nizhni Novgorod, Russia
gergel@unn.ru, evgeny.kozinov@itmm.unn.ru

Abstract. In the present paper, an efficient method is proposed for parallel solving of the multicriterial optimization problems with non-convex constraints, where the optimality criteria could be the multiextremal ones and computing the values of the criteria and constraints could require a large amount of computations. The developed approach is based on the reduction of the multicriterial problems to the nonlinear programming ones by means of the minimax convolution of the partial criteria, on the dimensionality reduction with the use of Peano space-filling curves, and on the application of efficient information-statistical global optimization methods with a novel index scheme of the constraints handling instead of the penalty functions applied usually. When performing the parallel computations, the maximum utilization of the whole search information obtained in the course of the search process is provided. The results of the computational experiments have demonstrated such an approach to allow reducing the computational costs of solving the multicriterial optimization problems essentially – tens and hundred times.

Keywords: Decision making · Multicriterial optimization · Global optimization with Non-convex constraints · High performance computations · Dimensionality reduction · Criteria convolution · Global search algorithms · Computational complexity

1 Introduction

The multicriterial optimization (MCO) problems are among the most general problem statements for the decision-making problems – the statement of MCO problems covers many classes of optimization problems, including unconstrained optimization, nonlinear programming, global optimization, etc. The opportunity to specify several criteria is very useful in formulating the complex decision-making problems, and is used in the applications widely. The practical importance has caused a high research activity in the field of the MCO problems. As a result of intensive research, a plenty of efficient methods for solving the MCO problems have been proposed, and many practically important problems have been solved - see, for example, the monographs [1–3, 19] and reviews of scientific and practical results [4, 5, 7, 20, 32, 33].

Among key features of the multicriterial optimization problems is a potential contradiction between the partial efficiency criteria. This makes impossible to achieve

© Springer International Publishing AG 2017
V. Voevodin and S. Sobolev (Eds.): RuSCDays 2017, CCIS 793, pp. 121–135, 2017.
https://doi.org/10.1007/978-3-319-71255-0_10

the optimum (the best) values with respect to all partial criteria simultaneously. Consequently, the finding of some compromised (effective, non-dominated) decisions, when the achieved values of particular partial criteria are consistent with each other is understood as a solution of a MCO problem usually. It is important to note that the viewpoint on an expedient compromise can be changed in the course of computations that could require finding several different compromised decisions.

Among the developed approaches for solving the MCO problems, one can outline the methods of lexicographic optimization, when some arrangement of the criteria according to the importance of these ones is made, and the optimization of the partial criteria is performed successively according to the decreasing of their importance – see, for example, [3]. Another approach is represented by the iterative methods [4, 17], when the researcher (the decision-maker) takes an active part in the process of selecting the decisions. One more direction developed extensively consists in the development of the evolutionary algorithms based on the simulation of some natural phenomena and the application of these ones to solving the MCO problems [17, 18, 22, 23]. The scalarization, when some methods for the convolution of the partial criteria into a single criterion are applied, is an approach used widely – see, for example, [2, 6].

The present work is devoted to the solving of the MCO problems, which are used for formulating the decision-making problems in the computer-aided design of the complex technical objects and systems. In these applications, the partial criteria can have a multiextremal form, and the domain of feasible decisions can be defined by non-convex constraints. The presence of constraints can result in a partial computability, when the computations of some criteria and constraints are impossible if even one constraint is not satisfied. Also, it was supposed that the computations of the values of criteria and constraints could require a large amount of computations. In these conditions, the finding of even one compromised decision requires a considerable amount of computations whereas the finding of several effective decisions (or of the complete set of these ones) becomes a problem of a huge computational complexity.

The properties of the considered class of the MCO problems listed above determine the key feature of these ones – a high computational complexity. One of the promising directions of the search for the methods of solving such problems consists in the use of the model-based approach, when after a small number of computations of the values of the computation-costly criteria and constraints, the fast-computed approximation functions are constructed [25, 26]. Such an approach is efficient enough, however, the construction of good approximations is difficult at the essentially multiextremal behavior of the optimized criteria and constraints.

The approach to solving the computational-costly class of the MCO problems proposed in the present paper is based on the following key statements. First of all, the scalarization of the vector criterion is used that allows reducing the solving of a MCO problem to the solving of a series of global optimization problems [2, 6]. Next, an efficient global search algorithm developed in the framework of the information-statistical theory of the multiextremal optimization [9, 10] is applied for solving the constrained global optimization (CGO) problems with the non-convex constraints. The parallelization methods developed for this algorithm provide high indicators of efficiency of the parallel computations allowing full utilization of the great computational potential of modern supercomputer systems. Finally, the whole search information obtained in the course of solving a MCO problem is utilized in full amount when performing all necessary computations. In general, the developed approach allows reducing the amount of computations performed for the searching of the next efficient decisions essentially – down to the execution of several iterations only.

Further structure of the paper is as follows. In Sect. 2, the statement of a multi-criterial optimization problem with non-convex constraints is given. In Sect. 3, the basics of the developed approach are presented. In Sect. 4, the global search algorithm for solving the reduced scalar nonlinear programming problems is described. In Sect. 5, the issues of the parallel computations with the reuse of the search information obtained in the course of computations are discussed. Section 6 presents the results of numerical experiments. In Conclusion, the obtained results are discussed and main directions of further investigations are outlined.

2 Problem Statement

A problem of multicriterial optimization with non-convex constraints can be stated in the following form:

$$f(y) = (f_1(y), f_2(y), \ldots, f_s(y)) \rightarrow min, y \in Q,$$
$$Q = \{y \in D : g_i(y) \le 0, 1 \le i \le m\}, \tag{1}$$
$$D = \{y \in R^N : a_i \le y_i \le b_i, 1 \le i \le N\}$$

where

- $y = (y_1, y_2, \ldots, y_N)$ is the vector of varied parameters,
- N is the dimensionality of the multicriterial optimization problem being solved,
- $f(y) = (f_1(y), f_2(y), \ldots, f_s(y))$ is the vector criterion of efficiency,
- $g(y) = (g_1(y), g_2(y), \ldots, g_s(y))$ is the vector function of the constraints,
- Q is the domain of feasible solutions, D is the search domain and $a, b \in R^N$ are given constant vectors.

In further consideration, the following notations will be used also:

$$g_{m+1}(y) = f_1(y), g_{m+2}(y) = f_2(y), \ldots, g_{m+s}(y) = f_s(y), M = s + m.$$

Without any loss in generality, the partial criteria values in the problem (1) are supposed to be non-negative, and the decrease of these ones corresponds to increasing efficiency of the considered decisions $y \in D$.

Usually, the partial criteria of the MCO problem (1) contradict to each other, and there is no decision $y \in D$, which would provide the optimal (minimal) values for all criteria simultaneously. In such cases, the decisions $y^* \in D$, where the values of particular partial criteria cannot be improved without worsening the efficiency values with respect to other criteria, are considered as the solutions of the MCO problem. Such unimprovable decisions are called the effective or Pareto-optimal ones. Any effective decision can be considered as a *partial solution*, and the set of all unimprovable decisions represent a *complete solution* of the MCO problem.

As it has been already mentioned above, in the present paper, the problem (1) will be considered in application to the most complex decision-making problems where the partial criteria $f_i(y)$, $1 \leq i \leq s$ could be multiextremal, the constraints could be non-convex, and obtaining the values of the criteria and constraints at the points of the search domain $y \in D$ could require a large amount of computations. Let us suppose also the partial criteria $f_i(y)$, $1 \leq i \leq s$ and the constraints $g_i(y)$, $1 \leq i \leq m$ to satisfy the Lipschitz condition

$$|g_i(y') - g_i(y'')| \leq L_i \|y' - y''\|, \ y', y'' \in D, 1 \leq i \leq M, \tag{2}$$

where L_i are the Lipschitz constants for the functions $g_i(y)$, $1 \leq i \leq M$ and $\|*\|$ denotes the Euclidean norm in R^N.

3   The Basics of the Approach

3.1   The Reduction of the MCO Problems to the Global Optimization Problems with the Non-convex Constraints

The approach applied in the present work is based on the scalarization of the vector criterion by means of the minimax convolution scheme that allows reducing the solving of the problem (1) to solving a nonlinear programming problem

$$\min[F(\lambda, y) = \max(\lambda_i f_i(y), \ 1 \leq i \leq s)], \ y \in Q,$$
$$\lambda \in \Lambda \subset R^s : \sum_{i=1}^{s} \lambda_i = 1, \lambda_i \geq 0, 1 \leq i \leq s. \tag{3}$$

The necessity and sufficiency of this approach for solving the MCO problem is a key property of the minimax convolution scheme: the result of the minimization of $F(\lambda, y)$ leads to the obtaining of an effective decision[1] for the MCO problem and, vise versa, any effective decision of the MCO problem can be obtained as a result of the minimization of $F(\lambda, y)$ at the corresponding values of the convolution coefficients λ_i, $1 \leq i \leq s$ – see, for example, [4].

The coefficients λ_i, $1 \leq i \leq s$ in (3) can be understood as the indicators of importance of the partial criteria – the larger the value of the coefficient λ_i of a particular partial criterion, the more the contribution of this partial criterion in the scalar criterion $F(\lambda, y)$. As a result, a method of solving the MCO problems can be formulated in a step-by-step manner. At every step, the decision maker chooses the desired values of the coefficients λ_i, $1 \leq i \leq s$. Then, the solving of the formed problem (3) is performed. Afterwards, the decision maker analyzes the obtained effective decisions and corrects the chosen coefficients λ_i, $1 \leq i \leq s$ if necessary. Such a multistep method corresponds to the practice of the choice of the compromised decision in the complex decision-making problems to much extent. And the possibility to determine several

---

[1] More exactly, the minimization of $F(\lambda, y)$ can lead to the obtaining of the weakly – effective decisions (the set of the weakly effective decisions includes the Pareto domain).

effective decisions (or the whole set of these ones) at reasonable computational costs becomes a key problem in solving the complex multicriterial optimization problems.

It is worth noting that the scalar criterion $F(\lambda, y)$ satisfies the Lipschitz condition also:

$$|F(\lambda, y') - F(\lambda, y'')| \leq L\|y' - y''\|, \ y', \ y'' \in D. \tag{4}$$

3.2 The Dimensionality Reduction for the Multidimensional Global Optimization Problems

The use of the global search algorithms developed within the framework of the information-statistical theory of global optimization [8–11] for solving the multiextremal optimization problems (3) is one more key statement of the approach developed in the present work. This theory has served as the basis for the development of a large number of optimization algorithms, which have been substantiated mathematically and have demonstrated a high efficiency, and have allowed solving many complex optimization problems in various fields of application [11, 28–31, 34].

The reduction of the dimensionality of the problems being solved with the use of Peano space-filling curves or evolvents $y(x)$ mapping the interval $[0, 1]$ onto an N-dimensional hypercube D unambiguously is a distinctive feature of the information-statistical global optimization algorithms – see, for example, [9–11]. As a result of such reduction, the initial multidimensional global optimization problem (3) is reduced to a one-dimensional problem:

$$F(\lambda, y(x^*)) = min\{F(\lambda, y(x)) : g_i(y(x)) \leq 0, 1 \leq i \leq m, x \in [0, 1]\}. \tag{5}$$

It is important to note that the one-dimensional functions obtained as a result of the reduction satisfy the uniform Hölder condition (see [9, 10]) i.e.

$$|F(\lambda, y(x')) - F(\lambda, y(x''))| \leq H|x' - x''|^{\frac{1}{N}}, x', x'' \in [0, 1],$$
$$|g_i(y') - g_i(y'')| \leq H_i|x' - x''|, x', x'' \in [0, 1], 1 \leq i \leq m \tag{6}$$

where the Hölder constant H (H_i) is defined by the relation $H = 4L\sqrt{N}$ $(H_i = 4L_i\sqrt{N})$, $1 \leq i \leq m$, $L(L_i)$ is the Lipschitz constant from (2) and (4) and N is the dimensionality of the optimization problem (1).

4 An Efficient Method for Solving the Global Optimization Problems with the Non-convex Constraints

The basics of the approach presented in Sect. 3 allow reducing the solving of the MCO problem (1) to the solving of a series of the reduced multiextremal problems with the constraints (5). And, thus, the global search algorithms can be applied for solving the MCO problems [8, 12–16].

It is worth noting that the presence of the non-convex constraints complicates solving the global optimization problems considerably – the obtained solutions should belong to the feasible domain Q. The situation becomes even more complicated in the case of the partial computability, when the computing of some criteria and constraints is impossible if there is even one unsatisfied constraint. Often, for solving the constrained optimization problems, more simple cases are selected – for example, the problems with the linear or quadratic constraints are considered. Various methods of approximation of the complex constraints using the constraints of simpler forms (linear, convex, etc.) are applied as well. However, the most often applied method is the penalty function method. The approach used in the present work is based on a novel method of the constraint handling. This approach was developed within the framework of the information-statistical theory of global search [10]. The idea of the approach consists in the construction of a scalar unconstrained objective function, the solving of which leads to the solving of the initial problem (5) – more detailed description of the approach is given below.

Within the framework of this approach, the algorithm of constrained global search (ACGS) for the multiextremal optimization problems with the non-convex constraints[2] makes the basis of the developed optimization methods. The general computational scheme of the algorithm can be represented in the following form [9, 10].

Let us introduce a simpler notation for the one-dimensional problems (5) as

$$min\{\varphi(x) : g_i(y(x)) \leq 0, 1 \leq i \leq m, x \in [0, 1]\},$$
$$\varphi(x) = g_{m+1}(x) = F(\lambda, y(x)). \tag{7}$$

The problem (7) can be considered in the partial computability form, when each function g_j, $1 \leq j \leq m+1$ is defined and computable in the corresponding subdomain $\Delta_j \subset [0, 1]$ only, where

$$\Delta_1 = [0, 1], \Delta_{j+1} = \{x \in \Delta_j : g_j(y(x)) \leq 0\}, 1 \leq j \leq m. \tag{8}$$

Taking into account the condition (8), the initial problem (7) can be represented as follows

$$\varphi(x^*) = min\{g_{m+1}(y(x)) : x \in \Delta_{m+1}\}. \tag{9}$$

This form of the problem (7) allows defining an *index* $v = v(x)$ for the points x from the search domain $[0, 1]$ where $v - 1$ is the number of constraints, which are satisfied at this point. The index v is defined by the conditions

$$g_v(y(x)) > 0, g_j(y(x)) \leq 0, 1 \leq j \leq v - 1, 1 \leq v = v(x) \leq m+1. \tag{10}$$

where the last inequality is insufficient if $v = m+1$ Computing the index v can be provided by the sequential computation of the values $g_j(y(x))$, $1 \leq j \leq v = v(x)$, i.e. the next value $g_{j+1}(x)$ is computed in the case, when $g_j(x) \leq 0$ only. The process of computations is terminated either as a result of the fulfillment of the inequality

---

$g_j(x) > 0$ or as a result of the achievement of the value $v(x) = m+1$ (this procedure is called hereafter a *trial*).

The main idea of such index scheme consists in the reduction of the constrained problem (7) to an unconstrained problem

$$\Phi(x^*) = min\{\Phi(x) : x \in [0,1]\}, \tag{11}$$

where

$$\Phi(x) = \begin{cases} g_v(y(x))/H_v, & v < m+1, \\ (g_{m+1}(y(x)) - g^*_{m+1})/H_{m+1}, & v = m+1. \end{cases}$$

It is worth noting that the values of the Lipschitz constants L_v, $1 \le v < m+1$ and the value g^*_{m+1} are unknown. However, when performing the computations, one can use the adaptive estimates of these values obtained in the course of solving the optimization problem (see the description of the algorithm below) instead.

The general computational scheme of the ACGS method consists in the following.

The first trial is performed at an arbitrary point $x_1 \in (0,1)$. The choice of the point x_{k+1}, $k \ge 1$ of any next trial is determined by the following rules.

Rule 1. Renumber the points of preceding trials $x_1, \ldots, x_k$ by the lower indices in the order of increasing of the coordinate values i.e.

$$0 = x_0 < x_1 < \ldots < x_i < \ldots < x_k < x_{k+1} = 1, \tag{12}$$

and juxtapose these ones to the values $z_i = g_v(x_i)$, $v = v(x_i)$, $1 \le i \le k$ from (10) computed at these points. The points $x_0 = 0$ and $x_{k+1} = 1$ are introduced additionally for convenience of further notations (the values z_0 and z_{k+1} are undefined).

Rule 2. Subdivide the indices i, $1 \le i \le k$ of the points from (12) with respect to the number of constraints of the problem fulfilled at these points by constructing the sets

$$I_v = \{i : 1 \le i \le k, v = v(x_i)\}, 1 \le v \le m+1 \tag{13}$$

containing the indices of all points x_i, $1 \le i \le k$ having the indices equal to the same value v. The boundary points $x_0 = 0$ and $x_{k+1} = 1$ are interpreted as the ones having the zero indices, and are juxtaposed to an auxiliary set $I_0 = \{0, k+1\}$.

Determine the maximum value of the index

$$M = max\{v = v(x_i), 1 \le i \le k\}. \tag{14}$$

Rule 3. Compute the current estimates

$$\mu_v = max\left\{ |z_i - z_j| / \sqrt[N]{(x_i - x_j)}, i,j \in I_v, i > j \right\} \tag{15}$$

for the Hölder constants H_v of the functions g_v, $1 \le v \le m+1$ from (6). If the set I_v contains less than two elements or if μ_v from (15) appears to equal zero, then accept $\mu_v = 1$.

Rule 4. Compute the estimates z_v^*, $1 \leq v \leq M$ for all nonempty sets I_v, $1 \leq \leq m+1$ from (13),

$$z_v^* = \begin{cases} 0, & v < M, \\ \min\{g_v(x_i) : i \in I_v\}, & v = M. \end{cases} \tag{16}$$

Rule 5. Compute the characteristics $R(i)$ for each interval (x_{i-1}, x_i), $1 \leq i \leq k+1$ where

$$R(i) = \begin{cases} \rho_i + \frac{(z_i - z_{i-1})^2}{r_v^2 \mu_v^2 \rho_i} - 2\frac{(z_i + z_{i-1} - 2z_v^*)}{r_v \mu_v}, & v = v(x_{i-1}) = v(x_i), \\ 2\rho_i - 4\frac{(z_i - z_v^*)}{r_v \mu_v}, & v = v(x_i - 1) < v(x_i), \\ 2\rho_i - 4\frac{(z_{i-1} - z_v^*)}{r_v \mu_v}, & v = v(x_{i-1}) > v(x_i), \end{cases} \tag{17}$$

$$\rho_j = \sqrt[N]{(x_i - x_{i-1})}, \ 1 \leq j \leq k+1$$

where z_v^*, $1 \leq v \leq M$ from (16), M from (14).

The values $r_v > 1$, $1 \leq v \leq m+1$ are the parameters of the algorithm. The appropriate values r_v allows using the products $r_v\mu_v$ as the estimates of the Hölder constants H_v, $1 \leq v \leq m+1$.

Rule 6. Determine the interval (x_{t-1}, x_t) with the maximum characteristic:

$$R(t) = max\{R(i) : 1 \leq i \leq k+1\}. \tag{18}$$

Rule 7. Execute the next trial at the point of the interval $x^{k+1} \in (x_{t-1}, x_t)$ determined according to the expression

$$x^{k+1} = \begin{cases} \frac{x_t + x_{t-1}}{2}, & v(x_{t-1}) \neq v(x_{t-1}), \\ \frac{x_t + x_{t-1}}{2} + sign(z_t - z_{t-1})\frac{1}{2r_v}\left[\frac{|z_t - z_{t-1}|}{\mu_v}\right]^N, & v = v(x_{t-1}) = v(x_{t-1}). \end{cases} \tag{19}$$

The iterations of the algorithm are terminated if the stopping condition is satisfied

$$\rho_t \leq \varepsilon, \tag{20}$$

where t is from (18), and $\varepsilon > 0$ is the predefined accuracy.

Various modifications of this algorithm and the corresponding theory of convergence are presented in [9, 10].

5   Parallel Computations for the Time-Consuming Multicriterial Constrained Optimization Problems

The proposed approach for parallel computations when solving the computation-costly multicriterial optimization problems is based on the simultaneous computing of the values of partial criteria and constraints of the initial problem (1) at several different points of the search domain D. Such an approach provides the parallelization of the most time-consuming part of the global search, and is a general one – it can be applied for many global search methods for various global optimization problems. Besides, an essential speedup of the computations can be provided by means of full utilization of the whole search information obtained in the course of optimization.

5.1   The Reuse of the Search Information for Accelerating the Computations

The numerical solving of the optimization problems consists in the sequential computation of the values of the partial criteria $f^i = f(y^i)$ and constraints $g^i = g(y^i)$ at the points $y^i, 1 \leq i \leq k$ of the search domain D. The search information obtained can be represented in the form of the *search information set* (SIS):

$$\Omega_k = \left\{ (y^i, f^i, g^i)^T : 1 \leq i \leq k \right\}. \tag{21}$$

The availability of SIS allows reducing the results of previous computations to the values of any next optimization problem (11) being solved without any time-consuming computations of the values of partial criteria and constraints of the initial problem (1) at any new values of the convolution coefficients $\lambda \in \Lambda$.

And, thus, all search information can be utilized for continuing the computations in full amount. In general, the reuse of the search information will require less and less amount of computations for solving every next optimization problem downto performing several iterations only to find the next effective decision (see Sect. 6 for the results of the numerical experiments).

As a result of the dimensionality reduction, the search information Ω_k from (21) can be transformed into the matrix of search state (MSS)

$$A_k = \left\{ (x_i, z_i, v_i, l_i)^T : 1 \leq i \leq k \right\}, \tag{22}$$

where $x_i, 1 \leq i \leq k$ are the reduced trial points of the executed global search iterations, $z_i, 1 \leq i \leq k$ are the values of scalar criterion of current reduced optimization problem (11) being solved, $v_i, 1 \leq i \leq k$ are the indices of the scalar criterion values, and $l_i, 1 \leq i \leq k$ are the indices of the global search iterations, where the points $x_i, 1 \leq i \leq k$ have been computed.

The ACGS algorithm improved by the possibility to use the search information A_k from (22) will be called hereafter the Algorithm of Multicriterial Constrained Global Search (AMCGS).

5.2   Parallel Algorithm of the Multicriterial Global Search

The choice of the points in the search domain D for the simultaneous execution of several trials (computing the values of the criteria and constraints of initial MCO problem (1)) can be provided by means of the following parallel generalization of the ACGS method – see, for example, [10, 34].

Let p be the number of employed parallel computing nodes (processors or cores) of the computational system with shared memory. The rules of the parallel algorithm correspond to the computational scheme of the ACGS method except the steps of computing the points of the next global search iteration. The modified rules for the parallel algorithm can be presented as follows.

Rule 6 (updated). Arrange the characteristics of the intervals (x_{i-1}, x_i), $1 \leq i \leq k+1$ obtained in (17) in the decreasing order

$$R(t_1) \geq R(t_2) \geq \ldots \geq R(t_{k-1}) \geq R(t_k) \tag{23}$$

and select p intervals with the indices t_j, $1 \leq j \leq p$ having the maximum values of the characteristics.

Rule 7 (updated). Perform new trials at the points x^{k+j}, $1 \leq j \leq p$ placed into the intervals with the maximum characteristics from (23) according the expression (19).

The stopping condition (20) of the parallel algorithm, which terminates the trials, should be checked for all intervals, where the scheduled trials are performed, i.e.

$$\rho_{t_j} \leq \varepsilon, \quad 1 \leq t_j \leq p.$$

The ACGS algorithm updated by the opportunity of the parallel computations for the computing nodes with shared memory will be named hereafter the Parallel Algorithm of Multicriterial Constrained Global Search (PAMCGS).

6   Results of Numerical Experiments

The numerical experiments have been carried out using the Lobachevsky supercomputer at State University of Nizhni Novgorod (the operating system – CentOS 6.4, the supercomputer management system – SLURM). Each supercomputer node had 2 Intel Sandy Bridge E5-2660 2.2 GHz, 64 Gb RAM processors. The central processor units (CPUs) had 8 cores (i.e., total 16 CPU cores were available per a node).

First, let us consider the results of the comparison of the developed PAMCGS algorithm with several other multicriterial optimization algorithms. A bi-criterial test problem proposed in [21]:

$$f_1(y) = (y_1 - 1)y_2^2 + 1, f_2(y) = y_2, 0 \leq y_1, y_2 \leq 1. \tag{24}$$

was used for this experiment. The construction of a numerical approximation of Pareto domain was understood as the solution of the problem (24). To evaluate the quality of approximation, the completeness and uniformity of coverage of the Pareto domain were evaluated with the use of the following two indices [21, 24]:

- The hypervolume index (HV). This index features the completeness of approximation of the Pareto domain (a larger value corresponds to a denser coverage of the Pareto domain).

- The distribution uniformity index (DU). This index features the uniformity of coverage of the Pareto domain (a lower value corresponds to more uniform coverage of the Pareto domain).

Within the framework of this experiment, five multicriterial optimization algorithms were compared: the Monte-Carlo (MC) method, the genetic algorithm SEMO from the PISA library [20, 24], the Non-Uniform Coverage (NUC) method [21], the Bi-objective Lipschitz Optimization (BLO) method proposed in [24], and the serial variant of the PAMCGS algorithm proposed in the present paper. The results of solving the problem (24) for all methods listed above were obtained in [24].

For the AGCS method, 50 problems (3) have been solved at various values of the convolution coefficients λ distributed in Λ uniformly. The results of performed experiments are presented in Table 1.

Table 1. Comparison of the efficiency of the multicriterial optimization algorithms

Method	MC	SEMO	NUC	BLO	ACGS
Iterations	500	500	515	498	370
Number of points in the Pareto domain approximation	67	104	29	68	100
HV index	0.300	0.312	0.306	0.308	0.316
DU index	1.277	1.116	0.210	0.175	0.101

The results of the performed experiments have demonstrated that the ACGS algorithm have a considerable advantage as compared to the considered multicriteria optimization methods even when solving the relatively simple MCO problems.

In the second series of the numerical experiments, the solving of the bi-criterial two-dimensional MCO problems with two constraints i.e. $N = 2$, $s = 2$, $m = 2$ has been performed. The multiextremal functions obtained with the use of the GKLS generator [27] were used as the problem criteria. In the course of experiments, the solving of 100 multicriterial problems of this class has been performed. In every problem, the search of the Pareto-optimal decisions for 50 convolution coefficients λ from (3) distributed in Λ uniformly has been performed. The obtained results were averaged over the number of solved MCO problems. In Fig. 1 an example of two criteria as well as the result of convolution of the criteria and the feasible domain are presented.

The numerical experiments have been performed with stopping upon achievement the method accuracy. For the checking, the points of the solution found by the method have been compared to the points of the Pareto domain approximation computed taking into account the selected convolution coefficients λ. The accuracy of method $\varepsilon = 0.02$ and the reliability parameter $r = 5.6$ were used when solving the series of problems. The results of the numerical experiments are presented in Table 2.

In the first column of Table 2, the number of computing cores employed for solving the problems from the considered series of experiments is given. In the second and fourth columns, the averaged number of iterations executed by the PAMCGS algorithm

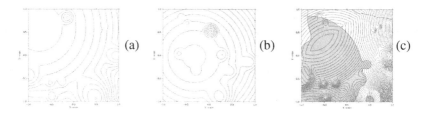

Fig. 1. Contour plots of two criteria obtained with the use of the GKLS generator (a, b); the problem to be solved obtained by the convolution of the criteria $\lambda = \{0.5, 0.5\}$ (c). The feasible domain is highlighted by green

for the solving of the optimization problem are presented. The third and fifth columns contain the percentage of the solved problems at given parameters of the method. The last two columns contain the information on the obtained speedup. The column (S1) shows the effect of the reuse of the accumulated search information. The column (S2) contains the information on the total speedup achieved as compared to the initial algorithm without the use of the search information.

The obtained results of experiments (Table 2) demonstrate that the reuse of the search information to allow reducing the total amount of computations by the factor of 18.4 without employing any additional computational resources. When using 25 computer cores, the maximum speedup reaches 295.6 times.

Table 2. The results of the series of experiments on solving the two-dimensional bi-criterial constrained MCO problems

Number of computing cores	Search information				S1	S2
	Not used		Used			
	Number of iterations	Problems solved	Number of iterations	Problems solved		
1	26191.8	88%	1420.5	93%	18.4	18.4
2	12146.1	85%	653.3	90%	18.6	40.1
5	5019.3	85%	285.7	91%	17.6	91.7
10	2141.5	85%	152.5	93%	14.0	171.8
25	1022.4	88%	88.6	94%	11.5	295.6

In the third series of the numerical experiments, the solving of the three-criterial four-dimensional MCO problems with five constraints (i.e. $N = 4$, $s = 3$, $m = 5$) has been performed. The criteria and constraints of the MCO problems to be solved were generated with the use of the GKLS generator [27] as in the previous experiment. When solving the problem series, the accuracy of the method $\varepsilon = 0.01$ and the reliability parameter $r = 5.6$ were used. The results of the numerical are presented in Table 3. As it can be noted, for example, the speedup achieved when using 25 computing cores was 244.6 times.

Table 3. The results of the series of experiments on solving the four-dimensional three-criterial constrained MCO problems

Number of computing cores	Search information				S1	S2
	Not used		Used			
	Number of iterations	Problems solved	Number of iterations	Problems solved		
1	49 988 246.5	91%	6 153 261.0	90%	8.1	8.1
2	20 369 550.2	90%	2 400 575.1	89%	8.5	20.8
5	8 228 684.5	90%	709 672.4	92%	11.6	70.4
10	5 582 125.8	92%	702 522.4	91%	7.9	71.2
25	1 704 359.8	91%	204 342.8	90%	8.3	244.6

7   Conclusion

In the present paper, an efficient parallel method is proposed for solving complex multicriterial optimization problems with non-convex constraints, where the criteria of optimality could be the multiextremal ones, and the computing of the criteria values could require a large amount of computations. The proposed approach is based on the reduction of the multicriterial problems to the nonlinear programming ones by means of the minimax convolution of the partial criteria, on the dimensionality reduction with the use Peano space-filling curves, and of application of the efficient information-statistical global optimization methods with a novel index scheme of the constraints handling instead of the penalty functions applied usually.

The key aspect of the developed approach is the overcoming of a large computational complexity of the global search of the set of effective decisions when solving the multicriterial optimization problems. A considerable increase in the efficiency and an essential reduction of the amount of computations was provided by means of the maximum possible use of the search information obtained in the course of computations. To do so, it was necessary to provide the possibilities of storing large amounts of the search information, of its efficient processing, and of using the search data in the course of solving the multicriterial optimization problems. Within the framework of the developed approach, the methods for converting all available search information to the values of current scalar problem of nonlinear programming being solved have been proposed. The search information transformed to current optimization problem was used by the applied optimization methods for the adaptive planning of the global search iterations performed. The availability of the search information allows also executing the parallel computation efficiently providing the choice of the most promising points of the search domain when searching the effective decisions for the MCO problems.

The results of the numerical experiments have demonstrated the developed approach to allow reducing the computational costs of solving the multicriterial optimization problems with the non-convex constraints essentially – tens and hundreds times.

As a conclusion, one can note that the developed approach is a promising one and needs continuing the investigations further. First of all, it is necessary to continue carrying out the numerical experiments on solving the multicriterial optimization problems with larger number of partial criteria and constraints for lager dimensionalities of the optimization problems to be solved. Also, a possibility of parallel computations for the high-performance systems with distributed memory should be explored.

Acknowledgements. This research was supported by the Russian Science Foundation, project No 16-11-10150 "Novel efficient methods and software tools for time-consuming decision making problems using supercomputers of superior performance."

References

1. Miettinen, K.: Nonlinear Multiobjective Optimization. Springer, New York (1999)
2. Ehrgott, M.: Multicriteria Optimization, 2nd edn. Springer, Heidelberg (2010)
3. Collette, Y., Siarry, P.: Multiobjective Optimization: Principles and Case Studies (Decision Engineering). Springer, Heidelberg (2011)
4. Marler, R.T., Arora, J.S.: Survey of multi-objective optimization methods for engineering. Struct. Multi. Optim. **26**, 369–395 (2004)
5. Figueira, J., Greco, S., Ehrgott, M. (eds.): Multiple Criteria Decision Analysis: State of the Art Surveys. Springer, New York (2005)
6. Eichfelder, G.: Scalarizations for adaptively solving multi-objective optimization problems. Comput. Optim. Appl. **44**, 249–273 (2009)
7. Zavadskas, E.K., Turskis, Z., Kildienė, S.: State of art surveys of overviews on MCDM/MADM methods. Technol. Econ. Dev. Econ. **20**, 165–179 (2014)
8. Pintér, J.D.: Global Optimization in Action (Continuous and Lipschitz Optimization: Algorithms, Implementations and Applications). Kluwer Academic Publishers, Dortrecht (1996)
9. Strongin, R.G.: Numerical Methods in Multiextremal Problems: Information-Statistical Algorithms. Nauka, Moscow (1978). (in Russian)
10. Strongin, R., Sergeyev, Y.: Global Optimization with Non-convex Constraints. Sequential and Parallel Algorithms. Kluwer Academic Publishers, Dordrecht (2nd ed. 2013, 3rd ed. 2014)
11. Sergeyev, Y.D., Strongin, R.G., Lera, D.: Introduction to Global Optimization Exploiting Space-filling Curves. Springer, New York (2013)
12. Floudas, C.A., Pardalos, M.P.: Recent Advances in Global Optimization. Princeton University Press, Princeton (2016)
13. Horst, R., Tuy, H.: Global Optimization: Deterministic Approaches. Springer, Berlin (1990)
14. Locatelli, M., Schoen, F.: Global Optimization: Theory, Algorithms, and Applications. SIAM (2013)
15. Törn, A., Žilinskas, A.: Global Optimization. Lecture Notes in Computer Science, vol. 350. Springer, Berlin (1989)
16. Zhigljavsky, A.A.: Theory of Global Random Search. Kluwer Academic Publishers, Dordrecht (1991)
17. Branke, J., Deb, K., Miettinen, K., Slowinski, R. (eds.): Multi-Objective Optimization—Interactive and Evolutionary Approaches. Springer, Berlin (2008)
18. Yang, X.-S.: Nature-Inspired Metaheuristic Algorithms. Luniver Press, Frome (2008)

19. Marler, R.T., Arora, J.S.: Multi-Objective Optimization: Concepts and Methods for Engineering. VDM Verlag, Saarbrucken (2009)
20. Hillermeier, C., Jahn, J.: Multiobjective optimization: survey of methods and industrial applications. Surv. Math. Ind. 11, 1–42 (2005)
21. Evtushenko, Y., Posypkin, M.A.: A deterministic algorithm for global multi-objective optimization. Optim. Method. Softw. 29(5), 1005–1019 (2014)
22. Deb, K.: Multi-Objective Optimization using Evolutionary Algorithms. Wiley, Chichester (2001)
23. Tan, K.C., Khor, E.F., Lee, T.H.: Multi-objective Evolutionary Algorithms and Applications. Springer, London (2005)
24. Žilinskas, A., Žilinskas, J.: Adaptation of a one-step worst-case optimal univariate algorithm of bi-objective Lipschitz optimization to multidimensional problems. Commun. Nonlinear Sci. Numer. Simulat. 21, 89–98 (2015)
25. Jones, D.R.: A taxonomy of global optimization methods based on response surfaces. J. Glob. Optim. 21, 345–383 (2001)
26. Voutchkov, I., Keane, A.: Multi-objective optimization using surrogates. Comput. Intell. Optim. Adapt. Learn. Optim. 7, 155–175 (2010)
27. Gaviano, M., Kvasov, D.E.: Lera, D., and Sergeyev, Ya.D.: Software for generation of classes of test functions with known local and global minima for global optimization. ACM Trans. Math. Softw. 29(4), 469–480 (2003)
28. Barkalov, K., Gergel, V., Lebedev, I.: Use of Xeon Phi coprocessor for solving global optimization problems. In: Malyshkin, V. (ed.) PaCT 2015. LNCS, vol. 9251, pp. 307–318. Springer, Cham (2015). https://doi.org/10.1007/978-3-319-21909-7_31
29. Gergel, V.P., Grishagin, V.A., Gergel, A.V.: Adaptive nested optimization scheme for multidimensional global search. J. Global Optim. 66(1), 35–51 (2016)
30. Barkalov, K., Gergel, V.: Parallel global optimization on GPU. J. Glob. Optim. 66(1), 3–20 (2016)
31. Gergel, V., Lebedev, I.: Heterogeneous parallel computations for solving global optimization problems. Procedia Comput. Sci. 66, 53–62 (2015)
32. Gergel, V.P., Kuzmin, M.I., Solovyov, N.A., Grishagin, V.A.: Recognition of surface defects of cold-rolling sheets based on method of localities. Int. Rev. Autom. Control 8(1), 51–55 (2015)
33. Kvasov, D.E., Sergeyev, Y.D.: Deterministic approaches for solving practical black-box global optimization problems. Adv. Eng. Softw. 80, 58–66 (2015)
34. Sergeyev, Y.D., Famularo, D., Pugliese, P.: Index branch-and-bound algorithm for Lipschitz univariate global optimization with multiextremal constraints. J. Glob. Optim. 21(3), 317–341 (2001)

Increasing Performance of the Quantum Trajectory Method by Grouping Trajectories

Alexey Liniov[1(✉)], Valentin Volokitin[1], Iosif Meyerov[1], Mikhail Ivanchenko[1], and Sergey Denisov[1,2]

[1] Lobachevsky State University of Nizhni Novgorod, Nizhny Novgorod, Russia
`alin@unn.ru`, `valyav95@mail.ru`, `meerov@vmk.unn.ru`, `ivanchenko.mv@gmail.com`
[2] Institut für Physik, Universität Augsburg, 86135, Augsburg, Germany
`sergey.denisov@physik.uni-augsburg.de`

Abstract. The quantum trajectory method is the most popular and widely used algorithm to simulate the evolution of an open N-dimensional quantum system. The key idea is to unravel Markovian equation describing evolution of the system density operator (a $N \times N$ Hermitian matrix) into a set of independent stochastic realizations obtained by propagating system wave function (a complex N vector). Since the method decreases the scaling of the computational problem from N^2 to N, it is especially efficient for the systems of large dimensions. Intrinsic parallelism that is characteristic to all Monte Carlo schemes allows for efficient implementations of quantum trajectories on a high-performance computational cluster. One of the core mathematical operations involved into the method is the matrix-vector multiplication. We propose to improve the algorithm by grouping trajectories into matrices and substituting a set of matrix-vector multiplications with a single matrix-matrix multiplication. By using a testbed model with 1024 states, we demonstrate that, even in the presence of intrinsic asynchrony between different trajectories, this step leads to a 17-fold acceleration on the 4-socket 96-core Intel Broadwell CPU.

Keywords: Open quantum systems · Quantum trajectory method · High-performance computing · Supercomputing technologies · Parallel computing · Performance analysis and optimization

1 Introduction

Physics of open quantum systems attracts a lot of attention during the last decade. This is because it considers quantum systems in their natural habitats, i.e., when the former interact with their environments [1]. The growing interest to open systems was initiated by the rise of quantum technologies and is maintained by ever-increasing number of real-life applications of quantum systems that a decade ago existed on paper only. It is evident that in order to blueprint a realistic quantum device, effects of its interaction with environment should be taken into account.

© Springer International Publishing AG 2017
V. Voevodin and S. Sobolev (Eds.): RuSCDays 2017, CCIS 793, pp. 136–150, 2017.
https://doi.org/10.1007/978-3-319-71255-0_11

The most elaborated, both from mathematical and physical point views, approach to model open quantum system is the Lindblad formalism, which is based on the idea of quantum dynamical semi-groups and culminates into the Lindblad equation [2]. This approach is very popular in such fields as quantum optics, optomechanics, cavity quantum electrodynamics, and cold atom physics [3]. Straightforward numerical solution of the Lindblad equation (and thus obtaining the asymptotic state of the model of interest) is not feasible when the model dimension N – that is the dimension of the Hilbert space the model lives in – is larger than 500. When the model Hamiltonian is explicitly time-periodic, i.e., the system is additionally modulated in time [4], evaluation of the system non-equilibrium asymptotic state involves numerical integration of the Lindblad equation in time. It is hardly doable even when $N \simeq 400$.

Model with $N = 400$ states may still be too small to describe real-life quantum systems. It is possible to go beyond this limit by unraveling the Lindblad equation into a set of stochastic realizations, called "quantum trajectories" [3,5]. This method allows transform the problem of the numerical solution of the Lindblad equation into a task of statistical sampling over quantum trajectories, with every trajectory specified by a complex vector of the size N. The price to pay for the reduction from N^2 to N is that one now has to sample over many realizations.

In our work [6] we presented an implementation of the quantum trajectory method that allowed us to resolve non-equilibrium asymptotic states (which we called "quantum attractors") of a periodically modulated quantum model. We demonstrated that a regular high-performance cluster (with up to 512 computational cores) is enough to sample such attractors with high accuracy for the model of the dimension $N \approx 2000$. The aim of this paper is to investigate the potential for the further optimization of the implementation and improvement of its performance.

Like in a number of other numerical software, a substantial advance can be potentially reached by increasing parallelism on each level of computing (processes, threads, SIMD, instruction level parallelism), as well as by improving memory usage efficiency. Since the quantum trajectory method belongs to the Monte Carlo family, it should possess high intrinsic parallelism. Note that due to the nature of the method, intrinsic stochastic steps – quantum jumps – can occur at different random times for different trajectories, and their number (for a fixed time interval) can vary too. Nevertheless, numerical experiments indicate the absence of a substantial variation, which together with the opportunity of merging a set of trajectories into a single computational task for parallel computing, leads to a small imbalance of computational load, order 5% only. At the same time, an empiric choice of the ratio between the number of the employed processes and threads of the hybrid MPI + OpenMP parallelization scheme, on the contrary, significantly affects the total computation time. The next level of parallelism is related with vectorization of computing and effective usage of wide vector registries and AVX2-instructions in modern processors. It is essential that the main computational core of the method that consumes $> 95\%$ of the total time is the dense matrix-vector multiplication, which can be vectorized.

In that respect, our code makes use of the high performance implementation of the matrix-vector multiplication from the Intel Math Kernel Library (MKL). Summarizing the above, there is a considerable room for exploiting parallelism in quantum trajectory method.

Improving performance of computations here could be achieved by reducing the number of calls to the main memory and by a more efficient usage of the cache. The most straightforward idea along these lines would be a substitution of the groups of matrix-vector multiplications to single matrix-matrix multiplications. In [7] it was demonstrated, that such optimization can substantially reduce computational cost, at least for the addressed class of problems. There, merging was achieved by substituting propagation of separate vectors with propagation of a matrix composed of them. In the case of quantum trajectories, there is an asynchrony between different trajectories in times of jumps and such merging is not so straightforward. Here we propose a solution to the problem that allows to attain the same results but at a smaller time, as a rule. We will demonstrate the way to organize computing, focusing on a matrix multiplication, where appropriate.

The paper is organized as follows. In Sect. 2 we give a mathematical model – a system of indistinguishable interacting bosons hopping between the sites of a periodically rocked dimer. In Sect. 3 the description of the quantum trajectory method is given. In Sect. 4 we present the optimized method. Numerical results and performance analysis are given in Sect. 5. Section 6 concludes the paper.

2   Model

The Lindblad equation is described by its generator $\mathcal{L}$, which has a universal structure [2]:

$$\dot{\varrho} = \mathcal{L}(\varrho) = -i[H(t), \varrho] + \sum_{k=1}^{K} \gamma_k(t) \cdot \mathcal{D}_k(\varrho),$$

$$\mathcal{D}_k(\varrho) = A_k \varrho V_k^\dagger - \frac{1}{2}\{A_k^\dagger A_k, \varrho\}. \tag{1}$$

Here ϱ is the system density matrix, while the set of quantum jump operators, A_k, $k = 1, ..., K$, captures the action of the environment on the system. Namely, it acts through K 'channels' with time-dependent (in general) rates γ_k. Finally, $[.,.]$ and $\{.,.\}$ denote commutator and anti-commutator, respectively.

As a testbed model we use a system of $N - 1$ indistinguishable interacting bosons hopping over a periodically rocked dimer [8]. The system Hamiltonian is

$$H(t) = -J\left(b_1^\dagger b_2 + b_2^\dagger b_1\right) + \frac{U}{2(N-1)}\sum_{g=1,2} n_g\left(n_g - 1\right)$$

$$+ \varepsilon(t)\left(n_2 - n_1\right) \tag{2}$$

where J is the tunneling amplitude, U is the interaction strength, and $\varepsilon(t)$ presents the modulation of the on-site potential difference. In particular, we choose $\varepsilon(t) = \varepsilon(t+T) = \mu_0 + \mu_1\theta(\omega t)$, where μ_0 and μ_1 are static and dynamic energy offsets between the two sites, respectively, and $\theta(x)$ is a step-like periodic function of period one. Here, b_g and $b_g^\dagger$ are the annihilation and creation operators on site $g \in \{1,2\}$, while $n_g = b_g^\dagger b_g$ is the particle number operator. The system Hilbert space has dimension N and can be spanned with the Fock basis vectors, labeled by the number of boson on the first site t, $\{|t+1\rangle\}$, $t = 0,...,N-1$. So, the model has N states and its size is controlled by the total number of bosons. Hamiltonian (2) has been used for theoretical studies and was already implemented in experiments [8]. On top, this is a nicely scalable model; its dimension N can be incremented by simply adding one boson.

We use a single jump operator [9,10],

$$A = (b_1^\dagger + b_2^\dagger)(b_1 - b_2), \qquad (3)$$

which tries to 'synchronize' the dynamics on the sites by constantly recycling anti-symmetric out-phase modes into the symmetric in-phase ones. The coupling constant $\gamma = (N-1)\gamma_0$ is assumed to be time-independent.

3 Quantum Trajectory Method

3.1 Base Algorithm

Solution to the Lindblad equation (1) for the density matrix of an open system can be unraveled into an ensemble of quantum trajectories, which are governed by the equation

$$|\dot\psi\rangle = -i\tilde{H}|\psi\rangle, \qquad (4)$$

where $|\psi\rangle$ is the state vector of dimension N, and $\tilde{H} = H - \frac{i}{2}\sum_k \gamma_k A_k^* A_k$ is the non-Hermitian Hamiltonian, constructed from the original system Hamiltonian and jump operators.

The method is implemented as follows (see Algorithm 1). Initially, the code loads the model and method parameters: system size N, number of quantum trajectories for sampling L, end time T_{max}, the matrices for exponential operators $expM$ (lines 12). Initial conditions $|\psi_0\rangle$ are chosen such that the norm of the vector equals 1; this corresponds to the initial condition $\varrho_0 = |\psi_0\rangle\langle\psi_0|$ for the original Lindblad equation (1). The main computational cycle (lines 312) contains numerical propagation of L vectors $\{|\psi_l(t)\rangle; l = 1,..,L\}$ in time within the interval $[0; T_{max}]$.

The details of propagation follow (lines 411).

1. Choose a random number η from the uniform distribution on $[0,1]$ (line 5).
2. Perform propagation in time τ (lines 69), until the following is satisfied: $\||\psi(t)\rangle\|^2 = \eta$. Reaching it is ensured by the special form of the Hamiltonian $\tilde{H}$ that monotonously decreases the norm.
3. Make quantum jump (line 10).

   3.1 Normalize the state vector again $|\psi(t)\rangle = \frac{|\psi(t)\rangle}{\||\psi(t)\rangle\|}$.

Algorithm 1. Quantum trajectory method

1: load the system and method parameters (N, L, T_{max}, ExpM)
2: **for** $l = 1$ to L **do**
3:     **for** $t = 0$ to T_{max} **do**
4:         generate $\eta = U[0, 1]$
5:         **while** $\||\psi(t)\rangle\|^2 > \eta$ **do**
6:             calculate t_{next}
7:             propagate $|\psi(t)\rangle$ on $[t_{cur}, t_{next}]$
8:         **end while**
9:         makes quantum jump
10:     **end for**
11: **end for**
12: calculate the final density matrix
13: release memory

3.2 Calculate probabilities of selecting quantum jump channels $[p_1, .., p_K]$, $p_k = \frac{\gamma_k \|A_k|\psi(t)\rangle\|^2}{\sum_{i=1}^{K} \gamma_i \|A_k|\psi(t)\rangle\|^2}$. Therefore, we split a unit interval in K parts of the lengths $p_1, .., p_K$, respectively.

3.3 Choose random ξ from a uniform distribution on $[0, 1]$. Determine the corresponding m-th quantum channel such that $\xi \in p_m$, and complete the quantum jump according to: $|\psi(t)\rangle = \frac{A_m|\psi(t)\rangle}{\|A_m|\psi(t)\rangle\|}$.

As a result, we obtain an ensemble of quantum trajectories $\{|\psi_l(t)\rangle; t \in [0, T_{max}]; l = 1, .., L\}$. The density matrix, approximating an exact solution to Eq.(1) at an arbitrary time $t \in [0; T_{max}]$, can be unraveled by averaging over the trajectories:

$$\tilde{\rho}_L(t) = \frac{1}{L} \sum_{l=1}^{L} \frac{|\psi_l(t)\rangle\langle\psi_l(t)|}{\||\psi_l(t)\rangle\|^2} \tag{5}$$

It is proved that $\lim_{L \to \infty} \tilde{\rho}_L(t) = \rho(t)$ [2].

3.2 Exponential Operators

The most computationally intensive part of the described algorithm is propagating a vector $|\psi(t)\rangle$ until the condition of a jump is met (line 8). This step can be substantially accelerated, taking into account that $\tilde{H}$ is constant between switching of $\theta(t)$. There, propagation is explicitly described by an exponential operator [6]:

$$B(\Delta t) = e^{-i\tilde{H}\Delta t},$$

which does not depend on a particular state $|\psi(t)\rangle$, and therefore allows for calculating evolution of an arbitrary vector over time Δt. To implement a high-precision approach to resolution of the jump moments, one can pre-calculate a set of exponential operators, for different time steps. The maximal time step should be chosen to be in an integer ratio to the time interval T, where the Hamiltonian $\tilde{H}$ is constant: $T = k\Delta t_1$, $k \in Z$. For finer timescales one defines:

$$\{B(\Delta t_i)|\Delta t_i = 2\Delta t_{i+1} : i = \overline{1, m-1}\}$$

The particular value of Δt_1 should be chosen as to decrease the computation time between quantum jumps, and so it is model and parameter dependent. In our experience, a "rule of thumb" defines the best value of Δt_1 as that makes $|||\psi(t)\rangle||$ decreasing by $30-60\%$ under action of $B(\Delta t_1)$. The depth of timescale hierarchy, m, should be chosen with regard to the required precision for the resolution of jump moments, which is $\leq \Delta t_m$.

Propagation of the trajectory untill the next jump implements bisection method (Algorithm 2). It is initialized with the a current time t, state vector $|\psi(t)\rangle$ and random η. The first step takes $s = 1$, $\delta t = \Delta t_1$ (line 1), next steps take values of s and δt, obtained in the end of the preceding step. The moment of jump is found as follows:

1. The main cycle of the algorithm (lines 2–13) implements propagation with a given time step until quantum jump conditions are fulfilled, $|||\psi(t)\rangle||^2 \leq \eta$.
2. The value of the state vector at $t + \delta t$ is calculated (line 3).
3. If $|\psi(t + \delta t)\rangle$ fulfills the jump condition, the time of the jump is resolved with higher precision. This is achieved by taking a smaller time step (line 5), if the minimal one, t_m, has not been reached yet. Otherwise, the time of the jump is determined with the maximal possible precision.
4. If the jump condition is not fulfilled, or its moment is found with maximal precision, then the current time and state vector values are renewed (lines 7–8). Then, the maximal time step, Δt_s, to be used for the next iteration of the algorithm (lines 9–11), is chosen such that the time to the next switch of the Hamiltonian is the multiple of Δt_s.

4   Optimized Algorithm

Algorithm 2 that propagates the state vector $|\psi(t)\rangle$ to the next time moment $t + \delta t$, involves the multiplication of the vector with the matrix of the exponential operator. Our idea of accelerating the algorithm is to cluster (group) the trajectories so that multiple independent matrix-vector multiplications are substituted with a single matrix-matrix multiplication, which, for example, for matrix dimension of the order of 10^3 reduces computational time by several-fold. The main challenge lies in clustering of the vectors in groups; that is because moments of next quantum jumps for the trajectories are different and independent. It should also be noted that while Hamiltonian matrices for quantum dynamics are often sparse, the matrices for exponential operators are not.

Below we present the detailed description of our solution. The task is formulated as the propagation of the group of vectors $|\psi_l(t)\rangle$, $l = 1, .., L'$ over the

Algorithm 2. Finding the moment of a quantum jump with the set of exponential operators

---

1: $\delta t = \Delta t_1$
2: $s = 1$
3: **while** $\||\psi(t)\rangle\|^2 > \eta$ **do**
4:     $|\mu\rangle = B(t_s)|\psi(t)\rangle$
5:     **if** $\||\mu\rangle\|^2 \leq \eta$ & $s < m$ **then**
6:         $s = s + 1$
7:     **else**
8:         $|\psi(t)\rangle = |\mu\rangle$
9:         $t = t + \delta t$
10:        **while** $s > 1$ & $\delta t = k * \Delta t_{s-1}, k \in \mathbb{Z}^+$ **do**
11:            $s = s - 1$
12:            $\delta t = \Delta t_s$
13:        **end while**
14:    **end if**
15: **end while**

---

time interval $[t_{\tilde{H}}, t_{\tilde{H}} + T_{\tilde{H}}]$, where the Hamiltonian $\tilde{H}$ is constant. As a result, all vectors have to be propagated to the end of the specified time interval.

For simplicity of description and without loss of generality, we consider the set of operators $\{B(\Delta t_i) | k\Delta t_1 = T_{\tilde{H}}, k \in Z; \Delta t_i = 2\Delta t_{i+1} : i = \overline{1, m-1}\}$. The group of vectors is stored in the matrix form $V \in C^{L' \times N}$, where each vector is given by a separate row. Here L' is the number of vectors in a group, N is the dimension (number of states) of the model quantum system, $V[l]$ is the l-th state vector, and $\|V[l]\|$ is its norm.

Each l-th vector is given additional characteristics.

- $\eta[l]$, a random number from a uniform distribution, $U[0, 1]$, which determines the value of the norm, when a quantum jump occurs.
- $d[l]$ is an integer number that sets a current time within the propagation interval, when Hamiltonian remains constant, $[t_{\tilde{H}}, t_{\tilde{H}} + T_{\tilde{H}}]$. $d[l] = 0$ corresponds to the beginning of the interval. Increasing $d[l]$ by 1 corresponds to increasing current time by Δt_m. $d[l] = k * 2^{m-1}$ corresponds to the end of the time interval, $t_{\tilde{H}} + T_{\tilde{H}}$.

Matrix V has a specific structure (Table 1) and contains 4 blocks of rows of distinct classes. In course of the run, vectors are moved from one block to another (change their class), and the sizes of blocks change.

Start of the propagation from the left boundary of the interval $[t_{\tilde{H}}, t_{\tilde{H}} + T_{\tilde{H}}]$ is accompanied by the following initialization.

- All vectors belong to class A.
- If $t_{\tilde{H}} = 0$, then for all vectors the values $\eta[l] = U[0, 1]$ are calculated. Otherwise $\eta[l]$ is brought forward from the previous time interval.
- For all vectors $d[l] = 0$ is set.

Table 1. The structure of the matrix

Class/Block	Description
VA, rows $(1) - (A)$	Vectors, which could experience a jump until the right boundary of the interval is reached $[t_{\tilde{H}}, t_{\tilde{H}} + T_{\tilde{H}}]$, but an estimate of the time of jump is missing
VB, rows $(A+1) - (A+B)$	Vectors, which will experience a jump during the time interval Δt_1.
VC, rows $(A+B+1)-$ $(A+B+C)$	Vectors, which cannot experience a jump before the right boundary of the time interval is reached $[t_{\tilde{H}}, t_{\tilde{H}} + T_{\tilde{H}}]$
VD, rows $(A+B+C+1)-$ $(A+B+C+D = L')$	Vectors, propagated to the right boundary of the time interval $[t_{\tilde{H}}, t_{\tilde{H}} + T_{\tilde{H}}]$

Propagation is performed in the forward step (Algorithm 3) and backward step (Algorithm 4). The forward step is repeated until there remain trajectories from VA class. Otherwise, the backward step is performed once.

The forward step of the algorithm finds the next moment for the quantum jump for all vectors of class A. It is specified in Algorithm 3.

1. The first part of the algorithm (lines 1–17) propagates each vector VA by a maximally possible number of steps Δt_1, which does not lead to the quantum jump yet. For that a vector is sequentially multiplied by B_1, but before it is renewed, one of the following is checked.

   1.1 Next product produces the quantum jump condition. Then, the vector is moved to block VB (lines 4–6).

   1.2 Next product does not lead to quantum jump and will not allow the current time to reach the right boundary of the interval $[t_{\tilde{H}}, t_{\tilde{H}} + T_{\tilde{H}}]$. Then, the value of the vector should be renewed, and propagation continued (rows 7–9).

   1.3 Next product does not lead to quantum jumps condition, but the current time becomes equal to the right boundary of the interval $[t_{\tilde{H}}, t_{\tilde{H}} + T_{\tilde{H}}]$. In that case, the vector should be renewed and moved to block VD (lines 10–12).

   1.4 Next product does not lead to quantum jump, but the current time becomes greater then the right boundary of the interval $[t_{\tilde{H}}, t_{\tilde{H}} + T_{\tilde{H}}]$. Then, the vector should be moved to block VC (rows 13–15).

   On completion of the first part of the algorithm block A becomes empty.

2. The second part of the algorithm (rows 18–34) searches for the quantum jump times for vectors from block VB, provided that it occurs within $[t_{\tilde{H}}, t_{\tilde{H}} + T_{\tilde{H}}]$. Namely, each vector from VB is multiplied by the matrices of the exponential operators, B_i, $i = 2..m$, one by one. Processing the result $VBnext[l]$ of the multiplication of each vector by each matrix is determined by the following.

Algorithm 3. Forward step of the algorithm

1: **while** $A > 0$ **do**
2:     $VAnext = B_1 \times VA$
3:     **for** $l = 1; l < A; l = l + 1$ **do**
4:         **if** $\|VAnext[l]\|^2 < \eta[l]$ **then** move $VA[l]$ to VB
5:         **end if**
6:         **if** $\|VAnext[l]\|^2 \geq \eta[l]$ & $d[l] + 2^{m-1} < k * 2^{m-1}$ **then**
7:             $VA[l] = VAnext[l]d[l] = d[l] + 2^{m-1}$;
8:         **end if**
9:         **if** $\|VAnext[l]\|^2 \geq \eta[l]$ & $d[l] + 2^{m-1} == k * 2^{m-1}$ **then**
10:             $VA[l] = VAnext[l]$; $d[l] = 0$;
11:             move $VA[l]$ to VD
12:         **end if**
13:         **if** $\|VAnext[l]\|^2 \geq \eta[l]$ & $d[l] + 2^{m-1} > k * 2^{m-1}$ **then**
14:             move $VA[l]$ to VC
15:         **end if**
16:     **end for**
17: **end while**
18: **for** $i = 2; i \leq m; i = i + 1$ **do**
19:     $VBnext = B_i \times VB$
20:     **for** $l = 1; l \leq B; l = l + 1$ **do**
21:         **if** $\|VBnext[l]\|^2 < \eta[l]$ & $d[l] + 2^{m-i} < k * 2^{m-1}$ **then**
22:             $VB[l] = VB[l]$;
23:         **end if**
24:         **if** $\|VBnext[l]\|^2 \geq \eta[l]$ & $d[l] + 2^{m-i} < k * 2^{m-1}$ **then**
25:             $VB[l] = VBnext[l]$; $d[l] = d[l] + 2^{m-i}$
26:         **end if**
27:         **if** $\|VBnext[l]\|^2 \geq \eta[l]$ & $d[l] + 2^{m-i} == k * 2^{m-1}$ **then**
28:             $VB[l] = VBnext[l]$; $d[l] = 0$; move $VB[l]$ to VD
29:         **end if**
30:         **if** $\|VBnext[l]\|^2 \geq \eta[l]$ & $d[l] + 2^{m-i} > k * 2^{m-1}$ **then**
31:             move $VB[l]$ to VC
32:         **end if**
33:     **end for**
34: **end for**
35: $VB = B_m \times VB$
36: Make quantum jump for all vectors in VB
37: **for** $l = 1; l \leq B; l = l + 1$ **do**
38:     generate $\eta[l] = U[0, 1]$
39:     $d[l] = d[l] + 1$
40:     **if** $d[l] == k * 2^{m-1}$ **then**
41:         $d[l] = 0$; move $VB[l]$ to VD
42:     **else**
43:         move $VB[l]$ to VA
44:     **end if**
45: **end for**

  2.1 If a quantum jump occurs for $VBnext[l]$, the result is not saved (lines 21–23). In this case, we can either find a moment for the quantum jump to a higher precision or make the last move towards time Δt_m in the third part of the algorithm.

  2.2 If the quantum jump condition is not fulfilled for $VBnext[l]$, and the time has not reached the end of the interval $[t_{\tilde{H}}, t_{\tilde{H}} + T_{\tilde{H}}]$, then the vector $VB[l]$ is renewed, together with its current time $d[l]$ (lines 24–26).

  2.3 If the quantum jump condition is not fulfilled for $VBnext[l]$, but the right boundary of the interval $[t_{\tilde{H}}, t_{\tilde{H}} + T_{\tilde{H}}]$ is reached, then the result is saved, the vector is moved to block VD (lines 27–29). Propagation step is finished.

  2.4 In the quantum jump condition is not fulfilled for $VBnext[l]$, and the current time has gone beyond the interval $[t_{\tilde{H}}, t_{\tilde{H}} + T_{\tilde{H}}]$, then the result is not saved, and the vector is moved to block VC (lines 30–32).
On completion of the second part of the algorithm, block VB contains only those vectors, for which quantum jump occurs only after propagation to the time Δt_m.

3. In the third part of the algorithm (lines 35–45), vectors from block VB are multiplied by matrix B_m, and undergo quantum jumps (lines 35–36). Then for each vectors from block VB, there is a new value of $\eta[l]$ is generated, the current time $d[l]$ is renewed, and the vector is moved to VA or VD (lines 37–45).

The backward step of the algorithm brings all vectors from VC to the right end of the interval $[t_{\tilde{H}}, t_{\tilde{H}} + T_{\tilde{H}}]$. It is implemented in the case, when the class VA becomes empty after an iteration of the forward step. The backward step is organized as follows (Algorithm 4).

1. Vectors from block VC are propagated to the right boundary of the time interval $[t_{\tilde{H}}, t_{\tilde{H}} + T_{\tilde{H}}]$. Block VC is multiplied by exponential operator matrices $B_i, i = m, .., 2$, one by one (lines 1–11). Processing of the result of the multiplication of each vector of the block by a matrix, $VCnext[l]$, is determined by the following.
   - In propagation over Δt_i is required to reach an exact boundary of the time interval, $[t_{\tilde{H}}, t_{\tilde{H}} + T_{\tilde{H}}]$, then both the current time and vector are renewed (lines 4–6).
   - If after propagation over time Δt_i we reach the right end of the interval $[t_{\tilde{H}}, t_{\tilde{H}} + T_{\tilde{H}}]$, then the current time and vector are renewed, the vector is moved in block VD (lines 7–9).
2. All vectors from VD move to VA (line 12).

5   Numerical Results

5.1   Computational Infrastructure

We used a node of the Intel Endeavor cluster with 4 high-end 24-core Intel Xeon E7-8890v4 CPUs (2.2 GHz, codename Broadwell). We employed the Intel

Algorithm 4. Backward step of the algorithm

1: **for** $i = m; i \geq 2$ & $C > 0; i = i - 1$ **do**
2:     $VCnext = B_i \times VC$
3:     **for** $l = 1; l \leq C; l = l + 1$ **do**
4:         **if** $d[l]$ & $2^{m-i} = 1$ /* & means bitwise AND here */ **then**
5:             $VC[l] = VCnext[l]; d[l] = d[l] + 2^{m-i}$
6:         **end if**
7:         **if** $d[l] == k * 2^{m-1}$ **then**
8:             $d[l] = 0$; move $VC[l]$ to VD
9:         **end if**
10:     **end for**
11: **end for**
12: move all vectors from VD to VA

Math Kernel Library, Intel MPI, and Intel C++ Compiler from the Intel Parallel Studio XE Cluster Edition 2017.

5.2 Methodology

The Goal of the Experiments and the Model Problem. The main scientific contribution of this paper is the algorithmic optimization of the quantum trajectory method described in the previous section. In this section we empirically show the advantages of the optimized algorithm compared to the baseline implementation. We also identify the most promising run modes by means of trying different combinations of MPI processes, OpenMP threads, and MKL threads. The quantum dimer with $N = 1024$ states described in Sect. 2 is chosen as a testbed problem. For this number of states the run time of the baseline algorithm is acceptable and so we can run and analyze extensive performance tests. Besides, this system size is big enough to highlight the advantage of the optimized algorithm.

Correctness Tests. First, we check the correctness of the optimized implementation. Note that since in both cases we are dealing with stochastic algorithms, the results will not be exactly the same. Given that the correctness of the baseline implementation has been verified in our earlier work [6], we take its results as a basis for further comparison. Correctness evaluation of the optimized version consisted of two stages. At the first stage, we generated a sequence of pseudorandom numbers and used it for both algorithms. Then we compared the resulting density matrices. For the time periods considered, the relative difference did not exceed 10^{-14}, which can be explained by the different order of floating-point operations. At the second stage we used the time intervals and numbers of trajectories sufficient to reach the attractor, which is of great interest for researchers of this kind of problems. We found that the results of the optimized algorithm were in the 95% confidence interval computed for the baseline version. Thus, our experiments demonstrate that the results of the optimized code match the expectations.

Performance Evaluation. A straightforward choice of the performance metric is to compare the run time to reach the attractor with the given accuracy. However, the problem considered is computationally intensive and testing many configurations is somewhat wasteful in terms of CPU-hours consumed. Therefore, we employ the number of trajectories processed per second, while propagating the system for one time period, as the performance metric. Since our performance evaluation experiments are done for a single period, the metric corresponds to simply trajectories per second.

The plan for performance evaluation takes into account the following features. The hotspot of the baseline implementation is the dense matrix-vector multiplication routine. On the contrary, the optimized version spends most time on dense matrix-matrix multiplication. Workload imbalance is under 5% even for small numbers of trajectories.

The baseline implementation employs MPI + OpenMP parallelism on the level of trajectories with sequential MKL matrix-vector multiplication. Parallel matrix-vector multiplication is not beneficial because of a small workload per invocation combined with good balancing on MPI + OpenMP level. The optimized version uses the same MPI + OpenMP scheme, but we additionally study efficiency of internal parallelism in MKL matrix-matrix multiplication.

Based on the above-mentioned considerations, we fixed the integration time to be equal to one period and varied the number of processes P, the number of threads T and, for the optimized version, the number of MKL threads M. For both versions the total number of threads in each configuration was equal to the number of cores (96). Previously we have checked that using all 96 cores indeed yields better performance (in terms of the metric used) compared to smaller numbers of cores. All experiments were performed on a single node of the Endeavor system, since the scaling efficiency on distributed memory is close to linear due to a small workload imbalance and virtually absent communications between nodes.

5.3   Results and Discussion

First, we found the empirically best combination of processes and threads for the baseline version (Table 2). The value of the performance metric varies between 0.8 and 1.69 trajectories per second, with the optimal configuration being 4 processes with 24 threads per process. In this mode each process is run on a separate 24-core CPU with an affinity mask used to pin OpenMP threads to cores.

The next series of experiments concerns the optimized version with varying numbers of MPI processes, OpenMP threads and MKL threads. The results are presented at Table 3. Same as for the baseline version, the configurations with external parallelism and sequential MKL routines are superior. The best configuration is again 4 processes with 24 threads per process, scoring 29.09 trajectories per second. In this configuration the optimized version outperforms the baseline version by a factor of 17.21, which proves efficacy of the proposed approach to optimization. Increasing the problem size will likely further increase

Table 2. Comparison of MPI + OpenMP configurations for the baseline version.

# processes	# threads	Trajectories per second
1	96	0.80
2	48	0.63
4	**24**	**1.69**
8	12	1.49
12	8	1.44
24	4	1.39
48	2	1.36
96	1	1.35

Table 3. Comparison of MPI, OpenMP and MKL configurations for the optimized version.

# processes	# OpenMP threads	# MKL threads	Trajectories per second
1	1	96	5.57
1	96	1	27.60
2	1	48	11.69
2	48	1	28.07
4	1	24	21.58
4	**24**	**1**	**29.09**
8	1	12	16.32
8	12	1	25.75
12	1	8	27.03
12	8	1	21.02
24	1	4	27.33
24	4	1	18.48
48	1	2	13.99
48	2	1	16.87
96	1	1	13.78

the speedup due to the growing advantage of matrix-matrix multiplication over a set of matrix-vector multiplications.

To assess the hardware usage efficiency we apply the Roofline model [11]. Presented several years ago, this method of analysis is widely used to compare the achieved and theoretically attainable performance on specific computing systems. The main advantage of this model is a visual representation of the achieved performance and its theoretical upper bounds. In our experiments we collected the data traffic through L1 cache and arithmetic intensity (AI) using the Roofline Analysis of Intel Advisor. The baseline and optimized implementations were run

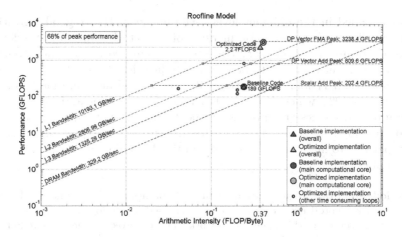

Fig. 1. Roofline model for the baseline and optimized algorithms

using the best combination of MPI processes and OpenMP threads. The resulted Roofline model is presented below as a log-log plot (Fig. 1). The arithmetic intensity, computed as the number of floating point operations related to the data traffic through L1 cache, is shown on the horizontal axis. The vertical axis corresponds to the achieved and attainable floating-point performance in double precision. Four dotted sloping lines show the peak performance as a function of arithmetic intensity with fixed memory bandwidth of L1, L2, L3, and DRAM. Three horizontal dotted lines show peak performance for double precision floating point computations in the scalar, vector and vector fused multiply-add (vector FMA) modes, respectively. The collected performance data is presented as follows. The red triangle represents the overall baseline version performance 184.69 GFLOPS with $AI = 0.239$ FLOP/Byte. The green triangle represents the overall optimized version performance 2211.92 GFLOPS with $AI = 0.37$ FLOP/Byte. The red and green circles correspond to the main hotspots of both implementations. Thus, the largest red circle corresponds to the MKL dense matrix multiplication routine. It achieved 3128.504 GFLOPS that is very close to the peak hardware performance. Overall, the optimized version achieved 68% of the 4-socket 96-core Intel Broadwell CPU peak performance which is quite well for state-of-the-art scientific applications.

6   Conclusions

We proposed and validated the optimized version of the quantum trajectory method, which allows to find asymptotic states of open quantum systems. The central idea is the clustering (grouping) of trajectories into matrices and substitution of multiple matrix-vector multiplication operations with a single matrix-matrix multiplication. This modification significantly increases efficiency of the multi-level hierarchic memory usage due to the potential of re-using the data, previously loaded in the different level cache memory. The original algorithm [6] did not allow for an automatic merging of trajectories due to the different times of quantum jumps on every trajectory. It required a substantial rewriting of the code, which proved to be completely justified. Computational results showed more than 17-fold acceleration with the testbed quantum model of the dimension $N = 1024$, which demonstrated a possibility of substantial economy of computational resources or/and time of calculations. The obtained results open the door to studying systems of even greater dimension. We expect our approach to be applicable to many models, actual and timely in different fields of modern quantum physics.

Acknowledgements. We are grateful to Intel Corporation and Lobachevsky University of Nizhni Novgorod for access to the systems used for performing computational experiments presented in this paper. This work was supported by the Russian Science Foundation grant No. 15-12-20029.

References

1. Breuer, H.-P., Petruccione, F.: The Theory of Open Quantum Systems. Oxford University Press, Oxford (2002)
2. Lindblad, G.: On the generators of quantum dynamical semigroups. Commun. Math. Phys. **48**(2), 119–130 (1976)
3. Daley, A.: Quantum trajectories and open many-body quantum systems. J. Adv. Phys. **63**(2), 77–149 (2014)
4. Bukov, M., D'Alessio, L., Polkovnikov, A.: Universal high-frequency behavior of periodically driven systems: from dynamical stabilization to Floquet engineering. Adv. Phys. **64**(2), 139–226 (2015)
5. Carmichael, H.J.: An Open Systems Approach to Quantum Optics. Springer, Berlin (1993)
6. Volokitin, V., Liniov, A., Meyerov, I., Hartmann, M., Ivanchenko, M., Hänggi, P., Denisov, S.: Towards quantum attractors: Sampling asymptotic states of modulated open systems with quantum trajectories. arXiv:1612.03848 (2016)
7. Laptyeva, T.V., Kozinov, E.A., Meyerov, I.B., Ivanchenko, M.V., Denisov, S., Hänggi, P.: Calculating Floquet states of large quantum systems: A parallelization strategy and its cluster implementation. Comp. Phys. Comm. **201**, 85–94 (2016)
8. Hartmann, M., Poletti, D., Ivanchenko, M., Denisov, S., Hänggi, P.: Asymptotic state of many-body open quantum systems under time-periodic modulations. arXiv:1606.03896 (2016)
9. Diehl, S., Micheli, A., Kantian, A., Kraus, B., Büchler, H.P., Zoller, P.: Quantum states and phases in driven open quantum systems with cold atoms. Nature Phys. **4**(11), 878–883 (2008)
10. Diehl, S., Tomadin, A., Micheli, A., Fazio, R., Zoller, P.: Dynamical phase transitions and instabilities in open atomic many-body systems. Phys. Rev. Lett. **105**, 015702 (2010)
11. Williams, S., Waterman, A., Patterson, D.: Roofline: an insightful visual performance model for multicore architectures. Commun. ACM **52**(4), 65–76 (2009)

Tensor Train Global Optimization: Application to Docking in the Configuration Space with a Large Number of Dimensions

A.V. Sulimov[1,2], D.A. Zheltkov[3], I.V. Oferkin[1], D.C. Kutov[1,2], E.V. Katkova[1,2], E.E. Tyrtyshnikov[2,3,4,5], and V.B. Sulimov[1,2(✉)]

[1] Dimonta, Ltd., Moscow 117186, Russia
{as,io,dk,katkova}@dimonta.com, vladimir.sulimov@gmail.com
[2] Research Computer Center, Lomonosov Moscow State University, Moscow 119992, Russia
eugene.tyrtyshnikov@gmail.com
[3] Institute of Numerical Mathematics of Russian Academy of Sciences, Moscow 119333, Russia
dmitry.zheltkov@gmail.com
[4] Faculty of Computational Mathematics and Cybernetics,
Lomonosov Moscow State University, Moscow 119992, Russia
[5] Siedlce University of Natural Sciences and Humanities, Siedlce, woj. Mazowieckie, Poland

Abstract. The novel docking algorithm is presented and it is applied to the docking problem with flexible ligand and moveable protein atoms. The energy of the protein-ligand complex is calculated in the frame of the MMFF94 force field in vacuum. The conformation space of the system coordinates is formed by translations and rotations of the ligand as a whole, by the ligand torsions and also by Cartesian coordinates of the selected target protein atoms. The algorithm is realized in the novel parallel docking SOL-P program and results of its performance for a set of 30 protein-ligand complexes are presented. It is shown that mobility of the protein atoms improves docking positioning accuracy. The SOL-P program is able to perform docking of a flexible ligand into the active site of the target protein with several dozen of protein moveable atoms – up to 157 degrees of freedom.

Keywords: Docking · Tensor train · Protein-ligand complex · Protein moveable atoms · Flexible ligand · Drug design

1 Introduction

Search of molecules-inhibitors of a given target protein is the key stage of the new drug development. Inhibitors block the active site of the protein associated with a disease and the disease is cured. Molecular modeling on the base of supercomputer simulation by docking and molecular dynamics programs should increase effectiveness of new inhibitors development [1, 2]. On the base of such calculations it is possible to predict inhibition activity of new compounds. The reliable prediction is defined by the accuracy of these programs. Docking programs perform positioning of a compound (a ligand) in the active site of the target protein. Computed poses of the ligand are used for the calculation

© Springer International Publishing AG 2017
V. Voevodin and S. Sobolev (Eds.): RuSCDays 2017, CCIS 793, pp. 151–167, 2017.
https://doi.org/10.1007/978-3-319-71255-0_12

of the protein-ligand binding free energy which is directly connected with the inhibition constant. Compounds with higher binding energy are better inhibitors because the same inhibition effect can be reached with smaller concentration of the compound. The accuracy of binding energy calculations should be better than 1 kcal/mol [3] for the reliable prediction of the inhibitory activity. However, the accuracy of binding energy calculations for arbitrary target proteins and ligands is too bad now. This accuracy depends on many factors and simplifications: the force field choice for modeling intra- and intermolecular interactions instead of the use of quantum chemical methods, the solvent model, target protein and ligand models, the docking algorithm, the free energy calculation method, respective approximations and computer resources required for docking of one ligand. Main simplifications of many existing docking programs, e.g. the SOL [4] program, is the rigid protein approximation and the use of the grid of preliminary calculated potentials of ligand probe atoms interactions with the protein (the grid approximation) which restrict strongly performance of docking programs and make worse the docking accuracy. However proteins are flexible and some protein atoms near the ligand binding pose relax from their initial positions in the process of protein-ligand binding – a difference between bound and unbound protein's structures is often observed [5]. In this study we describe the novel docking algorithm which makes it possible to reject the rigid protein as well as the grid approximations, to take into account many proteins' degrees of freedom and to increase the docking accuracy.

The protein-ligand binding free energy ΔG_{bind} can be calculated as the difference between the free energy of the protein-ligand complex G_{PL} and the sum of free energies of the unbound protein G_P and the unbound ligand G_L:

$$\Delta G_{bind} = G_{PL} - G_P - G_L \qquad (1)$$

Free energies of the protein, the ligand and their complex are described by respective energy landscapes and they can be calculated through the configuration integrals over the respective phase space. In the thermodynamic equilibrium the molecular system occupies its low energy minima. The configuration integral will come to the sum of configuration integrals over the separate low energy minima if these minima are separated by sufficiently high energy barriers [6, 7]. So, the docking accuracy is defined by the completeness of the low energy minima finding and by the accuracy of the configuration integral calculation in each of these minima.

Docking without the preliminary calculated energy grid requires much more computational resources because the protein-ligand energy has to be computed in the frame of the whole given force field for each system conformation appearing in the minima search algorithm. Such docking programs, FLM [7] and SOL-T [8], have been developed for the rigid target protein and the flexible ligand. The parallel FLM program can perform the comprehensive minima search either in vacuum or with the rigorous implicit solvent model [7] but at the expense of too large supercomputer resources – about 20000 CPU*h per one complex. The parallel SOL-T program employs the novel tensor train global (TT) optimization algorithm and it requires much less supercomputer resources than FLM. The docking positioning accuracy of FLM and SOL-T in vacuum for the rigid protein are comparable with one another at least for some test complexes [8]. Also it is demonstrated [7] that the ligand positioning accuracy is much better when the recent quantum chemical semiempirical methods, PM7 [9] and PM6 [10], are used instead of classical force fields and water solvent is taken into account.

Algorithms of most modern docking programs are based on the docking paradigm [7–9]. This paradigm assumes that the ligand binding pose in the active site of the target protein corresponds to the global minimum of the protein-ligand energy or is near it and the docking problem is reduced to the global optimization problem on the multi-dimensional protein-ligand energy surface. The dimensionality of this surface (d) is defined by the number of protein-ligand system degrees of freedom and commonly used docking algorithms, e.g. the genetic algorithm, are not able to perform docking for $d \geq 25$. Therefore docking of a flexible ligand into a flexible target protein requires more effective global optimization algorithms. The present study demonstrates that it is possible to perform successfully such docking employing the novel tensor train global optimization algorithm [8, 11]. We describe here main features of this novel algorithm, the respective program SOL-P for docking flexible ligands into target proteins with moveable atoms [12, 13] and the results of validation of the ligand positioning accuracy for a test set of 30 protein-ligand complexes [8].

2 Materials and Methods

For the realization of the novel docking algorithm we use the MMFF94 force field [14] in vacuum. The results will be much better, if either MMFF94 is used with the solvent model or PM7 is used with the solvent model [7, 9]. While looking for low-energy minima, ligands are considered to be fully flexible and some of protein atoms are moveable. The force field determines energy of the protein-ligand complex for its every conformation. The MMFF94 force field combines sufficiently good parameterization based on ab initio quantum-chemical calculations of a broad spectrum of organic molecules and the well-defined procedure of atom typification applicable to an arbitrary organic compound. MMFF94 is implemented in the SOL docking program [4] used successfully for new inhibitors development, e.g. see [15].

2.1 TT-docking

The novel docking algorithm (TT-docking) [8, 11] utilizes the TT global optimization method. It is based on the novel methods of tensor computations.

If d is the number of degrees of freedom of the protein-ligand complex, then we can introduce a grid in the configuration space with n_i nodes in each direction $i = 1, 2 \ldots d$. If the grid is fine enough, then the solutions of continuous and discrete problems are expected to be close.

The basis of this consideration is the Tensor Train (TT) decomposition [16, 17] of a tensor $A \in \mathbb{R}^{n_1 \times \ldots \times n_d}$ in the form:

$$A(i_1, \ldots, i_d) \approx \sum_{\alpha_1=1, \ldots, \alpha_{d-1}=1}^{r_1, \ldots, r_d} G_1(i_1, \alpha_1) G_2(\alpha_1, i_2, \alpha_2) \ldots G_{d-1}(\alpha_{d-2}, i_{d-1}, \alpha_{d-1}) G_d(\alpha_{d-1}, i_d) \quad (2)$$

The numbers $r_1, \ldots, r_{d-1}$ are called TT-ranks of the tensor; for convenience, dummy ranks $r_0 \equiv r_d \equiv 1$ are also introduced. The 3-dimensional tensors $G_i \in \mathbb{R}^{r_{i-1} \times n_i \times r_i}$ are called cores or carriages of the tensor train. If TT-ranks are reasonably small, then the TT decomposition possesses several very useful properties [16, 17]. However, we cannot afford computing or storing all the elements for large tensors. Therefore, it becomes crucial to have for tensors a fast approximation method utilizing only a small number of their elements. Such a method was proposed and called the TT-Cross method [18]. It heavily exploits the matrix cross interpolation [19–23] algorithm applied cleverly, although heuristically, to selected submatrices in the unfolding matrices of the given tensor. The matrix $A_k \in \mathbb{R}^{n^k \times n^{d-k}}$, $A_k(i_1 \ldots i_k, i_{k+1} \ldots i_d) = A(i_1, i_2, \ldots, i_d)$ is called the k-th unfolding matrix of the tensor A. Such matrices are intrinsically linked with the TT-decomposition, TT-rank r_k is just the rank of the matrix A_k.

To explain the idea of the global optimization method consider a rank-1 matrix $A = uv^T \in \mathbb{R}^{m \times n}$. It is evident that the largest magnitude element of the matrix could be easily found in $m + n$ operations: if i and j are positions of the largest magnitude element in vectors u and v, respectively, then the required element is A_{ij}. Moreover even if factors u and v are unknown, such element could be found in $m + n$ evaluations of matrix elements. For this purpose, select any nonzero column of the matrix and find its largest magnitude element. Then select the row containing that element. The largest magnitude element of the matrix is the largest magnitude element of that row.

It was noticed, that latter strategy finds the largest magnitude element of the matrix with high probability even if matrix is not a rank-one matrix (though in this case more than $m + n$ elements should be evaluated, the search continues until the element is of the largest magnitude in both its row and column). Of course, this is evident if matrix is very close to a rank-one matrix. But such a strategy works with high probability even if the error in the optimal rank-one approximation of the matrix is quite large (as is proved by A. Osinsky, a good approximation exists if the error is even $1/8$ of the matrix Frobenius norm).

Moreover, consider a rank-2 matrix, for which a rank-one approximation is not very accurate. Apply the above strategy to the original matrix, perform the Gauss elimination with the selected element as a pivot and then apply the search strategy again. The largest in magnitude element with high probability is within evaluated elements of the matrix.

This is just how the matrix cross approximation method [20, 21] works. This method performs the search of the largest in magnitude matrix element, uses the found element to perform the Gauss elimination (constructing its factors but not performing elimination for all matrix elements) and repeats operations with the obtained matrix until the stopping criteria is met. Great advantage of the method is that it does not evaluate all matrix elements but only $O((m + n)r)$ of them, where r is the approximation rank. Also it has low complexity: $O((m + n)r^2)$ arithmetic operations. Moreover, the approximation obtained by this method is quasioptimal, *i.e.* its accuracy is close (by a not very large factor) to the accuracy of the optimal rank-r approximation, especially when the rank is small.

So, the matrix cross interpolation method could be used as a simple global optimization method as it finds the largest in magnitude element among all evaluated elements.

More sophisticated variants of such optimization method use the local optimization of interpolation points because these points with high probability are near to local optima with large values. Such methods find in practice a global optimum with the ranks much less than those which are needed for a good approximation. Usually the parameter r_{max} limiting the rank from above is introduced to reduce the number of operations.

Complexity of the global optimization is $O((m + n)r)$ function evaluations, $O((m + n)r^2)$ arithmetic operations and $O(r)$ local optimizations.

The TT-Cross approximation method applies to tensors (multi-dimensional arrays) [18]. It uses the matrix cross approximation method and pursues the same goal, which is to construct the approximation of a tensor in such a way that only small number of its elements are picked up. For tensors this is even by far more important than for matrices because the number of elements of many practical tensors is so huge that it cannot be computed or stored in any memory we may have at our disposal. If approximation ranks are reasonably low, the method evaluates only the logarithmic number of the total amount of tensor elements.

The idea of TT-cross approximation method is based on the following fact. Let a set I consist of r row indicies of A and a set J contain r column indicies, and let $A(I, J)$ be a submatrix of volume (modulus of the determinant) that is close to the maximal one among all submatrices of order r. Then a sufficiently good approximation is as follows:

$$A \approx A(:, J)A(I, J)^{-1}A(I, :) \tag{3}$$

Here $A(:, J)$ means the matrix consisting of columns of the A matrix with indicies from J, similarly $A(I, :)$ means the matrix consisting of rows of the A matrix with indices from I.

To facilitate explanation, let us introduce some special matrices associated with tensors. For the tensor $T \in \mathbb{R}^{n_1 \times \cdots \times n_d}$ the matrix $T_k \in \mathbb{R}^{n_1 \cdots n_k \times n_{k+1} \cdots n_d}$ is k-th *unfolding matrix* of this tensor if its elements are just reordered elements of the given tensor:

$$T_k(i_1 \ldots i_k, i_{k+1} \ldots i_d) = T(i_1, \ldots, i_d) \tag{4}$$

Let us consider a T_1 matrix. If we know r_1 rows I_1 and columns J_1 for which $T_1(I_1, J_1)$ has large enough volume then using (3) we obtain:

$$T_1 \approx T_1(:, J_1)T_1(I_1, J_1)^{-1}T_1(I_1, :), \tag{5}$$

Denote $T_1(:, J_1)T_1(I_1, J_1)^{-1}$ as a matrix G_1 of size $n_1 \times r_1$ and rewrite Eq. (5) element-wise taking in account that elements of matrices T_1, T_2 and tensor T are the same:

$$T(i_1, \ldots, i_d) \approx \sum_{\alpha_1=1}^{r_1} G_1(i_1, \alpha_1)T(\alpha_1, i_2, \ldots, i_d) = \sum_{\alpha_1=1}^{r_1} G_1(i_1, \alpha_1)T_2(I_1(\alpha_1)i_2, i_3, \ldots, i_d) \tag{6}$$

Note that $T_2(I_1(\alpha_1)i_2, i_3 \ldots, i_d)$ are elements of the T_2 submatrix with rows selected in a special way. Denote this submatrix by $\tilde{T}_2$. Assuming good enough sets of rows I_2 and columns J_2 of size r_2, for this matrix we can obtain:

$$\tilde{T}_2 \approx \tilde{T}_1(:,J_2)\tilde{T}_2(I_2,J_2)^{-1}\tilde{T}_2(I_2,:) \tag{7}$$

Denote by $G_2 \in \mathbb{R}^{r_1 \times n_2 \times r_2}$ the tensor for which the matrix $\tilde{T}_1(:,J_2)\tilde{T}_2(I_2,J_2)^{-1}$ is the first unfolding matrix and substitute (7) to (6) using those elements of matrices $\tilde{T}_2, T_3$ which are elements of tensor T:

$$
\begin{aligned}
T(i_1,\dots,i_d) &\approx \sum_{\alpha_1=1,\alpha_2=1}^{r_1,r_2} G_1(i_1,\alpha_1)G_2(\alpha_1,i_2,\alpha_2)T_2(I_2(\alpha_2),i_3\dots i_d) \\
&= \sum_{\alpha_1=1,\alpha_2=1}^{r_1,r_2} G_1(i_1,\alpha_1)G_2(\alpha_1,i_2,\alpha_2)T_3(I_2(\alpha_2),i_3,i_{4,}\dots,i_d).
\end{aligned}
\tag{8}
$$

Now $T_3(I_2(\alpha_2),i_3,i_{4,}\dots,i_d)$ are elements of the submatrix of the T_3 matrix with rows selected in a special way. Denote it by $\tilde{T}_3$ and continue the procedure.

After repeating the procedure described above for $\tilde{T}_3, \dots \tilde{T}_{d-1}$ and denoting $\tilde{T}_d$ by G_d the approximation of the tensor in the TT format is obtained:

$$T(i_1,\dots,i_d) \approx \sum_{\alpha_1=1,\dots,\alpha_{d-1}=1}^{r_1,\dots,r_d} G_1(i_1,\alpha_1)G_2(\alpha_1,i_2,\alpha_2)\dots G_{d-1}(\alpha_{d-2},i_{d-1},\alpha_{d-1})G_d(\alpha_{d-1},i_d) \tag{9}$$

Note that the approximation error may grow exponentially with d, but in practice it is not large even for d of several hundreds.

The problem of the procedure described above is that matrices $T_1, \tilde{T}_k$ have a lot of columns, especially those considered first. So, finding a submatrix of the large volume in this matrix is a nontrivial task. But, if some small sets of columns which contain the large volume are known (at the start such columns could be selected by some mathematical assumptions or randomly), then a submatrix of large volume could be found in a fast way by the matrix cross approximation method. When the whole procedure is completed, we obtain a tensor approximation in the TT-format (may be not good enough) and rows containing submatrices of large enough volume. Next we can start this procedure in the reversed order (or, equivalently, for the tensor with the reversed order of indicies). After the completion of this procedure the new approximation of the tensor and columns containing large volume submatrices are obtained. Such iterations could be repeated, for example, until the difference between two consequent approximations (fast calculation of this difference is possible in the TT format) becomes sufficiently small.

The TT-Cross method is transformed into a global optimization strategy by the same way as it is done for the matrix cross approximation method, in the result the largest in magnitude evaluated element is close to the largest in magnitude element of tensor. More fast convergence could be obtained using the local optimization of pivots obtained by the matrix cross method. To reduce the number of evaluations, the maximal rank is bounded by r_{max}. After the rank limitation iterations could possibly never converge and the *maximal iterations number* parameter is introduced.

Note that such a global optimization strategy allows us to find only the largest in magnitude element. For other optimization problems, e.g. for the docking problem which is the global minimization problem, the problem should be transformed into an equivalent problem of the largest magnitude search. It could be easily done by some monotonic continuous function. Selection of such a function is a nontrivial task as this function must separate optimums as good as possible. For the docking problem it is convenient to apply the TT magnitude maximization to the functional $f(x, E_*) = exp\{100 arccot[E(x) - E_*]\}$, where $E(x)$ is the dimensionless MMFF94 energy for the given configuration x of the protein-ligand complex, E_* is the global minimum found on the previous iteration.

For continuous problems, such as the docking problem, at first the grid must be introduced to obtain a tensor. For such problems, some additional artificial tensorisation might be very useful: instead of applying the method to the d-dimensional tensor with size n in each direction, the $D = md$-dimensional tensor with the size of 2 in every dimension is used. In this case, since the method complexity depends linearly on the tensor dimensionality and the size in each direction, the complexity grows logarithmically with the grid size and it is possible to use very fine grids. Note that artificial tensorisation may increase values of parameters of the TT-optimization method (r_{max} and the maximal iterations number) which are needed to find optimum robustly. However in practice for most of global optimization problems these parameters stay almost the same.

The TT-docking iteratively performs the following steps:

1. Generation of submatrices of unfolding matrices using sets of tensor elements.
2. Interpolation of submatrices using TT-Cross method with the rank $\leq$ r_max.
3. A set of interpolation points for each submatrix contains elements with large values in modulus.
4. Rough local optimization of interpolation points (protein-ligand poses) by the simplex method, addition of optimized point projections to the tensor and to the interpolation point sets.
5. Updating of each set of interpolation points of the unfolding matrix by merging the interpolation points of the previous unfolding matrix and ones of the subsequent unfolding matrix.
6. Addition of the best points (protein-ligand poses) to the interpolation point set of the unfolding matrix, and transition to step 1 using the obtained point set as the tensor elements.

The complexity of the TT global optimization method is $O(dnr_{max}^2)$ functional evaluations, $O(dr_{max})$ local optimizations and $O(dnr_{max}^3)$ arithmetic operations.

2.2 SOL-P Docking Program

The parallel SOL-P docking program is constructed on the base of the TT-docking algorithm (see above). The SOL-P program is developed for finding the low energy local minima spectrum of protein-ligand complexes, proteins or ligands including the respective global energy minimum. The energy of each molecule conformation is calculated directly in the frame of the MMFF94 force field [14] in vacuum without any simplification or fitting parameters. The conformation space of the system coordinates is formed by translations and rotations of the ligand as a whole, by the ligand torsions and also by Cartesian coordinates of the selected target protein atoms. The parallel MPI (message passing interface) based SOL-P program is written on C++ with usage of BLAS and LAPACK libraries. Main SOL-P parameters are: the maximal rank r_{max} of the TT-Cross approximation method, the power m of the discretization degree of the search space (there are $n = 2^m$ nodes along one dimension) and the number of iterations of the TT global optimization algorithm.

As it is mentioned in the previous section there is a rough local energy optimization in the TT-docking algorithm by the Nelder-Mead simplex method [24] within the Subplex algorithm [25] implemented as Sbplx program in NLOpt library [26].

2.3 Moveable Atoms

In the present consideration a protein atom is moveable when it is close to at least one of reference ligand poses. The protein atom is close to a ligand pose when the distance between this protein atom and at least one ligand atom is less than a given threshold. In the present work we took three ligand poses as reference ones: the ligand pose corresponding to the global protein-ligand energy minimum found by the FLM program [8] for the rigid protein, the locally optimized native ligand pose and the nonoptimized native ligand pose. Such choice of the reference ligand poses is taken here only for the uniformity of the consideration of all different proteins and ligands of the test set. Determination of moveable protein atoms is carried out by the specially written our original program Mark-PMA (Mark Protein Moveable Atoms) with the MLT (Moveable Layer Thickness) parameter defining the threshold distance. The MLT parameter is taken up to 3 Å in the present investigation.

2.4 Docking Procedure

The molecular data of the ligand and the protein with the marked moveable atoms are the input of the SOL-P program (shown in I stage in Fig. 1). The SOL-P program uses a cube centered in the geometrical center of the native ligand position in the crystallized protein-ligand complex as the spatial region for the low-energy minima search: all found ligand positions have their geometrical centers inside this cube (the docking cube). Each of moveable protein atoms can move inside its own small cube centered in the initial atom position taken from the crystallized protein-ligand complex. In this work we set the docking cube edge equal to 10 Å and the small cube edge equal to 1 Å. The SOL-P program performs MPI-parallelized search for the low-energy minima of protein-ligand complexes by TT-docking algorithm containing the rough local optimization by the simplex method. The ligand has six rotational-translational degrees of freedom as a whole rigid body plus torsional degrees of freedom for each single non-cyclic bond; each of the protein moveable atoms has three degrees of freedom – its Cartesian coordinates. Data about all found low-energy minima including protein-ligand configurations is too large to be saved in the molecular data format. These configurations are saved as the binary data (shown in Fig. 1 as "Binary data of all non-optimized minima").

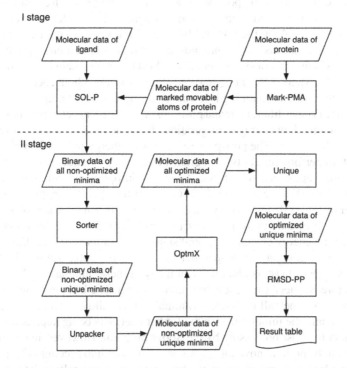

Fig. 1. Flowgraph of the program complex for low energy local minima search with flexible ligand and moveable target protein atoms. I stage: the data preparation and TT global energy minima search with the SOL-P program. II stage: the analysis of binary data with the "non-optimized minima" obtained from the SOL-P program and preparation of the table with the results and the final minima set.

2.5   Analysis of Local Minima

Docking of a flexible ligand into the target protein with moveable protein atoms differs strongly for docking into the rigid protein. In the former case we obtain after docking much larger volume of information about low energy minima than in the latter case. Different minima found in docking with moveable protein atoms are described by different protein-ligand conformations containing different ligand poses as well as different protein conformations. When docking is performed with a large number of degrees of freedom, e.g. with a flexible ligand and moveable protein atoms, the local energy optimization is too laborious and it is performed in the TT-docking not very precisely by the simplex method. All these peculiarities of the docking with a large number of degrees of freedom lead to importance of post-docking processing (post-processing): elimination of equivalent minima, more accurate local energy optimization and elimination of equivalent minima again. All these operations are performed at the stage II (see Fig. 1).

At the stage II in Fig. 1 the post-processing of low energy configurations stored in the binary data is performed with the Sorter program. The Sorter program sorts the "nonoptimized minima" by their MMFF94 energies in vacuum and excludes minima with equal ligand positions – only one minimum with the lowest energy is being kept. Two ligand positions are considered equal if RMSD between them is less than a given threshold (0.1 Å), where RMSD is calculated atom-to-atom without chemical symmetry accounting. Thus, all the remaining low-energy configurations ("unique non-optimized minima" in Fig. 1) have different ligand positions. Then, the Unpacker program performs exporting all unique low-energy configurations from the binary file to the file with the MOL2 molecular format. The post-processing of low energy protein-ligand configurations consists of performance of two programs: OptmX and Unique (Fig. 1). The OptmX program locally optimizes all of the "unique non-optimized minima". For these purposes, the OptmX program uses L-BFGS algorithm [27, 28] applied to the local optimization of the MMFF94 energy function in vacuum with variations of Cartesian coordinates of all ligand atoms and moveable protein atoms. Each local optimization stopped when the energy change at several steps was less than 10^{-8} kcal/mol. Optimization of different minima is MPI-parallelized. After this optimization, the "all optimized minima" (Fig. 1) set is obtained. But many of these minima may become equal again. Therefore, we need to re-exclude similar minima. The Unique program excludes equal minima from the "all optimized minima" set as follows. Among several equal configurations only the minimum with the lowest energy is being kept as it is made in the binary data file post-processing by the Sorter program. However, in contrast to the Sorter program the protein moveable atoms are also taken into account in RMSD calculation, and the RMSD is calculated with chemical symmetry analysis. The decrease of the number of minima at the post-processing stage can be very large comparing with the number of minima found at the docking stage. For example, after the processing with the Sorter program there are 30365 and 28166 minima for the protein-ligand complexes with PDB ID 1MRW (with 30 moveable atoms) and 5BT3 (with 27 moveable atoms), respectively; however, after the precise local energy optimization with the OPTM-X program and filtering the obtained minima with the Unique program the numbers of different local energy minima decrease down to 7580 and 5891 minima for 1MRW and 5BT3, respectively.

Analysis of the local minima remaining after post-processing is carried out by the RMSD-PP program which calculates RMSD (with respect to all ligand atoms) between the ligand pose in a certain energy minimum of the protein-ligand complex and the ligand pose in the energy minimum corresponding to the native ligand position obtained after the local optimization from its configuration in the crystallized complex. The RMSD here is calculated taking into account the approximate chemical symmetry analysis [13] and it is a good metric to estimate geometrical difference between two configurations of a protein-ligand complex; it can correctly discard geometrical pseudo-differences such as phenyl residue flip, comparing to the native atom-to-atom RMSD calculation.

As a result the RMSD-PP program creates in its output (Fig. 1) the resulting table containing: the minimum index, the minimum energy, RMSD from the optimized native configuration and the distance from the ligand geometric center in the given minimum to the ligand geometric center in the optimized native configuration. The energy minima are sorted by their energy in the ascending order; that is, every minimum gets its own index equal to its number in this sorted list of minima. The lowest energy minimum has the index equal to 1.

Some minima from the list might be close in space to the optimized native ligand position. We designate the index of the minimum having RMSD from the optimized native ligand position less than 2 Å as "Index of the minimum Near Optimized Native" or "INON." If there are several such minima which are close to the optimized native ligand position, we will choose the minimum with the lowest energy (with the lowest index) as "INON". When INON = 1 the docking paradigm is satisfied: the global minimum of the protein-ligand energy is near the native configuration. If there are no minima with the ligand pose near the optimized native configuration among all minima found by the SOL-P program, we use notation INON = inf.

In the present consideration we compare the energy minima found by the SOL-P program with ones obtained by the FLM program [7] with the same target function – energy in the frame of the MMFF94 force field in vacuum, for the same test set of 30 protein-ligand complexes [8] which are taken from the Protein Data Bank [29].

2.6   Parallel Performance of SOL-P

In docking problem (and many others) the evaluation of any tensor element has almost the same complexity. So, the parallelization is considered for this case. The parallel implementation of the matrix cross method is available [30] for such case. However, matrices which are used by TT global optimization strategy are relatively small, especially in the case when the additional artificial tensorisation is used. So, this parallel resource is very limited.

But for the TT global optimization and even for TT-cross approximation methods submatrices of different unfolding matrices are not necessary to be considered consequently and rows or columns used for the approximation are not necessary to be nested. In the case of the approximation this will lead to some additional (but independent for all unfolding matrices with the single communication between unfoldings number k and $k + 1$ prior) computations for the construction the tensor approximation. In the global optimization strategy the approximation of the tensor is not constructed explicitly, so such additional computations are not needed.

To balance computations, all submatrices of unfolding matrices are selected of the same size (maximal amongst all original submatrices) and the approximation is performed till the same rank. As a positive side effect this leads to faster convergence and better robustness. Moreover, in the case of the global optimization, especially when the additional tensorisation is used, original sizes of these submatrices are very close to each other due to the rank limitation and to the equal size of each dimension.

Finally, in the parallel algorithm such operations are done for every unfolding submatrix independently at every iteration:

1. From the set of tensor elements P_k, which are obtained on the previous iteration, construct the set of unfolding matrix columns and rows. Rows are constructed in the same way as in the TT-cross procedure in the normal order, columns – as in the reversed order. The number of rows and columns will be approximately n times larger than the number of elements in P_k.

2. Using random rows and columns extend their number to $7nr_{max}$ each.

3. Perform the parallel matrix cross interpolation of the unfolding matrix submatrix of the order $7nr_{max}$ with the rank bounded by r_{max}.

4. Perform in parallel a small number of local optimization steps for obtained pivots and project them back to tensor elements.

5. Every unfolding has approximately $2r_{max}$ tensor elements now – matrix cross pivots and elements obtained by the projection of locally optimized points. Send positions of these elements to unfoldings number $k - 1$ and $k + 1$ and receive positions from them. Also, by the parallel reduction find r_{max} points with the best functional values amongst all unfoldings. After these operations each unfolding has about $7r_{max}$ elements which are used at the next iteration.

Note, that only steps 3 and 4 have high computational cost.

Parallel efficiency of such algorithm is highly dependent on the number of unfoldings (denote it as K) and the number of processors (p). If p is less than K then the parallel efficiency is the best when K is divisible by p. In the other case, the parallel efficiency is higher when p is divisible by K and it is even more better when additionally Kr_{max} or $7Knr_{max}$ is divisible by p. The maximal number of processors which is reasonable to use is $7Knr_{max}$.

The multi-processor performance of SOL-P is investigated at the Lomonosov-2 supercomputer [31]. The results of SOL-P performance for the first set of TT-docking parameters ($r_{max} = 4$, $n = 2^{16}$ and the number of iterations equal to 15) are demonstrated in Fig. 2 for the 1SQO complex with 6 protein moveable atoms, the ligand consisting of 34 atoms (4 torsions) and with different numbers of cores. We see the non-monotonic behavior of the parallel efficiency $(T_{14} \times 14)/(T_N \times N)$ on the number of cores which are used for the calculations (at the nodes containing 14 cores per node). The detailed plots are different for different complexes but their general behavior is the same. For example, for the 1SQO complex $K = 435$ unfolding matrices were considered for constructed tensor. It is easy to see, that efficiency is maximal when the number of used cores p is close to the number which is divisible by K (for numbers of cores larger than K) or when the fractional part of K/p is close to zero.

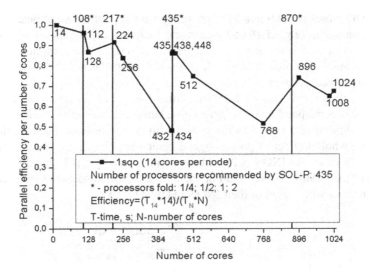

Fig. 2. Parallel efficiency of the SOL-P program for different numbers of core of the Lomonosov-2 supercomputer [31] for the 1SQO test complex with 6 protein moveable atoms, the ligand consisting of 34 atoms (4 torsions).

3   Results

Performance of SOL-P is investigated for different values of the maximal rank $r_{max} = \{4, 8, 16\}$, the initial grid size $n = \{2^8, 2^{12}, 2^{16}\}$ and the number of iterations. Results of this testing demonstrate that for the higher initial grid size even the lowest tested maximal rank $r_{max} = 4$ is enough to find the optimum reliably and precisely. However, the increase of the initial grid size leads to slower convergence of the method and the iteration number must be larger (for $n = 2^{16}$ from 10 to 15 iterations are needed). The high grid size $n = 2^{16}$ for ranks 8 and 16 makes computations significantly slower, thus the initial grid size of $n = 2^{12}$ is used for ranks 8 and 16. For such initial grid size the computation time is reduced by 1.5 times and the number of iterations decreases. SOL-P with three parameter sets: $\{r_{max} = 16, n = 2^{12}\}$, $\{r_{max} = 8, n = 2^{12}\}$ and $\{r_{max} = 4, n = 2^{16}\}$ demonstrates similar ability to find the global energy minimum near the optimized native ligand position for several test complexes but the fastest performance is observed for $\{r_{max} = 4, n = 2^{16}\}$, and we choose the latter set with 15 iterations as the optimal parameters for the present investigation.

It is found that for some complexes (e.g. 1SQO: 4 ligand internal torsions and 34 ligand atoms) the docking paradigm is satisfied for the rigid protein as well as for the protein up to 35 moveable atoms. For some complexes (e.g. 3CEN: 7 torsions and 50 ligand atoms) the docking paradigm is satisfied only for sufficiently large number (13, 26, 48) of protein moveable atoms when INON is equal to 1 or 2. SOL-P finds the global energy minimum for this complex when 48 protein atoms are moveable and the dimensionality of the energy surface is equal to 157 = 144 (protein) + 13 (ligand). For such docking SOL-P uses about 9 h at 512 core of the Lomonosov supercomputer [31, 32]. For other complexes (e.g. 4FT9: 5 torsions and 32 ligand atoms) the MMFF94 force field energy in vacuum is not adequate and the energy surface is so complicated that for the too large number of protein moveable atoms (42) SOL-P is not able to find minima near the native configuration.

The validation shows that SOL-P finds either the global minimum or one of low energy minima corresponding to the ligand pose being near the optimized native ligand pose for the rigid protein and/or for the protein with moveable atoms for more than two thirds of the whole test set of protein-ligand complexes (for 22 out 30) for these 22 complexes INON = 1 or INON ≤ 25 and the docking paradigm is fulfilled for them in the frame of the MMFF94 force field in vacuum. The test complexes are collected in groups in respect with values of their INON index in Fig. 3.

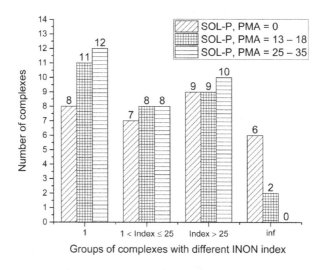

Fig. 3. Numbers of complexes with different values of INON index. PMA indicates the range of protein moveable atoms for the SOL-P program. INON is the index of the minimum having RMSD from the optimized native ligand position less than 2 Å; if there are several such minima, the minimum with the lowest energy (with the lowest index) should be taken.

Protein atoms mobility is crucial for 4 complexes (1J01, 1K1J, 1MQ6 and 3CEN) out of 30 ones: SOL-P does not find any minima near the optimized native ligand pose for docking into the rigid protein (INON = inf) but, when mobility of protein atoms is taken into account, docking finds near the optimized native ligand pose either the global minimum (INON = 1) or one of the lowest energy minima (INON ≤ 25). On the other hand, for rigid proteins SOL-P and FLM cannot find such minima (INON = inf) for 6 and 5 complexes, respectively. It is worth to note that SOL-P is able to find the minimum near the optimized native ligand pose for all 5 complexes where FLM is not able to do this. In tote, we can say that SOL-P, with and without protein moveable atoms, works not worse than the FLM program and much faster than the latter.

Our observation that neither SOL-P nor FLM can find any minimum near the optimized native ligand pose for 11 complexes (out of 30) is connected with inadequacy of the energy target function calculated in the frame of the MMFF94 force field in vacuum. It has been previously demonstrated [7] that protein-ligand energy calculation in the frame of the MMFF94 force field in solvent (with an implicit model) improves docking performance of the FLM program for the rigid proteins and with such target energy function SOL-P should also work better.

4   Conclusions

The novel algorithm is realized in the supercomputer SOL-P docking program where protein and ligand atoms mobility is taken into account simultaneously and equally. Energies of low-energy minima found in the docking procedure and respective ligand poses are carefully analyzed.

It is shown that the program is able to perform docking of a flexible ligand into the active site of the target protein taking mobility of assigned protein atoms into account: up to 157 degrees of freedom in the conformation space using about 9 h at 512 cores of the Lomonosov supercomputer [31, 32]. This is the first time when the docking program is able to perform successfully the global energy minimum search in the conformational space with such a large dimensionality. This result is achieved due to the novel docking algorithms (TT-docking) which is based on the so-called tensor train decomposition of multi-dimensional arrays and the TT global optimization method [8, 11].

The SOL-P docking performance is comparable with one of the FLM program [7] which executes the massive parallel local energy minima for rigid target proteins due to employment of much larger computing resources.

It is demonstrated that the docking paradigm is fulfilled for the target energy function calculated in the frame of the MMFF94 force field in vacuum for a flexible ligand and for a target proteins with 25–35 moveable atoms for two thirds of the whole test set of protein-ligand complexes. Interaction with solvent should increase this number. It is demonstrated that in some cases docking results are being improved even when small movements of protein atoms is taken into account in the docking procedure.

The present investigations became possible due to computing resources of M.V. Lomonosov Moscow State University supercomputer Lomonosov [32].

Acknowledgements. The work was financially supported by the Russian Science Foundation, Agreement no. 15-11-00025.

References

1. Sliwoski, G., Kothiwale, S., Meiler, J., Lowe, E.W.: Computational Methods in Drug Discovery. Pharmacol. Rev. **66**, 334–395 (2014). https://doi.org/10.1124/pr.112.007336
2. Sadovnichii, V.A., Sulimov, V.B.: Supercomputing technologies in medicine. In: Voevodin, V.V., Sadovnichii, V.A., Savin, G.I. (eds.) Supercomputing Technologies in Science, Education, and Industry, pp. 16–23. Moscow University Publishing (2009). (in Russian)

3. Mobley, D.L., Dill, K.A.: Binding of small-molecule ligands to proteins: "what you see" is not always "what you get". Structure **17**(4), 489–498 (2009). https://doi.org/10.1016/j.str.2009.02.010

4. Sulimov, A.V., Kutov, D.C., Oferkin, I.V., Katkova, E.V., Sulimov, V.B.: Application of the docking program SOL for CSAR benchmark. J. Chem. Inf. Model. **53**(8), 1946–1956 (2013). https://doi.org/10.1021/ci400094h

5. Antunes, D.A., Devaurs, D., Kavraki, L.E.: Understanding the challenges of protein flexibility in drug design. Expert Opin. Drug Discov. **10**(12), 1301–1313 (2015). https://doi.org/10.1517/17460441.2015.1094458

6. Chen, W., Gilson, M.K., Webb, S.P., Potter, M.J.: Modeling protein-ligand binding by mining minima. J. Chem. Theor. Comput. **6**(11), 3540–3557 (2010)

7. Oferkin, I.V., Katkova, E.V., Sulimov, A.V., Kutov, D.C., Sobolev, S.I., Voevodin, V.V., Sulimov, V.B.: Evaluation of docking target functions by the comprehensive investigation of protein-ligand energy minima. Adv. Bioinf. **2015**, 12 (2015). https://doi.org/10.1155/2015/126858. Article ID 126858

8. Oferkin, I.V., Zheltkov, D.A., Tyrtyshnikov, E.E., Sulimov, A.V., Kutov, D.C.: Evaluation of the docking algorithm based on tensor train global optimization. Bull. South Ural State Univ. Ser. Math. Model. Program. Comput. Softw. **8**(4), 83–99 (2015). https://doi.org/10.14529/mmp150407

9. Sulimov, A.V., Kutov, D.C., Katkova, E.V., Sulimov, V.B.: Combined docking with classical force field and quantum chemical semiempirical method PM7. Adv. Bioinf. **2017**, 6 (2017). https://doi.org/10.1155/2017/7167691. Article ID 7167691

10. Pecina, A., Meier, R., Fanfrlík, J., Lepšík, M., Řezáč, J., Hobza, P., Baldauf, C.: The SQM/COSMO filter: reliable native pose identification based on the quantum-mechanical description of protein-ligand interactions and implicit COSMO solvation. Chem. Commun. **52**, 3312–3315 (2016)

11. Zheltkov, D.A., Oferkin, I.V., Katkova, E.V., Sulimov, A.V., Sulimov, V.B.: TTDock: docking method based on tensor train. Vychislitelnie metody i programmirovanie (Numer. Meth. Program.) **14**, 279–291 (2013). (in Russian). http://num-meth.srcc.msu.ru/english/zhurnal/tom_2013/v14r131.html. Accessed 10 April 2017

12. Sulimov, A., Zheltkov, D., Oferkin, I., Kutov, D., Tyrtyshnikov, E.: Novel gridless program SOL-P for flexible ligand docking with moveable protein atoms. In: 21st EuroQSAR Where Molecular Simulations Meet Drug Discovery. Aptuit Conference Center, Verona Italy, Abstract book, OC15, p. 52, 4–8 September 2016. www.euroqsar2016.org

13. Sulimov, A.V., Zheltkov, D.A., Oferkin, I.V., Kutov, D.C., Katkova, E.V., Tyrtyshnikov, E.E., Sulimov, V.B.: Evaluation of the novel algorithm of flexible ligand docking with moveable target protein atoms. Comput. Struct. Biotechnol. J. **15**, 275–285 (2017). https://doi.org/10.1016/j.csbj.2017.02.004

14. Halgren, T.A.: Merck molecular force field. I. basis, form, scope, parameterization and performance of MMFF94. J. Comput. Chem. **17**, 490–519 (1996)

15. Sinauridze, E.I., Romanov, A.N., Gribkova, I.V., Kondakova, O.A., Surov, S.S.: New synthetic thrombin inhibitors: molecular design and experimental verification. PLoS ONE **6**(5), e19969 (2011). https://doi.org/10.1371/journal.pone.0019969

16. Oseledets, I.V., Tyrtyshnikov, E.E.: Breaking the curse of dimensionality, or how to use SVD in many dimensions. SIAM J. Sci. Comput. **31**(5), 3744–3759 (2009). https://doi.org/10.1137/090748330

17. Oseledets, I.V.: Tensor-train decomposition. SIAM J. Sci. Comput. **33**(5), 2295–2317 (2011). https://doi.org/10.1137/090752286

18. Oseledets, I.V., Tyrtyshnikov, E.E.: TT-Cross approximation for multidimensional arrays. Linear Algebra Appl. **432**(1), 70–88 (2010). https://doi.org/10.1016/j.laa.2009.07.024
19. Goreinov, S.A., Tyrtyshnikov, E.E., Zamarashkin, N.L.: Pseudo-skeleton approximations of matrices. Rep. Russ. Acad. Sci. **342**(2), 151–152 (1995). https://doi.org/10.1016/S0024-3795(96)00301-1
20. Goreinov, S.A., Tyrtyshnikov, E.E., Zamarashkin, N.L.: A theory of pseudo-skeleton approximations. Linear Algebra Appl. **261**, 1–21 (1997). https://doi.org/10.1016/S0024-3795(96)00301-1
21. Tyrtyshnikov, E.E.: Incomplete cross approximation in the mosaic-skeleton method. Computing **64**(4), 367–380 (2000). https://doi.org/10.1007/s006070070031
22. Goreinov, S.A., Tyrtyshnikov, E.E.: The maximal-volume concept in approximation by low-rank matrices. Contemp. Math. **208**, 47–51 (2001)
23. Goreinov, S.A., Oseledets, I.V., Savostyanov, D.V., Tyrtyshnikov, E.E., Zamarashkin, N.L.: How to find a good submatrix. Research Report 8-10, ICM HKBU, Kowloon Tong, Hong Kong (2008). https://doi.org/10.1142/9789812836021_0015
24. Nelder, J.A., Mead, R.: A simplex method for function minimization. Comput. J. **7**, 308–313 (1965)
25. Rowan, T.: Functional stability analysis of numerical algorithms. Ph.D. thesis, Department of Computer Sciences, University of Texas at Austin (1990)
26. Steven, G.J.: The NLopt nonlinear-optimization package. http://ab-initio.mit.edu/nlopt
27. Byrd, R.H., Lu, P., Nocedal, J., Zhu, C.: A limited memory algorithm for bound constrained optimization. SIAM J. Sci. Comput. **16**(5), 1190–1208 (1995). https://doi.org/10.1137/0916069
28. Zhu, C., Byrd, R.H., Lu, P., Nocedal, J.: Algorithm 778: L-BFGS-B: fortran subroutines for large-scale bound-constrained optimization. ACM Trans. Math. Softw. **23**(4), 550–560 (1997). https://doi.org/10.1145/279232.279236
29. Berman, H.M., Westbrook, J., Feng, Z.: The protein data bank. Nucleic Acids Res. **28**(1), 235–242 (2000). http://www.rcsb.org/pdb/home/home.do
30. Zheltkov, D.A., Tyrtyshnikov, E.E.: Parallel Implementation of Matrix Cross Method. Vychislitelnye metody i programmirovanie (Numer. Meth. Program.) **16**, 369–375 (2015). (in Russian)
31. MSU Supercomputers: Lomonosov-2. http://hpc.msu.ru/?q=node/159. Accessed 30 May 2017
32. Sadovnichy, V.A., Tikhonravov, A.V., Voevodin, V.V., Opanasenko, V.: "Lomonosov": supercomputing at Moscow State University. In: Contemporary High Performance Computing: From Petascale toward Exascale, pp. 283–307. CRC Press (2013)

On the Parallel Least Square Approaches in the Krylov Subspaces

V.P. Il'in[1,2(✉)]

[1] Institute of Computational Mathematics and Mathematical Geophysics RAS,
6, pr. Lavrentieva, 630090 Novosibirsk, Russia
ilin@sscc.ru
[2] Novosibirsk State University, Novosibirsk, Russia

Abstract. We consider different parallel versions of the least squares methods in the Krylov subspaces which are based on computing various basis vectors. These algorithms are used for solving very large real, non-symmetric, in gerenal, sparse systems of linear algebraic equations (SLAEs) which arise in grid approximations of multi-dimensional boundary value problems. In particular, the Chebyshev acceleration approach, steepest descent and minimal residual, conjugate gradient and conjugate residual are applied as preliminary iterative processes. The resulting minimization of residuals is provided by the block, or implicit, orthogonalization procedures. The properties of the Krylov approaches proposed are analysed in the "pure form", i.e. without preconditioning. The main criteria of parallelezation are estimated. The convergence rate and stability of the algorithms are demonstated on the results of numerical experiments for the model SLAEs which present the exponential fitting approximation of diffusion-convection equations on the meshes with various steps and with different coefficients.

Keywords: Large sparse systems of linear algebraic equations · Non-symmetric matrices · Block implicit least squares methods · Krylov subspaces · Parallel technologies · Numerical experiments

1 Introduction

The mathematical modeling in real extremal interdisciplinary problems includes the solution of the multi-dimensional direct and inverse tasks, linear and non-linear, stationary and non-stationary, which are approximated by various order numerical schemes on the non-structured grids in the complicated computational domains. In any case, at a low level of these procedures, the multi-fold solution to the systems of linear algebraic equations (SLAEs) is required. The practical high resolution demands very large degrees of freedom (dof). So, the solution of the corresponding ill-conditioned SLAEs is the bottle-neck of the general numerical process, because necessary computational resources grow nonlinearly at this stage if the dimension of the system increases (for example, 10^{10} and higher).

© Springer International Publishing AG 2017
V. Voevodin and S. Sobolev (Eds.): RuSCDays 2017, CCIS 793, pp. 168–180, 2017.
https://doi.org/10.1007/978-3-319-71255-0_13

In this case the road map to provide a high performance consists in parallel implementation of modern multi-preconditioned iterative processes in the Krylov subspaces based on the domain decomposition methods (DDM) (see [1,2] and the references therein). The main achivements are based on the combination of efficient mathematical discoveries and scalable parallel technologies on the multi-processor systems (MPS) with distributed and hierarchical shared memory.

This paper deals with just one particular side of the general problem. Namely, we consider the possibility of parallel "implicit" construction of the iterative methods in the Krylov subspace "in the pure form", i.e. without preconditioning, which is supposed to be a separate problem.

Let us consider the solution of the SLAE

$$Au = \left\{ \sum_{l' \in \omega_l} a_{l,l'} u_{l'} \right\} = f, \quad A = \{a_{l,l'}\} \in \mathcal{R}^{N,N},$$
$$u = \{u_l\}, \quad f = \{f_l\} \in \mathcal{R}^N \tag{1}$$

with a large real sparse matrix resulting from grid approximations of multi-dimensional boundary value problems by finite element, finite volume, or other methods. In general, this matrix is non-symmetric and ill-conditioned. In Eq. (1), ω_l denotes a set of indices of nonzero entries in the ℓ-th row of the matrix A, whose number N_ℓ is assumed to be much smaller than N. The algorithms considered below can easily be extended to the case of complex SLAEs.

In [3], the authors have offered special procedures for accelerating the convergence of the Jacobi method as an "efficient alternative" to the classical Krylov methods. In order to solve a linear system, they have used the Anderson acceleration, which had been originally proposed in [4] for solving systems of nonlinear algebraic equations, A comparative experimental analysis presented in [3] has demonstrated a considerable superiority of the original alternating Anderson-Jacobi (AAJ) method over the popular generalized minimal residual method (GMRES) as concerns the solution time. The idea of the AAJ method consists in periodical (after a prescribed number of stationary iterations) use of an acceleration method based on solving an auxilary least squares problem not involving successive orthogonalization of the direction vectors, which is typical of the Krylov variational type methods.

The present paper aims at generalization and experimental study of the similar approaches. We apply several non-stationary iterative algorithms as a preliminary tool for constructing some basis vectors in the Krylov subspaces and further minimization of the residual vector norm by means of the least squares method. In this context, parallel implementation of the approaches proposed is considered.

This paper is organized as follows. In Sect. 2, we present the idea of implicit, or block, least squares method in the Krylov subspaces which uses a preliminary construction of the basis vectors. Section 3 is devoted to analyzing the efficiency of parallel versions of the iterative algorithms considered in comparison with the classical variational method of semi-conjugate residuals in the Krylov subspaces. Section 4 discusses the results of numerical experiments obtained for the

algorithms offered on a series of the test SLAEs, resulting from the grid approximation of two-dimensional boundary value problems for the convection-diffusion
equation. In conclusion, we observe the efficiency of the algorithms presented and
discuss some plans for future studies.

2   Versions of the Least Squares Methods in the Krylov Subspaces

The wide class of iterative processes for solving SLAE (1) can be written in the
form

$$
\begin{aligned}
u^{n+1} &= u^n + \alpha_n p^n = u^0 + \alpha_0 p^0 + \dots + \alpha_n p^n, \\
r^{n+1} &= r^n - \alpha_n A p^n = r^0 + \alpha_0 A p^0 + \dots + \alpha_n A p^n.
\end{aligned}
\tag{2}
$$

Here u^0 and $r^0 = f - Au^0$ are the initial guess and the corresponding residual
vector, and p^n, α^n are some direction vectors (usually $p^0 = r^0$) and the iterative
parameters which are defined from the additional conditions in the different
approaches.

If A is a symmetric positive definite (spd) matrix, then the following conjugate direction (CD) methods [1,5]:

$$
\begin{aligned}
p^{n+1} &= r^{n+1} + \beta_n^{(s)} p^n, \\
\alpha_n^{(s)} &= \frac{(A^s r^n, r^n)}{(Ap^n, A^s p^n)}, \quad \beta_n^{(s)} = \frac{(A^s r^{n+1}, r^{n+1})}{(A^s r^n, A^s r^n)},
\end{aligned}
\tag{3}
$$

for $s = 0, 1$ present the classical conjugate gradient (CG) and conjugate resid-
ual (CR) algorithms, respectively, which minimize the functionals $\Phi_n^{(s)}(r^0) = (A^{-s} r^{n+1}, r^{n+1})$ in the Krylov subspaces

$$
\mathcal{K}_n(r^0, A) = \text{span}\,(r^0, Ar^0, \dots, A^n r^0).
\tag{4}
$$

The residual and direction vectors in these approaches for all k, n satisfy the
orthogonal properties

$$
(A^s r^k, r^n) = (A^s r^n, r^n)\delta_{k,n}, \quad (A^s p^k, Ap^n) = (A^s p^n, Ap^n)\delta_{k,n}
\tag{5}
$$

where $\delta_{k,n}$ is the Kronecker symbol.

However, if A is a non-symmetric matrix, then these methods have no such
variational and orthogonal properties. In such cases, the global minimization of
the functionals $\Phi_n^{(s)}$ is provided by the general minimized residual type (GMRES)
approaches or by the equivalent, in some sense, semi-conjugate direction (SCD)
methods [6]

$$
p^{n+1} = r^{n+1} - \sum_{k=0}^{n} \beta_{n,k}^{(s)} p^k, \quad \beta_{n,k}^{(s)} = (Ap^k, A^s r^{n+1})/(Ap^n, A^s p^n).
\tag{6}
$$

Let us remark that the formulas (6) realize the orthogonal properties (5) by
Gram–Schmidt procedure. It fact, this procedure should be changed by more

stable modified Gram–Schmidt (MGS) orthogonalization [7]. If $\alpha_n^{(s)}$ are defined by (3) then for $s = 0, 1$ from (6) we provide the extremum conditions

$$\partial \Phi_n^{(s)} / \partial \alpha_n = 0, \quad \Phi_n^{(s)} = (r^{n+1}, A^{s-1} r^{n+1}), \tag{7}$$

and for $s = 1$ the functional $\Phi_n^{(s)}$ has the minimum in the Krylov subspace (4).

In this case the resulting residual vectors are not conjugate, but semi-conjugate only, i.e.,

$$(A^s r^n, r^k) = \begin{cases} 0, & k < n, \\ \sigma^n, & k = n, \end{cases}$$

and for $s = 0, 1$ we have a semi-conjugate gradient and a semi-conjugate residual (SCG and SCR) methods, respectively.

Let us remark, that for spd - matrix A, the CD methods (both CG and CR), as well as SCD approaches (SCG and SCR) have the same theoretical number of iterations, see [1,5]:

$$n(\varepsilon) \approx 0.5 |\ell n(\varepsilon/2)| (cond A)^{-1/2},$$

where $cond A$ is the condition number of A and $\Phi_n^{(s)} \le \varepsilon^2 \Phi_{n-1}^{(s)}, \; 0 < \varepsilon \ll 1$. But if A is non-symmetric, the same estimate is valid for SCD but not for CD methods.

In the general case, to compute the vectors u^n and r^n using (2)–(6), it is necessary to store all the vectors $p^n, p^{n-1}, ..., p^0$ and $Ap^n, Ap^{n-1}, ..., Ap^0$. In practice, these methods are realized with periodic restarts every m iteration. This means that the residual vector is computed from the original equation

$$r^{ml} = f - Au^{ml}, \quad \ell = 0, 1, ..., \tag{8}$$

rather than using (2), and the subsequent approximations are computed "from the beginning", i.e., for $n > m$ one should change n for $n = ml$ in the formulas. Here, it is necessary to store only the last $m + 1$ vectors $p^n, p^{n-1}, ..., p^{n-m}$, and $Ap^n, Ap^{n-1}, ..., Ap^{n-m}$. The restarted versions of SCD methods, similar to restarted GMRES, have lower convergence rate, but this is the cost for the memory saved.

The most expensive stage of the SCD methods consists in successive computations of the direction vectors p^{n+1} by means of long recursions (6). In accord with the Anderson acceleration approach, we can simplify (6) and use in the sum the last direction vector p^n only (but save the vectors $p^n, ..., p^{n-m}$ and $Ap^n, ..., Ap^{n-m}$). In these cases, the minimization of the residual norm $||r^{n+1}||_2 = (r^{n+1}, r^{n+1})^{1/2}$ in the Krylov subspace

$$K_{n,m}(r^n, A) = \text{span } (r^n, Ar^n, ..., A^m r^n) \tag{9}$$

can be provided by the following least squares method:

$$r^{n+m} = r^n - W_{n,m} \bar{\gamma}_{n,m} \approx 0, \quad W_{n,m} = (w_n w_{n+1} ... w_{n+m}) \in \mathcal{R}^{N,m+1}, \tag{10}$$
$$w_{n+k} = A^k p^n, \quad \bar{\gamma}_{n,m} = (\gamma_n, \gamma_{n+1}, ..., \gamma_{n+m})^T \in \mathcal{R}^{m+1}.$$

The coefficient vector $\bar{\gamma}_{n,m}$ can be computed from the over-determined SLAE

$$W_{n,m}\bar{\gamma}_{n,m} = r^n, \tag{11}$$

which can be solved, for example, by means of the singular value decomposition (SVD) or an other approach (see [7]). In particular, the left-hand Gauss transformation procedure

$$B_{n,m}\bar{\gamma}_{n,m} = g_{n,m}, \ B_{n,m} = W_{n,m}^T W_{n,m} \in \mathcal{R}^{m+1,m+1}, \ g_{n,m} = W_{n,m}r^n \in \mathcal{R}^{m+1} \tag{12}$$

can be here efficiently applied.

In fact, the computing vectors p^k, Ap^k in such algorithms can be realized by formulas (2), (3), and we call them CD-LSM-ℓ (CG-LSM-ℓ and CR-LSM-ℓ for $s = 0, 1$, respectively) where the integer $\ell = 1, 2$ corresponds to application of formulas (11) or (12).

If the coefficient vector $\bar{\gamma}_{n,m}$ is known, the improved numerical solution can be computed by the formulae

$$u^{n+m} = u^n + \gamma_n p^n + \dots + \gamma_{n+m}p^{n+m}. \tag{13}$$

The considered algorithms can be simplified even to a greater extent if we use instead CG or CR method, the two-terms formulas of the steepest descent (SD) or the minimal residual (MR) method, which can be formaly described (for $s = 0, 1$ respectively) as follows, see [1,5]:

$$\alpha_n^{(s)} = (A^s r^n, r^n)/(Ar^n, A^s r^n), \ \beta^n = 0, \ p^n = r^n. \tag{14}$$

For the spd-matrices, these approaches provide the local variational properties only, i.e. for just one iteration, but minimization of the functional $\Phi_{n,m}^{(s)} = (A^{s-1}r^{n+m}, r^{n+m})$ in the Krylov subspaces $\mathcal{K}_{n,m}(r^n, A)$ can be achieved by the LSM-ℓ approaches (11) or (12). Such methods will be called SD-LSM-ℓ and MR-LSM-ℓ, $\ell = 1, 2$. Of course, for SD and MR methods with local variational properties, the convergence rates of iterations are worse as compared to the previous algorithms ($n(\varepsilon) \sim condA$ only), but let us remind that it is just the way to obtain the basis vector for LSM optimization.

In all the approaches considered above, we use the least squares methods, based on the direction vectors p^n with weak orthogonal, or variational, properties. Instead of this, we can construct the basis vectors by application of the some spectral iterative process. If the matrix A has real positive eigenvalues $\lambda \in [0 < \lambda_1, \lambda_N]$, then the optimal convergence rate of iterations is provided by the Chebyshev acceleration [1,5,8]. Such approaches can be implemented in different forms, and we use the two-terms recurrent representation, which consists of the following relations:

$$\begin{aligned}
p^0 &= r^0 = f - Au^0, \\
u^n &= u^{n-1} + \alpha_{n-1}p^{n-1}, \\
r^n &= r^{n-1} - \alpha_{n-1}Ap^{n-1}, \\
p^n &= r^n + \beta_n p^{n-1}.
\end{aligned} \tag{15}$$

Here we use the restarted procedures which also suppose applying the LSM approaches by (11) or (12) after each m iteration. The coefficients in (15) are defined via three terms description of the Chebyshev acceleration presented in [8]:

$$
\begin{aligned}
&u^1 = u^0 + \tau\, r^0, \quad \tau = 2/(\lambda_1 + \lambda_N), \quad r^n = f - Au^n, \\
&u^{n+1} = u^n + \tau_n \tau\, r^n + (\tau_n - 1)(u^n - u^{n-1}), \quad \tau_0 = 2, \\
&\tau_n = 4(4 - \tau_{n-1}\gamma)^2) - 1, \quad \gamma = (1 - c)/(1 + c), \quad c = \lambda_1/\lambda_N.
\end{aligned}
\tag{16}
$$

The values of α_n, β_n from (15) provide the equivalence to reccurences (16) by the formulas

$$
\alpha_0 = \tau, \quad \alpha_n = \tau_n\tau, \quad \beta_n = (\tau_n - 1)\alpha_{n-1}/\alpha_n.
\tag{17}
$$

After each m iterations by formulas (15)–(17) we can apply the acceleration procedures according to (10)–(13). The corresponding algorithms we will call the Chebyshev least squares methods (CHEB-LSM-1 and CHEB-LSM-2). We conclude this section with the following two remarks. First, it is easy to check that from theoretical viewpoint, LSM-1 and LSM-2 coincide because, in exact arithmetic, by solving Eqs. (11) and (12) one obtains one and the same vector $\bar{\gamma}_{n,m}$. Second, an approach similar to the one considered above was applied by P.L. Montgomery in [9] (see [10] also) in solving special systems of linear algebraic equations over a finite field and was referred to as the block Lanczos method.

3   Properties of Parallel Implementation

As is seen, the implementation of the optimal SCR method includes at each iteration the following main stages:

- one matrix-vector multiplications (MV-operations);
- $2m + 3$ vector-vector (VV) operations, i.e. linear combinations of the vectors;
- computing the $m + 2$ inner vector products.

It is important that all these operations are fulfilled successively. The idea of parallel implementation of the methods proposed with LSM-2 approaches consists in the simultaneous computation of the entries of the matrix

$$
B_{n,m} = \{b_{k,\ell}^{(n,m)} = (w_k, w_\ell); \; k, \ell = n, ..., n - m\}.
$$

And for $m \ll N$, we can neglect the costs for solving SLAEs (12) and compute the vector $\bar{\gamma}_{n,m}$ by formula (13) on the all processor units simultaneously.

Now we compare parallel realizations of a cycle of m iterations in the methods LSM and SCR. This will suffice for a qualitative comparison of the performances of the algorithms in question because they minimize the same functional in the same Krylov subspace and, consequently, are theoretically equivalent with respect to the convergence rate. Concerning the methods considered, we assume that they are applied to a block system of linear equations of the form (1), and the block rows $A_k = \{A_{k,\ell}, \; \ell = 1, .., P\} \in \mathcal{R}_{N_k,N}, \; N_k \cong N/P, \; N_1 + ... + N_p = N$

of the coefficient matrix A are distributed in the memory of the corresponding MPI processes used for the first level of parallelizing the algorithms, as is done in the domain decomposition methods (where every block row corresponds to a subdomain, see [11]). Note that in fact to different MPI processes different computer processors correspond (though this is not formally necessary). In the SCR method, the direction vectors $p^n, p^{n-1}, ..., p^{n-m}$ and also the current vectors u^n and r^n are partitioned into subvectors of lengths N_k, each being stored in the corresponding k-th MPI process. As the iterations proceed, data exchanges among processes are needed, and their volumes should be minimized. When arithmetic operations are performed in the k-th MPI process using a multicore processor, "inner" parallelization (of the second level) can be effected based on multi-thread computations (here, we omit the details). A similar distributed data structure is formed in the least squares methods, in which case the block partition is used for the vectors w_k, $k = 1, ..., m$. We assume that in all the algorithms the standard double-precision computer arithmetics is used. For a comparative analysis of the performances of the methods considered, we estimate the time T_P of performing a cycle of m iterations on P MPI processes based on the following simple model of the computation process:

$$T_P = T_P^a + T_P^c \approx \tau_a N_a + (\tau_0 + \tau_c V_c) N_c. \qquad (18)$$

Here, T_P^a and T_P^c are the times for performing arithmetic and communication operations, respectively;; τ_a is the average time of a single arithmetic operation, and N_a is the number of such operations (for one processor); N_c is the total number of data transmittings; τ_0 is the delay (tuning) time of a single transaction; τ_c is the average time of transmitting a real number, and V_c is the average volume of one package of data transmitted. Note that in view of the relations $\tau_0 \gg \tau_c \gg \tau_a$, it is natural to attempt to minimize not only the total volume of information to be transmitted but also the number of exchanges. This is important not only from the viewpoint of the time of data transmissions but also in view of high energy costs of communication operations.

It is easy to check that in CG-LSM-2 or CR-LSM-2 for $n \neq m$ we need to compute by formulas (2), (3) just 2 inner products and 3 VV-operations. And if we use SD or MR approaches by (2), (14) with local variational properties, then we must perform 2 inner products and 2 vector linear combination, i.e. the difference is not significant as compared with CG or CR methods.

Let us now consider the combination of the Chebyshev acceleration (15)–(17) and the LSM approach. These algorithms do possess orthogonal or variational properties, but have the same optimal estimation of $n(\varepsilon)$. And what is important: the spectral iterations do not need computation of inner products!

The last circumstance is highly valuable in terms of the implementation of the iterative process at the MPS, because these operations obviously need data communications. But this approach demands the knowledge of the spectrum boundaries of the matrix. Of course, this is too strong requirement, but in many practical problems the necessary estimations can be obtained.

It should be remarked that the implementation of the LSM with different preliminary iterative approaches does not need the computation of the vectors u^n, because at the end of any algorithms considered, the resulting vector is realized by (13). Of course, this operation can also be parallelized efficiently.

4   Discussion of Numerical Experiments

Let us consider the Dirichlet problem [8] for the convection-diffusion equation

$$-\frac{\partial^2 u}{\partial x^2} - \frac{\partial^2 u}{\partial y^2} + p\frac{\partial u}{\partial x} + q\frac{\partial u}{\partial y} = f(x, y), \ (x, y) \in \Omega, u|_\Gamma = g(x, y), \quad (19)$$

in a square computational domain $\Omega = (0, 1)^2$ with the boundary Γ and the convection coefficients p, q, which for simplicity are assumed to be constant. This boundary value problem is approximated on a square grid with the step size $h = 1/(L + 1)$ and the total number of interior nodes $N = L^2$,

$$x_i = ih, \ y_j = jh, \ i, j = 0, 1, ..., L + 1, \quad (20)$$

using the five-point finite-volume monotone approximations of exponential type [12]

$$(Au)_l = a_{l,l}u_l + a_{l,l-1}u_{l-1} + a_{l,l+1}u_{l+1} + a_{l,l-L}u_{l-L} + a_{l,l+L}u_{l+L} = f_l, \quad (21)$$

having the second order of accuracy. Here, ℓ is the "global" number of a grid node in the natural node ordering, $\ell = i + (j - 1)L$. Generally speaking, formulas for the coefficients in equations (20) may be different, and we use the following ones:

$$a_{l,l\pm 1} = e^{\pm ph/2}/h, \ a_{l,l\pm L} = e^{\pm qh/2}/h,$$
$$a_{l,l} = a_{l,l-1} + a_{l,l-L} + a_{l,l+1} + a_{l,l+L}. \quad (22)$$

Equations (21) are written for the interior nodes of the grid, but for the near-boundary nodes with the subscripts i = 1, L or j = 1, L the values of the solution on the boundary should be substituted into the system of equations and moved to the right-hand side; here, the corresponding coefficients of the left-hand side can be formally set to zero. In our experiments, we have actually solved the normalized equations, which are obtained by the following transformations with the diagonal matrix $D = \text{diag} \{a_\ell, \ell\}$:

$$D^{-1/2}AD^{-1/2}D^{1/2}u = D^{-1/2}f,$$
$$\bar{A}\bar{u} = \bar{f}, \ \bar{A} = D^{-1/2}AD^{-1/2}, \ \bar{u} = D^{1/2}u, \ \bar{f} = D^{-1/2}f. \quad (23)$$

The numerical experiments have been carried out using the standard double-precision arithmetic for computing the values of the functions $f(x, y) = 0$ and $g(x, y) = 1$ corresponding to the exact solution $u(x, y) = 1$ of problem (19). Since the convergence rate of iterations depends on the initial error $u - u^0$, its influence has been analyzed by comparing the results for the initial guesses $u^0 = 0$ and $u^0 = P_2(x, y) = x^2 + y^2$. The stopping criterion used has been of the from $(r^n, r^n) \leq \varepsilon^2(f, f)$, with $\varepsilon = 10^{-7}$. The computations have been carried out on grids with $N = 7^2, 15^2, 31^2, 63^2$, and 127^2 nodes and for the restart parameter $m = 8, 16, 32, 64$, and 128. In the tables below, we present the results obtained in solving problem (19) with the convection coefficients $p = q = 0$ and $p = q = 4$ on the grids with $N = 7^2, 15^2, 31^2, 63^2, 127^2$ nodes and for different initial guesses. The algorithms applied differ in the method of forming the auxiliary linear system for finding the coefficient vector of correction (to be exact, the systems obtained in LSM-1 and LSM-2 have been solved using the SVD program (the singular value decomposition algorithm) from LAPACK, included into the program library MKL Intel [13]). Let us remark that the matrix $B_{n,m}$ from SLAE (12), which corresponds to LSM-2, has a bigger condition number, as compared to the matrix $W_{n,m}$ from (11). So, LSM-1 is more preferable, from the stability point of view. But in our experiments, the resulting errors are approximately equal as for LSM-1 and LSM-2. So, in the following tables we present the numerical results for LSM-2 only.

The main goal of our experimental research consists not in demonstration of the high performance of algorithms for very large SLAEs, but in study of the stability and convergence rate of LSM approaches with preliminary cheap iterative processes. All the calculations have been carried out on the Siberian Super Computing Center cluster (http://www2.sscc.ru).

In the each cell of the following tables we present two values: the upper is the number of iterations, and the lower is the resulting maximal error $\delta = \max_{i,j}\{|1 - u_{i,j}^n|\}$. In our experiments the results are approximately the same for different initial guesses, and we present data for $u^0 = x^2 + y^2$ only.

In the Tables 1 and 2 we give the results for CHEB-LSM-2 algorithm for symmetric and non-symmetric SLAEs. In both cases the boundaries λ_1, λ_N of matrix spectrum in formylas (16), (17) were taken for $p = q = 0$, but the presented results are close to each other enough. The columns with $m = \infty$ correspond to "pure" Chebyshev acceleration without LSM. It is evident from these tables, that in all cases considered there is an optimal value m.

The Tables 3 and 4 demonstrate the similar results for CR-LSM-2 algorithm. The symmetric case ($p = q = 0$) show that conjugate residual is optimal for such SLAEs, and least squares approach is not reasonable here. But for non-symmetric algebraic systems the application of LSM gives the considerable improvement of the iterative process. Let us remark, that the resulting numbers of iteration and errors δ are approximatelly the same in CR and CHEB.

In the Tables 5 and 6, we present the results for CG-LSM, which confirm that the efficiency of conjugate residual method, in combination with the least squares approach is approximately the same that of CR algorithm.

At last, in the Table 7 we give the similar results for the minimal residual method with local variational properties. This approach presents a big disadvantage in efficiency, as compared to the previous algorithms, even with application of the least squares methods. The close effect is demonstrated for steepest decent (SD) method, both for symmetric and non-symmetric matrices.

Table 1. CHEB-LSM-2, $p = q = 0$, $u^0 = x^2 + y^2$

$N \setminus m$	8	16	32	64	128	∞
7^2	34	29	32	41	41	41
	$2.4 \cdot 10^{-7}$	$7.8 \cdot 10^{-8}$	$9.9 \cdot 10^{-16}$	$1.3 \cdot 10^{-7}$	$1.3 \cdot 10^{-7}$	$1.3 \cdot 10^{-7}$
15^2	90	75	63	64	82	82
	$1.2 \cdot 10^{-6}$	$5.3 \cdot 10^{-7}$	$5.3 \cdot 10^{-8}$	$5.9 \cdot 10^{-9}$	$2.0 \cdot 10^{-7}$	$2.0 \cdot 10^{-7}$
31^2	281	197	140	127	128	163
	$3.6 \cdot 10^{-6}$	$3.5 \cdot 10^{-6}$	$1.3 \cdot 10^{-6}$	$1.6 \cdot 10^{-7}$	$3.1 \cdot 10^{-8}$	$3.0 \cdot 10^{-7}$
63^2	960	586	390	267	251	327
	$1.0 \cdot 10^{-5}$	$1.0 \cdot 10^{-5}$	$9.6 \cdot 10^{-6}$	$6.8 \cdot 10^{-6}$	$2.3 \cdot 10^{-6}$	$3.1 \cdot 10^{-7}$
127^2	3429	1991	1148	734	528	653
	$2.9 \cdot 10^{-5}$	$2.9 \cdot 10^{-5}$	$2.9 \cdot 10^{-5}$	$2.7 \cdot 10^{-5}$	$2.2 \cdot 10^{-5}$	$3.5 \cdot 10^{-7}$

Table 2. CHEB-LSM-2, $p = q = 4$, $u^0 = x^2 + y^2$

$N \setminus m$	8	16	32	64	128	∞
7^2	34	31	32	45	45	45
	$7.5 \cdot 10^{-8}$	$2.6 \cdot 10^{-8}$	$4.6 \cdot 10^{-15}$	$8.2 \cdot 10^{-8}$	$8.2 \cdot 10^{-8}$	$8.2 \cdot 10^{-8}$
15^2	67	75	71	64	91	91
	$5.0 \cdot 10^{-7}$	$2.6 \cdot 10^{-7}$	$3.4 \cdot 10^{-7}$	$9.8 \cdot 10^{-9}$	$1.6 \cdot 10^{-7}$	$1.6 \cdot 10^{-7}$
31^2	210	158	142	149	128	184
	$2.9 \cdot 10^{-6}$	$3.4 \cdot 10^{-7}$	$1.3 \cdot 10^{-6}$	$8.6 \cdot 10^{-7}$	$4.3 \cdot 10^{-8}$	$2.2 \cdot 10^{-7}$
63^2	740	421	348	285	271	363
	$7.9 \cdot 10^{-6}$	$6.6 \cdot 10^{-6}$	$3.7 \cdot 10^{-6}$	$3.6 \cdot 10^{-6}$	$2.7 \cdot 10^{-6}$	$1.8 \cdot 10^{-7}$
127^2	2654	1531	884	662	543	719
	$2.4 \cdot 10^{-5}$	$2.3 \cdot 10^{-5}$	$2.1 \cdot 10^{-5}$	$1.8 \cdot 10^{-6}$	$7.6 \cdot 10^{-6}$	$1.7 \cdot 10^{-7}$

Table 3. CR-LSM-2, $p = q = 4$

$N \setminus m$	8	16	32	64	128
7^2	34	31	63	127	255
	$1.3 \cdot 10^{-7}$	$7.9 \cdot 10^{-8}$	$1.6 \cdot 10^{-9}$	$4.6 \cdot 10^{-12}$	$5.1 \cdot 10^{-13}$
15^2	74	64	94	127	255
	$8.8 \cdot 10^{-7}$	$9.7 \cdot 10^{-7}$	$2.2 \cdot 10^{-7}$	$3.1 \cdot 10^{-9}$	$3.9 \cdot 10^{-12}$
31^2	236	149	129	190	255
	$2.9 \cdot 10^{-6}$	$2.0 \cdot 10^{-6}$	$8.1 \cdot 10^{-7}$	$3.0 \cdot 10^{-8}$	$7.3 \cdot 10^{-7}$
63^2	592	472	305	331	382
	$8.1 \cdot 10^{-6}$	$8.0 \cdot 10^{-6}$	$4.3 \cdot 10^{-6}$	$4.6 \cdot 10^{-6}$	$1.9 \cdot 10^{-7}$
127^2	2612	1347	897	539	659
	$2.4 \cdot 10^{-5}$	$2.3 \cdot 10^{-5}$	$2.1 \cdot 10^{-5}$	$1.3 \cdot 10^{-5}$	$1.2 \cdot 10^{-5}$

Table 4. CR-LSM-2, $p = q = 0$, $u^0 = x^2 + y^2$

$N \setminus m$	8	16	32	64	128
7^2	37	20	20	20	20
	$2.7 \cdot 10^{-7}$	$4.2 \cdot 10^{-8}$	$5.4 \cdot 10^{-9}$	$5.4 \cdot 10^{-9}$	$5.4 \cdot 10^{-9}$
15^2	99	75	42	40	40
	$6.9 \cdot 10^{-7}$	$8.3 \cdot 10^{-7}$	$3.1 \cdot 10^{-7}$	$8.8 \cdot 10^{-8}$	$8.8 \cdot 10^{-8}$
31^2	314	199	145	83	83
	$3.6 \cdot 10^{-6}$	$304 \cdot 10^{-6}$	$2.6 \cdot 10^{-6}$	$1.2 \cdot 10^{-6}$	$2.4 \cdot 10^{-7}$
63^2	1084	626	390	283	160
	$1.0 \cdot 10^{-5}$	$1.0 \cdot 10^{-5}$	$9.2 \cdot 10^{-6}$	$8.4 \cdot 10^{-6}$	$2.6 \cdot 10^{-6}$
127^2	3860	2119	1185	746	538
	$2.9 \cdot 10^{-5}$	$2.9 \cdot 10^{-5}$	$2.8 \cdot 10^{-5}$	$2.8 \cdot 10^{-5}$	$2.1 \cdot 10^{-5}$

Table 5. CG-LSM-2, $p = q = 0$, $u^0 = x^2 + y^2$

$N \setminus m$	8	16	32	64	128
7^2	38	20	20	20	20
	$1.8 \cdot 10^{-7}$	$4.7 \cdot 10^{-8}$	$5.4 \cdot 10^{-9}$	$5.4 \cdot 10^{-9}$	$5.4 \cdot 10^{-9}$
15^2	99	76	43	41	41
	$5.4 \cdot 10^{-7}$	$5.5 \cdot 10^{-7}$	$1.5 \cdot 10^{-7}$	$2.6 \cdot 10^{-8}$	$2.6 \cdot 10^{-8}$
31^2	316	211	156	86	81
	$3.0 \cdot 10^{-6}$	$1.3 \cdot 10^{-6}$	$9.9 \cdot 10^{-7}$	$5.7 \cdot 10^{-7}$	$1.7 \cdot 10^{-7}$
63^2	1086	631	404	316	167
	$9.8 \cdot 10^{-6}$	$8.6 \cdot 10^{-6}$	$5.1 \cdot 10^{-6}$	$2.3 \cdot 10^{-6}$	$1.2 \cdot 10^{-6}$
127^2	3865	2131	1210	757	614
	$2.9 \cdot 10^{-5}$	$2.7 \cdot 10^{-5}$	$2.1 \cdot 10^{-5}$	$2.2 \cdot 10^{-5}$	$3.6 \cdot 10^{-6}$

Table 6. CG-LSM-2, $p = q = 4$, $u^0 = x^2 + y^2$

$N \setminus m$	8	16	32	64	128
7^2	34	31	63	127	255
	$1.5 \cdot 10^{-7}$	$1.6 \cdot 10^{-8}$	$2.5 \cdot 10^{-10}$	$3.8 \cdot 10^{-12}$	$2.1 \cdot 10^{-13}$
15^2	78	69	94	127	455
	$1.3 \cdot 10^{-7}$	$2.2 \cdot 10^{-7}$	$2.2 \cdot 10^{-8}$	$1.1 \cdot 10^{-9}$	$6.6 \cdot 10^{-11}$
31^2	239	151	156	190	255
	$2.1 \cdot 10^{-6}$	$1.5 \cdot 10^{-6}$	$7.5 \cdot 10^{-7}$	$1.9 \cdot 10^{-8}$	$5.2 \cdot 10^{-8}$
63^2	596	481	311	337	382
	$7.8 \cdot 10^{-6}$	$5.8 \cdot 10^{-6}$	$1.7 \cdot 10^{-6}$	$7.1 \cdot 10^{-7}$	$9.9 \cdot 10^{-8}$
127^2	2612	1351	900	568	736
	$2.4 \cdot 10^{-5}$	$1.9 \cdot 10^{-5}$	$1.9 \cdot 10^{-5}$	$2.9 \cdot 10^{-6}$	$3.9 \cdot 10^{-6}$

Table 7. MR-LSM-2, $p = q = 0, u^0 = x^2 + y^2$

$N \setminus m$	8	16	32	64	128	∞
7^2	37	21	32	64	128	185
	$2.7 \cdot 10^{-7}$	$5.5 \cdot 10^{-8}$	$7.0 \cdot 10^{-9}$	$2.8 \cdot 10^{-9}$	$2.8 \cdot 10^{-9}$	$4.8 \cdot 10^{-7}$
15^2	99	76	67	82	128	703
	$5.4 \cdot 10^{-7}$	$5.5 \cdot 10^{-7}$	$4.3 \cdot 10^{-7}$	$2.1 \cdot 10^{-7}$	$2.6 \cdot 10^{-8}$	$1.3 \cdot 10^{-6}$
31^2	316	202	187	253	267	2614
	$3.0 \cdot 10^{-6}$	$3.1 \cdot 10^{-6}$	$1.3 \cdot 10^{-6}$	$1.9 \cdot 10^{-7}$	$1.4 \cdot 10^{-6}$	$3.7 \cdot 10^{-6}$
63^2	1086	631	559	505	636	9622
	$9.8 \cdot 10^{-6}$	$8.6 \cdot 10^{-6}$	$6.1 \cdot 10^{-6}$	$2.1 \cdot 10^{-6}$	$4.8 \cdot 10^{-6}$	$1.0 \cdot 10^{-5}$
127^2	3860	2123	1427	1702	1906	35050
	$2.9 \cdot 10^{-5}$	$2.9 \cdot 10^{-5}$	$2.6 \cdot 10^{-5}$	$2.2 \cdot 10^{-5}$	$1.7 \cdot 10^{-5}$	$2.9 \cdot 10^{-5}$

5 Conclusion

We consider the generalization of Anderson acceleration, for parallel solving
non-symmetric large SLAEs with sparse matrices, on the base of least squares
methods applied to some preliminary "cheap" iterative process, which is used
just for computing basis vectors for implicit, or block, implementation of the
Krylov type algorithms with periodically minimization of the residual vector
before restarts. The comparative experimental analysis of the variational conjugate gradient and conjugate residual methods, as well as spectral Chebyshev
acceleration demonstrates reasonable stability and convergence rate of the iterations the methods proposed. The idea of increasing parallelism consists in the
simultaneous computations of big number of inner products, in contrast to successive computations in the conventional Krykov algorithms. The performance
of the proposed approaches at the real multi-processor systems with distributed
and hierarchical shared memory is the topic of further research.

This work was supported by the Russian Science Foundation (project N 14-
11-00485) and the Russian Foundation for Basic Research (project N 16-29-
15122).

References

1. Saad, Y.: Iterative Methods for Sparse Linear Systems. PWS Publ., New York
   (2002)
2. Il'in, V.P.: Problems of parallel solution of large systems of linear algebraic equations. J. Mathem. Sci. **216**(6), 795–804 (2016)
3. Pratara, P.P., Suryanarayana, P., Pask, J.E.: Anderson acceleration of the Jacobi
   iterative method. An efficient alternative to Krylov methods for large, sparse linear
   systems. J. Comput. Phys. **306**, 43–54 (2016)
4. Anderson, D.G.: Itaerative procedures for nonlinear integral equations. J. Assoc.
   Comput. Mach **12**, 547–560 (1965)

5. Il'in, V.P.: Methods and Technologies of Finite Elements. ICM&MG SB RAS Publ., Novosibirsk (2007). (in Russian)
6. Il'in, V.P.: Methods of semiconjugate directions. Russ. J. Numer. Anal. Math. Model. **23**(4), 369–387 (2008)
7. Lawson, C.L., Hanson, R.Z.: Solving Least Squares Problems. Prentice-Hall Inc., Englewood Cliffs (1974)
8. Samarskii, A.A., Nikolaev, E.S.: Mehtod for Solving Grid Equations. Nauka, Moscow (1978)
9. Montgomery, P.L.: A block Lanczos algorithm for finding dependencies over GF(2). In: Guillou, L.C., Quisquater, J.-J. (eds.) EUROCRYPT 1995. LNCS, vol. 921, pp. 106–120. Springer, Heidelberg (1995). https://doi.org/10.1007/3-540-49264-X_9
10. Zamarashkin, N., Zheltkov, D.: Block Lanczos–Montgomery method with reduced data exchanges. In: Voevodin, V., Sobolev, S. (eds.) RuSCDays 2016. CCIS, vol. 687, pp. 15–26. Springer, Cham (2016). https://doi.org/10.1007/978-3-319-55669-7_2
11. Gurieva, Y.L., Il'in, V.P.: Parallel approaches and technologies of domain decomposition methods. J. Math. Sci. **207**(5), 724–735 (2015)
12. Il'in, V.P.: On exponential finite volume approximations. Rus. J. Num. Anal. Math. Model. **18**(6), 479–506 (2003)
13. Intel®Mathematical Kernel Library from Intel. http://www.software.intel.com/en-us/intel-mkl

Supercomputer Simulation

Simulation of Seismic Waves Propagation in Multiscale Media

Impact of Cavernous/Fractured Reservoirs

Vladimir Tcheverda[1]([✉]), Victor Kostin[1], Galina Reshetova[2],
and Vadim Lisitsa[1]

[1] Institute of Petroleum Geology and Geophysics SB RAS, 3, Prosp. Koptyug,
Novosibirsk 630090, Russia
{cheverdava,kostinvi,lisitsavv}@ipgg.sbras.ru
[2] Institute of Computational Mathematics and Mathematical Geophysics SB RAS,
6, Prosp. Lavrentiev, Novosibirsk 630090, Russia
kgv@nmsf.sscc.ru

Abstract. In order to simulate the interaction of seismic waves with cavernous/fractured reservoirs, a finite-difference technique based on locally refined time-and-space grids is used. The need to use these grids is due primarily to the differing scale of heterogeneities in the reference medium and the reservoir. Domain Decomposition methods allow for the separation of the target area into subdomains containing the reference medium (coarse grid) and reservoir (fine grid). Computations for each subdomain can be carried out in parallel. The data exchange between each subdomain within a group is done using MPI through nonblocking iSend/iReceive commands. The data exchange between the two groups is done simultaneously by coupling the coarse and fine grids.

The results of a numerical simulation of a carbonate reservoir are presented and discussed.

Keywords: Finite-difference schemes · Local grid refinement · Domain decomposition · MPI · Group of processor units · Master processor unit

1 Introduction and Motivation

One of the key challenges in modern seismic processing is to use the surface and/or borehole data to restore the microstructure of the hydrocarbon reservoir. This microstructure can have a significant impact on oil and gas production. In particular, in many cases the carbonate reservoir's matrix porosity contains the oil but the permeability is mainly through the fracture corridors. In some carbonate reservoirs the in-place oil is contained in karstic caves. Because of this, the ability to locate these microstructures precisely and to characterize their properties is of a great importance. Recently various techniques have been developed to perform this analysis with the help of scattered seismic waves. Among them,

© Springer International Publishing AG 2017
V. Voevodin and S. Sobolev (Eds.): RuSCDays 2017, CCIS 793, pp. 183–193, 2017.
https://doi.org/10.1007/978-3-319-71255-0_14

the scattering index presented by Willis et al. ([8]) or a variety of the imaging techniques recently developed under the generic name of interferometry (see e.g. book of G. Schuster [7]).

The first step in the development of any inversion/imaging procedure is to simulate accurately the wave field scattered by fractures and caves. The numerical and computer constraints even on very large clusters place limitations on the resolution of the model described. Really, a reservoir beds typically at a depth of 2000÷4000 m, which is about 50÷70 dominant wavelength. The current practice for the finite-difference simulation of seismic waves propagation at such distances is to use grid cells of 0.05–0.1 of a dominant wavelength, usually between 5–10 m. So, one needs to upscale heterogeneities associated with fracturing on a smaller scale (0.01–1 m) and to transform them to an equivalent/effective medium. This effective medium will help reproduce variations in the travel-times and an average change of reflection coefficients but absolutely cancels the scattered waves that are a subject of the above mentioned methods for characterizing fracture distributions.

Thus, the main challenge with a full scale simulation of cavernous/fractured (carbonate) reservoirs in a realistic environment is that one should take into account both the macro- and microstructures. A straightforward implementation of finite difference techniques provides a highly detailed reference model. From the computational point of view, this means a huge amount of memory required for the simulation and, therefore, extremely high computer cost. In particular, a simulated model of dimension 10 km $\times$ 10 km $\times$ 10 km, which is common for seismic explorations, with a cell size of 0.5 m claims 8×10^{12} cells and needs in $\approx 350 Tb$ of RAM.

The popular approach to overcome these troubles is to refine a grid in space only and there are many publications dealing with its implementation (see [6] for a detailed review), but it has at least two drawbacks:

- To ensure stability of the finite-difference scheme the time step must be very small everywhere in the computational domain;
- Unreasonably small Courant ratio in the area with a coarse spatial grid leads to a noticeable increase in numerical dispersion.

Our solution to this issue is to use a mutually agreed local grid refinement in time and space: spatial and time steps are refined by the same factor.

2   Numerical and Parallel Implementation

In our considerations propagation of seismic waves is simulated with help of an explicit finite-difference scheme (FDS) on staggered grids approximating elastic wave equations (velocity-stress formulation):

$$\varrho \frac{\partial \boldsymbol{u}}{\partial t} - A \frac{\partial \boldsymbol{\sigma}}{\partial x} - B \frac{\partial \boldsymbol{\sigma}}{\partial y} - C \frac{\partial \boldsymbol{\sigma}}{\partial z} = 0;$$

$$D \frac{\partial \boldsymbol{\sigma}}{\partial t} - A^T \frac{\partial \boldsymbol{u}}{\partial x} - B^T \frac{\partial \boldsymbol{u}}{\partial y} - C^T \frac{\partial \boldsymbol{u}}{\partial z} = \boldsymbol{f};$$

written for vectors of the velocity $\boldsymbol{u} = (u_x, u_y, u_z)^T$ and the stress $\boldsymbol{\sigma} = (\sigma_{xx}, \sigma_{yy}, \sigma_{zz}, \sigma_{xz}, \sigma_{yz}, \sigma_{xy})$.

Staggered grid finite difference scheme updates values of unknown vectors in two steps:

1. from velocities at t to stresses at $t + \Delta t/2$;
2. from stresses at $t + \Delta t/2$ to velocities at $t + \Delta t$.

In view of the local spatial distribution of the stencil used in this finite difference scheme to update the vector at some point M and time $(t + \Delta t/2)$, the previous time level (t) corresponding values should be known in a neighborhood of this point.

Parallel implementation of this FDS is based on the decomposition of the computational domain to elementary subdomains, being assigned to its individual Processor Unit (PU) (Fig. 1). Update unknown vectors while moving from a time layer to the next one requires two adjacent PU to exchange unknown vectors values in the grid nodes along the interface. Necessity of this exchange negatively impacts scalability of the method. However, the impact is less visible on 3D Domain Decomposition (DD) than in one- and two-dimensional ones (see theoretical estimates of acceleration for different versions of DD in Fig. 2)). In our implementation we choose 3D Domain Decomposition, moreover, in order to reduce the idle time, the asynchronous computations based on nonblocking MPI procedures iSend/iReceive are used.

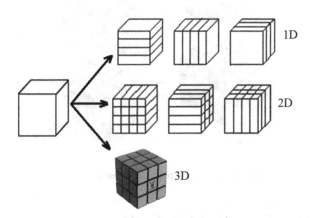

Fig. 1. Domain decomposition. From top to bottom: 1D, 2D, 3D.

In order to carry out the numerical simulation of seismic waves propagation through a multiscale medium we represent it as a superposition of the reference medium given on a coarse grid and the reservoir on a fine grid (see Fig. 3). Each of these grids is again decomposed to elementary subdomains being assigned to individual PU. Now these PU are combined into two groups for coarse and fine grids, and special efforts should be applied in order to couple these groups.

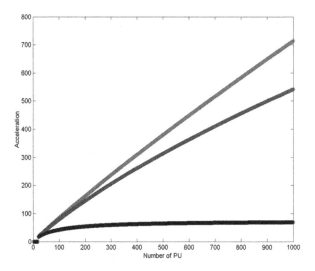

Fig. 2. Theoretical estimation of acceleration for different implementations of Domain Decomposition (top down): 3D, 2D and 1D (see Fig. 1).

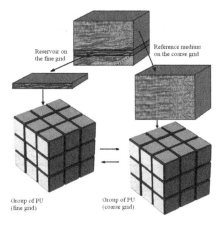

Fig. 3. Two groups of Processor Units.

2.1 Coupling of Coarse and Fine Grids

First of all, let us explain how a coarse and a fine grids are coupled to each other. The necessary properties of the finite difference method based on a local grid refinement should be its stability and an acceptable level of artificial reflections. Scattered waves we are interested in have an amplitude of about 1% of the incident wave. Artifacts should be at least 10 times less, that is about 0.1% of the incident wave. If we refine the grid at once in time and space stability of the FDS on this way (see [1,2]) can be provided via coupling coarse and fine grids on the base of energy conservation, which leads to an unacceptable level (more than 1%)

of artificial reflections (see [3,4]). We modify the approach so that the grid is refined by turn in time and space on two different surfaces surrounding the target area with microstructure. This allows decoupling temporal and spatial grid refinement and to implement them independently and to provide the desired level of artifacts.

Refinement in Time. Refinement in time with a fixed 1D spatial discretization is clearly seen in Fig. 4 and does not need any explanations. Its modification for 2D and 3D media is straightforward (see [3,4] for more detail).

Refinement in Space. In order to change spatial grids, the Fast Fourier Transform (FFT) based interpolation is used. Let us explain this procedure for a 2D problem. The mutual disposition of a coarse and a fine spatial grids is presented in Fig. 4b, which corresponds to updating the stresses by velocities (updating stresses by velocities is implemented in the same manner). As can be seen, to update the stresses on a fine grid it is necessary to know the displacement at the points marked with small (red) triangles, which do not exist on the given coarse grid. Using the fact that all of them are on the same line (on the same plane for 3D statement), we seek the values of missing nodes by FFT based interpolation. Its main advantages are an extremely high performance and exponential accuracy. It is this accuracy allows us to provide the required low level of artifacts (about 0.001 with respect to the incident wave) generated on the interface of these two grids. For 3D statement we again perform the FFT based interpolation but this time 2D.

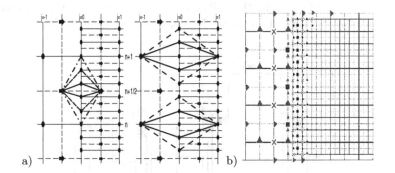

Fig. 4. From a coarse to a fine grid: (a) refinement in time (left - displacement, right - stresses) (b) refinement in space. (Color figure online)

2.2   Implementation of Parallel Computations

Our objective is to analyze the impact of cavernous-fractured reservoirs in the seismic waves for realistic 3D heterogeneous media. Therefore, parallel computations are necessary both in the reference medium, described by a coarse mesh, and in the reservoir itself, determined on a fine grid. The simultaneous use of a coarse and a fine grids and the need for interaction between them makes it difficult to ensure a uniform load of Processor Units under parallelization of computations based on Domain Decomposition. Besides, the user should be allowed to locate the reservoir anywhere in the background.

This problem is resolved through the implementation of parallel computations on two groups of Processor Units. One of them is fully placed 3D heterogeneous referent environment on a coarse grid, while the fine mesh describing the reservoir is distributed among the PU in the second group (Fig. 3). Thus, there is a need for both exchanges between processors within each group and between the groups as well. The data exchange within a group is done via faces of the adjacent Processor Units by non-blocking iSend/iReceive MPI procedures. Interaction between the groups is much more complicated. It is carried out not so much for data sending/receiving only, but for coupling a coarse and a fine grids as well. Let us consider the data exchange from the first group (a coarse grid) of PU to the second (a fine grid) and backwards.

From coarse to fine. First are found Processor Units in the first group which cover the target area, and are grouped along each of the faces being in contact with the fine grid. At each of the faces there is allocated the Master Processor (MP), which gathers the computed current values of stresses/displacements and sends them to the relevant MP on a fine grid (see Fig. 5). All the subsequent data processing providing the coupling of a coarse and a fine grids by the FFT based interpolation is performed by the relevant Master Processor in the second group (a fine grid). Later this MP sends interpolated data to each processor in its subgroup.

Interpolation performed by the MP of the second group essentially decreases the amount of sent/received data and, hence, the idle time of PU.

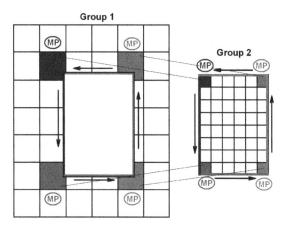

Fig. 5. Processor Units for a coarse (left) and a fine (right) grids. Relevant MP from different groups have the same color. (Color figure online)

From fine to coarse. As in the previous case, primarily there are identified PU from the second group which perform computations on the faces covering the target area. Next, again for each face Master Processor is identified. This MP as its partner from the coarse grid collects data from the relevant face and performs their preprocessing before sending to the first group of PU (a coarse grid). Now we do not need all data in order to move to the next time, but only those of them which fit the coarse grid. Formally, these data could be thinned out, but our experiments have proved that this way generates significant artifacts due to the loss of smoothness. Therefore for this direction (from fine to coarse) we also use the FFT based interpolation implemented by the relevant MP of the second group (a fine grid). The data obtained are sent to the first group.

3   Reservoir Simulation

3.1   2D Statement: Karstic Layer

In order to estimate the accuracy of the method, we first consider a 2D statement for a thin layered reservoir with karst intrusions presented in Fig. 6(a). In order to describe the microstructure of karstic intrusions we should use a grid with $h_x = h_z = 0.5$ m, while for the reference medium the dispersion analysis gives $h_x = h_z = 2.5$ m. In Fig. 6a one can see an area with the fivefold grid refinement in time and space. Let us compare now the results of simulation for a uniform fine grid and a grid with the local refinement in time and space. In Fig. 6b, one can see a free surface seismogram (horizontal displacement) generated by the vertical point force with a Ricker pulse of a dominant frequency 25 Hz and simulated on the uniform fine grid. Figure 7 represents a comparison of synthetic traces computed on a uniform fine grid and a grid with local refinement in time and space. As can be seen, there is an excellent coincidence of scattered PP-waves and rather good agreement of PS ones.

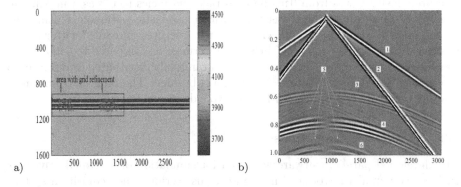

a)   b)

Fig. 6. (a) Karstic layer (b) Surface seismogram (horizontal component). 1 - direct P-wave, 2 - direct S-wave coupled with surface Rayleigh wave, 3 and 4 - reflected PP- and PS-waves, 5 - scattered PP- and PS-waves, 6 - reflected SP-wave.

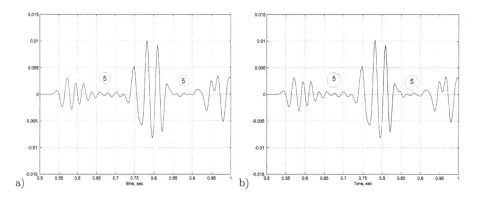

Fig. 7. Traces computed on a uniform fine grid (a) and on a grid with local refinement in time and space.

3.2   3D Statement: Fracture Corridors

Now we present the results of numerical simulation for some realistic model of a carbonate reservoir with fracture corridors. The reservoir is embedded into a homogeneous background with elastic properties equivalent to an average carbonate rock:

$$V_p = 4500 \, \text{m/s}, \; V_s = 2500 \, \text{m/s}, \; \text{density} \; \varrho = 2500 \, \text{kg/m}^3$$

The reservoir is treated as a horizontal layer 200 m thick and corresponds to a slightly softer rock with the elastic waves propagation velocities $V_p = 4400 \, \text{m/s}$, $V_s = 2400 \, \text{m/s}$ and the density $\varrho = 2200 \, \text{kg/m}^3$ and contains two fractured layers 30 m thick each. The fracturation is of a corridor type, that is, we have included into each layer a set of randomly distributed parallel fracture corridors. The fracture density varies from 0 in the non-fractured facies to 0.3 as a maximum. Finally, the fracture density was transformed to elastic parameters using the second order Hudson theory following [5]. Since fractures were filled with gas, the velocity diminishes down to 3600 m/s the lowermost as compared to 4400 m/s in the matrix. The fracture corridors were then randomly distributed into fractured layers until the desired fracture density was obtained. The final distribution of fracture corridors can be seen in Fig. 8 (two side views).

3.3   Synthetic Seismograms

The developed parallel software was used for simulation of scattered waves for the reservoir model introduced in the previous section. The acquisition system can be seen in Fig. 9. Three-component seismograms are presented in Fig. 10. There is a visible difference between the seismograms along the parallel and the perpendicular lines with respect to fracture corridors.

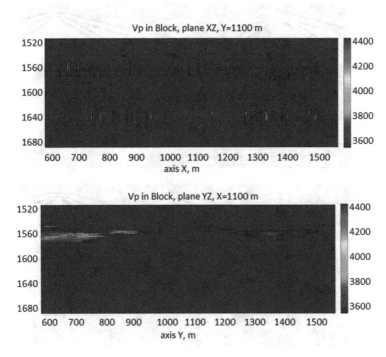

Fig. 8. Side view of fracture corridors within reservoir: orthogonal (top) and parallel (bottom) to the corridor direction

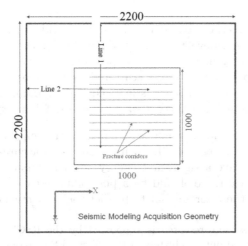

Fig. 9. Acquisition system. The source is at the intersection of Line 1 and Line 2.

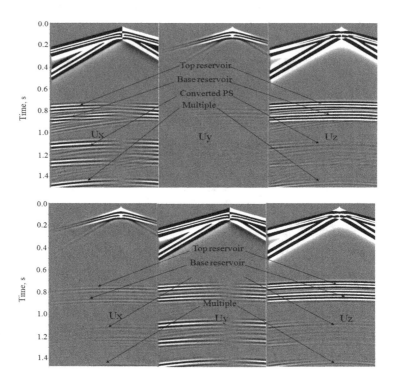

Fig. 10. 3C seismograms along Line 2 (top) and Line 1 (bottom). From left to right: X, Y and Z-displacements.

4   Conclusion

A finite difference method based on the use of grids with local space-time refinement is proposed, developed and verified. Implementing its parallel software opens up a fundamentally new opportunity to study the processes of formation and propagation of waves scattered by a microstructure of the cavernous/fractured reservoir for a realistic geological environment. The very first simulations carried out using this software, allow the following conclusions:

- Modeling techniques make possible to simulate the impact of fine-scale heterogeneities within a realistic 3D environment in an accurate manner;
- Scattered waves have a significant energy and can be acquired by the field observations, hence there should be a possibility not only to reveal cavities and fractures in the reservoir but to predict their orientation as well.

Acknowledgements. The research is supported by the RSCF grant 17-17-01128. Simulations were carried out on clusters of the Siberian Supercomputer Center of the Siberian Branch of RAS (Novosibirsk) and the Joint Supercomputer Center of the Russian Academy of Sciences (Moscow).

References

1. Collino, F., Fouquet, T., Joly, P.: A conservative space-time mesh refinement method for 1-D wave equation. Part I, II. Numer. Math. **95**, 197–251 (2003)
2. Diaz, J., Grote, M.J.: Energy conserving explicit local time stepping for second-order wave equations. SIAM J. Sci. Comput. **31**(3), 1985–2014 (2009)
3. Lisitsa, V., Tcheverda, V., Reshetova, G.: Finite-difference algorithm with local time-space grid refinement for simulation of waves. Comput. Geosci. **16**(1), 39–54 (2012)
4. Kostin, V., Lisitsa, V., Tcheverda, V., Reshetova, G.: Local time-space mesh refinement for simulation of elastic wave propagation in multi-scale media. J. Comput. Phys. **281**, 669–689 (2015)
5. Mavko, G., Mukerji, T., Dvorkin, J.: Rock Physics Handbook. Cambridge University Press, Cambridge (2009)
6. Kristek, J., Moczo, P., Galis, M.: Stable discontinuous staggered grid in the finite-difference modelling of seismic motion. Geophys. J. Int. **183**(3), 1401–1407 (2010)
7. Schuster, G.: Seismic Interferometry. Cambridge University Press, Cambridge (2009)
8. Willis, M., Burns, D., Rao, R., Minsley, B., Toksoz, N., Vetri, L.: Spatial orientation and distribution of reservoir fractures from scattered seismic energy. Geophysics **71**, O43–O51 (2006)

Computational Modeling of Turbulent Structuring of Molecular Clouds Based on High Resolution Calculating Schemes

Boris Rybakin[1(✉)], Valery Goryachev[2], and Stepan Ageev[1]

[1] Department of Gas and Wave Dynamics, Moscow State University, Moscow, Russia
rybakin@vip.niisi.ru
[2] Department of Mathematics, Tver State Technical University, Tver, Russia
gdv.vdg@yandex.ru

Abstract. The article submits the results of 3D computational modeling of the adiabatic interaction between a shock wave and molecular clouds, central impact and glancing collision between them, in the case of counter movement. According to the problem set in the first case, two spherical clouds with pre-established density fields interact with the post-shock medium of supernova blast remnants. It is demonstrated that the collision give rise to the supersonic turbulence in a cloud mixing zone, the formation of cone-like filamentous structures, the significant stratification of gas density and the disruption of clouds. Problems of vortex filaments origination in clouds wakes are analyzed after simulation of supersonic forward and glancing collision of two molecular clouds.

Keywords: Parallel computing · Supersonic turbulence · Shock waves · Small molecular clouds

1 Introduction

Giant molecular clouds (GMC) are a huge accumulation of interstellar gas and dust, composed mostly of molecular hydrogen. They are the coolest and densest portions of the interstellar medium (ISM). MCs are generated from this matter, a part of which falls under strong shock wave compression in extended filaments and globules that eventually collapse. Filaments formations are wide spread in the universe. Swellings of filaments and clumps occurring inside MCs are results of shock strong compression and cloud's self-gravitation, magnetic hydrodynamics after-effect [1, 2]. Molecular clouds may evolve to structure of interlacing and connecting filaments. This web of spatial voids, highly compressed gas fibers or enclosures, depends on the external influence on the interstellar medium, and has an indirect action on the condensation of cores formed into more massive intersected filament clumps that later could become protocores – embryos of future stars [3, 4].

One of possible scenarios of filament structuring initiation is a collision of MCs with strong shock waves (SW) of another gas formations propagated in the space. Others

© Springer International Publishing AG 2017
V. Voevodin and S. Sobolev (Eds.): RuSCDays 2017, CCIS 793, pp. 194–206, 2017.
https://doi.org/10.1007/978-3-319-71255-0_15

cases are initiated in different collisions between MCs of different mass with originating shock wave multiformity in mixing zones.

The shock wave interaction generated by proliferation of supernova explosion remnants and the molecular clouds leads to supersonic turbulence of gas in MCs, deep density stratification and their destruction followed by formation of filamentous structures. The shock wave impact accompanied by intense dynamic interaction of filaments with each other leads to the significant redistribution of gas density in MCs.

The present simulation has been performed to study vortex structuring in the molecular clouds formations disbalanced after collisions with a strong shock wave and between two MCs in different cases of collision, and associated with turbulization and filaments formation during these processes. The occurring filament formations governed by the vortex slipstream flow after MCs and a shock wave interact with each other and develop in the gradient regions of the gas density fields. It is shown that at specified points of time there occur ultra-dense regions, with the density contrast being an order of magnitude higher than the initial contrast in clouds. Evolution of such objects, running from the formation of filament rudiments to the moment, they reach the stellar densities, covers a vast range of spatial and time scales.

Main of numerical methods used in astrophysical hydrodynamics can be divided into classical continuum approach and Smoothed Particle Hydrodynamics (SPH) methods. A significant part of continuum solvers uses a high resolution regular or AMR grids. The most of them work on parallel HPC. SPH methods are challenges due to several benefits over traditional grid-based techniques (flexible parallel computing realization, multiphase fluid simulation, etc.). This method, however, has essential limitation, for example, particles leaved a domain with high gradient parameters (velocity, pressure) will be compensated, and this leads to loss of computational accuracy. Recent related publications using last approach can be found in [5–9].

Author's code used is a continual. The article submits the results of 3D computational modeling of the adiabatic interaction between a shock wave and molecular clouds, central impact and glancing collision between them. The article analyses the density fragmentation, investigates the process of filament formation and gas density stratification as a result of SW/MCs or MC/MC collision. Applying HPC technologies we have realized a numerical simulation of complicated gas dynamics task using calculation grids with more than two billion nodes.

2   Problem Definition

2.1   Initial Conditions and Parameters

We study three scenarios of MCs collision in ISM. Case I: strong shock wave and SW/MCs interaction in a configuration where a plane of shock frontal of supernova remnant gas runs onto systems of two clouds of spherical form. Case II: a central concussion of two molecular clouds (MC/MC) of initially spherical form moving in opposite direction. Case III: a glancing collision of MC to the side of another one, in reverse moving.

The schema of simulated collisions is shown in Fig. 1. Two MCs have different radial distribution of density (illustrated by different diagrams on schema).

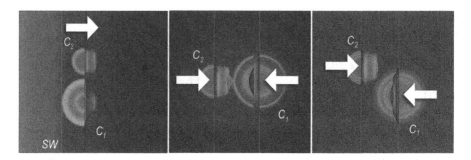

Fig. 1. Computation schemes for three case of simulation.

In the case I, at the moment of collision a shock wave contacts with outer MCs boundaries. Special rule of density radial distribution is used to represent more realistically the density smoothing profile on the border between the clouds and the outer medium. Appropriate functions were taken according to recommendations given in [10–12].

The density radial distribution formulas for clouds C_1 and C_2 are the following:

$$\rho(r) = \rho_{ism} + \frac{\rho_{cl} - \rho_{ism}}{1 + \left(r/R_{cl}\right)^{2.7128}}, \tag{1}$$

$$\rho(r) = \rho_{ism}\left(\chi + \frac{\alpha}{\alpha + 1}(1 - \chi)\right), \text{ where } \chi = \rho_{cl}/\rho_{ism} - \text{density contrast.} \tag{2}$$

Form factor α in (2) can be calculated by formula

$$\alpha = exp\left\{ min\left[20.0,\ 10 \cdot \left(\left(\frac{r}{R_{cl}}\right)^2 - 1\right)\right]\right\}$$

Several parameters in these formulas controlling the steepness of clouds border were changed to improve density smoothing.

The key physical parameters and assumptions of the present tasks were correlated by data from [17]. The interstellar medium consists of relatively warm matter with $T_{ism} = 10^4$ K, the temperature of colder MC gas $T_{cl} = 10^2$ K. The ambient gas density of the outer cloud medium $\rho_{ism} = 2.15 \, 10^{-25}$ g·cm^{-3}, the gas density in the undisturbed cloud centers $\rho_{cl} = 1.075 \, 10^{-22}$ g·cm^{-3}.

In the case I, characteristic (conventionally diffused) radius of each cloud R_{cl} is equal roughly to 0.1 pc. For system of two clouds (C_1, C_2), the mass of each is (approximately, considering fuzzy boundary) equal to 0.005 $M_\odot$ or 0.01 $M_\odot$ respectively, in solar mass fractions. The initial density contrast between the MCs centers and the interstellar medium is $\chi = 500$.

Mach number M_{sw} of incident shock wave is equal to seven, post-shock plasma density $\rho_{sw} = 8.6 \times 10^{-25}$ g·cm^{-3}, temperature $T_{sw} = 1.5 \times 10^5$ K, velocity of shock wave $U_{sw} = 104$ km·s^{-1}. The thickness of post-shock wave front is ~ 2–5 pc, which is much greater than the radius of a cloud. The period of time the shock wave propagates the upper cloud diameter is about 2000 years. This value is used as a scale for non-dimensional time.

In the case II, III the mass of each cloud C_1, C_2 is equal to 0.32 $M_\odot$ or 1.05 $M_\odot$ respectively. The velocity of each MC is 5 km·s^{-1}, the oncoming velocity is equal 10 km·s^{-1}. In the case of the glancing strike centers of MCs are displaced, linear shift is 0.2R_{cl}. The initial density contrast between the MCs centers and the interstellar medium is $\chi = 500$ and 100 accordingly.

2.2 Equations and Numerical Realization

Gas movement is described with a set of Euler equations which are conservation laws for mass, momentum, and energy

$$\frac{\partial U}{\partial t} + \nabla \cdot T = 0, \ U = \begin{pmatrix} \rho \\ \rho u \\ e \end{pmatrix}, \ T = \begin{pmatrix} \rho u \\ \rho uu + pI \\ (e+p)u \end{pmatrix}^T, \ e = \frac{p}{\gamma - 1} + \frac{|u|^2}{2}, \tag{3}$$

ρ denote the gas density, $\mathbf{u} = (u, v, w)$ is the velocity vector. The total energy density e and gas pressure p are related through the ideal gas closure, where adiabatic index - $\gamma = c_p/c_v$ is equal to 5/3.

Accuracy of numerical solution of multivariable problems in supersonic gas dynamics is eminently important in astrophysics phenomenon simulation. High-order accurate difference schemes have guaranteed monotonicity preservation of conservation laws. They are based on Godunov approach [18] to proof linear, first-order upwind schemes. The nonlinear, second-order accurate total variation diminishing (TVD) approach provides high resolution capturing of shocks and prevents unphysical oscillations, therefore it describes the local discontinuity preserving hyperbolic conservation laws [19, 20]. The TVD maintains a nonlinear stability condition. The total variation of a discrete solution defined as a measure of the overall amount oscillation in velocity $\mathbf{u}$ is

$$TV(u^t) = 2 \left(\sum u_{max} - \sum u_{min} \right) \tag{4}$$

The flux assignment scheme with condition $TV\left(u^{t+\Delta t}\right) \le TV(u^t)$ can guarantee that the amount of total oscillations will have a limit. Different TVD limiters are used: *minmod, superbee, vanleer*. The *vanleer* limiter proves to be preferential in our solution. TVD scheme is an approbated and robust method to solve systems of Euler equations. Such an approach and a sampling of physical coordinates [22] have enables us to take an appropriate parallelization and accelerate computing.

Numerical experiment is performed using different spatial resolution of physical description of SW/MCs collision in space. Grid sizes from $512 \times 512 \times 512$ (case II and III) to $2048 \times 1024 \times 1024$ (case I) units were used in systematic calculations. Minimal spherical clouds radius corresponds to 128 grid nodes. Last number exceeds the spatial resolution level, which is necessary to resolve correctly density and velocity turbulence fluctuations over energy high gradient gas layers and shock waves. The level of zonal discretization is more that used in [14–16]. The computing areas used for problem under consideration are parallelepipeds with dimensions $1.6 \times 0.8 \times 0.8$ pc for the case using high resolution mesh, and $1.6 \times 1.6 \times 1.6$ pc for MC/MC collisions. The lateral and outlet computational domain edges are determined as open boundary conditions for primitive variables.

We use author's parallel code allowing computations to be done with OpenMP. The setting of a code performance is done with Intel VTune Amplifier XE. Some computations are done with graphics accelerators NVIDIA K40 and CUDA for PGI Fortran. To compute with graphical accelerators the computation program has been retargeted so that some subprograms should be directed to GPU and the others - to CPU.

The numerical simulation procedure and peculiarities of 3D hydro code were detailed in [22, 23]. The wide set of CFD utilities and postprocessing systems were used to analyze a big data output after a numerical experiment. Simulated filamentous structures and fragmentation process observed are analyzed using computer visualization technique of author's program - HDVIS.

3   Analysis of SW/MCs and MC/MC Collision

Numerical simulation has been performed to study the morphology and vortex coherent structures in the molecular cloud formations disbalanced after collisions with a shock wave, shift and wake reformation in the situation with MCs forward and glancing impact.

The computations have shown that the formation of filaments and gas density stratification depend significantly on several factors, but the primary one is a shock wave compression of clouds matter near the sheet (inners and outers) layers of gas mixed.

3.1   SW/MCs Interaction

In first case of SW/MSs collision the density gas fragmentation is associated with supersonic turbulization during these processes. The evolution of transient coherent structures in MCs after passing a shock wave goes through three representative stages. At initial time, when a bow shock wave rounds clouds, a wave is formed behind its front. It moves

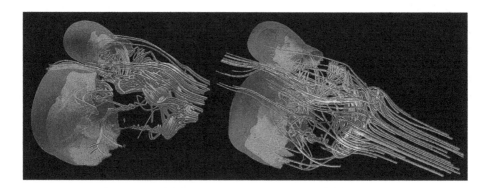

Fig. 2. Initial compression in SW/MCs interaction (t = 40, 60). (Color figure online)

towards the flow and forms vortical structures, as it is shown in Fig. 2. Iso-surfaces of density contrast $\chi = 10$ and 2000 and twisting pathlines are given in green and red colors.

At the next stage a SW continues to extend and initiates the Richtmyer-Meshkov instability (RMI) - disturbance of gas at outer boundaries of clouds. There occur convective acceleration of flow and whirling of the boundary layers of the conventional border between MCs and a surrounding matter, in zones with large and small gas density. The Kelvin–Helmholtz instability (KHI) can increase here additionally.

At the third stage of cloud transformation, the conical-like sheets commence to stretch. The flow streams accelerate in high gradient density layers and can initiate whirling of layers at the conventional MCs/ISM borders. Filament rudiments look like the elongated conical folded sheets (Figs. 3 and 4). The global circulation of a gas flow in the mixing zone begins to appear after cloud C_1 being rounded by a shock wave and finds its source in two vortex lines born inside the cloud at the back side. The flow swirl occurs in accordance with the scheme of spatial twin vortex.

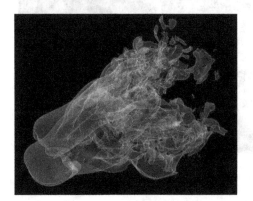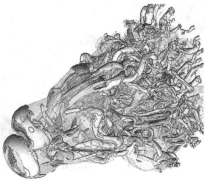

Fig. 3. Gas density stratification and vortical structure of MCs at $t = 330$.

Vortex sheets start to deflect and twist and become filamented practically after origination of instability. The observable vortices are illustrated in Fig. 3 by showing of vorticity magnitude $|\omega| = 30$ distribution. In the evolution process, the vortex lines elongate, kink, take the form of hairpins, and expand in a bend region. The visualization inset shows the formation of a hairpin structure near the outer gas layer that precedes the long streaks. Transition happens via streaky and misshapen structures behind shock layers on the leeward side of MCs. Hairpin vortices observed are similar to those found in plasma flows for high Mach boundary layers in supersonic transition, investigated in experiment.

Q-criterion - the second invariant of a velocity gradient tensor being used to identify the regions of non-uniformly scaled vortex concentrations and to differentiate peculiarities of the flow structure. Vortices have smaller distribution density within the mixing region, at the boundaries and surfaces of elongated filamentous film rudiments the vortex distribution densities are significantly higher and show a local velocity slope in different cloud regions. Figure 3 shows typical vortex formation: with elongated loops and helical deformations inside MCs at the moment of shell forming.

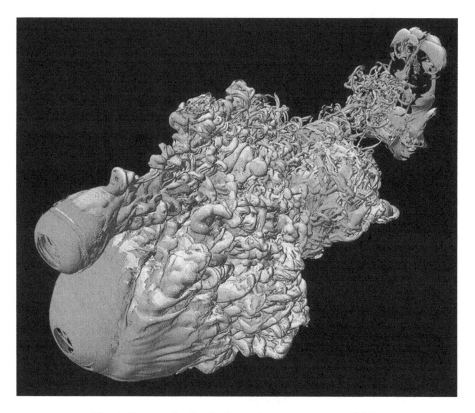

Fig. 4. Denstrophy distribution and vortex structure of MCs.

To reinforce the role of density interleaving and turbulent supersonic transfer in analyzing of MCs stratification the denstrophy characterization is used. The local dens-trophy is defined as a value: $\Omega_{1/2} = \frac{1}{2}\left|\nabla \times \left(\rho^{1/2}\boldsymbol{u}\right)\right|^2$, which is an indicator of compressible turbulent velocity fluctuation [24]. A view of iso-surface of denstrophy $\Omega_{1/2} = 1000$ at t = 300 is shown on Fig. 4.

Scanning of denstrophy distribution iso-surface one can emphasize the fractal recurrence of cone-like filament envelopes. Supersonic flow perturbation leads to a considerable grow of the denstrophy over filament in stripping phase of MCs transformation. Compressed gas sheets assume funneled form. The low-density gas is removed to center zone of cloud, and occupy low-pressure regions previously created by rarefaction waves. Stochastic void swelling is typical for shock-induced MCs. Generated vortices grow over time, slip and roll through the newly-formed sheets, eventually to be expelled outside.

One of the extreme forms of gas stratification in MCs is conventionally hollow filament. Envelopes of such formation are shown in Fig. 5 using selected display of iso-surfaces for contrast density $\chi = 1$ (opaque red-green-yellow), 5 (translucent rose). The map of local denstrophy is distributed along the surface of $\chi = 1$. Denstrophy color legend conforms to $100 < \Omega_{1/2} < 10000$. One can see that separate zones are practically eddy-free; they have the close to ISM density of gas. Hollow channels and caverns with low denstrophy and turbulent pulsation are intrinsic for giant molecular clouds.

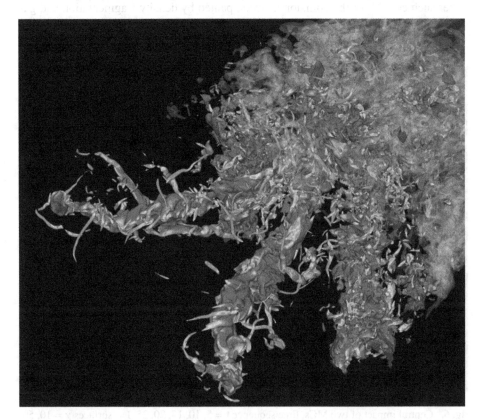

Fig. 5. Iso-surfaces $\chi = 1, 5$ with denstrophy contours at t = 300. (Color figure online)

Supersonic turbulence drives the fragmentation of dense cores and multiformity of pro-filamentary structures taking the original shell-like and clump forms. It is possible to establish some relationship between energy and density gradients strips on shell edges which is high-correlated. Gas currents near edges are accelerated by oblique collisions of secondary shock fronts that can arise from the initial supersonic shock fluctuations, either over the cloud recirculation zone or inside it, or behind the shock wave (primary or secondary) intersection lines and discs.

3.2 MC/MC Collisions

Collision of SMCs or molecular clumps in GMC can be realized in different ways. Outcomes of impact depend on initial parameters: velocities, mass ratio, impulse direction, matter inhomogeneity and another thing. In our study results of central and shifted collision have been explored as first. The parameters and initial conditions for formulated tasks assigned in Sect. 2.1.

Central impact between two molecular clouds studied with oncoming velocity of objects equal 10 km·s^{-1}. The mass ratio M_{c1}/M_{c2} in this collision is much less of three. Under such conditions the collision is accompanied by density fragmentation and gas scattering from the center to periphery outside. Time sequence of collision compressing is shown on Fig. 6.

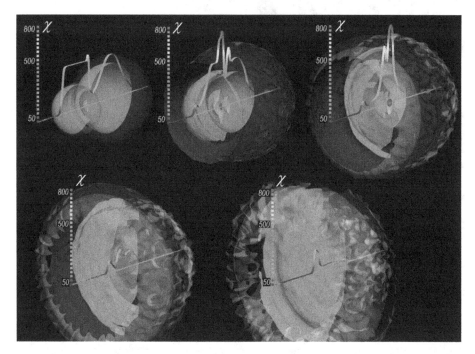

Fig. 6. Central impact of two MCs, time sequence t = 5, 10, 15, 20, 25. Iso-surfaces χ = 10, 50, 500, 800 and density contrast profile on the central line.

Arising from forward compression a density "splash" is similar to concave lens of asymmetrical form. Substance of cloud C_2, of smaller diameter having much larger mass penetrates into cloud C_1 – lightweight, of a greater diameter.

Density contrast diagram, shown on Fig. 6 indicates a fast (relatively to space time scale) spatial intermittency of supersonic flow, accompanied amplified KH instability and disturbance of gas at outer boundaries of clouds. During cloud deformation two-three compressed density lens-like clumps and rolled rings arise here (Figs. 6 and 7).

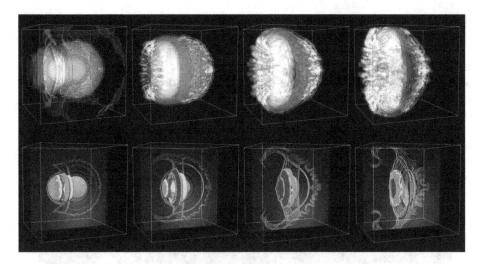

Fig. 7. Density contrast fields $\chi = 1, 5, 40, 100$ in time evolution for central impact of MCs. Isosurfaces of $\chi = 100$ and schlierens in the middle plane. Time sequence from $t = 10$ to $t = 25$.

The impulse does not lead to appearance of angular momentum. In work [7], using SPH modeling, a small rotation around central axis of clumps was discovered. Possibly

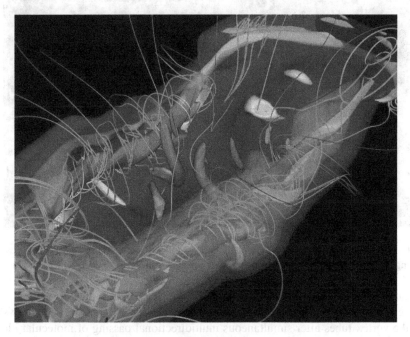

Fig. 8. Rolled filament layers of density contrast $\chi = 4$, wake vortices, indicated by $Q = 3$, and helix pathlines for case of MC/MC glancing collision at time $t = 15$.

it is due by mathematical setup to initial conditions in smooth particle approach and initial perturbation in solution.

The hot gas of colliding MCs is cooling rapidly. It can significantly affect the possible collapse of cool and heavily compressed protocores after collision. On figures given above one can see that from time moment t = 15 mutable density clumps began to break up leading to density fragmentation. Results of simulation are correlated well with observed data [11, 12].

In the case of glancing collision more heave molecular cloud penetrates into more light and friable side of another one. MC$_2$ cuts gas "hollow" in side of MC$_1$, boundary layers and conditional edges of which begin to roll. Time evolution of MCs separation process is shown in Figs. 8 and 9.

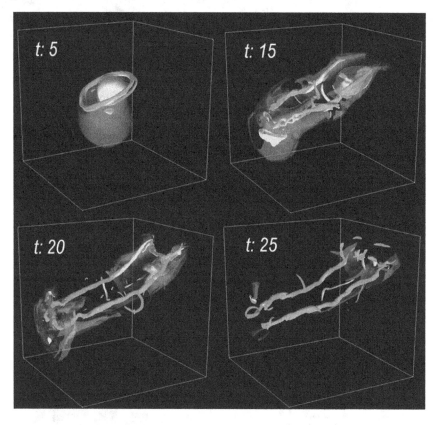

Fig. 9. Envelope layers of density contrast χ = 5, 10, 50 and vortex indicator Q = 5 for case of glancing collision of MCs. Origination of vortex tubes over filament sheet edges is shown.

Bound of mixing zone becomes film-like with curvilinear profile of their outer surfaces. Initially closed edges of cutting concavity become twisted and stretch. System of wake vortex tubes after simultaneous multidirectional passing of molecular clouds fairly stable on time scales examined in simulation.

Lens-like and cone-like clumps can be fragmented and ablated. Strong shock waves, dynamical supersonic collisions and local high compression zones shaking can be amplified by self-gravitation and MHD deformation of filaments originated in MCs.

The simulation SW/MCs/MC collisions taking into account self-gravitation process and magnetic condensation requires sharp increase of solution exposure, maybe even ten times over. To clarify possible developments of formed structures with more real time and spatial scales, it is necessary to carry out the numerical research using more power HPC systems.

4 Conclusions

1. Filaments forming and molecular clouds crushing were simulated using the HPC numerical modeling with high spatial resolution grids and parallelization codes developed.
2. The MCs dynamical transformation for different scenario of molecular clouds collision - between shock wave and MCs and impact between them - were analyzed in terms of supersonic perturbations over shocked sheets as the outcome of local strong shock compression.
3. The research has shown the ways the shock interaction initiates supersonic turbulence in mixed clouds, its effect on the filament origin and stratification of gas density, as well as on the transformation of emerging structures.

Acknowledgements. The work has been funded by the Russian Foundation for Basic Research grants No. 16-29-15099, 17-07-00569.

References

1. Ferriere, K.M.: The interstellar environment of our galaxy. Rev. Mod. Phys. **73**, 1031–1066 (2001)
2. Vázquez—Semandeni, E., Ostriker, E.C., Passot, T., Gammie, C.F., Stone, J.M.: Compressible MHD turbulence: implications for molecular cloud and star formation. In: Mannings, V. et al. (eds.), Protostars and Planets IV, University of Arizona, Tucson, pp. 3–28 (2000)
3. Beuther, H., Ragan, S.E., Johnston, K., Henning, T., Hacar, A., Kainulainen, J.T.: Filament fragmentation in high-mass star formation. A&A **584**(A67), 1–12 (2015)
4. Truelove, J.K., Klein, R.I., McKee, C.F., Holliman, J.H., Howell, L.H., Greenough, J.A., Woods, D.T.: Self-gravitational hydrodynamics with 3-D adaptive mesh refinement: methodology and applications to molecular cloud collapse and fragmentation. ApJ **495**, 821 (1998)
5. Bate, M.R., Bonnell, I.A., Price, N.M.: Modeling accretion in protobinary systems, monthly notices of the royal astronomical society. MNRAS **277**(2), 362–376 (1995)
6. Price, D.J., Monaghan, J.J.: An energy conserving formalism for adaptive gravitational force softening in SPH and N-body codes. MNRAS **374**, 1347–1358 (2007)
7. Vinogradov, S.B., Berczik, P.P.: The study of colliding molecular clumps evolution. Astron. Astrophys. Trans. **25**(4), 299–316 (2006)

8. Arreagga-Garcia, G., Klapp, J., Morales, J.S.: Simulations of colliding uniform density H_2 clouds. Int. J. Astr. Astrophys. **4**, 192–220 (2014)

9. Lucas, W.E., Bonnel, I.A., Forgan, D.H.: Can the removal of molecular cloud envelopes by external feedback affect the efficiency of star formation? MNRAS **469**(2) (2017)

10. Pittard, J.M., Falle, S.A.E.G., Hartquist, T.W., Dyson, J.E.: The turbulent destruction of clouds. MNRAS **394**, 1351–1378 (2009)

11. Elmegreen, B.G.: Star formation in a crossing time. Astrophys. J. **530**(277), 281 (2000)

12. Dawson, J.R., Ntormousi, E., Fukui, Y., Hayakawa, T., Fierlinger, K.: The Astrophysical Journal, **799**(64) (2015)

13. Johansson, E.P.G., Ziegler, U.: Radiative interaction of shocks with small interstellar clouds as a pre-stage to star formation. Astrophys. J. **766**, 1–20 (2011)

14. Nakamura, F., McKee, C.F., Klein, R.I., Fisher, R.T.: On the hydrodynamic interaction of shock waves with interstellar clouds. II. The effect of smooth cloud boundaries on cloud destruction and cloud turbulence. Astrophys. J. **164**, 477–505 (2006)

15. Pittard, J.M., Parkin, E.R.: The turbulent destruction of clouds – III. Three dimensional adiabatic shock-cloud simulations. MNRAS **457**(4), 1–30 (2015)

16. Pittard, J.M., Goldsmith, K.J.A.: Numerical simulations of a shock-filament interaction. MNRAS **458**(1), 1–25 (2015)

17. Melioli, C., de Gouveia Dal Pino, E., Raga, A.: Multidimensional hydro dynamical simulations of radiative cooling SNRs-clouds interactions: an application to starburst environments. Astronomy Astrophys. **443**, 495–508 (2005)

18. Godunov, S.K.: A difference scheme for numerical solution of discontinuous solution of hydrodynamic equations. Math. Sbornik. **47**, 271–306 (1959)

19. Harten, A.: High resolution schemes for hyperbolic conservation laws. J. Comp. Phys. **49**, 357–393 (1983)

20. Toro, E.F.: Riemann Solvers and Numerical Methods for Fluid Dynamics, 727 p. Springer, Heidelberg (1997)

21. Strang, G.: On the construction and comparison of difference schemes. SIAM J. Numer. Anal. **5**, 506–517 (1968)

22. Rybakin, B.P., Stamov, L.I., Egorova, E.V.: Accelerated solution of problems of combustion gas dynamics on GPUs. Comput. Fluids **90**, 164–171 (2014)

23. Rybakin, B., Smirnov, N., Goryachev, V.: Parallel algorithm for simulation of fragmentation and formation of filamentous structures in molecular clouds. In: Voevodin, V., Sobolev, S. (eds.) CCIS 687, RuSCDays 2016, vol. 687, pp. 146–157 (2016)

24. Kritsuk, A.G., Norman, M.L., Padoan, P., Wagner, R.: In turbulence and nonlinear processes in astrophysical plasmas. American Institute of Physical Conference Series, Shaikh, D., Zank, G.P. (eds.), vol. 932, pp. 393–399, ApJ, 665, 416 (2007)

25. Shu, F.H., Adams, F.C., Lizano, S.: Star formation in molecular clouds - observation and theory. ARA&A, pp. 25–23 (1987)

The Combinatorial Modelling Approach to Study Sustainable Energy Development of Vietnam

Aleksey Edelev[1(✉)], Valeriy Zorkaltsev[1], Sergey Gorsky[2], Doan Van Binh[3], and Nguyen Hoai Nam[3]

[1] ESI SB RAS, Irkutsk, Russia
{flower,zork}@isem.irk.ru
[2] ISDCT SB RAS, Irkutsk, Russia
gorsky@icc.ru
[3] IES VAST, Hanoi, Vietnam
{doanbinh,nhnam}@ies.vast.vn

Abstract. The article describes the combinatorial modelling approach to the research of energy sector development. The idea of the approach is to model a system development in the form of a directed graph which nodes correspond to the possible states of a system at certain moments of time and arcs characterize the possibility of transitions from one state to another. The combinatorial modelling is a visual representation of dynamic discrete alternatives and permits to simulate the long-term process of system development at various possible external and internal conditions, to determine an optimal development strategy of the system under study. The formation and analysis procedures of energy development options are implemented in the Corrective software package. The heterogeneous distributed computing environment is needed to compute an energy sector development graph. In 2015 Institute of Energy Science of Vietnamese Academy of Science and Technology performed the study of Vietnam sustainable energy development from 2015 to 2030. Based on data of this study the combinatorial modelling methods are applied to the formation and analysis of Vietnam energy development options taking into account energy security requirements. The created Vietnam energy sector development graph consists of 531442 nodes. It is computed on the cluster located at Institute for System Dynamics and Control Theory of Siberian Branch of Russian Academy of Science (Irkutsk) under control of the Orlando Tools software package. The found optimal path of Vietnam sustainable energy development provides the minimum costs of energy sector development and operation.

Keywords: Combinatorial modelling · Energy sector · Decision support · Distributed computing environment

1 Introduction

The study of long-term energy development with regard to uncertainty (ambiguity) of the initial information and development conditions [1] should be conducted on the basis of general energy research approaches [2, 3] with the use of special methods, models,

V. Voevodin and S. Sobolev (Eds.): RuSCDays 2017, CCIS 793, pp. 207–218, 2017.
https://doi.org/10.1007/978-3-319-71255-0_16

databases and software. The models should consider rather long time period (30-40 years) and distinguish several stages in the development and operation of energy systems. Also models should explicitly consider discreteness of the energy facilities development options. Tools to generate and analyse energy development options must be well-founded and flexible. They should be established on some general organizing research, algorithms to create and choose energy development options.

It is impossible to describe and test all distinctive combination of external conditions and energy development options within frames of an energy sector model taking into account uncertainty, energy security threats and other factors. It leads to a huge number of possible energy sector states and takes a lot of time to generate and analyse using usual methods of research. To deal with this issue the combinatorial modelling approach is used. The combinatorial modelling is a visual representation of dynamic discrete alternatives and permits to simulate the long-term process of system development at various possible external and internal conditions, to determine an optimal development strategy of the system under study.

This article describes the software that implements some combinatorial modelling approach procedures and considers their application to study some problems of sustainable energy development of Vietnam.

2   The Energy Sector Model

The balance economic-mathematical model [4] evaluates the energy sector state at a certain time period with regard to energy security (ES) requirements [5–7]. The model possibilities are quite close to MARKAL [8], MESSAGE [9], EFOM-ENV [10], TIMES [11], Balmorel [12] and others. The model allows:

- Considering a whole energy sector from the production of energy resources to final consumption in the various economic sectors including all stages of energy transformation;
- Investigating energy technological and territorial structure development.

The energy sector model is the following linear programming problem:

$$AX - \sum_{t=1}^{T} Y^t = 0, \tag{1}$$

$$0 \leq X \leq D, \tag{2}$$

$$0 \leq Y^t \leq R^t, \tag{3}$$

where t is a category of consumers; X – the decision vector whose components represent the intensity of energy facilities usage (storage, production, transformation and transmission of energy resources); Y_t – the decision vector whose components characterize the energy resource consumption for different categories t; A – the matrix of facilities technology factors (production, transformation) and transmission of energy resources;

D – the vector that determines technically possible capacities of production, transformation and transmission facilities; R_t – the vector that defines energy resources demands of the category t.

The objective function is as follows:

$$(C, X) + \sum_{t=1}^{T} \left(r^t, g^t \right) \rightarrow min \tag{4}$$

The first component of this objective function reflects the operation costs of the energy sector. The vector C contains unit functioning costs for the existing, reconstructed, upgraded and newly built production, transformation and transmission facilities.

The second component represents the losses due to the energy resources deficit for the different consumer categories. The energy resources deficit g^t of the category t is equal to the difference between R_t and Y_t. Vector r^t consists of the components called "specific losses" for consumer of category t.

3   The Combinatorial Modelling Approach

The procedures of formation and analysis of energy development options are based on the representation of components belong to an investigated system in the form of a directed graph. The graph nodes correspond to the possible states of components in the certain moments or cuts of time. The graph arcs define the admissibility of transitions between states. The research of the whole system development is performed by analyzing various combinations of states and transitions of particular components. This approach is known as combinatorial modelling [13].

A component is a structural unit of the system under research. It may be a factory, power plant, set of the similar energy sources or a consumer category. The degree of aggregation of the energy production or consumption facilities depends mostly on the goals of study and data base possibilities.

The first step of the combinatorial modelling approach is to describe the basic scenario of energy development to investigate as a graph with one node for each cut of time (see Fig. 1). These nodes contain the essential information to create new possible states of the energy sector.

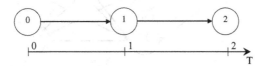

Fig. 1. Basic scenario of energy sector development

At second step the infrastructure of energy sector is separated into several components by territorial or industrial criteria. For the each component a development graph is built. It contains changes of energy facility parameters at the considered time period. The development graphs of two energy facilities are shown on Fig. 2. The source nodes corresponding to time 0 do not have numbers because they will not participate the next generation of the energy sector graph.

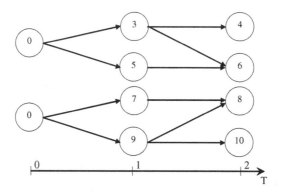

Fig. 2. Development graphs of 2 energy facilities

The third step is combining data of reference graph with information of different component graphs belonged to the same moment in time. This results in the set of possible states of energy sector for each moment in time. The created states (nodes) are linked by transitions (arcs) to form an energy sector development graph.

An energy sector development graph shown on Fig. 3 is created by means of combination of nodes and edges of the graphs on Figs. 1 and 2. The number of generated possible energy sector state is shown inside circle on Fig. 3. The numbers above a circle are combination of the graph nodes on Figs. 1 and 2. The beginning of all paths in the generated energy sector development graph is common initial node at moment 0.

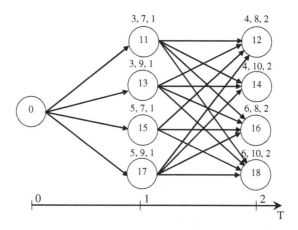

Fig. 3. Energy sector development graph

The forth step is to check the validity of nodes and arcs of the energy sector development graph since not all possible energy sector states and transitions can be valid.

For this purpose, there are system-wide constraints in the combinatorial modeling. Among them it can be distinguished two types:

1. Logical conditions. Some development alternatives of a component can depend on the implementation of certain development variants of other components.
2. Balance and other design constraints. These are restrictions on the available raw resources and products at every cut of time and transition. They can be defined in the form of balance equations or inequalities.

Lists of pairs of incompatible nodes are used to implement logical conditions. A couple of incompatible nodes is a pair of nodes of the different component graphs and their combination in a possible system state is not possible or does not make sense for some reasons.

The model of energy sector described above is of the second type of system-wide constraints. The admissibility of an energy sector state depends on the correctness of the decision results.

If ES requirements exist then ES status of a possible energy sector state is estimated by means of ES indicators. ES indicator value is calculated on the basis of the economic-mathematical model of energy sector decision results. The ES status is determined by comparison of ES indicators values and thresholds.

The energy sector development graph shown on Fig. 3 has four nodes that did not pass the validity check (see Fig. 4).

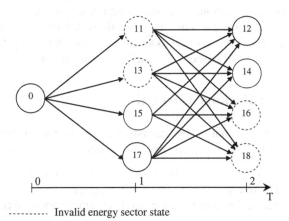

-------- Invalid energy sector state

Fig. 4. The validity check of energy sector states

The fifth stage is to build a graph containing valid states and transitions. States and transitions that are unreachable from the initial state are determined during the passage from the initial node to the end nodes. After that blind states and transitions are determined during the reverse passage. It is impossible to build a path from the initial node to the nodes at last time moment with blind states and transitions. The invalid, unreachable and blind states and transitions are removed from the graph which contains possible energy sector states and transitions.

At the last stage a set of optimal and close to optimal paths is determined by finding shortest paths from the initial node to end nodes with a criterion.

The graph which consists of valid energy sector states and transition is shown on Fig. 5. It was made from the graph shown on Fig. 4 where an optimal way to ensure minimum costs of energy sector development and operation is presented by the bold lines.

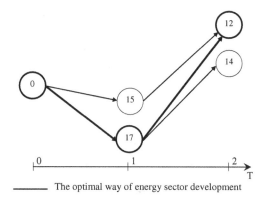

Fig. 5. The development graph with valid energy sector states and transitions

The main issue of the combinatorial modeling implementation is to deal with large number of the simulated system states and transitions. It grows exponentially with the increasing number of system components and their states. That is why the combinatorial modeling approach is usually used with distributed computation technologies [14].

4   The Software Package Corrective

The above procedure of formation and analysis of energy sector development are implemented in the software package Corrective. It consists of the following modules:

1. module m_1 to design basic scenario of energy sector development to study,
2. module m_2 to create energy sector development graph,
3. module m_3 to check the validity of a possible energy sector state (node of development graph),
4. module m_4 to support expert analysis of energy sector development paths.

The scheme of information and logical links between modules of software package Corrective is shown on Fig. 6 in the form of bipartite directed graph where modules m_1, m_2, m_3, m_4 are black ovals.

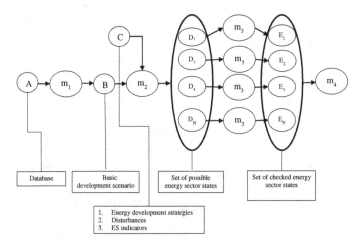

Fig. 6. The logical scheme of software package Corrective

5 Making Scalable Computations with the Software Package Orlando Tools

The Corrective software works in a heterogeneous distributed computing environment (HDCE) by means of the software package Orlando Tools [15]. It is a set of tools to create scalable applications. The architecture of Orlando Tools consists of the next main components: the user interface, the model designer, the knowledge base, the executive subsystem and the computations database (see Fig. 7).

The interface is implemented as a Web application and provides access to other components. The aim of the model constructor is to make the declarative specification of computational knowledge about modules of an application that solves the domain specific tasks, the knowledge of modular structure of a domain specific model and algorithms, the knowledge to support the decision making to choose the optimal computational algorithms depending on the HDCE conditions as well as the software and hardware parameters and administrative characteristics of the HDCE nodes. The model constructor provides the textual (as an XML document) and graphical (as diagrams) notations of the domain specific model. The model is stored in the knowledge base.

The executive subsystem includes an interpreter of the schemes to solve the domain specific tasks and a scheduler. The interpreter processes the control structures and executes the schemes to solve the domain specific tasks. The scheduler performs the schemes decomposition for the better HDCE communication optimization and load balancing. The decomposition can be made before the computation starts or immediately during the computation process. The initial data and the solution are stored in the computations database.

The parameters and operations of the Corrective package domain area as well as their interrelationships are shown on Fig. 8. The rounded rectangles are modules m_2, m_3, m_4. The ovals inside a rectangle are the module input and output parameters. The folder parameter is a temporary folder to keep the intermediate data. Square brackets in the

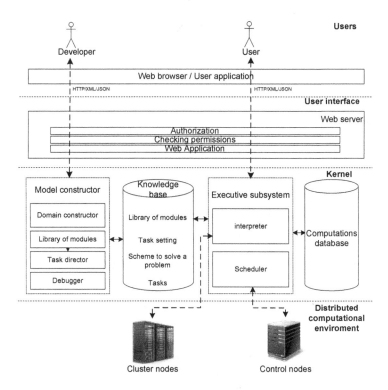

Fig. 7. The architecture of Orlando Tools

name denote that the operations with this parameter or module can be done in parallel. The output of the module m_4 is a result archive file.

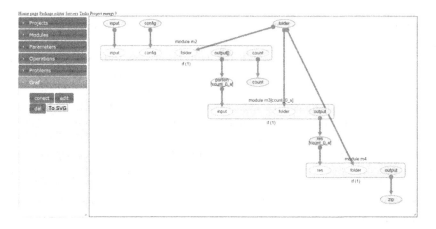

Fig. 8. The Orlando Tools model constructor with the Corrective package domain scheme

6   Modelling Sustainable Development of Vietnam Energy Sector

The Vietnam energy sector model was developed from 2011 to 2015 on the basis of the energy sector model presented above during the joint research conducted by the Melentiev Energy Systems Institute of Siberian Branch of the Russian Academy of Science (ESI SB RAS) and the Institute of Energy Science of Vietnamese Academy of Science and Technology (IES VAST) [16].

In order to analyse the characteristics of the key socio-economic regions the Vietnam energy sector structure and some other related issues the supply and demand balance are calculated according to eight regions: Red River Delta, Northeast, Northwest, North Central, South Central Coast, Highlands, Southeast and Mekong Delta. Input data includes energy supply (costs and value of production, import and export), conversion and transportation of energy, energy consumption by types of energy - fuel including coal, oil and gas and power system. Specifically, the regional parameters of production capacity, costs of production, transport capacity, transport costs are built on the basis of data from the individual production and transportation facilities. The data on regional energy consumption is extracted and calculated on energy consumption of five key economic sectors: industry, agriculture, transportation, commerce- service and residential.

In 2015 IES VAST with the help of module m_1 [17] investigated the sustainable energy development of Vietnam from 2015 to 2030 with regard to the energy security requirements. The energy development scenarios are assessed on energy security and sustainable development criteria. These scenarios should meet the national energy demand for the socio-economic development; apply the suitable and efficient energy technology, minimize the environmental impacts from the energy system, and achieve the cost effective energy system development.

The different energy development scenarios for the period 2020-2030 were built on the basis of capacity fluctuation of the following energy facilities: domestic coal production capacity (baseline, increase by 10%, decrease by 10%), domestic natural gas production capacity (baseline, increase by 10%, decrease by 10%) and domestic hydropower generation capacity (baseline, increase by 10%, decrease by 10%).

In the optimal way, natural gas capacity increases by 2020 to meet the national energy demand than follows the base scenario by 2025 and 2030. Hydropower capacity remains stable for the whole period 2020–2030, while the coal capacity reduces by 10% by 2020.

Below the algorithms for combinatorial modeling were applied using the same assumptions and data for the formation and analysis of Vietnam energy development [18, 19].

At the first stage, the basic energy sector development graph was created. At the second stage the component development graphs were built for the pairs of industries and regions of Vietnam which marked with "+" in the Table 1. A typical component development graph is shown on Fig. 9 where the component capacity fluctuation is shown in the circles.

Table 1. Energy industries and regions of Vietnam

Region	Domestic coal production	Domestic natural gas production	Domestic hydropower generation
Red River Delta	+		
North East	+		+
North West			+
North Central Coast		+	+
South Central Coast			+
Central Highland			+

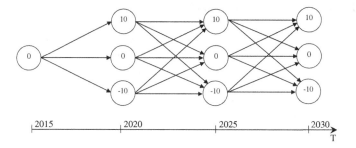

Fig. 9. A typical development graph of Vietnam energy industry

At the third stage, an energy sector development graph that consists of 531 442 nodes is created. At the next stage, the computational experiment on the new energy development graph is performed with HDCE which includes nodes of the high-performance cluster Academician V.M. Matrosov [20]. The cluster is located at Institute for System Dynamics and Control Theory of Siberian Branch of Russian Academy of Science (Irkutsk) and includes 60 computational nodes T-Blade V205S. Each computational node contains two AMD Opteron 6276 «Interlagos» processors with 16 cores.

At the fifth stage, the optimal path is found with minimum costs of development and operation criterion where the natural gas production increases and the coal production reduces for all time moments.

The creation of the energy sector development graph and its computation on 40 cores took 7 h 4 min. The computational time of one possible energy sector state was about 0.0365 s.

7   Conclusions

Traditionally while comparing development options by multiple-criteria decision analysis such as the analytic-hierarchical approach used by IES VAST experts the researchers usually compare the rather small number of options. Usually a choice depends on intuition and experience of the researchers. However, such a limited choice always reflects some subjectivity which reduces the evidence level of the obtained results.

The advantages of the combinatorial modelling are the clarity and compactness of representation of a modelled system development options in the form of a directed graph. A graph clearly illustrates as differences of various system development paths as their common states and transitions.

By the advantages are achievable with the completeness of their description. The traditional approaches to compare the development options based on the multi-criteria methods usually permit a researcher to choose a few options only. A choice depends on the researcher's intuition and experience. Such a choice even if it is right always reflects certain subjectivity and thus depreciates the level of result proof.

A result set of the admissible system development paths can be used in many forecasting tasks where, for example, it's necessary to take into account the uncertainty issue. Among the admissible system paths one can choose not only the best way but also paths close to it according to specified criteria.

Acknowledgments. The research was partially supported by Russian Foundation of Basic Research, project no. 15-07-07412a.

References

1. Saneev, B.G., Sokolov, A.D., Agafonov, G.V.: Methods and models of the development of regional energy programs, 140 p. Nauka, Novosibirsk (2003)
2. The Eastern vector of Russia's energy strategy: the current state of opinion in the future, 368 p. Novosibirsk Academic Publishing House "Geo" (2011)
3. Makarov, A.A., Mitrova, T.A., Malakhov, V.A.: World energy forecast and consequences for Russia. Stud. Russ. Econ. Dev. **24**(6), 511–519 (2013)
4. Zorkaltsev V.I.: The methods of forecasting and analysis of the efficiency of the fuel supply system operation, 144 p. M., Nauka (1988)
5. Voropai, N.I., Klimenko, S.M., Krivorutsky, L.D., Pyatkova, N.I., Rabchuk, V.I., Senderov, S.M., Trufanov, V.V., Cheltsov, M.B.: Comprehensive substantiation of the adaptive development of energy systems in terms of changing external conditions. Int. J. Global Energy Issue **20**(4), 416–424 (2003)
6. Senderov, S.: Energy security of the largest asia pacific countries: main trends. Int. J. Energy Power **1**(1), 1–6 (2012)
7. Ang, B.W., Choong, W.L., Ng, T.S.: Energy security: definitions, dimensions and indexes. Renew. Sustain. Energy Rev. **42**, 1077–1093 (2015)
8. Fishbone, L.G., Abilock, H.: MARKAL, a linear-programming model for energy systems analysis: technical description of the BNL version. Int. J. Energy Res. **5**, 353–375 (1981)
9. Gerking, H.: Modeling of multi-stage decision making process in multi-period energy models. Eur. J. Oper. Res. **32**(2), 191–204 (1987)
10. Van der Voort, E., et al.: Energy supply modelling package. In: EFOM-12C Mark I, Mathematical Description, 429 p., Louvain-La-Neuve (1984)
11. Loulou, R., Labriet, M.: ETSAP-TIAM: the TIMES integrated assessment model part I: model structure. CMS **5**(1), 7–40 (2008)
12. Ravn, H.F., Munksgaard, J., Ramskov, J., Grohnheit, P.E., Larsen, H.V.: Balmorel: a model for analyses of the electricity and CHP markets in the Baltic Sea Region. Appendices (No. NEI-DK–3934), Elkraft System (2001)

13. Zorkaltsev, V.I., Hamisov, O.V.: The equilibrium model of the economy and energy, 221 p. Nauka Publ., Novosibirsk (2006)

14. Edelev, A., Sidorov, I.: Combinatorial modeling approach to find rational ways of energy development with regard to energy security requirements. In: Dimov, I., Faragó, I., Vulkov, L. (eds.) NAA 2016. LNCS, vol. 10187, pp. 317–324. Springer, Cham (2017). http://doi.org/10.1007/978-3-319-57099-0_34

15. Feoktistov, A.G., Gorskij, S.A.: The language of computational models specification in large-scale program packages. Mod. Sci.-Intensive Technol. **7**(1), 84–88 (2016)

16. Edelev, A.V., Ninh, N.Q., Van The, N., Hung, T.V., Tu, L.T., Duong, D.B., Nam, N.H.: Developing "Corrective" software: 3-region model. In: International Conference Green Energy and Development, Hanoi, Vietnam, pp. 41–52 (2012)

17. Edelev, A.V., Pyatkova, N.I., Chemezov, A.V., Nam, N.H.: Corrective software to study long-term development of energy sector of Vietnam. Softw. Syst. **4**, 211–216 (2014)

18. Edelev, A.V., Sidorov, I.A., Van Binh, D., Nam, N.H.: The approach to find rational energy development ways in terms of energy security requirements. In: International Conference of Science and Technology. 50-th Anniversary of Electric Power University, Hanoi, pp. 548–556 (2016)

19. Edelev, A.V., Pyatkova, N.I., Sidorov, I.A., Van Binh, D., Nam, N.H.: An approach to studying sustainable energy development and energy security problems of Vietnam. Sci. Bull. NSTU **64**(3), 175–193 (2016)

20. Irkutsk Supercomputer Centre of SB RAS, http://hpc.icc.ru. Accessed 14 Apr 2017

Ani3D-Extension of Parallel Platform INMOST and Hydrodynamic Applications

Vasily Kramarenko[1,2(✉)], Igor Konshin[1,3], and Yuri Vassilevski[1,2]

[1] Institute of Numerical Mathematics of the Russian Academy of Sciences,
Moscow 119333, Russia
kramarenko.vasiliy@gmail.com, igor.konshin@gmail.com,
yuri.vassilevski@gmail.com
[2] Moscow Institute of Physics and Technology,
Dolgoprudny, Moscow Region 141701, Russia
[3] Dorodnicyn Computing Centre, FRC CSC RAS, Moscow 119333, Russia

Abstract. The paper is devoted to an extension of the parallel platform INMOST by finite element and meshing libraries of the Ani3D software package. The extension allows us to develop parallel finite element solvers of boundary value problems and, in particular, hydrodynamic problems. The Ani3D package allows one to build, refine, locally adapt and improve the quality of tetrahedral meshes, perform finite element discretizations of partial differential equations for various types of finite elements, solve the appearing algebraic systems, and visualize the discrete solutions. The INMOST software platform provides tools for creating and storing distributed general conformal grids with arbitrary polyhedral cells, parallel assembling and parallel solution of arising distributed linear systems. We present the integration of two libraries from Ani3D into INMOST platform and demonstrate the functionality of the joint software on the solution of two model hydrodynamic problems on multiprocessor systems.

Keywords: Parallel computing · Finite element method · Parallel solvers · Hydrodynamic problems

1 Introduction

We consider an extension of the parallel platform INMOST by finite element and meshing libraries of the Ani3D software package. The extension provides a tool for developing parallel finite element solvers of boundary value problems and, in particular, hydrodynamic problems. The Ani3D package [1] offers advanced finite element discretizations on tetrahedral meshes and various options of tetrahedral mesh generation, refinement, and adaptation. The parallel platform INMOST [2] provides tools for creating and storing distributed general conformal grids with arbitrary polyhedral cells, parallel assembling and parallel solution of arising distributed linear systems. Integration of two libraries from Ani3D into the INMOST platform offers a new technology of parallel solution of boundary

© Springer International Publishing AG 2017
V. Voevodin and S. Sobolev (Eds.): RuSCDays 2017, CCIS 793, pp. 219–228, 2017.
https://doi.org/10.1007/978-3-319-71255-0_17

value problems. Functionality of the joint software is demonstrated on the solution of two model hydrodynamic problems on multiprocessor systems.

The paper is organized as follows. Sections 2 and 3 contain brief descriptions of Ani3D and INMOST packages, respectively. Section 4 provides technical details of merging the packages. Section 5 demonstrates the parallel solution of two model hydrodynamic problems.

2   Ani3D Package

The package Ani3D [1] is developed for generation of unstructured tetrahedral meshes, adaptation of these meshes isotropically or anisotropically, discretization of PDE systems, solution of linear and nonlinear systems, visualization of meshes and associated solutions. It consists of a set of independent libraries oriented to the solution of the specific tasks. All these libraries allow a user to operate with data in sequential mode only. We consider the Ani3D-extension of the INMOST platform by two Ani3D libraries, Ani3D-MBA and Ani3D-FEM.

The main purpose of the Ani3D-MBA library is generation of conformal tetrahedral meshes which are quasi-uniform in a given metric. Additionally, the library provides tools to read/write a tetrahedral mesh from/to the disk in a specific Ani3D format and to perform its uniform mesh refinement by splitting each tetrahedron into 8 sub-tetrahedra.

The Ani3D-FEM library provides a flexible interface to generate a local finite element discretization (local matrix and right-hand side vector) on a mesh tetrahedron and to assemble the local discretizations into a global system of grid equations. Importantly, the local discretization may involve different types of finite elements: for instance, the local matrix for the Stokes problem may exploit quadratic basis functions for velocity and linear basis functions for pressure unknowns. Our finite element extension of the INMOST platform uses a user-defined subroutine FEM3Dext where the local finite element matrix is generated. The library Ani3D-FEM is equipped with a great number of examples of this subroutine for various applications. The rules for creation of the subroutine FEM3Dext and respective examples can be found in Ani3D documentation [1]. In particular, the user should specify explicitly the order of cell elements collocating the user's finite element basis functions. For instance, quadratic basis functions have four degrees of freedom collocated at the vertices of the tetrahedron and six degrees of freedom collocated at the mid-edges of the tetrahedron. Also, the user may transfer user data to the subroutine FEM3Dext with the help of special working arrays. This feature can be used for acquiring the solution from the previous time step which is inevitable for unsteady time stepping implementations.

3   INMOST Platform

The INMOST software platform [2] is instrumented for creating and storing distributed general conformal grids with arbitrary polyhedral cells, parallel assembling of systems of grid equations and their parallel solution. However, INMOST does not provide software for generation and assembling of local finite element discretizations. In order to add to INMOST the finite element environment from Ani3D-FEM, we take advantage of Mesh class, Solver class, and Sparse::Matrix class of INMOST.

The Mesh class is designed for storing distributed grids. It contains a number of cells consisting of nodes, faces, edges. In parallel mode each processor has a sets of "owned" and "shared" elements and a set of "ghost" elements. Each "ghost" element in fact is the copy of an element owned by another processor and marked as "shared" there. These elements are used for the construction of overlapping communication layers between processors. We note that some cell can be "ghost" for Processor A, but its node/edge/face can be owned by Processor B.

The important data structure of INMOST is Tag which is used to connect any data with a mesh element, i.e. cell, face, edge, or node. The simplest case of tagged data is a real or integer array associated with every mesh element of particular type (e.g. every edge). The main function of Tag is to provide automatic exchanges of tagged data between neighboring processors.

The important feature of Sparse::Matrix class is that it stores the matrix by rows in parallel regime. Processor B cannot add entries to a row owned by Processor A. In order to assemble local matrices in parallel, one has to use special numbering of rows.

No special features of Solver class will be used for our purposes. By this reason, any of five inner linear solvers or five external linear solvers from PETSc, Trilinos, and SuperLU can be exploited in the same interface.

The detailed description of INMOST software platform can be found in [3–5].

4   Parallelization Technology

In this section we present technological details of integration of the Ani3D uniform mesh refinement and the Ani3D local finite element matrix generation into the INMOST platform.

The first part of the Ani3D–INMOST technology allows us to generate huge meshes on multi-processor systems. The major steps are reading the initial mesh, its partitioning, refinement, and merging the refined submeshes into a global distributed mesh. Reading the mesh is performed by a standard Ani3D-MBA library routine on the root processor. The result of the reading, the object INMOST::Mesh, is processed by one of INMOST partitioning algorithms (e.g., ParMETIS [6] partitioning) and redistributed among available processors. Each submesh is refined independently. Parallel multilevel mesh refinement requires mesh conformity control. The uniform refinement function in Ani3D guaranties this property provided that the initial (coarse) submeshes form a conformal global mesh and the number of refinement levels is the same on all processors. Merging the fine submeshes removes duplicate nodes, edges, and faces in the global fine mesh. Once the fine submeshes are generated and processed to constitute the global conformal mesh, we construct an additional layer of "ghost" cells. This layer is needed for correct assembling of the global finite element matrix.

The second part of the integrated Ani3D–INMOST software initializes the INMOST tags and data which will be used in assembling of local finite element matrices generated by Ani3D-FEM library. First, all mesh elements involved in the discretization are numbered within the global mesh. INMOST function `AssignGlobalID` numbers the respective elements (cells and/or faces, edges, nodes) marked by appropriate mask. Second, for all globally numbered elements a special tag is created, the size of the tag in each cell being equal to the number of finite element degrees of freedom associated to the mesh element. Finally, the tags are synchronized between processors. Global numbering and synchronized tags provide easy recovering of the global matrix order as well as the order of matrices owned by processors. Importantly, flexibility for numbering degrees of freedom within INMOST allows the user to generate distributed matrices with desirable ordering. The proper ordering may improve the performance of INMOST linear solvers.

The third part of the Ani3D–INMOST technology assembles the local matrices generated on each tetrahedral cell by the user-defined Ani3D-FEM routine `FEM3Dext`. On each processor, INMOST-based assembling selects rows of each local matrix which correspond to owned (by the processor) mesh elements, and writes the respective entries to the global matrix and the right-hand side vector.

Once the global system is assembled, any INMOST parallel linear solver can be applied to the solution of the distributed global linear system.

5   Parallel Solution of Model Hydrodynamic Problems

5.1   Stokes Problem

We consider the finite element solution of the Stokes problem in a rectangular 3D channel with a backward step. We impose the non-homogeneous Dirichlet boundary condition (Poiseuille's profile) at the inflow boundary, the homogeneous Neumann boundary condition at the outflow boundary, and the homogeneous Dirichlet boundary condition (no-slip, no-penetration) on the channel walls. A sequence of three quasi-uniform tetrahedral meshes is considered. The coarsest mesh S0 with 25113 cells (see Fig. 1) is uniformly refined to get more finer meshes S1 and S2 with 200904 and 1607232 cells, respectively.

The minimal order Taylor–Hood finite elements are used for the discretization of the Stokes problem. The pressure p is approximated by continuous piecewise linear functions with nodal degrees of freedom, the velocity $\mathbf{v}$ is approximated by continuous piecewise quadratic functions with degrees of freedom collocated at nodes and mid-edges of the mesh (see Fig. 2). The discretization method results in a symmetric saddle-point matrix with zero diagonal pressure block:

$\mathbf{v}_{edge}$	$*$	$\times$	$\times$
$\mathbf{v}_{node}$	$\times$	$*$	$\times$
p_{node}	$\times$	$\times$	0

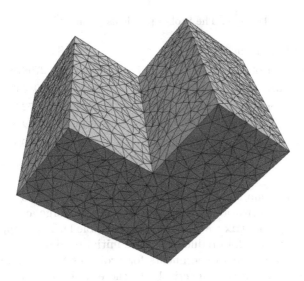

Fig. 1. The coarsest mesh S0

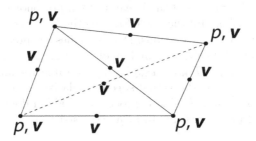

Fig. 2. Unknowns p and $\mathbf{v}$ associated with the vertices and edges of the tetrahedral cell for P1-P2 finite element

The numerical experiments were performed on the INM RAS cluster [7] in the x10core segment:

- Compute Node Arbyte Alkazar+ R2Q50;
- 20 cores (two 10-cores processors Intel Xeon E5-2670v2@2.50GHz);
- 64 GB RAM;
- SUSE Linux Enterprise Server 11 SP3 (x86_64).

Table 1 presents statistics for all three finite element problems.

To solve the linear system with the saddle-point matrix, we used the BiCGstab iterative solver preconditioned by the first order BIILU method [8,9]. In order to avoid zero pivots during ILU factorization, on each processor pressure unknowns are enumerated last so that the zero pressure block be the last diagonal block [10]. Parameters of the BIILU method are as follows: the threshold parameter for the conventional incomplete factorization ILU(τ) $\tau = 0.001$ and the number of overlap levels $q = 2$. The use of the conventional first order

Table 1. The Stokes problems parameters

Problem name	S0	S1	S2
Number of nodes	5187	36824	279903
Number of edges	31637	243079	1908542
Number of tetrahedra	25113	200904	1607232
Matrix size	115659	876533	6845238
Number of nonzeros	10751851	84374191	668849086

incomplete factorization is more beneficial compared to the more robust second order incomplete factorization ILU2(τ_1, τ_2) due to very large average number of nonzero elements per matrix row (about 100, see Table 1). The stopping criterion for the iterations is 10^{12}-fold reduction of the initial residual.

The results of numerical experiments for problems S0, S1, and S2 are presented in Tables 2, 3, and 4, respectively. In these tables p denotes the number of processors used, T_{ini}, T_{ass}, T_{prec}, and T_{iter} are the times for the preliminary data initialization, assembling of the linear system, preconditioner construction, and performing iterations by the BiCGStab method, respectively, N_{iter} stands for the number of BiCGstab iterations, Dens specifies the preconditioner density with respect to that of the original matrix of the system, PM is the number of pivot modifications, $T_{\mathrm{sol}} = T_{\mathrm{prec}} + T_{\mathrm{iter}}$ is the total linear system solution time, while $S = T_{\mathrm{sol}}(1)/T_{\mathrm{sol}}(p)$ is the actual speedup relative to the solution time on one (Tables 2 and 3) or four (Table 4) processors. Table 4 does not contain data for runs on 1 and 2 processors due to memory restrictions.

Table 2. The solution of S0 problem on $p = 1, ..., 32$ processors

p	T_{ini}	T_{ass}	T_{prec}	T_{iter}	N_{iter}	Dens	PM	T_{sol}	S
1	0.06	8.77	3.30	4.26	102	0.81	0	7.56	1.00
2	0.04	5.72	2.27	3.27	132	0.96	0	5.54	1.36
4	0.03	3.95	1.57	2.34	152	1.18	0	3.91	1.93
8	0.02	2.25	1.19	1.81	172	1.53	5	3.00	2.52
16	0.02	1.63	1.20	1.40	182	2.01	3	2.83	2.67
32	0.02	1.49	1.50	1.33	182	2.80	10	2.83	2.67

We do not observe a slowdown even for the solution of system with the smallest matrix S0 on 32 processors when approximately 5000 matrix rows are associated with each processor. For the moderate size matrix S1 the maximal speedup is 7.64, while for the largest matrix S2 the speedup is 5.15 when the number of processors increases from 4 to 32.

Table 3. The solution of S1 problem on $p = 1, ..., 32$ processors

p	T_{ini}	T_{ass}	T_{prec}	T_{iter}	N_{iter}	Dens	PM	T_{sol}	S
1	0.38	69.07	31.41	97.20	242	0.85	0	128.61	1.00
2	0.28	40.97	20.30	76.88	322	0.93	0	97.18	1.32
4	0.21	25.31	12.76	47.88	332	1.03	0	60.64	2.12
8	0.14	13.76	7.83	28.68	332	1.15	1	36.51	3.52
16	0.10	8.37	4.77	18.46	362	1.37	3	23.23	5.53
32	0.06	5.09	4.12	12.70	402	1.65	8	16.82	7.64

Table 4. The solution of S2 problem on $p = 4, ..., 32$ processors

p	T_{ini}	T_{ass}	T_{prec}	T_{iter}	N_{iter}	Dens	PM	T_{sol}	S
4	1.48	181.05	142.29	1484.02	722	0.97	0	1626.31	1.00
8	0.90	94.03	76.01	847.62	802	1.03	0	923.63	1.76
16	0.58	52.03	72.70	481.50	802	1.10	2	554.20	2.93
32	0.37	29.23	26.88	288.73	802	1.21	2	315.61	5.15

5.2 Unsteady Convection-Diffusion Problem

In the second test we consider the finite element solution of an unsteady convection–diffusion problem in a cubic domain. We impose homogeneous Dirichlet boundary conditions on all boundaries of the cube, except a patch centered at one cube face. In the patch the concentration is set to one. The diffusion coefficient is $D = 10^{-4}$, the convection field is the constant vector $\mathbf{v} = (1, 0, 0)$. The problem coefficients imply tongue-type propagation of the concentration in time and space (see Fig. 3). The initial quasi-uniform tetrahedral mesh L0 with 13952 cells is uniformly refined one and two times to produce meshes L1 and L2, respectively. The unknown concentration is approximated by continuous piecewise linear basis functions with nodal degrees of freedom. The finite element discretization of the convection operator is stabilized by the Streamline Upwind Petrov–Galerkin (SUPG) method. The second order implicit backward differentiation formula (BDF) scheme is used for time stepping. The problem is solved for the time period $[0; 0.5]$ with time step $\Delta t = 0.015$.

The numerical experiments were performed on the same computational system with the same computational method as in the previous test. Table 5 presents statistics for all three finite element problems on meshes L0, L1, and L2.

The time measurements for the solution of problems L0, L1, and L2 are presented in Tables 6, 7, and 8, respectively. In these tables p denotes the number of processors used, T_{ini}, T_{ass}, and T_{sol}, are the times for the preliminary data initialization, cumulative time of assembling the linear systems for all time steps, and cumulative time of the solution of all linear systems, respectively. In addition, T_{Σ} is the total problem solution time for all time steps and

$$S = \frac{T_{\text{sol}}(1) + T_{\text{ass}}(1) + T_{\text{ini}}(1)}{T_{\text{sol}}(p) + T_{\text{ass}}(p) + T_{\text{ini}}(p)} \qquad (1)$$

is the actual speedup relative to the solution time on one processor.

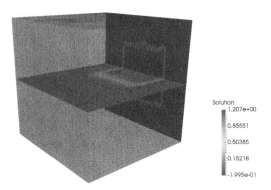

Fig. 3. The concentration at time $t = 0.5$

Table 5. The problems parameters

Problem name	L0	L1	L2
Number of nodes	20417	155905	1218561
Number of tetrahedra	111616	892928	7143424
Matrix size	20417	155905	1218561
Number of nonzeros	291393	2281217	18053121

Table 6. The solution of problem on mesh L0 on $p = 1, ..., 32$ processors

p	T_{ini}	T_{ass}	T_{sol}	T_{Σ}	S
1	2.87	62.68	3.27	68.82	1.00
2	1.93	40.45	2.05	44.43	1.54
4	1.29	26.94	1.24	29.47	2.33
8	0.82	16.76	0.76	18.34	3.75
16	0.58	11.91	0.53	13.02	5.28
32	0.45	8.89	0.55	9.89	6.95

Similarly to the solution of the Stokes problem, we do not observe a slowdown even for the smallest matrix L0 on 32 processors (when less than 700 matrix rows are associated to a processor). For the moderate size matrix L1 the maximal speedup is 11.77, while for the largest matrix L2 the speedup is 19.24. This test shows good scalability of the Ani3D-extension of the INMOST platform for unsteady problems.

Table 7. The solution of problem on mesh L1 on $p = 1, ..., 32$ processors

p	T_{ini}	T_{ass}	T_{sol}	T_Σ	S
1	26.16	552.87	33.86	612.89	1.00
2	16.48	327.93	21.31	365.72	1.67
4	9.83	197.89	12.19	219.91	2.78
8	5.75	114.81	7.28	127.84	4.79
16	3.56	73.57	4.03	81.16	7.55
32	2.31	47.28	2.45	52.04	11.77

Table 8. The solution of problem on mesh L2 on $p = 1, ..., 32$ processors

p	T_{ini}	T_{ass}	T_{sol}	T_Σ	S
1	436.82	5692.31	723.23	6852.36	1.00
2	169.17	2628.09	258.71	3055.97	2.24
4	94.33	1461.24	147.30	1702.87	4.02
8	53.62	874.21	91.24	1019.07	6.72
16	31.52	522.99	53.28	607.79	11.27
32	17.60	308.84	29.61	356.05	19.24

6   Conclusion

We presented the Ani3D-extension of the parallel platform INMOST. The extension widens the functionality of INMOST by the finite element and meshing libraries of the Ani3D software package. Two numerical examples demonstrated the efficiency of the presented approach for the parallel solution of two model hydrodynamic problems.

Acknowledgements. This work has been supported by RFBR grant 17-01-00886.

References

1. Lipnikov, K., Vassilevski, Y., Danilov, A., et al.: Advanced Numerical Instruments 3D. http://sourceforge.net/projects/ani3d. Accessed 15 Apr 2017
2. INMOST - a toolkit for distributed mathematical modeling. http://www.inmost. org. Accessed 15 Apr 2017
3. Vassilevski, Y., Konshin, I., Kopytov, G., Terekhov, K.: INMOST - A Software Platform and Graphical Environment for Development of Parallel Numerical Models on General Meshes. Lomonosov Moscow State University Publ., Moscow (2013). (in Russian)
4. Danilov, A.A., Terekhov, K.M., Konshin, I.N., Vassilevski, Y.V.: Parallel software platform INMOST: a framework for numerical modeling. Supercomput. Front. Innov. **2**(4), 55–66 (2015)

5. Konshin, I., Kapyrin, I., Nikitin, K., Terekhov, K.: Application of the parallel INMOST platform to subsurface flow and transport modelling. In: Wyrzykowski, R., Deelman, E., Dongarra, J., Karczewski, K., Kitowski, J., Wiatr, K. (eds.) PPAM 2015. LNCS, vol. 9574, pp. 277–286. Springer, Cham (2016). https://doi.org/10.1007/978-3-319-32152-3_26

6. ParMETIS - Parallel graph partitioning and fill-reducing matrix ordering. http://glaros.dtc.umn.edu/gkhome/metis/parmetis/overview. Accessed 15 Apr 2017

7. INM RAS cluster. http://cluster2.inm.ras.ru. Accessed 15 Apr 2017. (in Russian)

8. Kaporin, I.E., Konshin, I.N.: Parallel solution of large sparse SPD linear systems based on overlapping domain decomposition. In: Malyshkin, V. (ed.) PaCT 1999. LNCS, vol. 1662, pp. 436–446. Springer, Heidelberg (1999). https://doi.org/10.1007/3-540-48387-X_45

9. Kaporin, I.E., Konshin, I.N.: A parallel block overlap preconditioning with inexact submatrix inversion for linear elasticity problems. Numer. Linear Algebra Appl. **9**(2), 141–162 (2002)

10. Konshin, I., Olshanskii, M., Vassilevski, Y.: ILU preconditioners for nonsymmetric saddle-point matrices with application to the incompressible Navier-Stokes equations. SIAM J. Sci. Comp. **37**(5), A2171–A2197 (2015)

Numerical Simulation of Light Propagation Through Composite and Anisotropic Media Using Supercomputers

Roman Galev[1], Alexey Kudryavtsev[1,2(✉)], and Sergey Trashkeev[3,4]

[1] Khristianovich Institute of Theoretical and Applied Mechanics,
630090 Novosibirsk, Russia
{galev,alex}@itam.nsc.ru

[2] Novosibirsk State University, 630090 Novosibirsk, Russia

[3] Institute of Laser Physics, 630090 Novosibirsk, Russia
sitrskv@mail.ru

[4] Voevodsky Institute of Chemical Kinetics and Combustion,
630090 Novosibirsk, Russia

Abstract. Laser beam propagation through and absorption in composite and anisotropic media is simulated by solving numerically Maxwell's equations with the FDTD method. Laser treatment of materials, light beam transformation in micron-sized optical fiber systems and liquid crystalline materials, generation of optical vortices (beams with non-zero orbital angular momentum) due to interaction with liquid crystal disclinations are considered. Typical grids used for simulations consist of tens and hundreds of millions of cells. The numerical code is parallelized using geometrical domain decomposition and the MPI library for data transfer between nodes of a computational cluster.

Keywords: Computational electromagnetism · FDTD scheme · Laser treatment of materials · Metamaterials · Soft matter · Liquid crystals

1 Introduction

The development of coherent sources of optical radiation is often referred to as "optical revolution". Lasers have had a deep impact on telecommunications, industry, medicine, science and everyday life. Today we are witness to a new great progress in optical technologies which is resulted from a substantial body of pure and applied research in material science and soft matter physics. Optical properties of some soft matter materials (such as liquid crystals) can be substantially altered by weak external electromagnetic fields or thermal and mechanical stresses that opens unique opportunities for dynamic control of light propagation. Optical metamaterials, i.e. structured materials engineered to have optical properties that cannot be found in nature, enable us to manipulate radiation by blocking, absorbing, enhancing, or bending electromagnetic waves and achieve benefits that go far beyond what is possible with conventional materials.

© Springer International Publishing AG 2017
V. Voevodin and S. Sobolev (Eds.): RuSCDays 2017, CCIS 793, pp. 229–240, 2017.
https://doi.org/10.1007/978-3-319-71255-0_18

Numerical simulation is of great importance for better understanding of complex phenomena connected with light propagation through non-homogeneous, anisotropic and structured media as well as for development of new optical technologies. With the advent of modern supercomputers, numerical simulation of quite complicated optical systems and devices based on direct solving Maxwell's equations has become feasible. In the present paper we describe some examples of numerical simulations of problems connected with laser processing of materials, development of fiber-coupled liquid crystal systems and generation of "optical vortices", i.e. light beams with non-zero orbital angular moment, using liquid crystals.

2   Numerical Method and Its Computer Implementation

Maxwell's equations in an anisotropic inhomogeneous medium can be written as

$$\frac{\partial \mathbf{D}}{\partial t} = -(\mathbf{J} + \sigma_e \mathbf{E}) + \nabla \times \mathbf{H}, \qquad\qquad \mathbf{D} = \bar{\bar{\epsilon}}\,\mathbf{E}, \qquad (1)$$

$$\frac{\partial \mathbf{B}}{\partial t} = -(\mathbf{M} + \sigma_m \mathbf{H}) - \nabla \times \mathbf{E}, \qquad\qquad \mathbf{B} = \bar{\bar{\mu}}\,\mathbf{H}. \qquad (2)$$

Here t is time, $\mathbf{r} = (x, y, z)$ is the vector of spatial coordinates, $\nabla \equiv \partial/\partial\mathbf{r}$, $\mathbf{E}$ and $\mathbf{H}$ are the electric and magnetic fields, $\mathbf{D}$ and $\mathbf{B}$ are the densities of electric and magnetic fluxes, $\mathbf{J}$ and $\mathbf{M}$ are the densities of external electric and magnetic currents, $\bar{\bar{\epsilon}}$ and $\bar{\bar{\mu}}$ are the tensors of electrical permittivity and magnetic permeability, σ_e and σ_m are the electric and magnetic conductivities. In our simulations we consider nonmagnetic materials so that $\bar{\bar{\mu}}$ is the identity tensor and $\mathbf{B} \equiv \mathbf{H}$. The magnetic conductivity σ_m does not vanish only in a buffer zone surrounding the computational domain (see below). The electric conductivity σ_e and the positive definite symmetric tensor $\bar{\bar{\epsilon}}$ are assumed to be functions of $\mathbf{r}$.

Equations (1, 2) are solved numerically with the FDTD (Finite-Difference Time-Domain) method [1,2]. The FDTD is a simple but smartly devised and efficient second-order scheme that utilizes a computational grid staggered both in space and time so that electrical and magnetic fields are calculated in alternating time moments and all field components are determined in different points of the computational stencil — see Fig. 1. The FDTD scheme for x-components of Eqs. (1, 2) reads as

$$\frac{D_x\big|_{i,j',k'}^{n+1/2} - D_x\big|_{i,j',k'}^{n-1/2}}{\Delta t} = \frac{H_z\big|_{i,j'',k'}^{n} - H_z\big|_{i,j,k'}^{n}}{\Delta y} - \frac{H_y\big|_{i,j',k''}^{n} - H_y\big|_{i,j',k}^{n}}{\Delta z}$$
$$- J_x\big|_{i,j',k'}^{n} - \frac{1}{2}\sigma_e\big|_{i,j',k'}\left(E_x\big|_{i,j',k'}^{n-1/2} + E_x\big|_{i,j',k'}^{n+1/2}\right), \qquad (3)$$

$$\frac{B_x\big|_{i',j'',k''}^{n+1} - B_x\big|_{i',j'',k''}^{n}}{\Delta t} = \frac{E_y\big|_{i',j'',k'''}^{n+1/2} - E_y\big|_{i',j'',k'}^{n+1/2}}{\Delta z} - \frac{E_z\big|_{i',j''',k''}^{n+1/2} - E_z\big|_{i',j',k''}^{n+1/2}}{\Delta y}$$
$$- M_x\big|_{i',j'',k''}^{n+1/2} - \frac{1}{2}\sigma_m\big|_{i',j'',k''}\left(H_x\big|_{i',j'',k''}^{n} + H_x\big|_{i',j'',k''}^{n+1}\right). \qquad (4)$$

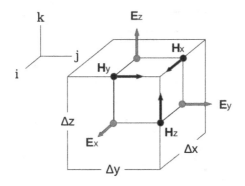

Fig. 1. FDTD computational stencil

Here $j' = j+1/2$, $j'' = j+1$, $j''' = j+3/2$ and similarly for other subscripts. Numerical approximations of remaining Maxwell's equations can be obtained from (3, 4) by a cyclic permutation of x, y, z and the corresponding change in subscripts.

The distributions of **J** and **M** are specified in such a way as to generate the incident electromagnetic wave or beam interacting with a material medium within the computational domain. In order to avoid or minimize false numerical reflections of scattered waves from boundaries of the computational domain, the perfectly matched layer (PML) technique [3,4] is employed. The computational domain is surrounded by a buffer zone filled with an artificial uniaxial anisotropic (with diagonal tensors $\bar{\bar{\epsilon}}$ and $\bar{\bar{\mu}}$) medium whose electric and magnetic conductivities grow rapidly with the distance from the computational domain boundary so that electromagnetic waves are absorbed virtually with no reflection.

In all cases considered below the medium is either anisotropic with a non-diagonal tensor $\bar{\bar{\epsilon}}$ but non-conductive or conductive but isotropic. So the FDTD scheme is only diagonally implicit and the solution on a new time level can be calculated non-iteratively. If the tensor $\bar{\bar{\epsilon}}$ is non-diagonal then, in order to calculate the electrical field **E**, all three components of **D** should be known in the same point that is not the case with the FDTD staggered grid. Thus, two missing components are determined by averaging over 4 neighboring grid nodes. More details about the FDTD method for anisotropic media can be found in [5].

Numerical simulations have been performed on the computational cluster of Novosibirsk State University. The code is written in Fortran-90 and parallelized using MPI. The computational domain is divided into rectangular blocks by planes parallel to the coordinate planes. Each block is assigned to one computational core. The domain decomposition in all three dimensions allows one to decrease the data transfer between processors in comparison with 1D or 2D decomposition. Tests performed on a grid containing $8 \cdot 10^6$ cells have shown that the computation with 16 cores is approximately 11 times faster than with a single core so that the efficiency of parallelization is close to 70%, see Fig. 2.

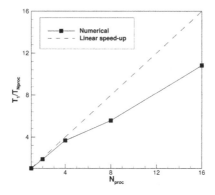

Fig. 2. Numerical speed-up

In different simulations presented below, spatial resolution was from 10 up 30 grid cells per a wavelength. Typically, a grid block assigned to each processor core consisted of $150^3 = 3.375 \cdot 10^6$ cells. The largest grid used contained $6 \cdot 10^8$ cells, 180 cores were employed to perform numerical simulation on this grid.

3   Laser Drilling

Today lasers are successfully applied for cutting, welding and drilling of materials [6,7]. Nevertheless, the problem of prediction of laser energy distribution over the cut surface is still of current interest because of the appearance of new materials and sources of laser radiation as well as growing requirements to the quality of laser treatment of materials.

Usually this problem is solved using geometrical optics approximation: the laser beam is represented as a set of single light rays, which are reflected and refracted on the metal surface according to Fresnel's laws [8,9]. However, in many cases it is not correct because small features of the treated surface can be comparable in size with the radiation wavelength. Below we compare results obtained using wave and geometrical optics approaches.

The problem under consideration is illustrated in Fig. 3a. A Gaussian beam of circularly polarized electromagnetic radiation propagates along the z axis. A cavity in a metal model is aligned coaxially with the beam. The cavity surface shape is specified analytically. The problem formulation corresponds to an initial stage of a laser drilling process. It is required to determine electromagnetic fields and calculate the volumetric density of absorbed radiation w in the metal surrounding the cavity surface. This density is equal to the time-averaged value of the divergence of the Poynting vector (with the opposite sign):

$$w = - \langle \nabla \cdot \mathbf{P} \rangle = -\nabla \cdot \langle \mathbf{E} \times \mathbf{H} \rangle . \tag{5}$$

Further, the dependence of absorbed energy on the cavity depth can be calculated as an integral of the volumetric density over the azimuthal angle and the distance from the axis.

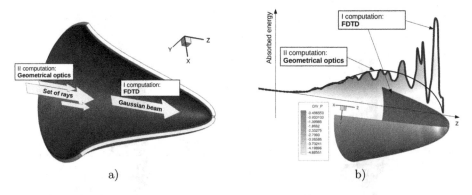

Fig. 3. Schematic of laser drilling problem (a) and distribution of the absorbed power (b): I — FDTD simulations, II — ray optics calculations

The FDTD simulations were performed on the computational grid containing $16 \cdot 10^6$ cells with 10 Gb of RAM and 16 processors used.

Figure 3b demonstrates a substantial qualitative difference of results obtained by solving Maxwell's equations and with geometrical optics approximation in a situation when the channel sizes are comparable with the wavelength. Additionally, the results of FDTD simulations point out that one possible reason for deterioration of laser drilling quality is a annular corrugation of the cavity bottom caused by nonuniform heating or vortex formation in the gas stream flowing over the region of melting.

More details about the results of numerical simulations of the laser drilling process can be found in [10].

4   Fiber-Coupled Liquid Crystal System

Liquid crystals (LC) is one of well-known examples of soft matter materials. Their unique optical properties are widely used in many devices. Along with other remarkable properties, they also have anomalously high values of nonlinear susceptibilities. In particular, the light-induced quadratic optical nonlinearity with a susceptibility index is several orders of magnitude higher than in solid crystals was observed in LCs experimentally [11].

The utilization of LCs as an active element of laser systems was reviewed in [12]. A very high efficiency allows LCs to be used in microscopic volumes for conversion and control of coherent radiation [13].

Recently a research team from Institute of Laser Physics (Novosibirsk, Russia), Novosibirsk State University and Aston Institute of Photon Technologies (UK) proposed to use a microscopic (2–8 μm) LC system placed inside the optical fiber as an optical trigger and a converter of electromagnetic radiation [14,15].

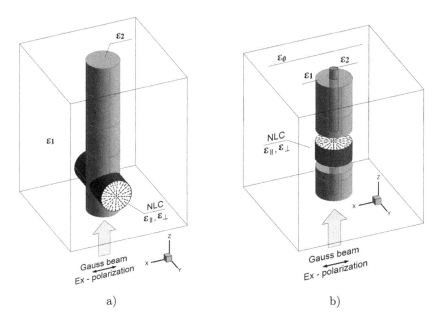

Fig. 4. Two configurations of fiber-coupled LC system: with cylindrical cavity filled with LC (a) and with plane layer of LC (b)

In the present paper the proposed integrated optical system is simulated numerically and the effects of cavity shape on light propagation are investigated. We simulate the interaction of a Gaussian laser beam propagating in the optical fiber with a nematic LC which fills a cavity whose size is comparable with the laser radiation wavelength. The laser beam is plane polarized.

Numerical simulations are performed for two different configurations of the proposed system (see Fig. 4). In the first case, the cavity is a cylindrical hole drilled across the optical fiber core. In the second case, the fiber is cut across and a plane layer of LC fills the gap formed. The first configuration is close to that was investigated in the experiment [14], the second — in the experiment [15]. The direction of preferred orientation of molecules at any point of LC is represented by a unit vector **n**, the director. In both the cases it is supposed that the director distribution is radially aligned and contains a linear singularity, the disclination of strength +1 [16], in the center. In the first case, the disclination coincides with the axis of the cylindrical hole, in the second case — with the axis of the optical fiber. In the experiments, the director alignment is forced using specially treated solid walls bounding the LC volume or by imposing an external heat flux.

The dielectric permittivity tensor for a LC medium is

$$\bar{\bar{\epsilon}} = \{\epsilon_{\alpha\beta}\} = \epsilon_\perp \delta_{\alpha\beta} + \left(\epsilon_\parallel - \epsilon_\perp\right) n_\alpha n_\beta, \tag{6}$$

where $\epsilon_\parallel = 2.82$ and $\epsilon_\perp = 2.28$ are the permittivities parallel and perpendicular to the director, $\delta_{\alpha\beta}$ is the Kronecker delta.

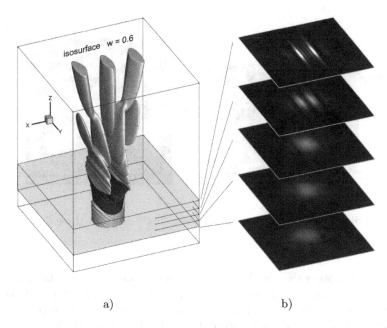

Fig. 5. Energy density of electromagnetic field: isosurface (a) and surface distributions in 5 successive cross-sections (b)

Numerical simulations for the first configuration (Fig. 4a) were performed on the grid of $27 \cdot 10^7$ cells using 80 processors and 80 Gb of RAM in total. They allows us to study how the intensity and directivity of laser beam change as a result of interaction with the transverse cylindrical cavity filled with a LC. Figure 5 shows the spatial distribution of energy density of electromagnetic field. It can be concluded that this configuration has serious drawbacks: the radiation is focused behind the cavity so that the optical fiber can burn out at high powers of the laser pulse, in addition a significant portion of the radiation is scattered outside the fiber core.

Numerical simulations for the second configuration when the laser beam passes through a layer of the LC filling a gap in a optical fiber (Fig. 4b) were performed on the grid of $48.5 \cdot 10^7$ cells using 144 processors and 145 Gb of RAM in total. A series of simulations with different values of the gap width h were conducted. In contrast to the first configuration, no significant scattering of radiation was observed (Fig. 6). The dependence of the reflection coefficient (the portion of energy reflected backwards) on the gap width is not monotonous having maximums and minimums at certain values of h. It can be explained by interference: the system behaves similar to a resonator with losses.

Thus, the configuration in which the laser beam passes through a plane layer of a LC material is preferable in comparison with that include a transverse cylindrical hole.

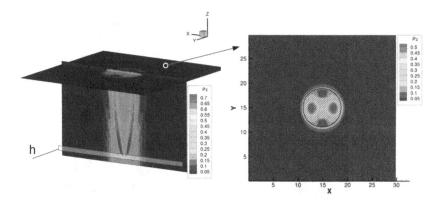

Fig. 6. Distributions of z-component of energy flux density in three different cross-sections

It is worth noting that in the present paper we neglect the influence of electromagnetic radiation on the properties of the medium so that the main subject of investigation is transformation of a laser beam in a non-homogeneous and anisotropic medium. A similar approach was used in many works on computational photonics, in particular devoted to electrodynamics of metamaterials — see, e.g., the recent paper [17] whose main subject is close to that of our investigation. It is clear that such approach is correct if the intensity of radiation is not very high so that one can neglect nonlinear effects.

5   Generation of Twisted Optical Beams

Generation, investigation and utilization of "optical vortices", i.e. light beams with helical dislocations of the wave front, is one of most rapidly developing area of modern optics [18]. Such "twisted" beams have not only the spin angular moment of photons but also an orbital angular momentum. Optical vortices find applications in many areas, from microscopy and manipulation of microscopic objects up to improvement of communication bandwidth and processing of images of astronomical objects [19,20].

Optical vortices can be effectively generated at the interaction of light with LCs. The advantage of this approach is the possibility to change the parameters of the output beam dynamically using weak external electromagnetic fields or mechanical and thermal stresses applied to the LC [21].

We performed numerical simulations of generation of optical vortices at propagation of a laser beam through a plane layer of nematic LC located in a gap between two ends of an optical fiber (Fig. 7). The laser beam propagating in the optical fiber represented the eigenmode HE_{11}. There was a disclination in the director distribution within the LC, which coincided with the fiber axis. The influence of the disclnation strength s and the gap width Δh on the angular moment of the transmitted beam was been investigated. The computational

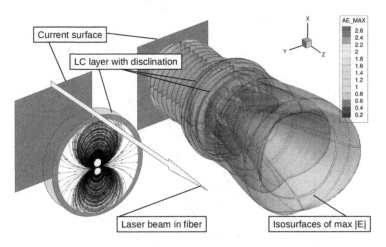

Fig. 7. Optical vortex generation at interaction of laser beam with LC. The disclination strength $s = 2$

domain was meshed with a grid containing $6 \cdot 10^8$ cells, 180 processors and 185 Gb of RAM were used for the computation.

A general view of laser beam transformation is shown in Fig. 7 where iso-surfaces of the maximum intensity of electric field are displayed. In the region ahead of the nematic LC layer the isosurfaces are corrugated because of the interference of incident and reflected beams. The interaction with a birefringent non-homogeneous medium leads to an extension of the beam and transferred an additional angular momentum to it.

The efficiency of angular momentum transfer can be characterized by the ratio of angular momentum of the transmitted beam (calculated with respect to the beam axis) to the energy flux:

$$LP_z(z) = \int\limits_{-\infty}^{+\infty}\!\!\!\int [xP_y(x,y,z) - yP_x(x,y,z)]\,dxdy \; \bigg/ \; \lambda \int\limits_{-\infty}^{+\infty}\!\!\!\int P_z(x,y,z)dxdy. \quad (7)$$

Here P_x, P_y, P_z are the components of the time-averaged Poynting vector, λ is the wavelength of incoming radiation.

The dependence of LP_z on the gap width for different strengths of the disclination is shown in Fig. 8. The non-monotonic, quasi-periodic behavior of $LP_z(z)$ deserves special attention: the beam is twisted and untwisted periodically. At the first glance, if light is twisted in a thin LC layer then an increase in the layer thickness should increase the twist. However, computations show that such conclusion is erroneous and the transferred angular momentum varies periodically with the gap width.

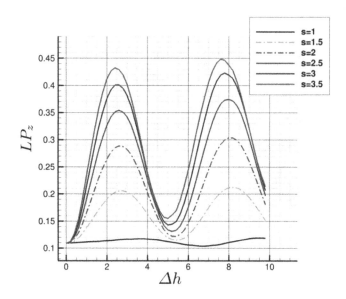

Fig. 8. Dependence of LP_z on the gap width at different disclination strengths

6   Conclusion

Numerical simulations of the interaction of electromagnetic radiation with a number of composite and anisotropic media were performed using supercomputers. The processes of laser treatment of materials such as laser drilling were simulated. It was demonstrated that, in many practical cases, the geometrical optics approximation (ray optics) is not sufficient for correct evaluation of the absorbed energy distribution, the latter can be achieved only by solving the full Maxwell's equations.

A fiber-coupled LC systems, which can be recently proposed for employing as an optical trigger and a converter of electromagnetic radiation, were simulated in two different configurations. It was shown that one of these configurations, containing a transverse cylindrical hole filled with a nematic LC, is impractical because of scattering of a significant portion of radiation outside the fiber core and focusing of another portion that can cause damage to the fiber. At the same time, the second configuration, in which the light beam passes through a LC layer, enable us to preserve most of radiation inside the fiber core and, thus, appears more suitable for a practical use.

The generation of light beams possessing an orbital angular momentum at propagation of laser radiation through a layer of a nematic LC containing disclinations of different strengths in the director field distribution was investigated numerically. It was found that the generated angular momentum grows as the disclination strength increases. A periodic dependence of the generated angular momentum on the layer thickness was observed that can be used to determine the optimal parameters for generation of such "twisted" beams.

Acknowledgements. This work was supported by Russian Foundation for Basic Research (joint Russia-India project No. 16-57-48007). Computational resources were kindly provided by Computational Center of Novosibirsk State University (nusc.nsu.ru).

References

1. Yee, K.S.: Numerical solution of initial boundary value problems involving Maxwell's equations in isotropic media. IEEE Trans. Antennas Propag. **14**, 302–307 (1966)
2. Taflove, A., Hagness, S.C.: Computational Electrodynamics, 3rd edn. Artech House, Boston & London (2005)
3. Berenger, J.-P.: Three-dimensional perfectly matched layer for absorption of electromagnetic waves. J. Comput. Phys. **127**, 363–379 (1996)
4. Berenger J.-P.: Perfectly Matched Layer (PML) for Computational Electromagnetics. Morgan & Claypool (2007)
5. Schneider, J., Hudson, S.: The finite-difference time-domain method applied to anisotropic material. IEEE Trans. Antennas Propag. **41**, 994–999 (1993)
6. Steen, W.M., Mazumder, J.: Laser Material Processing, 4th edn. Springer, London (2010)
7. Gladush, G.G., Smurov, I.: Physics of Laser Materials Processing: Theory and Experiment. Springer, Berlin (2011)
8. Duan, J., Man, H.C., Yue, T.M.: Modeling the laser fusion cutting process. I. Mathematical modeling of the cut kerf geometry for laser fusion cutting of thick metal. J. Phys. D: Appl. Phys. **34**, 2127–2134 (2001)
9. Kovalev, O.B., Zaitsev, A.V.: Modeling of the free-surface shape in laser cutting of metals. 2. model of multiple reflection and absorption of radiation. J. Appl. Mech. Tech. Phys. **46**, 9–13 (2005)
10. Kovalev, O.B., Galjov, R.V.: The application of Maxwell's equations for numerical simulation of processes during laser treatment of materials. J. Phys. D: Appl. Phys. **48**, 1–12 (2015)
11. Trashkeev, S.I., Klementyev, V.M., Pozdnyakov, G.A.: Highly efficient lasing at difference frequencies in a nematic liquid crystal. Quantum Electron. **38**, 373–376 (2008)
12. Coles, H., Morris, S.: Liquid-crystal lasers. Nat. Photonics **4**, 676–685 (2010)
13. Nyushkov, B.N., Trashkeev, S.I., Klementyev, V.M., Pivtsov, V.S., Kobtsev, S.M.: Generation of harmonics and supercontinuum in nematic liquid crystals. Quantum Electron. **43**, 107–113 (2013)
14. Nyushkov, B.N., Pivtsov, V.S., Trashkeev, S.I., Denisov, V.I., Bagayev, S.N.: Fiber laser systems for applied metrology and photonics. In: Russian-Chinese Workshop on Laser Physics and Photonics, Novosibirsk, Russia, 26–30 August 2015, pp. 69–70. Tech. digest (2015)
15. Trashkeev, S.I., Nyushkov, B.N., Galev, R.V., Kolker, D.B., Denisov, V.I.: Optical trigger based on a fiber-coupled liquid crystal. In: 17th International Conference "Laser Optics 2016", St. Petersburg, Russia, June 27–July 1 2016, pp. 69–70. Tech. digest (2016)
16. De Gennes, P.G., Prost, J.: The Physics of Liquid Crystals, 2nd edn. Clarendon Press, Oxford (1993)
17. Čančula, M., Ravnik, M., Žumer, S.: Generation of vector beams with liquid crystal disclination lines. Phys. Rev. E **90**, 022503 (2014)

18. Franke-Arnold, S., Allen, L., Padgett, M.: Advances in optical angular momentum. Laser & Photon Rev. **2**, 299–313 (2008)
19. Yao, A.M., Padgett, M.J.: Orbital angular momentum: origins, behavior and applications. Adv. Opt. Photon. **3**, 161–204 (2011)
20. Ritsch-Marte, M.: Orbital angular momentum light in microscopy. Phil. Trans. R. Soc. A **375**, 20150437 (2017)
21. Karimi, E., Piccirillo, B., Nagali, E., Marrucci, L., Santamato, E.: Efficient generation and sorting of orbital angular momentum eigenmodes of light by thermally tuned q-plates. Appl. Phys. Lett. **94**, 231124 (2009)

The Technology of Nesting a Regional Ocean Model into a Global One Using a Computational Platform for Massively Parallel Computers CMF

Alexandr Koromyslov[1,2], Rashit Ibrayev[2,3,4], and Maxim Kaurkin[2,4(✉)]

[1] M.V. Lomonosov Moscow State University, Moscow, Russia
alexandr.koromyslov@gmail.com
[2] Hydrometeorological Centre of Russia, Moscow, Russia
ibrayev@mail.ru, kaurkinmn@gmail.com
[3] Institute of Numerical Mathematics RAS, Moscow, Russia
[4] P.P. Shirshov Institute of Oceanology RAS, Moscow, Russia

Abstract. When developing regional ocean circulation model, the problem arises of providing the model with boundary conditions. An algorithm for one-way nesting (inclusion with boundary conditions) for a local model of an arbitrary ocean region in the model of the global ocean is proposed. Two problems are solved: (1) generation of a rectangular grid of the local model; (2) receive information. The nesting algorithm is developed within the framework of the CMF3.0 (Compact Modeling Framework) computing platform for massively parallel computers. Local and global models work as components of a coupled system running CMF3.0. Data nesting functions work as a CMF3.0 software service.

Keywords: Earth system modeling · Ocean modeling · Nesting · Coupler

1 Introduction

When modeling the World Ocean, it sometimes becomes necessary to obtain an accurate prediction only for a particular region, for example, a certain sea. At the same time, numerical integration of the entire high-resolution World Ocean dynamics model ($0.1°$ and more) is a computationally expensive task that sometimes requires several thousand processor cores [9]. To optimize the computational costs, it is reasonable to model the area under study using a high spatial-temporal resolution model (hereinafter, the local model), and to provide the model with the boundary and initial conditions for modeling the rest of the ocean using the World Ocean model of a more coarse resolution (global model). An important point that has a significant impact on the quality of the results is the way in which the boundary conditions are set in the local model. In real applications, several approaches are currently used.

One of these is nesting, a computational approach in which data is exchanged between global and local models. In this case, the global model usually has low resolution (for example, $1°$), and local - high or ultra-high resolution ($0.1°$ or more). This formulation of the problem requires considerable computational resources, since the global and local models sometimes use several hundred computing cores. But at the

© Springer International Publishing AG 2017
V. Voevodin and S. Sobolev (Eds.): RuSCDays 2017, CCIS 793, pp. 241–250, 2017.
https://doi.org/10.1007/978-3-319-71255-0_19

same time, this approach is more optimal from the point of view of computational costs than the modeling of the entire World Ocean with a high resolution, while the quality of the model forecast in the region under study with a correct formulation of the boundary conditions remains high.

The relevance of using the nesting method has previously been examined with examples of climate modeling in winter in South America [6], modeling the north-western Caribbean and the Scottish shelf in the northwestern Atlantic Ocean [4]. There is also an active research in the field of nesting algorithms [4, 5, 7].

The purpose of this work is to implement the nesting algorithm as a software service NST (abbr.NeSTing) of the CMF 3.0 computing platform for use in high spatial reso-lution models on massively parallel computers with distributed memory. This work continues a series of studies of authors published earlier in [2, 8] and devoted to the development of tools for modeling the Earth system on computers with distributed memory based on the compact modeling framework CMF 3.0.

2 Compact Modeling Framework CMF 3.0 and NST Service

Despite the fact that the logic of the operation of the coupler interpolation procedures in CMF 3.0 remained the same as in CMF 2.0 [3], the PGAS abstraction made it possible to greatly simplify the code. Now all the data necessary for the process from neighbors is obtained using the Communicator class.

The disadvantage of this approach is the performance degradation associated with the inability to use deferred MPI operations and the availability of GA own costs. However, the results of CMF 2.0 showed that we can sacrifice some of the performance to select a simpler (and possibly less efficient) abstraction to simplify communication algorithms [2]. Because, first, although the pure MPI-approach to communications has a high speed, it requires explicit work with data buffers, which greatly complicates the development and improvement of PCM. Secondly, the development of regional sub-models of the seas embedded in the grid of the global model, became rather complicated when using only MPI-procedures. On the other hand, this solution has made it possible to simplify the connection of new models, as well as to expand the functionality of the joint simulation system by adding new software services.

2.1 PGAS-Communicator

CMF3.0 contains the Communicator class, which encapsulates the logic of working with the Global Arrays library (GA) [6] and provides only an interface for put/get operations of parts of global data of model components. The GA library implements the PGAS (Partitioned Global Address Space) paradigm and allows working with distributed memory as a shared memory.

All exchanges between parts of the system are implemented using the Communicator class. It contains a hash table for storing all information about arrays, including their state and metadata. Each array of the component involved in exchanges contains a distributed copy, stored as a virtual global array GA. When a process needs to send data, it fills this copy with its current data. Due to the fact that the distribution of the global array completely repeats the decomposition of the component, this operation takes place locally.

2.2   The Architecture of the Coupling Model

Due to the growing complexity of the coupling system, a more convenient way of combining components - SOA (Service-Oriented Architecture) was used.

In CMF 3.0, all models send their general requests to a single message queue (Fig. 1). Service components take from this queue only messages that can process, take data from virtual global arrays, and perform appropriate actions. The architecture allows minimizing the connections between physical and service components and greatly simplifying the development. Moreover, since all services inherit the common base class Service, adding a new service is not difficult. Now CMF 3.0 contains the following independent

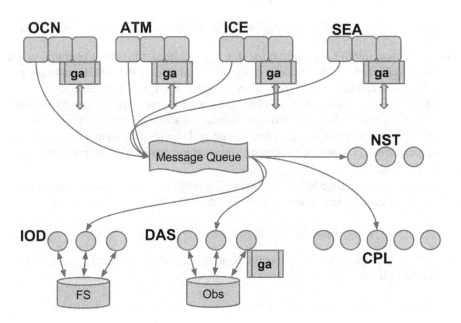

Fig. 1. The architecture of the compact framework CMF 3.0. There are four components in this example: ocean model (OCN), ice model (ICE), atmosphere model (ATM) and sea model (SEA). They send requests to the common message queue, where they are retrieved by coupler (CPL), data assimilation (DAS), input and output data (IOD), nesting (NST) services. The data itself is transferred through the mechanism of global arrays, which are also used for interprocessor communication in the components and services

parallel services: CPL (mapping operations), IOS/IOF (I/O Fast, I/O Slow - fast and slow file devices, DAS (Data Assimilation System) [8].

Service CPL is a CMF 2.0 coupler, which now deals exclusively with operations related to interpolation. It receives data using Communicator, performs interpolation, and places the data in the virtual global array of the recipient.

In CMF 3.0, a separate I/O service responsible for all I/O operations was allocated. It is interesting that one service solves only half of the problem, because, for example, a simultaneous request to write a large amount of information (a system check point) and a small model diagnostics will also take place sequentially. Therefore, the service was divided into a fast and slow recording device (thanks to the abstraction this division is performed by several lines of code). This mechanism provides a flexible and asynchronous mechanism for working with the file system.

3   Description of the NST Service

3.1   Mathematical Formulation One-Way Nesting Problem

We are interested in representing as accurately as possible the local ocean model in a domain Ω_{loc}. The circulation is supposed to be described on a time period [0, T] by a model which can be written symbolically

$$L_{\text{loc}} u_{loc} = f_{loc} \quad \text{in } \Omega_{loc} \times [0, \text{T}] \tag{1}$$

with convenient initial conditions at t = 0. L_{loc} is a partial differential operator, u_{loc} is the state variable, and f_{loc} the model forcing. The conditions at the solid boundaries will never be mentioned in this note, since they do not interfere with our subject.

Since Ω_{loc} is not closed, a portion of its boundary does not correspond to a solid wall, and has no physical reality. This artificial interface, also called open boundary (OB), is denoted Γ. The local solution u_{loc} is thus in interaction with the external ocean through Γ, and the difficulty consists in adequately representing this interaction in order to get a good approximation of u_{loc} in $\Omega_{\text{loc}} \times [0, \text{T}]$.

We also assume that we have at our disposal a (probably less accurate) representation of the global ocean, either under the form of some data u_{glob} or of a global model

$$L_{\text{glob}} u_{glob} = f_{glob} \quad \text{in } \Omega_{glob} \times [0, \text{T}] \tag{2}$$

where Ω_{glob} is an global oceanic domain. Note that, in our notations, Ω_{loc} and Ω_{glob} do not overlap (Fig. 2).

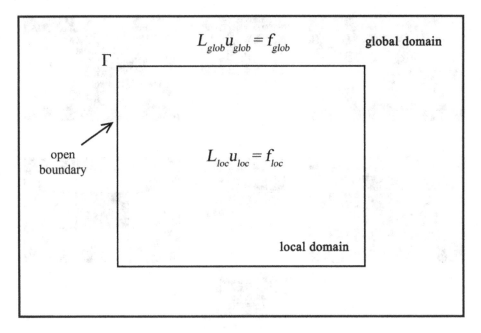

Fig. 2. A schematic view of nesting problem

The solution U_{glob} of an external model cover an area $\Omega_{glob} \cup \Omega_{loc}$ larger than $_{loc}$ is available. Therefore tis larger scale solution can be used to force the local model along Γ. The formulation of the problem which is solved in that approach is (3).

$$L_{glob}u_{glob} = f_{glob} \quad in \, \Omega_{glob} \cup \Omega_{loc} \times [0, T]$$

$$\begin{cases} L_{loc}u_{loc} = f_{loc} & in \, \Omega_{loc} \times [0, T] \\ Bu_{loc} = Bu_{glob} & on \, \Gamma \times [0, T] \end{cases} \tag{3}$$

where B denotes an open boundary operator (in particular cases are Dirichlet and Neumann conditions).

This interaction between the two models can be perfomed on-line using features CMF 3.0.

3.2 Description of the Algorithm

Suppose we have a global model with a rough spatial resolution (Fig. 3a), a local high-resolution model (Fig. 3b). The mutual arrangement of the local and global models is as follows (Fig. 3c). Let the decomposition between the processors local and global models differ as well (Fig. 3d and e). The global model needs to send the boundary cells of the area that covers the local model (Fig. 3f) to the nesting service. The nesting service interpolates the data received from the global model on the grid of the local model and then sends it to the boundary cells of the local model (Fig. 3g).

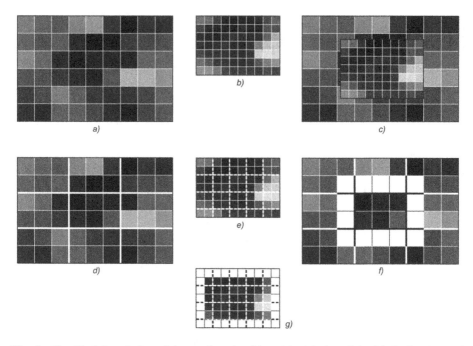

Fig. 3. Graphical description of the nesting algorithm: (a) global model grid; (b) local model grid; (c) the combination of the calculated grids of the global and local models; (d) processor decomposition of the global model area (shown in white bars); (e) processor decomposition of the region of the local model (shown in white bars); (f) white cells are highlighted in the global grid required for the nesting; (g) the cells of the local grid are highlighted in white, in which the boundary conditions are set by the nesting

3.3   Features of Parallel Implementation of the Service

Like any service of the CMF3.0 software complex, the nesting service is performed on separate computing cores. This approach allows us to obtain a transparent structure of the joint model, where each component is engaged in the solution of its task. At the same time, neither the global nor the local models practically participate in the nesting, only mesh and model masks are used, on the basis of which the nesting service builds the matrix of interpolation scales. Moreover, in the NST service, in addition to specifying the classical Neumann or Dirichlet conditions, the logic can be encapsulated to generate conditions for the sum of the flows (heat, salt, etc.) to be zero across the boundaries of the region under investigation or their correspondence to certain integral characteristics. The data from the global to the local model go without access to the file system, using the cluster interconnect through the GA library, which positively affects the performance of the system as a whole.

Problems and reasons for making a nesting in a separate service:

1. Global and local models have different resolutions and are calculated on a different number of cores.

2. The NST service can be used for other Earth system models (for example, the atmosphere model), which is also successfully used in conjunction with the CMF complex [10].
3. Interpolation matrices occupy a significant amount of memory, and the logic of calculating them can be very difficult, especially in the case of setting complex boundary conditions. Therefore, it is more advantageous to store them on separate computational cores from model components.

General algorithm of the local and global model using the NST service

- The initial values in the global and local models are initialized
- The NST service receives from the global and local models the configuration of grids, masks, dimensionality of models, decomposition of processors for each model, steps in time.
- Based on the received data, the NST service calculates the position of the local model relative to the global model.
- Based on the position of the local model, the NST service computes the elements of the arrays that need to be transferred from the global model to the local model.
- Based on the given boundary conditions for the local model and the obtained information about the model masks and grids, the matrices of the interpolation weights are calculated.
- When the transmission time is reached, the global model sends only the necessary parts of the model arrays to the NST service.
- The NST service interpolates the data from the global to the local grid using the previously obtained matrix of weights.
- The NST service sends the resulting boundary conditions to the local model.

3.4 Testing the NST Service

To test the computational efficiency of the parallel nesting algorithm using the GA software library, a global test model with a resolution of $0.25°$ (1440×720 grid) and a local model with a resolution of $0.1°$ (400×400 grid) were specified, in both 49 vertical z-levels. The global model used two-dimensional decomposition of the computational domain into 64 processor cores, and the local model used 32 cores, and the NST service was allocated from 1 to 8 cores. From the global to the local model, boundary conditions of the Dirichlet type were transmitted for model fields of temperature, salinity, and velocity [5]. The matrix of the interpolation weights was constructed by the method of bilinear interpolation, taking into account the land mask for the global and local models. The scalability of the parallel Nesting algorithm within the NST service is shown in the calculation of the boundary condition for 4 (the number of model fields) * 1600 (the extent of the local domain boundary) * 49 (the number of z-levels) ~ O (10^5) points in the local model. When constructing the parallel efficiency graph, the sum of the time spent on transferring data from the global model to the NST service, the time of parallel calculation of the boundary conditions by the NST service cores, and the time of data transfer to the local model (4) (Fig. 4).

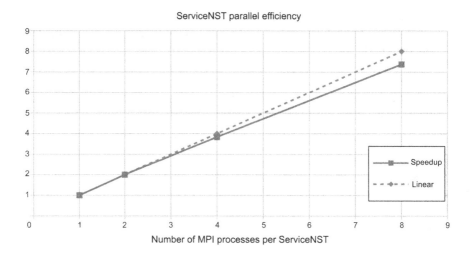

Fig. 4. Scalability of the parallel nesting method within the NST service for obtaining boundary conditions for O (10^5) points on the MVS-10P supercomputer. On the X axis, the number of processor cores per NST service. On the Y-axis, the parallel nesting method is speeded up based on $T_{nesting}$ time measurements. The global test model used a two-dimensional decomposition of the computational domain into 64 processor cores, and the local one - to 32 cores

$$T_{nesting} = T_{ocn\_to\_nst} + T_{nst} + T_{nst\_to\_sea} \qquad (4)$$

4   Example of Use

(Figure 5) shows the temperature field in the test region for the local model-the Barents Sea with a resolution of 0.1° (Fig. 5a) and a global model of the World Ocean [1] with a resolution of 0.5° (Fig. 5b). The size of the spatial grids for the World Ocean and the Sea is 720 × 360 and 300 × 220, in each 49 levels vertically. The model of the ocean uses 64 processor cores in calculations, and the Barents Sea model has 12 cores, 2 processor cores are allocated to the NST service for nesting boundary conditions.

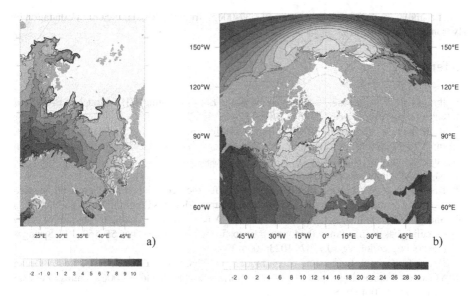

Fig. 5. Use of nesting technology for modeling the dynamics of the Barents Sea. (a) the surface temperature field in the Barents Sea model with a spatial resolution of $0.1°$; (b) the surface temperature field in the World Ocean model with a resolution of $0.5°$ for specifying the boundary conditions in the sea model. The Barents Sea model uses a two-dimensional processor decomposition of the calculated region into 12 processor cores, and a global one - into 64 cores

5   Conclusion

In this paper, a parallel implementation of nesting technology as a software service NST of the compact computing platform CMF3.0 is presented [2]. A parallel nesting algorithm is described in detail, an example of using the developed technology is given.

For the first time in Russia, a parallel technology for transferring boundary conditions (nesting) through a cluster interconnect (without the use of a file system) between ocean and sea dynamics models using CMF 3.0 was implemented. This implementation allowed the ocean and sea models to function as a single coupling model on massively parallel computers with distributed memory. The advantages of this approach include computing efficiency and speed, which will allow further use of this technology for operational forecast of the state of the ocean.

Nesting technology is implemented in other most advanced frameworks for Earth system modeling - CESM [11] and OASIS [12]. The nesting service (NST service) based on CMF 3.0, which is described in this article demonstrates similar results of computing efficiency.

Acknowledgements. To test the performance of the nesting service and the compact computing platform CMF 3.0, tests were conducted on the supercomputers Lomonosov and MVS-10P, which confirmed the numerical efficiency of the proposed software product.

The study was performed by a Grant #14-37-00053 from the Russian Science Foundation in Hydrometcentre of Russia.

References

1. Ibrayev, R.A., Khabeev, R.N., Ushakov, K.V.: Eddy-resolving 1/10° Model of the World Ocean. Izv. Atmos. Oceanic Phys. **48**(1), 37–46 (2012)
2. Kalmykov, V., Ibrayev, R.: CMF - framework for high-resolution Earth system modeling. In: CEUR Workshop Proceedings, Proceedings of the 1st Russian Conference on Supercomputing (RuSCDays 2015), Moscow, Russia, 28–29 September 2015, vol. 1482, pp. 34–40. MSU publishers, Moscow (2015)
3. Kalmykov, V.V., Ibrayev, R.A.: A framework for the ocean-ice-atmosphere-land coupled modeling on massively-parallel architectures. Numer. Methods Program. **14**(2), 88–95 (2013)
4. Sheng, J., Greatbatch, R.J., Zhai, X., Tang, L.: A new two-way nesting technique for ocean modeling based on the smoothed semi-prognostic method. Ocean Dyn. **55**, 162–177. Springer (2005). https://doi.org/10.1007/s10236-005-0005-6
5. Blayo, E., Debreu, L.: Nesting ocean models. In: Chassignet, E.P., Verron, J. (eds.) Ocean Weather Forecasting, pp. 127–146. Springer, Dordrecht (2006). https://doi.org/10.1007/1-4020-4028-8_5
6. Menéndez, C.G., Saulo, A.C., Li, Z.-X.: Simulation of South American windertime climate with a nesting system. Clim. Dyn. **17**, 219–231 (2001). Springer
7. Scheinert, M., Biastoch, A., Böning, C.W.: The agulhas system as a prime example for the use of nesting capabilities in ocean modelling. In: Resch, M., Roller, S., Benkert, K., Galle, M., Bez, W., Kobayashi, H. (eds.) High Performance Computing on Vector Systems 2009, pp. 191–198. Springer, Heidelberg (2010). https://doi.org/10.1007/978-3-642-03913-3_15
8. Kaurkin, M., Ibrayev, R., Koromyslov, A.: EnOI-Based Data Assimilation Technology for Satellite Observations and ARGO Float Measurements in a High Resolution Global Ocean Model Using the CMF Platform. In: Voevodin, V., Sobolev, S. (eds.) RuSCDays 2016. CCIS, vol. 687, pp. 57–66. Springer, Cham (2016). https://doi.org/10.1007/978-3-319-55669-7_5
9. Tolstykh, A., Ibrayev, R.: Models of the global atmosphere and the World Ocean: algorithms and supercomputer technologies. In: Tutorial, p. 144. Publishing House of Moscow University, Moscow (2013). (in Russian)
10. Fadeev, R., Ushakov, K., Kalmykov, V., et al.: Coupled atmosphere–ocean model SLAV–INMIO: implementation and first results. Russ. J. Numer. Anal. Math. Model. **31**(6), 329–337 (2016)
11. Craig, A.P., Vertenstein, M., Jacob, R.: A new flexible coupler for Earth system modeling developed for CCSM4 and CESM1. Int. J. High Perform. C **26**(1), 31–42 (2012)
12. Redler, R., Valcke, S., Ritzdorf, H.: OASIS4 - a coupling software for next generation earth system modeling. Geosci. Model Dev. **3**, 87–104 (2010)

Parallel Heterogeneous Multi-classifier System for Decision Making in Algorithmic Trading

Yuri Zelenkov[⊠] [ID]

National Research University Higher School of Economics, Moscow, Russia
yuri.zelenkov@gmail.com

Abstract. The most important factors of successful trading strategy are the decisions to sell or buy. We propose multi-classifier system for decision making in algorithmic trading, whose training is carried out in three stages. At the first stage, features set is calculated based on historical data. These can be oscillators and moments that used in technical analysis, other characteristics of time series, market indexes, etc. At the second stage, base classifiers are trained using genetic algorithms, and optimal feature set for each of them is selected. At the third stage, a voting ensemble is designed, weights of base classifiers are selected also using genetic algorithms. However, the usage of genetic algorithms requires considerable time for computing, so the proposed system is implemented in a parallel environment. Testing on real data confirmed that the proposed approach allows to build a decision-making system, the results of which significantly exceed the trading strategies based on indicators of technical analysis and other techniques of machine learning.

Keywords: Algorithmic trading, trading strategy · Multi-classifier system · Genetic algorithm

1 Introduction

Algorithmic Trading (AT) refers to any form of trading using sophisticated algorithms and programmed systems to automate all or some part of the trade cycle [1, 2]. The trade cycle and components of AT system are described in [1, 2]. The key stages in AT are the pre-trade analysis, signal generation, trade execution, post-trade analysis, risk management, and asset allocation.

The key factors of a successful trading strategy are the decisions to "buy" or "sell". These solutions are based on the alpha model, which is the mathematical model designed to predict the future behavior of the financial instruments that the algorithmic system is intended to trade [2]. A large number of studies related to the design of alpha models are known, including using machine learning methods. In this paper, we propose a method for designing the alpha model, based on multi-classifier system, whose training is carried out in three stages. At the first stage, features set is calculated based on historical data. At the second stage, base classifiers are trained using genetic algorithms, and optimal feature set for each of them is selected. At the third stage, a voting ensemble is designed, weights of base classifiers are selected also using genetic algorithms.

© Springer International Publishing AG 2017
V. Voevodin and S. Sobolev (Eds.): RuSCDays 2017, CCIS 793, pp. 251–265, 2017.
https://doi.org/10.1007/978-3-319-71255-0_20

To build an alpha model based on a multi-classifier system, the following actions should be performed: obtain and clean data that will drive AT; select base classifiers that mutually complementary; select architecture of their ensemble. Therefore, an organization of paper is following: after literature review in the second section, process of feature engineering is described. In next sections techniques of features wrapping and classifiers combination in ensemble are discussed. All proposed techniques are illustrated by practical examples. Obtained results are compared with other methods. In last section, parallel implementation of proposed algorithm is discussed.

2 Related Works

There are two financial instrument prediction methodologies:

- *Fundamental Analysis* is concerned more with the company and its macro-economic environment rather than the actual asset. The decisions are made based on the past performance of the company, the forecast of earning etc.
- *Technical Analysis* deals with the determination of the asset price based on the past patterns of the stock using time-series analysis.

When applying Machine Learning to stock data, Technical Analysis is the more applicable methodology, because it can learn the underlying patterns in the financial time series. The search of patterns is carried out in two main ways, the first is the identification of graphic figures that are formed by price charts, the second is the calculation of various indicators, the dynamics of which allows predicting asset price behaviour [3, 4]. Developers of alpha models based on machine learning usually use technical indicators. For example, Zhang and Ren [5] presented a trading strategy model that utilizes different technical indicators such as Moving Average (MA), Moving Average Convergence Divergence (MACD), Relative Strength Index (RSI), Slow Stochastic etc.

At the same time, various models of machine learning are used. A major difficulty in dealing with financial time series representing asset prices is their non-stationary behavior (or concept drift), i.e. the fact that the underlying data generating mechanism keeps changing over time. Therefore, in real-time forecasting and trading applications one is often interested in on-line learning, a situation where the prediction function is updated following the arrival of each new sample [6]. Various approaches for incremental learning have been proposed in the literature, for both classification [7] and regression problems [6, 8].

One of the most popular tools are artificial neural networks (ANNs) [9], which are often used together with evolutionary techniques, such as genetic algorithms (GA), because the combination of two or more techniques offers a better result [10, 11]. Scabar and Cloette [12] developed a hybrid prediction model based on an ANN and GA, which gives evidence that financial time series are not entirely random, and that— contrary to the predictions of the efficient markets hypothesis—a trading strategy based solely on historical price data can be used to achieve returns better than those achieved using a buy-and-hold strategy. Butler [13] developed an Evolutionary ANN (EANN) that makes future predictions based on macro-economic data. Tsai & Chiou [14] used technique that combines ANN with decision trees.

Peters [15, 16] proved that time series of stock prices are produced by systems with memory, he also determined cycles for different industries and stock markets. Therefore, the predictive tools that can model the memory effect, for example, recurrent neural networks, are of considerable interest. For example, feedforward networks and recurrent networks (Elman network) that can build "memory" in the evolution of neurons are reviewed in [17] with application to finance.

As follows from this brief survey, the use of statistical characteristics of time series and indicators of technical analysis is a widespread practice in AT. Researchers choose different techniques of machine learning and their combinations. At the same time, comparatively little attention has been paid to researching the possibilities of combinations of simple classifiers, such as k Nearest Neighbors (kNN), Logistic Regression (LR), Naive Bayes (NB), Decision Tree (DT), Support Vector Machine (SVM). This paper intends to fill this gap.

3   Problems of Trading Strategy Design

As it was stated above, we suggest use an ensemble of heterogeneous classifiers. When creating an effective method for design the AT system, several problems must be solved.

The first problem concerns the feature engineering and selection. A feature is a piece of information that might be useful for prediction (wikipedia.org). It can be structured attribute, combination of attributes and any unstructured information that relevant to the context. Feature engineering is the process of using domain knowledge to create features that make model works. Feature selection problem deals with selection of an optimal and relevant set of features that are necessary for the recognition and prediction [18, 19]. It helps reduce the dimensionality of the measurement space and facilitates the use of easily computable algorithms for efficient classification.

The second problem concerns hybrid multi classifier system (MCS) design [20]. It is: system topology (how to interconnect individual classifiers), ensemble design (how to drive the generation and selection of a pool of valuable classifiers) and fuser design (how to build a decision combination function).

We use parallel architecture because most MCS's reported in the literature are structured in a parallel topology [20, 21]. In this architecture, each classifier is feed the same input data, so that the final decision of the MCS is based on the individual classifiers outputs obtained independently. We use voting ensemble with majority voting rule. In that case MCS output is formed as the weighed sum of individual classifiers responses.

The design of hybrid ensemble should support involving of mutually complementary individual classifiers that provide high diversity and accuracy [21], but it is impossible to predict what classifiers can be complementary. Therefore, we suggest using the combination of several approaches: random sampling, features selection for each individual classifier on base of the GA; determination of each classifier weight in ensemble also through GA. Thus, the method proposed here includes three stages

(Fig. 1). The first one generates a set of features that can be used for forecasting; on the second one, the relevant set of features is selected using genetic algorithms for each individual classifier; on the third one, the weights of the voting ensemble are determined also using genetic algorithms.

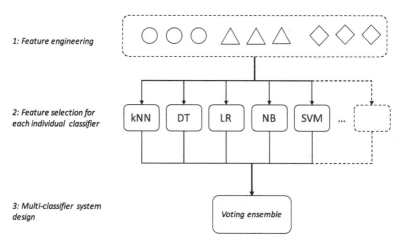

Fig. 1. Three stages of MCS training.

4   Features Engineering

4.1   Target Variable

A time series describing the dynamics of a financial instrument (for example, shares of a company) that can be downloaded from http://finance.yahoo.com includes the following variables: opening and closing prices, maximum and minimum prices, trading volume. The service finance.yahoo.com also provides a value of Adjusted Closing Price (ACP), which is a stock's closing price on any given day of trading that has been amended to include any distributions and corporate actions that occurred at any time prior to the next day's open. The ACP is often used when examining historical returns or performing a detailed analysis on historical returns.

Since we view trading strategy as a classification problem solved with the help of supervised learning, it is necessary to set the target variable. We define it as follows:

$$target(t) = \begin{cases} 1, & p(t)/p(t+1) < 1 \\ -1, & p(t)/p(t+1) \geq 1 \end{cases} \tag{1}$$

where $target(t)$ is target variable (trading signal) in the time t, $p(t)$ and $p(t+1)$ are the ACP's in current (t) and next $(t+1)$ trading intervals respectively.

This means that if the asset price in the next period increases, our strategy should generate a buy signal (1). If a financial instrument has already been acquired, it is necessary to retain it. If the price decreases in the next period, then the strategy should give a sell signal (−1) if the asset is already acquired. If the asset is not bought by this time, you should refrain from buying. In other words, our system has an ambitious goal - to predict based on historical data whether the price in the next period will increase or decrease.

We will assume that the initial capital of the investor is $x(0)$. If the 'buy' signal is received, he buys the maximum possible number of shares $d(t) = v(t)p(t)$, where $v(t)$ is the amount of acquired assets, which is restricted by condition $d(t) \leq x(t)$. If a 'sell' signal is received, he sells shares at the current price and his capital becomes $x(t) = x(t-1) + v(t)p(t)$. Thus, the goal of the strategy is to maximize the sum $x(t) + d(t)$, therefore, its effectiveness can be estimated as $e = [x(n) + d(n)]/x(0)$, where n is the number of trading periods.

Also, we use several assumptions that are typical for research of this kind:

- the volume of sales and purchases is quite small and does not affect the behavior of the market;
- Transaction costs for operations are zero.

4.2   Features Set

We suggest including in feature set most popular indicators of technical analysis: Moving Average (MA), Moving Average Convergence Divergence (MACD), Relative Strength Index (RSI), Stochastic and signals of trading strategies based on these indicators [3, 4].

Trading signals based on MA are generated as follows:

$$S_{MA}(t) = \begin{cases} -1, & MA_{n1}(t-1) > MA_{n2}(t-1) \wedge MA_{n1}(t) < MA_{n2}(t) \\ 1, & MA_{n1}(t-1) < MA_{n2}(t-1) \wedge MA_{n1}(t) > MA_{n2}(t) \end{cases}, \quad (2)$$

where $MA_n(t)$ is mean average in time t, n is widows size. Usually, $n_1 = 9$ and $n_2 = 50$, these values were determined empirically.

Trading signals based on MA are calculated as follows:

$$S_{RSI}(t) = \begin{cases} -1, & RSI(t) > UB \\ 1, & RSI(t) < LB \end{cases}, \quad (3)$$

where $RSI(t) = 100\left[1 - \frac{1}{1+RS(t)}\right]$, $RS(t) = \max_n AG / \min_n AL$, AG is average gain, AL is average loss, n is number of periods, UB and LB upper and lower limits respectively. Usually, in technical analysis $n = 14$, $UB = 70$, $LB = 30$.

Trading signals based on MACD are calculated as follows:

$$S_{MACD}(t) = \begin{cases} -1, & F(t-1) > S(t-1) \wedge F(t) \langle S(t) \wedge F(t) \rangle 0 \wedge S(t) > 0 \\ 1, & F(t-1) < S(t-1) \wedge F(t) > S(t) \wedge F(t) < 0 \wedge S(t) < 0 \end{cases}, \quad (4)$$

where $F(t) = EMA_{n1}(t) - EMA_{n2}(t)$ is fast MACD line, $S(t) = EMA_{n3}(F(t))$ is slow MACD line, EMA_n is exponential mean average with window size n. Typically,

$n_1 = 13$, $n_2 = 26$, and $n_3 = 9$. Useful information also can be extracted from MACD histogram: $Hist(t) = S(t) - F(t)$.

Stochastic signals are calculated as follows:

$$S_{stoch}(t) = \begin{cases} -1, & \%K(t) > UB \wedge \%D(t) > UB \\ 1, & \%K(t) < LB \wedge \%D(t) < LB \end{cases}, \qquad (5)$$

where $\%K(t) = 100[p(t) - \min_{n1} p(t)]/[\max_{n1} p(t) - \min_{n1} p(t)]$, and $\%D(t) = MA_{n2}$ $(\%K(t))$. Usually, $n_1 = 14$, $n_2 = 3$, $UB = 70$, $LB = 30$. Sometimes also parameter 'slow D' is used $D(t) = MA_{n3}(\%D(t))$ with $n_3 = 3$.

4.3   R/S Analysis

The first algorithm runs, performed on the historical data of Alphabet Inc. (ticker GOOG), showed that the above indicators are not enough to build an effective trading strategy. Therefore, to search for relevant characteristics, an R/S analysis of this time series was conducted using the methodology described by Peters [15, 16].

R/S analysis helps to determine the nature of price series by measuring its speed of diffusion. The speed of diffusion can be characterized by the variance [22]:

$$\left\langle |z(t+\tau) - z(t)|^2 \right\rangle \sim \tau^{2H}, \qquad (6)$$

where $z(t) = \log p(t)$ is the log prices (ACP) at the time t, τ is the arbitrary time lag, and $\langle \cdots \rangle$ is average over all τ's. The $\sim$ means that this relationship turns into equality with some proportionality constant, is the Hurst exponent. For a price series exhibiting geometric random walk, $H = 0.5$, for a mean reverting series, $H < 0.5$, and for a trending series $H > 0.5$. In last case, a future data point is likely to be like a data point preceding it, i.e. logarithms $\log[p(t)/p(t-1)]$ and $\log[p(t+1)/p(t)]$ likely will have the same signs. So, value of Hurst exponent is very valuable domain knowledge that can help to design effective algorithm.

According to Peters [15, 16] Hurst exponent, H, is calculated as asymptotic approximation of the rescaled range as a function of the time span of a time series as follows

$$E\left[\frac{R(n)}{S(n)}\right] = Cn^{H} \text{ as } n \to \infty, \qquad (7)$$

where $R(n)$ is the rescaled range of first n values of $z(t)$, $S(n)$ is their standard deviation, $E[\cdot]$ is the expected value, n is the number of data points in time series, and C is a constant. Therefore, to find H, it is enough to find a regression $\log[R(n)/S(n)] = \log C + H \log n$.

The results of the Hurst exponent estimation for a series of daily closing prices (ACP) of Alphabet Inc. for 10 years (2007–2016) are shown in Fig. 2a, the dependence of its values on period length is shown in the Fig. 2b. As already mentioned, the value

$H = 0.597 > 0.5$ means that the logarithms of successive price changes are likely to have the same sign.

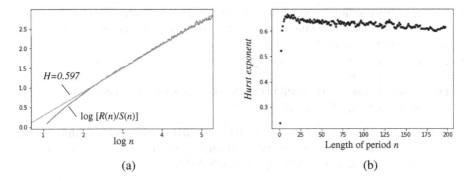

(a)   (b)

Fig. 2. Hurst exponent for the GOOG ACP time series.

Figure 3a shows the dependence of $\log[p(t+1)/p(t)]$ on $\log[p(t+1)/p(t-1)]$, a significant part of the points is concentrated in areas II and IV. These values do not change sign. Figure 3b shows the probability of changing the sign of the logarithm of the price increment, depending on the duration of the period:

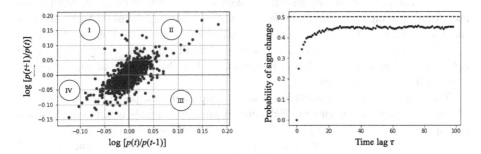

Fig. 3. Probability of sign of logarithm of price ratio change.

$$P(\tau) = P\left[\log\frac{p(t+\tau)}{p(t)} \Big/ \log\frac{p(t)}{p(t-1)} < 0\right] \tag{8}$$

It follows from the Fig. 3b that this probability does not never reach the value 0.5, corresponding to complete uncertainty. Thus, the investigated price range is not entirely random. Moreover, for a one day period we have $P(\log\frac{p(t+1)}{p(t)} \big/ \log\frac{p(t)}{p(t-1)} < 0) = 0.25$, i.e. the probability that the sign of logarithm of the price ratio in the next day will not change is 0.75. It is very important domain knowledge, so we must include sign of $\log[p(t)/p(t-1)]$ in feature set.

Other useful information may be extracted form market indices, as well as logarithms of their changes. Therefore, we will also include them in the features set.

As result 41 features where selected for algorithm training including:

- Values of price series (open, close, min and max prices, volume and ACP);
- Technical indicators described in Sect. 4.2: $MA_9(t)$, $MA_{50}(t)$, $RSI(t)$, $F(t)$, $S(t)$, $Hist(t)$, $\%K(t)$, $\%D(t)$, $D(t)$;
- Trading signals described in Sect. 4.2: $S_{MA}(t)$, $S_{MACD}(t)$, $S_{RSI}(t)$, $S_{stoch}(t)$;
- Sign of $\log[p(t)/p(t-1)]$;
- Dow-Jones (ticker ^DJI), NASDAQ (^NDX) and S&P500 (^GSPC) indexes and signs of logarithms of their ratios.

To train algorithm we used two-year (2015 and 2016) daily prices of Alphabet Inc. (ticker GOOG). This training set contains 504 samples of 41 features. Test set contains 61 samples of daily data from January to March 2017.

5   Features Selection for Individual Classifiers

We use wrapping method based on genetic algorithm for features selection. The program code was developed on the Python language and based on machine learning library scikit-learn [23]. Therefore, only the classifiers available in this library were used as basic (kNN, NB, LR, DT and SVM).

Each classifier is coded by an array G with length N, which describes features set used for its training (N – quantity of features in the researched dataset). Array elements can take values 0 or 1. If the element is equal to 0, the corresponding feature is excluded from training set. Value 1 is assigned to all elements of a genotype G of individuals when initial population is generated. Thereby training of each classifier begins with full range of features.

The best individual in population is always copied in new population without any changes (the principle of elitism). Selection of other individuals is rank-based. Mutation operation is applied with probability p_m to randomly selected G-genotype element of selected individual, with its value replaced by opposite, i.e. 0 becomes 1, and 1 becomes 0. Crossover operation is applied with probability of p_c, it is implemented as exchange of randomly selected substring between two individuals.

Fitness is calculated as average classification accuracy value:

$$accuracy = \frac{TP + TN}{TP + FP + TN + FN} \tag{9}$$

where TP is true positives, FP is false positives (negatives classified as positives), TN is true negatives, and FN is false negatives (positives classified as negatives).

On this stage of training the researcher determines a set of classifiers types which will be used for ensemble design (denote the number of types by M). The classifier of each type is trained according to the algorithm described above. The set of the trained classifiers is transferred to the following stage.

Results of individual classifiers training are given in Table 1. For all classifiers, the following parameters were used: the population size 40, number of generations 20, $p_m = 0.5$, $p_c = 0.5$. Accuracy before training is calculated before start of a genetic algorithm (the features set includes all 41 features). Widely known models kNN, LR, NB, DT and SVM are used as the basic classifiers with the default parameters of scikit-learn library. The average accuracy and subsequent confidence interval after training are given in the column with caption "Accuracy after wrapping". Also, Table 1 lists the number of selected features and precision/recall values.

Table 1. Results of base classifiers training.

Classifier	Accuracy		Number of selected features	Precision/recall
	Before wrapping (N = 41)	After wrapping		
KNN	0.526	0.558 ± 0.100	21	0.581/0.632
LR	0.536	0.585 ± 0.061	14	0.589/0.624
NB	0.542	0.577 ± 0.020	16	0.598/0.614
DT	0.530	0.530 ± 0.050	17	0.565/0.543
SVM	0.552	0.552 ± 0.059	18	0.595/0.612

As it follows from Table 1, features wrapping improves performance of classifiers, but their accuracy remains low, slightly bigger than 0.5.

6 Multi-classifier System Training

At the third stage ensemble with majority voting rule is designed from the set of the classifiers trained at the previous stage. The GA is used again. The ensemble is coded by an array w of M real numbers, $w_i \geq 0$. They set value of weight coefficient to corresponding classifier. During creation of initial population, the elements w_i are initialized as random numbers with the normality condition $\sum w_i = 1$.

Selection rules are the same as at the previous stage: elitism and rank selection. Mutation operation is applied to all elements w_i of the selected individual, their values are randomly changed by the uniformly distributed number $(-0.1; 0.1)$. If negative w_i is received as the result, it is replaced by 0. At the same time the normality condition isn't satisfied. These parameters were determined during experimental launches of the algorithm. Crossover operation is like the crossover at the first stage. Ensemble fitness is calculated as the accuracy. Object class C_E is calculated as the weighed sum of outcomes c_i of individual classifiers $C_E = \sum_{i \in M} w_i c_i$.

If the sign of C_E matches the object type ($C_E > 0$ means that next price will growth, $C_E < 0$ means that this price will go down), then an object is considered as recognized correctly. Absolute value $|C_E|$ corresponds with confidence of classification.

After training, ensemble with accuracy 0.744, precision 0.698, and recall 0.880 was received. This accuracy value notably outperforms the accuracy of individual classifiers.

7   Trading Strategy Results

The results of back testing of generated strategy on daily prices of Alphabet Inc. shares (ticker GOOG) in two-year period 2015–2016 is presented on Fig. 4a. Proposed algorithm gives the return $e = [x(n) + d(n)]/x(0) = 1.781$, this result outperforms market growth, which is 1.48. To check real possibility of proposed strategy to generate profit, another test was conducted on test set (Fig. 4b), return is $e = 1.104$.

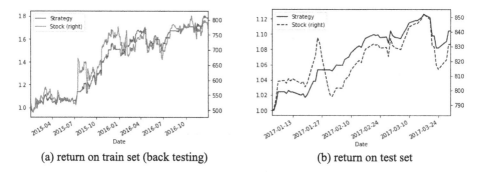

(a) return on train set (back testing)   (b) return on test set

Fig. 4. Tests of generated trading strategy.

To check prediction performance of proposed method, few other well-known ensemble methods were tested on the same training and test set (Table 2). We tested five Bagging algorithms (on base DT, kNN, NB, LR, and SVC), three Adaptive Boosting algorithms (on base DT, SVC, and NB) and three other methods (Gradient Boosting, Random Forest, and ExtraTrees).

As follows from Table 2, some techniques outperform proposed method on training set, but it shows better results on the test set. It means that proposed method more effectively avoids overfitting. Better results on test set are shown by Bagging on base NB and Adaboosting on base SVC. Comparison of trading strategies based on these two techniques and proposed algorithm is presented on Fig. 5. Presented data show that proposed algorithm provides better results (return of Bagging + NB is 1.071, return of AdaBoost + SVC is 1.057, return of proposed method is 1.104).

To check capability of proposed algorithm to generate profit for assets of different companies and industries, test on securities of other companies was carried. We used two-years (2015–2016) daily data of 10 companies from 5 industries for training and three months' data (January–March 2017) for testing. Results are presented in Table 3, including accuracy of ensemble on train and test sets, stock prices changes $p(n)/p(0)$, where n is the length of price series, and return e as it define above.

Table 2. Comparison of different ensemble techniques.

Method	Training set			Test set		
	Accuracy	Precision	Recall	Accuracy	Precision	Recall
BAGGING						
DT	0.988	0.985	0.992	0.508	0.643	0.474
kNN	0.687	0.676	0.744	0.377	0.500	0.316
NB	0.575	0.585	0.585	0.590	0.651	0.737
LR	0.563	0.565	0.643	0.574	0.667	0.632
SVC	0.575	0.585	0.589	0.492	0.667	0.368
ADABOOST						
DT	1.000	1.000	1.000	0.459	0.609	0.368
SVC	0.512	0.512	1.000	0.623	0.623	1.000
NB	0.569	0.575	0.609	0.541	0.625	0.658
OTHER TECHNIQUES						
Gradient Boosting	0.978	0.981	0.977	0.459	0.619	0.342
Random Forest	0.980	0.992	0.969	0.492	0.684	0.342
ExtraTrees	1.000	1.000	1.000	0.443	0.577	0.395
Proposed Method	0.744	0.698	0.880	0.590	0.633	0.816

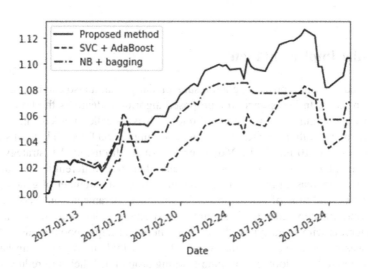

Fig. 5. Comparison of three trading strategies.

The obtained results confirm that the presented algorithm ensures successful trade irrespective of the type of industry both in the growing and falling markets. This means that it can be used as the alpha model in the Portfolio Construction Model [1, 2].

Table 3. Performance of proposed method on different assets.

Company	Industry	Ensemble accuracy		Back testing (train set)		Real testing (test set)	
		Test set	Train set	Price change	Return	Price change	Return
Alphabet	ITC	0.744	0.590	1.502	1.781	1.055	1.104
Amgen	Pharma	0.785	0.519	0.969	2.491	1.096	1.109
Apple	ITC	0.709	0.607	1.133	11.173	1.242	1.246
Exxon Mobile	Oil	0.714	0.541	0.932	8.300	0.799	1.168
General Electric	Manuf.	0.750	0.507	1.371	2.194	0.948	1.003
Gilead Sciences	Pharma	0.881	0.516	0.766	11.436	0.923	1.001
HSBC	Finance	0.889	0.508	0.997	7.296	1.024	1.042
JPMorgan Chase	Finance	0.775	0.514	1.497	10.453	1.013	1.040
Shell	Oil	0.765	0.581	0.981	2.913	0.972	1.013
United Techn.	Manuf.	0.877	0.505	1.019	1.970	1.018	1.027

8   Parallel Implementation

The proposed algorithm provides good results on daily data, it also can be used on data of shorter periods. In its essence, the proposed algorithm identifies the trading orders through reverse engineering of observed quotes. It is so called market microstructure trading, and many authors suggest that typical holding period for such kind of strategies should not exceed 10 min [24]. Moreover, return of discussed AT strategy can be improved, first, by including additional basic classifiers (e.g. different models of ANN), and second, by increasing the size of the population and the number of generations on training stages. But time of calculations with presented parameters on 2-core 1.5 GHz CPU is approximately 20 min, it should be extremely reduced to work on shorter trading intervals with extra types of base classifiers and larger populations.

As it was noted above, algorithm was realized on Python programming language, because there are lot of tools of machine learning around it. It helps to reduce time for algorithm design and testing, but as Python code does not compiled to native CPU code, there are possible performance problems.

The most applicable approach without code rewriting is usage of parallel capabilities of `ipython` library [25] and multi-core system. Several tests with different number of CPU cores were executed, to determine if it is possible to reach the required performance within the `ipython` framework. Figure 6 presents a test environment, which includes server with eight 3,5 GHz CPU cores and client computer, both connected to trusted network.

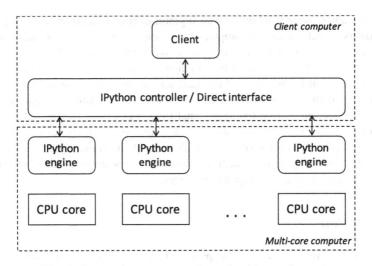

Fig. 6. Test environment to evaluate algorithm performance

On remote multi-core computer, several instances of IPyton engine were started, according with number of CPU cores used in test. The IPython engine is a regular Python interpreter that handles incoming and outgoing Python objects sent over a network connection. All program modules required for computation were located on local disks of server. IPython controller and client interface were ran on client computer.

As it follows from Fig. 1 there are a few opportunities to parallelize program code. First, it is possible to parallelize individual classifiers training, because they are trained independently. Second, there is possibility to parallelize genetic algorithms where they

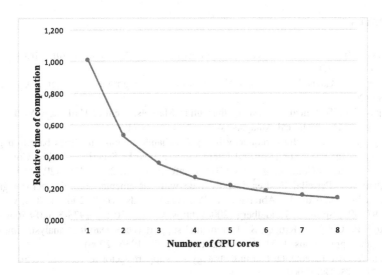

Fig. 7. Reduction in computation time as a function of the number of CPU cores

used. Both features were used on base of IPython engine direct interface that provides the possibility directly manage computation on each engine (without automatic load balancing), because it required a small correction of the source code.

Results of tests performed on 8-core system are presented on Fig. 7, which shows relative computational time (time of computation on one CPU core is 1). It is evident that the system with 4 cores provides the performance that satisfies the requirements of market microstructure trading (computational time is less than 10 min [24]).

However, from Fig. 7 it also follows that further possibilities for increasing performance with this approach are exhausted. To compute in shorter time intervals (1 min and less), it is necessary to implement the algorithm in the programming language that allows more efficient use of computer resources.

9 Conclusion

The presented results show that the proposed algorithm allows to build a trading strategy that stably generates positive return regardless of the behavior of the stock market (growth or decline). This can be explained by the two reasons. The first is the domain knowledge, which was used for features engineering. The second is the use of the multi-classifier system, which combines enough simple classifiers, it helps notable improve the prediction of price behavior.

Note, that the ways to improve this algorithm are obvious. It is the inclusion of additional classifier models in the ensemble, as well as an extension of the search space when using the genetic algorithm.

Using the parallel capabilities of the ipython allows to reduce the computation time to 10 min or less. However, further performance improvement will require a transition to another programming language.

References

1. Nuti, G., Mirghaemi, M., Treleaven, P., Yingsaeree, C.: Algorithmic trading. IEEE Comput. **44**(11), 61–69 (2011)
2. Treleaven, P., Galas, M., Lalchand, V.: Algorithmic trading review. Commun. ACM **56**(11), 76–85 (2013)
3. Murphy, J.J.: Technical Analysis of the Future Markets. Prentice Hall, New York (1986)
4. Schwager, J.D.: Technical Analysis. Wiley, New York (1996)
5. Zhang, H., Ren, R.: High frequency foreign exchange trading strategies based on genetic algorithms. In: Proceedings of 2nd International Networks Security Wireless Communications and Trusted Computing (NSWCTC) Conference, vol. 2, pp. 426–429 (2010)
6. Montana, G., Parrella, F.: Learning to trade with incremental support vector regression experts. In: Corchado, E., Abraham, A., Pedrycz, W. (eds.) HAIS 2008. LNCS, vol. 5271, pp. 591–598. Springer, Heidelberg (2008). https://doi.org/10.1007/978-3-540-87656-4_73
7. Laskov, P., Gehl, C., Kruger, S.: Incremental support vector learning: analysis, implementation and applications. J. Mach. Learn. Res. **7**, 1909–1936 (2006)
8. Wang, W.: An incremental learning strategy for support vector regression. Neural Process. Lett. **21**, 175–188 (2005)

9. de Oliveira, F.A., Nobre, C.N., Zarate, L.E.: Applying artificial neural networks to prediction of stock price and improvement of the directional prediction index–case study of PETR4, Petrobras, Brazil. Expert Syst. Appl. **40**(18), 7596–7606 (2013)

10. Evans, C., Pappas, K., Xhafa, F.: Utilizing artificial neural networks and genetic algorithms to build an algo-trading model for intra-day foreign exchange speculation. Math. Comput. Model. **58**(5), 1249–1266 (2013)

11. Yu, L., Wang, S., Lai, K.: A novel nonlinear ensemble forecasting model incorporating GLAR & ANN for foreign exchange rates. Comput. Oper. Res. **32**, 2523–2541 (2004)

12. Scabar, A., Cloete, I.: Neural networks, financial trading and the efficient market hypothesis. In: Oudshoorn, M. (ed.) XXV Australasian Computer Science Conference (ACSC2002), Melbourne, Australia, vol. 4 (2002)

13. Butler, M., Daniyal, A.: Multi-objective optimization with an evolutionary artificial neural network for financial forecasting. In: GECCO 2009 Proceedings of the 11th Annual Conference on Genetic and Evolutionary Computation, Montreal, Canada, pp. 1451–1458 (2009)

14. Tsai, C.F., Chiou, Y.J.: Earnings management prediction: a pilot study of combining neural networks and decision trees. Expert Syst. Appl. **36**(3), 7183–7191 (2009)

15. Peters, E.E.: Chaos and Order in the Capital Markets. Wiley, New York (1996)

16. Peters, E.E.: Fractal Market Analysis: Applying Chaos Theory to Investments and Economics. Wiley, New York (2003)

17. McNeils, P.D.: Neural Networks in Finance: Gaining Predictive Edge in the Market. Elseiver, Amsterdam (2005)

18. Han, J., Kamber, M., Pei, J.: Data Mining: Concepts and Techniques. Morgan Kaufmann, Waltham (2012)

19. Mukhopadhyay, A., Maulik, U., Bandyopadhyay, S.C.: A survey of multiobjective evolutionary algorithms for data mining: Part I. IEEE Trans. Evol. Comput. **18**(1), 4–19 (2014)

20. Wozniak, M., Grana, M., Corchado, E.: A survey of multiple classifier systems as hybrid systems. Inf. Fusion **16**(1), 3–17 (2014)

21. Kuncheva, L.: Combining Pattern Classifiers: Methods and Algorithms. Wiley, New York (2004)

22. Chan, E.P.: Algorithmic Trading: Winning Strategies and Their Rationale. Wiley, Hoboken (2013)

23. Pedregosa, F., et al.: Scikit-learn: machine learning in python. J. Mach. Learn. Res. **12**, 2825–2830 (2011)

24. Aldridge, I.: High-Frequency Trading: a Practical Guide to Algorithmic Strategies and Trading Systems. Wiley, Hoboken (2009)

25. Pérez, F., Granger, B.E.: IPython: a system for interactive scientific computing. Comput. Sci. Eng. **9**(3), 21–29 (2007)

Smoothed-Particle Hydrodynamics Models: Implementation Features on GPUs

Sergey Khrapov$^{(\boxtimes)}$ and Alexander Khoperskov

Volgograd State University, Volgograd, Russia
{khrapov,khoperskov}@volsu.ru

Abstract. Parallel implementation features of self-gravitating gas dynamics modeling on multiple GPUs are considered applying the GPU-Direct technology. The parallel algorithm for solving of the self-gravitating gas dynamics problem based on hybrid OpenMP-CUDA parallel programming model has been described in detail. The gas-dynamic forces are calculated by the modified SPH-method (Smoothed Particle Hydrodynamics) while the N-body problem gravitational interaction is obtained by the direct method (so-called Particle-Particle algorithm). The key factor in the SPH-method performance is creation of the neighbor lists of the particles which contribute into the gas-dynamic forces calculation. Our implementation is based on hierarchical grid sorting method using a cascading algorithm for parallel computations of partial sums at CUDA block. The parallelization efficiency of the algorithm for various GPUs of the Nvidia Tesla line (K20, K40, K80) is studied in the framework of galactic' gaseous halos collisions models by the SPH-method.

Keywords: Multi-GPU · OpenMP-CUDA · GPU-Direct · NVIDIA TESLA · SPH-method · Self-gravitating gas dynamics · Numerical simulation

1 Introduction

Research of astrophysical systems applies special demands on the properties of computational fluid dynamics models. Supersonic and hypersonic flows with the Mach number $\mathcal{M} \sim 1000$, turbulence including small-scale, and magnetic fields self-consistent accounting are essential for the extragalactic astronomy, cosmology or accreting relativistic objects.

To describe star formation we have to model a multicomponent system with chemical transformations (accounting for tens or even hundreds of chemical reactions) [8]. These processes occur in non-stationary and non-homogeneous gravitational fields on small spatial scales [6]. We should provide long-time calculations due to the problem rigidity taking into account fast processes and spatial small-scale inhomogeneities, when the total evolution time may exceed 10^7 integration time steps.

V. Voevodin and S. Sobolev (Eds.): RuSCDays 2017, CCIS 793, pp. 266–277, 2017.
https://doi.org/10.1007/978-3-319-71255-0_21

Let us particularly emphasize the presence of the dynamic boundaries between matter and vacuum. The same problem appears in the case of the free water surface modeling in reservoirs at Earth's conditions [9].

All these and many other factors are important for modern numerical astrophysical models.

The fast calculation methods' usage for the gravitational force calculation has disadvantages due to poorly controlled errors of the acceleration in the case of approximate numerical methods (TreeCode, Fast Fourier Transform [18], Fast Multipole Methods, etc.) that may require a large number of particles N.

We use the direct method (so-called Particle-Particle algorithm) for the gravitational force calculation. Due to the low prices on the new hardware based on GPU technologies [5] the direct method looks promising especially in models with a number of particles greater than 1 million.

All the features of parallel simulations are considered on the problem of gas halos collision around galaxies. The new precise estimations of the intergalactic gas density in observations makes problem relevant. These results are based on the observations of X-ray coronas around both elliptical and disk galaxies [17].

Initially the Smoothed Particle Hydrodynamics (SPH) method was proposed to simulate the astrophysical gas [3,12]. It has shown to be efficient for various applications as well as in other fields of physics and technology. The SPH approach occupies a significant market quota in astrophysical computational fluid dynamics and engineering applications. It should be specially distinguished the GASOLINE code [15], Weakly Compressible Smoothed Particle Hydrodynamics for multi-GPUs systems [4] and gpuSPHASE for the engineering calculations [16]. The aim of our research is the computational characteristics analysis of a parallel program for galaxies' gas self-gravitating subsystems modeling by the SPH and direct N-body methods using GPU technology. An additional positive aspect of GPUs usage is the visualization efficiency for such processors, which is very important for multidimensional non-stationary multicomponent flows.

2   Mathematical and Numerical Models

2.1   Basic Equations

Let us consider the collision process of two galactic systems each of which includes $N_g/2$ particles gas subsystem (SPH) and $N_h/2$ component (N-body) collisionless dark halo. The dynamics of gas particles is described by a system of differential equations:

$$\frac{d\mathbf{v}_i}{dt} = -\frac{\nabla p_i}{\rho_i} + \sum_{j=1, j\neq i}^{N} \mathbf{f}_{ij} \,, \tag{1}$$

$$\frac{d\mathbf{r}_i}{dt} = \mathbf{v}_i \,, \tag{2}$$

$$\frac{de_i}{dt} = -\frac{p_i}{\rho_i} \nabla \cdot \mathbf{v}_i \,, \tag{3}$$

where $N = N_g + N_h$, the radius-vector $\mathbf{r}_i(t)$ determines the position of the i-th particle in space, ρ_i, p_i, e_i, $\mathbf{v}_i$ are the mass density, gas pressure, specific internal energy, and velocity vector of the i-th particle, respectively. The gravitational interaction force between i-th and j-th particles is

$$\mathbf{f}_{ij} = -G \frac{m_j (\mathbf{r}_i - \mathbf{r}_j)}{|\mathbf{r}_i - \mathbf{r}_j + \delta|^3} \,, \tag{4}$$

where G is the gravitational constant, m_j is the mass of the particle, δ is the gravitational softening length at very short distances.

We use the quasi-isothermal model for the initial density distribution of dark matter in the halo and King model for the initial density profile in the bulge [7,17]. The equation of an ideal gas state is used to close the system of equations (1)–(3)

$$e_i = \frac{p_i}{(\gamma - 1)\rho_i} \,, \tag{5}$$

where γ is the adiabatic index.

2.2   The Numerical Scheme

For the numerical integration of the hydrodynamics equations (1) and (3) the spatial derivatives in these equations should be approximated. In accordance with the SPH-approach [12] for a finite number of particles N_g any medium characteristic $A = \{\rho, e, \mathbf{v}\}$ and its derivatives ∇A are replaced in the flow region Ω by their smoothed values:

$$\widehat{A}(\mathbf{r}) = \sum_{j=1}^{N_g} \frac{m_j}{\rho(\mathbf{r}_j)} A(\mathbf{r}_j) W(|\mathbf{r} - \mathbf{r}_j|, h) \,,$$

$$\nabla \widehat{A}(\mathbf{r}) = \sum_{j=1}^{N_g} \frac{m_j}{\rho(\mathbf{r}_j)} A(\mathbf{r}_j) \nabla W(|\mathbf{r} - \mathbf{r}_j|, h) \,, \tag{6}$$

where W is the smoothing kernel function, h is the smoothing length. The following conditions are imposed on the kernel W:

– the kernel finiteness;

– $\int_\Omega W(|\mathbf{r} - \mathbf{r}'|, h) \, d\mathbf{r}' = 1$ is the normalization condition;

– $\lim_{h \to 0} W(|\mathbf{r} - \mathbf{r}'|, h) = \delta(|\mathbf{r} - \mathbf{r}'|)$, δ is Dirac delta-function.

Different authors have used spline functions of different orders or Gaussian distribution for the smoothing kernel W [1,12,14]. In current paper a cubic spline

$$\rho_i = \rho(\mathbf{r}_i) = \sum_{j=1}^{N_g} m_j W(|\mathbf{r}_i - \mathbf{r}_j|, h_{ij}) \tag{7}$$

has been used to calculate mass-density of the i-th particle *Monaghan*:

$$W(\xi, h) = \frac{1}{\pi h^3} \begin{cases} 1 - \frac{3}{2}\xi^2 + \frac{3}{4}\xi^3, & 0 \le \xi \le 1; \\ \frac{1}{4}(2-\xi)^3, & 1 \le \xi \le 2; \\ 0, & \xi \ge 2; \end{cases} \tag{8}$$

where $\xi = |\mathbf{r}_i - \mathbf{r}_j| / h$ is the relative distance from the center of the i-th particle, $h_{ij} = (h_i + h_j)/2$ is the effective smoothing length. The smoothing length value for each particle depends on its mass and density as $h_i = \sigma \left(m_i/\rho_i\right)^{1/3}$, where σ is a constant $\sim 1.2 \div 1.3$ [10, 13].

If a smoothing core (8) is used to calculate the gas-dynamic forces (pressure gradient), then unphysical (numerical) particles clustering [1] will occur in the high-pressure regions. The latter is caused by the interaction force weakening between particles in the neighborhood of $0 < \xi < \frac{2}{3} \left(\lim_{\xi \to 0} \frac{\partial W}{\partial \xi} = 0 \right)$. A smoothing kernel W_p presenting in the following form [14]:

$$W_p(\xi, h) = \frac{15}{64\pi h^3} \begin{cases} (2-\xi)^3, & 0 \le \xi \le 2; \\ 0, & \xi \ge 2; \end{cases} \tag{9}$$

eliminates clustering of particles and increases the stability of the numerical algorithm. From equation (9) it follows that $\lim_{\xi \to 0} \frac{\partial W_p}{\partial \xi} = -\frac{45}{64\pi h^4}$.

Applying the SPH-approach (6)–(9) to equations (1) and (3) we finally get:

$$\frac{d\mathbf{v}_i}{dt} = -\sum_{j=1, j\neq i}^{N_g} m_j\, \Pi_{ij}\, \nabla W_p\left(|\Delta\mathbf{r}_{ij}|, h_{ij}\right) + \sum_{j=1, j\neq i}^{N} \mathbf{f}_{ij}, \tag{10}$$

$$\frac{de_i}{dt} = \frac{1}{2}\sum_{j=1, j\neq i}^{N_g} m_j\, \Pi_{ij}\, \Delta\mathbf{v}_{ij} \cdot \nabla W_p\left(|\Delta\mathbf{r}_{ij}|, h_{ij}\right), \tag{11}$$

where $\Delta\mathbf{r}_{ij} = \mathbf{r}_i - \mathbf{r}_j$, $\Delta\mathbf{v}_{ij} = \mathbf{v}_i - \mathbf{v}_j$, $\nabla W_p(|\Delta\mathbf{r}_{ij}|, h_{ij}) = \frac{\partial W_p}{\partial \xi} \frac{\Delta\mathbf{r}_{ij}}{|\Delta\mathbf{r}_{ij}|} \frac{1}{h_{ij}}$, $\Pi_{ij} = \frac{p_i}{\rho_i^2} + \frac{p_j}{\rho_j^2} + \nu_{ij}^a$ is the pressure force symmetric SPH-approximation ensuring Newton's third law fulfillment. The artificial viscosity ν_{ij}^a is expressed via

$$\nu_{ij}^a = \frac{\mu_{ij}\,(\beta\,\mu_{ij} - \alpha\, c_{ij})}{\rho_{ij}}, \qquad \mu_{ij} = \begin{cases} \frac{h_{ij}\,\Delta\mathbf{r}_{ij}\cdot\Delta\mathbf{v}_{ij}}{|\Delta\mathbf{r}_{ij}|^2 + \eta\, h_{ij}^2}, & \Delta\mathbf{r}_{ij}\cdot\Delta\mathbf{v}_{ij} < 0; \\ 0, & \text{else}; \end{cases}$$

where $\rho_{ij} = (\rho_i + \rho_j)/2$, $c_{ij} = \left(\sqrt{\gamma p_i/\rho_i} + \sqrt{\gamma p_j/\rho_j}\right)/2$ are the density and sound velocity average values for i-th and j-th interacting particles, respectively. The empirical constants α, β and η determine the intensity of artificial viscosity (in our calculations their reference values are $\alpha = 0.5$, $\beta = 1$ and $\eta = 0.1$).

A second-order accuracy method of the predictor-corrector type (the so-called leapfrog method) is used for the numerical integration of differential equations (10), (11) and (2). The main steps of the leapfrog method for self-gravity SPH-models are:

(I) The velocity $\mathbf{v}_i$ and internal energy e_i predictor calculations at time $t + \Delta t$:

$$\widetilde{\mathbf{v}}_i(t + \Delta t) = \mathbf{v}_i(t) + \Delta t \, \mathbf{Q}_i[\mathbf{r}(t), \mathbf{v}(t), e(t)], \tag{12}$$

$$\widetilde{e}_i(t + \Delta t) = e_i(t) + \Delta t \, E_i[\mathbf{r}(t), \mathbf{v}(t), e(t)], \tag{13}$$

where Δt is the time step, $\mathbf{Q}_i$ and E_i are the right-hand side of equations (10) and (11), respectively.

(II) particles' spatial position calculation $\mathbf{r}_i$ at time $t + \Delta t$:

$$\mathbf{r}_i(t + \Delta t) = \mathbf{r}_i(t) + \frac{\Delta t}{2} \left[\widetilde{\mathbf{v}}_i(t + \Delta t) + \mathbf{v}(t) \right]. \tag{14}$$

After the particles' new positions $\mathbf{r}_i(t+\Delta t)$ to be defined the density $\rho[\mathbf{r}_i(t+\Delta t)])$ is refined according to equation (7).

(III) During the corrector step the velocity, $\mathbf{v}_i$, and internal energy, e_i, values are recalculated at time $t + \Delta t$:

$$\mathbf{v}_i(t + \Delta t) = \frac{\mathbf{v}_i(t) + \widetilde{\mathbf{v}}_i(t + \Delta t)}{2} + \frac{\Delta t}{2} \, \mathbf{Q}_i[\mathbf{r}(t + \Delta t), \widetilde{\mathbf{v}}(t + \Delta t), \widetilde{e}(t + \Delta t)], \tag{15}$$

$$e_i(t + \Delta t) = \frac{e_i(t) + \widetilde{e}_i(t + \Delta t)}{2} + \frac{\Delta t}{2} \, E_i[\mathbf{r}(t + \Delta t), \widetilde{\mathbf{v}}(t + \Delta t), \widetilde{e}(t + \Delta t)]. \tag{16}$$

In general, the right-hand sides in the predictor-corrector scheme (12)–(16) are calculated twice at the same time layer. Since the gravitational interaction of particles (4) depends only on the particles' positions $\mathbf{r}_i$ the calculation of the force between the particles in this approach is performed once per integration time step. The latter allows increasing of the calculations performance about 2 times keeping the same order of accuracy for the method.

To increase the stability of the numerical method during the modeling of supersonic self-gravitating gas flows, we modified the standard SPH stability condition [12] as follows:

$$\Delta t = C_{CFL} \min_i \left[\frac{s_{ij}^{min}}{c_{ij}^{max}(1 + 1.2\alpha) + 1.2\beta\mu_{ij}^{max} + \sqrt{s_{ij}^{min}(|\mathbf{Q}_i| + |E_i|)}} \right], \tag{17}$$

where $s_{ij}^{min} = \min_j |\mathbf{r}_i - \mathbf{r}_j|$, $c_{ij}^{max} = \max_j c_{ij}$, $\mu_{ij}^{max} = \max_j \mu_{ij}$. We added the third term with the square root in the denominator of (17) and replaced $h_{ij} \to s_{ij}^{min}$ in the numerator. Using relation (17), a stable calculation can be performed with larger Courant number ($0.5 \lesssim C_{CFL} < 1$) and lower artificial viscosity value.

3   Parallel Algorithm Design

A parallel implementation of the numerical algorithm (12)–(16) for multiple GPUs has been performed using OpenMP-CUDA and GPU-Direct technologies. Figure 1a presents a two-level OpenMP-CUDA parallelization scheme for $k \times$ GPUs, and Fig. 1b shows a data transfer scheme between GPUs based on GPU-Direct technology for NVIDIA graphics processors.

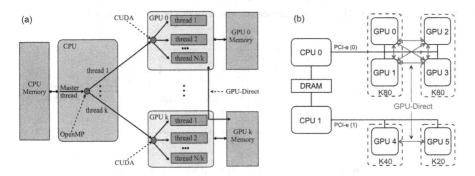

Fig. 1. (a) The two-level scheme of parallelization with OpenMP–CUDA. (b) Architecture 2×CPU+6×GPU.

The two-level parallelization scheme OpenMP-CUDA (Fig. 1a) is more suitable for shared memory systems type CPU + $k \times$ GPU. Using OpenMP technology to create k-threads on the CPU allows us to run CUDA kernels on $k \times$ GPUs on each of which we calculate the dynamics of N/k particles [2]. GPU-Direct technology provides the fast data exchange between GPUs via PCI-e bus. This technology is only applicable to graphics processors that connect to PCI-e buses under the control of one CPU (Fig. 1b).

The numerical algorithm consists of five major Global CUDA Kernels being run from CPU on multiple GPUs using the OpenMP parallel programming model:

- The Sorting Particles (SP) is a set of CUDA Kernels to determine the particles' numbering and number of particles in three-dimensional grid cells. Further this information is used to define the particles' neighbor list during the calculation of SPH sums in equations (7), (10) and (11). The computational complexity of the kernel is $\sim O(N)$.
- The Density Computation (DC) is a CUDA Kernel for density calculation using (7). It has the similar computational complexity $\sim O(N)$.
- The Hydrodynamics Force Computation (HFC) is a CUDA Kernel for the hydrodynamic forces calculation in (10) and (11). The kernel has two states {predictor, corrector} and its computational complexity is $\sim O(N \cdot \overline{N}_{pc})$, where $\overline{N}_{pc}$ is the average number of particles in the cells.

- The Gravity Force Computation (GFC) is a CUDA Kernel for the gravitational forces calculation using (4). The computational complexity of the kernel is $\sim O(N^2)$, because of the direct N-body method.
- The System Update (SU) is a CUDA Kernel for the particle characteristics updating $(\mathbf{r}_i, \mathbf{v}_i, e_i)$ corresponding to equations (12)–(16). The kernel has two states {predictor, corrector}, and its computational complexity is $\sim O(N)$.

The Global CUDA Kernels execution sequence corresponds to the predictor-corrector scheme stages (12)–(16). It is shown at the diagram in Fig. 2. CUDA kernels SP, DC and HFC are skipped in the case of a collisionless system.

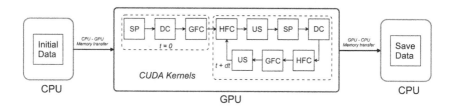

Fig. 2. Flow diagram for the calculation module.

An important factor affecting the efficiency of the parallel implementation of the SPH method is the sorting algorithm and building a particles neighbour list.

Let us consider the algorithm parallel implementation for the particles sorting on CUDA Kernels SP in details. The computational domain is covered by a grid $M_x \times M_y \times M_z$ with the total number of cells $M = M_x M_y M_z$. We use the following auxiliary arrays to build particles neighbor list in the SPH method:

- CellSPH[M] is a vector type array of int2, the components CellSPH[k].x and CellSPH[k].y contain a number of particles at the k-cell and a number of all the particles in the 0 to k cells, respectively;
- indexPC[N] is a vector type array of int2, the component indexPC[i].$x = k$ comprises a cell number k, where the i-th particle is located, while indexPC[j].$y = i$ links the initial number of the i-th particle with the sequential numeration of j particles in the cells;
- indexCell[M] is an integer type array specifying the current particle number at the corresponding k-th cell;
- maxPBC[$M/BlockSize$] is an integer array containing the number of particles at the CUDA block, where the $BlockSize$ and $M/BlockSize$ are the CUDA block number of threads and the CUDA grid number of CUDA blocks, respectively;
- hmaxCell[M] is a double type array comprising the smoothing length maximum value of the particle h_i at the k-th cell.

The entire particle sorting stage contains 5 separate CUDA Kernels:

- the kernelSortingSPH0$<<<M/BlockSize, BlockSize>>>$ is the NULL-initialization of sorting arrays.

- the kernelSortingSPH1$<<<N/BlockSize, BlockSize>>>$ includes numbers of cells where particles are located, a number of particles and maximum value of smoothing length in cells (indexPC$[i].x$, CellSPH$[k].x$ and hmaxCell$[k]$, where $k = 0, ..., M - 1$, $i = 0, ..., N - 1$.
- In the kernelSortingSPH2$<<<M/BlockSize, BlockSize>>>$ the total number of particles in all the cells from k to $k + BlockSize$ is defined for each CUDA block using the cascading algorithm of parallel partial sums finding. The latter is the analog of the sequential algorithm CellSPH$[k].y =$ CellSPH$[k - 1].y+$ CellSPH$[k].x$. The maxPBC$[]$ is evaluated next.
- Based on the total number of particles computed in the previous kernel at the CUDA block (maxPBC$[]$), in the kernelSortingSPH3$<<<M/BlockSize, BlockSize>>>$ the total number of particles in all cells from 0 to k is specified.
- In the kernelSortingSPH4$<<<N/BlockSize, BlockSize>>>$ the correspondence between the original i-th particle number and the sequential numbering of j-th particles in cells is determined (the value of indexPC$[j].x = i$ is calculated).

The fragment of the sorting algorithm code is listed below.
The code for CUDA-core: kernelSortingSPH2

```
__global__ void kernelSortingSPH2(int2 *CellSPH, int *maxPBC){
  __shared__ int sp[BlockSize], sp0[BlockSize];
  int ss, i, k = threadIdx.x + blockIdx.x * blockDim.x;
  sp[threadIdx.x] = CellSPH[k].x;
  sp0[threadIdx.x] = sp[threadIdx.x];        __syncthreads();
  for(i = 1; i < BlockSize; i*=2){
    if (threadIdx.x + i < BlockSize)
    sp[threadIdx.x+i] += sp0[threadIdx.x]; __syncthreads();
    sp0[threadIdx.x] = sp[threadIdx.x];       __syncthreads();}
  CellSPH[k].y = sp[threadIdx.x];
  if(threadIdx.x == 0){
    i = blockIdx.x; ss = sp[BlockSize - 1];
    while(i < gridDim.x){atomicAdd(&(maxPBC[i]), ss); i++;}}
}
```

The code for CUDA-core: kernelSortingSPH3

```
__global__ void kernelSortingSPH3(int2 *CellSPH, int *maxPBC){
  int k = threadIdx.x + blockIdx.x * blockDim.x;
  if (blockIdx.x > 0) CellSPH[k].y += maxPBC[blockIdx.x - 1];
}
```

The code for CUDA-core: kernelSortingSPH4

```
__global__ void kernelSortingSPH4(int2 *indexPC, int2 *CellSPH,
                      int *indexCell){
  int i = threadIdx.x + blockIdx.x * blockDim.x;
```

```
    int k = indexPC[i].x, j = (k>0) ? CellSPH[k - 1].y : 0;
    int ibk = atomicAdd(&indexCell[k], 1), indexPC[j + ibk].y = i;
}
```

In the CUDA kernels, DC and HFC, arrays indexPC[], CellSPH[] and hmaxCell[] are used to find the particles neighbor list upon SPH sums calculation.

4   The Principal Results and Discussions

We have studied the parallelization efficiency of our algorithm solving the relevant problem of galactic gaseous halos collisions modeling. The calculations have been carried out on GPU Nvidia Tesla processors: K20 (1GPU), K40 (1GPU), K80 (2GPU).

A different amount of gas $N_g = N/2$ and collisionless $N_h = N/2$ particles has been used in the calculations. The total number of particles $N = N_g + N_h$ has been set in the range from 2^{18} to 2^{23}.

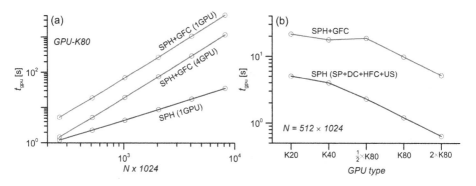

Fig. 3. The execution time of CUDA kernels SPH (SP, DP, HFC, US) and GFC on GPUs. The dependence of t_{gpu} on (a) the number of the particles N; (b) the GPU type.

Figure 3 represents the computation time of the hydrodynamic and gravitational interaction of the particles for different amount of N and GPUs types. For CUDA kernel GFC the calculation time dependence on the number of particles is almost quadratic which corresponds to the Particle-Particle algorithm complexity $O(N^2)$. The SPH calculation time has almost a linear dependence on the number of particles, which also corresponds to the kernel HFC CUDA algorithm complexity $\sim O(N \cdot \overline{N}_{pc})$ ($\overline{N}_{pc} \simeq$ const, since $h \sim N^{-1/3}$). The parallelization efficiency of the algorithm on two and four GPUs is 95% and 90%, respectively.

Table 1 shows some numerical values of the execution time of CUDA kernels SPH and GFC on different GPUs as a function of the number of particles N. Figure 3b and Table 1 show that the runtime of CUDA kernels SPH (SP + DC + HFC + US) on one K80 GPU is 1.7 times less than for the K40 GPU, but the

Table 1. The execution time of CUDA kernels SPH and GFC on GPUs.

	$K20$ (1GPU)		$K40$ (1GPU)		$\frac{1}{2} \times K80$ (1GPU)	
$N \times 1024$	t_{SPH}	t_{GFC}	t_{SPH}	t_{GFC}	t_{SPH}	t_{GFC}
256	2.54	4.13	2.02	3.51	1.20	4.10
512	5.04	16.50	3.99	13.72	2.30	16.37
1024	9.93	65.90	7.87	53.62	4.40	65.32

CUDA kernel GFC runs 1.2 times faster on the GPU K40. The SPH algorithm uses only global GPU memory, and the calculation of forces between the i-th and j-th particles in the CUDA kernel GFC is organized using shared memory of the GPU. Therefore, the different speed of CUDA kernels SPH and GFC execution on GPUs may be due to more efficient access to global memory on the K80.

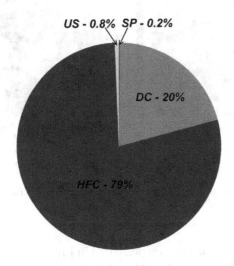

Fig. 4. The contributions of the different stages of the SPH numerical scheme at given time step.eps

Figure 4 demonstrates that the SP sorting time is borrowed only 0.2% of the total SPH simulation time. The sorting algorithm parallel implementation on GPUs proposed in current article requires less computational and memory resources in comparison with tree-based and hash-tables algorithms [11]. Note that the integration time step decreases ($\Delta t \sim h \sim N^{-1/3}$) with an increase in the particles number in accordance with the stability condition (17). Therefore, the total time for modeling the self-consistent dynamics of particles of the gas and collisionless subsystems has a stronger dependence on N than the one shown in Fig. 3a: $t_{all} = t_{SPH} + t_{GFC} = O(N^{4/3}) + O(N^{7/3}) = O(N^{7/3})$.

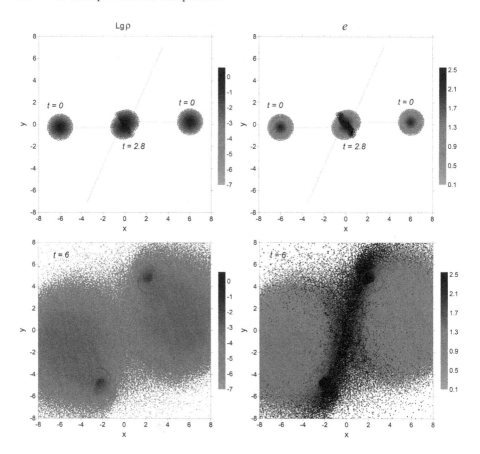

Fig. 5. The distribution of density (left) and internal energy (right) at different times. The dotted line and circles indicate the trajectory of the dark halo.

The results of our simulation are presented in Fig. 5. In the process of collision of galaxies, there is a mixing of matter of two galactic systems. An important factor in the interaction of galaxies is the formation of nonstationary shock waves in the collision of gas halos, leading to a substantial heating of the gas in the halo. After the passage of the gas halos, some of their matter is emitted into the surrounding space with the formation of clouds with a nonzero angular momentum.

Acknowledgments. The first author is thankful to the RFBR (grants 16-07-01037, 15-02-06204 and 16-02-00649). The second author has been supported by the Ministry of Education and Science of the Russian Federation (government task No.2.852.2017/4.6).

References

1. Desbrun, M., Cani, M.-P.: Smoothed Particles: a new paradigm for animating highly deformable bodies. In: Proceedings of the Eurographics Workshop on Computer Animation and Simulation 1996, pp. 61–76 (1996)
2. Dyakonova, T., Khoperskov, A., Khrapov, S.: Numerical model of shallow water: the use of NVIDIA CUDA graphics processors. J. Commun. Comput. Inf. Sci. **687**, 132–145 (2016)
3. Hwang, J.-S., Park, C.: Effects of Hot Halo Gas on star formation and mass transfer during distant galaxy-galaxy encounters. Astrophys. J. **805**(2), 19 (2015). article id. 131
4. Ji, Z., Xu, F., Takahashi, A., Sun, Y.: Large scale water entry simulation with smoothed particle hydrodynamics on single- and multi-GPU systems. J. Comput. Phys. Commun. **209**, 1–12 (2016)
5. Khan, F.M., Berentzen, I., Berczik, P., Just, A., Mayer, L., Nitadori, K., Callegari, S.: Formation and hardening of supermassive black hole binaries in minor mergers of disk galaxies. Astrophys. J. **756**(1), 10 (2012). article id. 30
6. Khoperskov, S.A., Vasiliev, E.O., Ladeyschikov, D.A., Sobolev, A.M., Khoperskov, A.V.: Giant molecular cloud scaling relations: the role of the cloud definition. J. Mon. Not. R. Astron. Soc. **455**(2), 1782–1795 (2016)
7. Khoperskov, S.A., Moiseev, A.V., Khoperskov, A.V., Saburova, A.S.: To be or not to be oblate: the shape of the dark matter halo in polar ring galaxies. J. Mon. Not. R. Astron. Soc. **441**(3), 2650–2662 (2014)
8. Khoperskov, S.A., Vasiliev, E.O., Khoperskov, A.V., Lubimov, V.N.: Numerical code for multi-component galaxies: from N-body to chemistry and magnetic fields. J. Phys. **510**(1), 13 (2014). article id. 012011
9. Khrapov, S., Pisarev, A., Kobelev, I., Zhumaliev, A., Agafonnikova, E., Losev, A., Khoperskov, A.: The numerical simulation of shallow water: estimation of the roughness coefficient on the flood stage. J. Adv. Mech. Eng. **5**, 1–11 (2013). article id. 787016
10. Lodato, G., Clarke, C.J.: Resolution requirements for smoothed particle hydrodynamics simulations of self-gravitating accretion discs. J. Mon. Not. R. Astron. Soc. **413**, 2735–2740 (2011)
11. Mokos, A., Rogers, B.D., Stansby, P.K., Dominguez, J.M.: Multi-phase SPH modelling of violent hydrodynamics on GPUs. J. Comput. Phys. Commun. **196**, 304–316 (2015)
12. Monaghan, J.: Smoothed particle hydrodynamics. J. Ann. Rev. Astron. Astrophys. **30**, 543–574 (1992)
13. Monaghan, J.J.: Smoothed particle hydrodynamics. J. Rep. Prog. Phys. **68**, 1703–1759 (2005)
14. Muller, M., Charypar, D., Gross, M.: Particle-based fluid simulation for interactive applications. In: Proceedings of 2003 ACM SIGGRAPH Symposium on Computer Animation, pp. 154–159 (2003)
15. Wadsley, J.W., Stadel, J., Quinn, T.: Gasoline: a flexible, parallel implementation of treeSPH. J. New Astron. **9**(2), 137–158 (2004)
16. Winkler, D., Meister, M., Rezavand, M., Rauch, W.: gpuSPHASE-A shared memory caching implementation for 2D SPH using CUDA. J. Comput. Phys. Commun. **213**, 165–180 (2017)
17. Zasov, A.V., Saburova, A.S., Khoperskov, A.V., Khoperskov, S.A.: Dark matter in galaxies. J. Phys. Usp. **60**(1), 3–39 (2017)
18. Zhang, X.: N-body simulations of collective effects in spiral and barred galaxies. J. Astron. Comput. **17**, 86–128 (2016)

The Integrated Approach to Solving Large-Size Physical Problems on Supercomputers

Boris Glinskiy, Igor Kulikov, Igor Chernykh, Alexey Snytnikov, Anna Sapetina[✉],
and Dmitry Weins

The Institute of Computational Mathematics and Mathematical Geophysics of SB RAS,
Novosibirsk, Russia
gbm@opg.sscc.ru, kulikov@ssd.sscc.ru, chernykh@parbz.sscc.ru,
snytav@gmail.com, afsapetina@gmail.com, wns.dmitry@gmail.com

Abstract. This paper presents the results obtained by the authors on applying an integrated approach to solving geoseismics, astrophysics, and plasma physics problems on high-performance computers. The concept of the integrated approach in the context of mathematical modeling of physical processes is understood as constructing a physico-mathematical model of a phenomenon, a numerical method, a parallel algorithm and its software implementation with the efficient use of a supercomputer architecture. With this approach, it becomes relevant to compare not only the methods of solving a problem but, also, physical and mathematical statements of a problem aimed at creating the most effective implementation of a chosen computing architecture. The scalability of algorithms is investigated using the multi-agent system AGNES simulating the behavior of computing nodes based on the current state of computer equipment characteristics. In addition, special attention in this paper is given to the energy efficiency of algorithms.

Keywords: Integrated approach · Co-design · Agent simulation · Energy efficiency of algorithms · Parallel algorithm · Supercomputers

1 Introduction

The modern stage of supercomputer development is characterized by the emergence of many projects on creation of an exascale-class supercomputer. Thus far, developments in the field of exascale supercomputers are conducted by different teams of developers in the United States. The collaboration in this direction is carried out, for example, by national laboratories of the US Department of Energy: Sandia and Oak Ridge. In Europe, there are also similar programs; seven European countries have signed the declaration of the Joint Project EuroHPC aimed at the creation of exascale supercomputers. In Japan (the RIKEN institution), the development of a supercomputer has already begun in, where it will be assembled and installed.

There are numerous international projects to develop the system and application software for exascale-class supercomputers with the participation of the United States, countries of the European Union, Japan, China, Russia (IESP, G8 EXASCALE,

© Springer International Publishing AG 2017
V. Voevodin and S. Sobolev (Eds.): RuSCDays 2017, CCIS 793, pp. 278–289, 2017.
https://doi.org/10.1007/978-3-319-71255-0_22

CRESTA, etc.). In [1–3], a review and various approaches to exascale scientific software design are given. In [4], the problems typical of exascale systems are listed, such as insufficient concurrent work available to maintain high utilization of all resources, time-distance delay intrinsic of parallel actions and resources on the critical execution path, which is not necessary in a sequential version, delay due to the lack of availability of oversubscribed shared resources. Also, in [4] various methods for overcoming the above-mentioned problems are discussed.

It should be pointed out that numerical algorithms are being developed slower than hardware. Therefore existing algorithms and programs for solving physical problems will be applied at the first stage of using exascale-class supercomputers.

We have offered the integrated approach to the development of algorithms and software for petascale- and exascale-class supercomputers [5]. It contains the three stages.

The first stage is the co-design, which is based on the development of a parallel computational technology, with allowance for all aspects of parallelism. The co-design of parallel methods for solving large-scale problems is difficult to formalize. It is impossible to make a "collection of recipes" for the efficient solution of any problem. However, some general approaches can be proposed. The co-design approach concept consists of the following steps, with allowance for the target hardware/software platform:

(1) Formulation of the physical statement of the problem;
(2) Mathematical formulation of the physical problem;
(3) Development of the numerical methods;
(4) Selection of data structures and parallel algorithms;
(5) Consideration of a supercomputer architecture;
(6) Usage of code optimization tools.

We use the extended definition of the co-design, in contrast to the common conception which consists in the joint development of software and hardware. In such an approach, not only the comparison of the problem solution methods is becoming relevant but also the comparison of the efficiency of using various physical and mathematical statements [5].

The second stage is the anticipated development of algorithms and software for the most promising exascale supercomputers. This stage is based on the simulation of the algorithm behavior within a certain supercomputer architecture. For the simulation of distributed systems it is best to use distributed simulation based on message passing.

The multi-agent approach [6] is used for the simulation of parallel programs run on a large number of cores due to such properties as decentralization, self-organization, and intelligent behavior [7]. Among many multi-agent simulation platforms the adaptable distributed simulation system AGNES [8] was chosen. It was successfully used in the scalability study of a series of parallel problems when executed on a large number of cores [5, 9].

The third stage is estimating the energy efficiency of the algorithm with different implementations for a single architecture or for different supercomputer architectures. In this paper, the term «energy efficiency for scientific HPC applications» means the most efficient use of each core, processor or computational accelerator; the minimization of communications between computational nodes; a good workload balancing of the program. The minimization of communications enables a decrease in the idle standing time for processors and accelerators. The good workload balancing enables a uniformly load of a computational system. The most energy-efficient algorithm gives the best FLOPS per Watts (Joules/sec) value.

The efficiency of this approach is illustrated on some examples of complex computing problems in seismology, plasma physics and astrophysics.

2   Using the Integrated Approach to Solving an Elastodynamic Problem

The numerical modeling of elastic wave propagation in heterogeneous 3D media with complex subsurface geometries is a complex problem in terms of computation, thus demanding the use of efficient methods of parallelization and scaling of algorithms. Quite often the topography of various real geophysical objects does not allow one to maintain an observational system. Therefore, constructing their 3D models requires solving the inverse problem by solving a set of direct problems: for different values of the elastic parameters of a heterogeneous medium; at various geometries of objects composing a model. This complicates the problem in the context of computation.

We apply the above-discussed approach to solving the problem of seismic wave propagation in a heterogeneous medium typical of magmatic volcanoes. Both active and sleeping volcanoes are potentially dangerous to the environment due to the possibility of sudden catastrophic eruptions. Using methods of the active vibroseismic monitoring of magmatic structures will allow predicting a probable time of eruption.

For the purpose of the co-design, we have made a comparison of the developed parallel implementations of solutions to the elastodynamic problem written in terms of the velocities of displacement and stress and in terms of displacements for the computational clusters, equipped with graphics cards. The simulation domain is considered to be an isotropic 3D-inhomogeneous elastic structurally complex medium which is a parallelepiped, one of whose sides is a free surface.

At the step of designing a numerical method, the most "flexible" and widespread technique for solving a three-dimensional elastodynamic problem is a finite difference method. Let us preliminarily notice that explicit finite difference schemes fit the architecture of graphics accelerator, because they are directly mapped on the topology of GPU architecture, and involve independent computations of values at each step and in each cell of the computational domain. In order to numerically solve elastodynamics equations in terms of the velocities of displacement and stress we apply the well-known Verrier finite difference scheme on a staggered grid [10]. The calculation of its difference coefficients is based on integral conservation laws. To solve the problem in terms of displacements, we use a similar finite difference scheme [11].

We should note the main difference between the algorithms, which can be constructed on the above-mentioned finite difference schemes. The calculation of the velocities of displacement and stress requires a larger memory size (at least, 18 3D arrays with the unknowns should be stored), but requires a smaller number of floating-point operations in total (57 operations for calculating the values in a cell for one time step).

The calculation of the displacements requires a smaller memory size (at least, 6 3D arrays with unknowns), but a larger number of operations (98 operations for one time step).

As software parallelizing tools we have chosen CUDA and MPI, which make possible to simultaneously use the largest number of parallel processes and ultimately to attain a maximum efficiency.

At the step of adaptation to a hybrid cluster equipped with GPU (for example, NKS-30T+GPU, which was installed in the Siberian Supercomputer Center and consists of 40 computational nodes, each one equipped with two six-core CPU Xeon X5670 and three NVIDIA Tesla M2090 graphics cards) we have implemented the next operation for the both statements implementations [11]. For carrying out the parallelization, we decompose the computational domain to layers along one of the coordinate axes. Each layer is calculated at a separate node, where, in turn, it is sub-divided into sub-layers along another coordinate axis (to attain a better scaling) according to the number of graphics accelerators at a node. In order to minimize the time of the data exchange, the data are transferred among nodes using appropriate non-blocking asynchronous functions of MPI, and exploiting the asynchronous copy function of CUDA for exchanges among the graphics cards. Let us note that the data for the exchange have an equal size in both approaches.

The numerical experiments have shown that the time of the calculations of the displacements and the time of the calculations of the velocities of displacement and stress at an equal number of nodes is roughly the same in spite of the fact that displacements calculation is performed with a larger amount of floating point operations at the each time step. Therewith the displacement calculations requires almost half as much of the GPU memory size. Based on the results obtained we prefer using the approach proposed to calculating the displacements.

Results of numerical simulation for the truncated model of the volcano Elbrus are presented in Fig. 1. For numerical simulation, a spatial grid of $1360 \times 701 \times 2600$ nodes

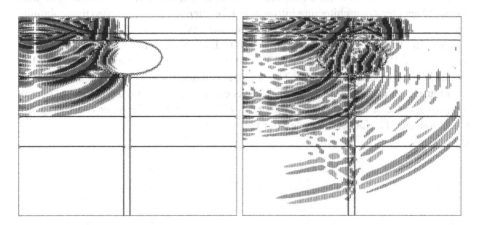

Fig. 1. Results of numerical simulation for the truncated model of the volcano Elbrus. In the snapshots of the wave field, the component u of the displacement velocities vector is presented in the plane Oxz at different time points.

and 15000 time steps were used. The calculation was carried out on 20 hybrid nodes of the NKS-30T+GPU cluster during 5 h. One can learn more about the geophysical model of the volcano and the results of numerical experiments in [12].

3   Using the Integrated Approach to Solving Astrophysical Problems

In this paper, we use a multi-component hydrodynamic model of galaxies considering the chemodynamics of molecular hydrogen and cooling in the following form. A detailed description of this model can be found in [13].

At the first stage of the co-design procedure, we define the main physical process of a problem. In the case of astrophysics, this process is hydrodynamics. For the description of hydrodynamics, the hyperbolic equations are used. There are many grid numerical methods for solving the hyperbolic equations [13, 14]. Some of these methods can be effectively realized by the computational domain decomposition. With adding the subgrid physics (e.g., cooling/heating, chemodynamics, a magnetic field), the structure of the equations remains hyperbolic. For the characterization of collisionless components, the first moments of the Boltzmann equation [14–16] can be used. In this case, a uniform numerical method can be used for solving hydrodynamic and collisionless components. It is possible to use the conjugate gradient method for the Poisson equation solution, which is successfully adopted in the HERACLES code [17]. The use of conformal mappings allows the construction of a moving mesh for solution detailing.

The numerical method of solving hydrodynamic equations is based on a combination of the operator splitting approach, the Godunov method with a modification of the Roe averaging, and a piecewise-parabolic method on a local stencil [18]. The redefined system of equations is used to ensure a non-decrease of entropy and for speed corrections. The detailed description of the numerical method can be found in [19].

Our AstroPhi code is based on the methods in question. We have taken the Intel Xeon Phi accelerator architecture into account for our code and controlled the scalability using the Intel Vectorization Advisor software tool. We use the RSC PetaStream architecture 8-node engineering prototype with 64x Intel Xeon Phi 7120D accelerators for the simulation. Our tests show that 10% of the total simulation time is spent on MPI/OpenMP send/receive operations. This value is suitable for massively parallel systems.

Figure 2 shows the expansion of two gas clouds after the galaxy collision. One of the possible scenarios is realized: one galaxy flying through the other formatting two gas clouds and H2 formation zone after the impact.

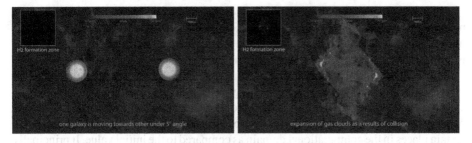

Fig. 2. The galaxy collision AstroPhi test with chemodynamics: Initial stage (on the left), Expansion of gas clouds after collision and H_2 formation zone (on the right). In the numerical simulation, cluster RSC PolyTechnik (32 Intel Xeon Phi 7150) with 1024^3 mesh during 48 h was used.

4 Using the Integrated Approach to Solving the Plasma Physics Problems

One of the most interesting problems in the fusion physics is the resonance interaction of a powerful electron beam with plasma. This problem has many practical applications such as studying the processes in the outer layers of the Sun, a fast ignition scheme in inertial fusion and plasma heating in tokamaks.

Let us consider the required computational resources to run a 3D kinetic plasma simulation that are necessary in the above-mentioned plasma physics problems. A rough estimate is a mesh with 100^3 nodes with 1000 particles each. A particle in a 3D case has 6 attributes: 3 double precision variables for 3 components of the coordinate vector and for the impulse vector, finally, 48 bytes. Then this requires about 0.05 terabyte for 1 billion model particles.

The number of flops of Particle-In-Cell method consumes about 250 floating point operations per particle (the number, of course, depends on the code details) during one time step. This results in 2.5 TFLOPS per one time step. Usually, from 10^4 to 10^6 time steps are required for a simulation, that is from 25 PFLOPS to 2500 PFLOPS total.

In the present paper, the following physical statement of the problem is used. The 3D computational domain has the shape of a cube with the following dimensions:

$$0 \leq x \leq L_X, 0 \leq y \leq L_Y, 0 \leq z \leq L_Z. \tag{1}$$

Within this domain there is model plasma. The model plasma particles (superparticles) are uniformly distributed within the domain. The density of plasma is set by the user as well as the electron temperature. The temperature of ions is considered to be zero. Electrons of the beam are also uniformly distributed along the domain. Thus, the beam is considered to be already present in the plasma, and the effects that occur while the beam is entering the plasma, are beyond the scope of this study.

A 3D kinetic study of the relaxation processes caused by the propagation of an electron beam in high-temperature plasma was carried out (Lotov et al., Phys.Plasmas, 2015) using the Vlasov-Maxwell equation system.

This problem has two different spatial scales: the Debye length of plasma and the beam-plasma interaction wavelength, that is, some 10 or 100 times larger, thus one needs high-performance computing to observe the two lengths at once. The numerical model is built based on the Particle-in-Cell (PIC) method.

Figure 3 shows the heat flow of plasma electron after the beam relaxation (that is, after beam electrons have given their energy to plasma and have mixed with plasma electrons). The values in Fig. 3 are normed to the initial heat flow value. It is seen in the picture that the resulting heat flow is one or two orders lower than the magnitude in certain places in the computational domain as compared to the initial value. It principally corresponds to the physics of the process.

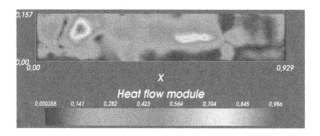

Fig. 3. The heat flow in plasma after beam relaxation in the simulation. The plasma parameters are typical of a magnetic mirror trap.

In the case under consideration the co-design begins at the stage of the physical consideration of the problem. The absence of dramatic density modulations makes it possible not to use the dynamic load balancing. The next stage is the numerical method design. Here, the FDTD method was chosen to provide the memory locality. At the stage of selecting a supercomputer architecture, the PIC method details are taken into account. Particle data and field data are stored in the same place in RAM. At the stage of selecting the software design tools the co-design is the following. For the PIC method, the use of the CUDA technology is highly efficient. Other parallel technologies for hybrid super-computers such as OpenCL, OpenMP, OpenACC could also be used, but it is CUDA that provides the possibility to employ the highest number of parallel processes and to gain the highest performance. The last stage of the co-design is the adaptation of the algorithm to the GPU architecture.

In order to attain the best scalability, the algorithm was parallelized by means of the mixed Lagrange-Eulerian domain decomposition; the details can be found in (Snytnikov A.V., Procedia Comp.Sci. 2009). The parallelization efficiency exceeds 90% for 500 nodes [5], the simulations were conducted with "Lomonosov" supercomputer in Moscow State University, the mesh size in plasma simulations $100 \times 4 \times 4$, 10 thousand model particles in each cell.

Considering the energy efficiency, one must notice that GPUs are not just faster, but they also consume less energy per Flops. In such a way, the Particle-In-Cell plasma simulation with GPUs is at least one order of magnitude better in terms of the energy efficiency. The algorithm was also implemented for GPU clusters. The particle push (the most time-consuming part of the algorithm) is computed 160 times faster with Nvidia

Kepler K40 and 4000 times faster with Nvidia P100 (Pascal) as compared to one core of the Intel Xeon E5450 processor. The simulations with Kepler architecture were conducted in the Siberian Supercomputer Center, the mesh size is 100^3, 100 model particles in each cell, and test runs with Pascal were conducted with Nvidia cluster.

The integrated approach applied to the plasma simulation results in more portable, better scalable and energy efficient codes because of bearing in mind all the three issues all the time when solving the problem.

5   The Results of Using the Integrated Approach

The development of power-efficient algorithms is one of important problems on the way to exascale computations. The CPU power modeling [20, 21] and the code level power efficiency optimizations [22] are well-studied issues. The results in [23] show that the computation performance is unaffected by a decrease in the CPU frequency, i.e. the execution time is independent of a change in the CPU frequency, but the power efficiency has been significantly improved with each frequency step as the CPU frequency changes from 2.67 to 1.60 GHz for the matrix multiplication algorithm on an ordinary Intel Core i7 CPU. However, for GPU computing, the paradigm of power modeling research and code optimization must change to incorporate such parameters as CPU efficiency, GPU efficiency and Bus efficiency between GPU and CPU.

Taking into account the above features for the geophysics code, we have attained 9 GFLOPS/W and 12 GFLOPS/W energy efficiency for the displacement problem and the stress problem tests, respectively, on Nvidia Tesla K40M GPU. We have also achieved 4.3 GFLOPS/W and 4.5 GFLOPS/W energy efficiency for the same problems on Nvidia Tesla 2090M GPU without changing the code. Figure 4 shows arithmetic and logic unit (ALU) utilization as well as memory utilization for the displacement problem solver of the geophysics code (on the left) and for the stress problem solver of the geophysics code (on the right).

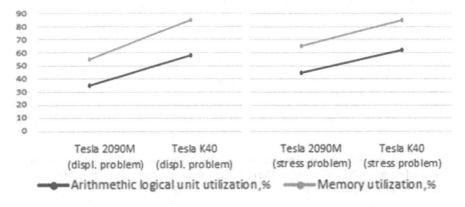

Fig. 4. The ALU and memory utilization for the geophysics code.

For carrying out the simulation of execution of a parallel algorithm, let us present the computational process as a set of threads that are executed in parallel at an individual node and which interact with each other through the exchange of values (messages). The main characteristic of threads is the execution time and the time of data exchange with another computing node. For the simulation of execution of computational processes on supercomputer a model of the interaction of certain process threads is composed. In the model, the calculation and communication blocks are separated for each thread.

Scalability is an important parameter with regard to the HPC-oriented algorithm efficiency. Scalability criteria are algorithm-dependent, for grid-based numerical simulations it keeps the execution times being constant with a simultaneous increase both in the number of node and simulation area size.

The thread interaction model is used for the multiagent simulation of the computational process. Threads are grouped according to their common behavior: the computations and the data exchange. For each thread group there is a class of software agents simulating computation blocks execution and message passing, each with a corresponding number of instances. It is worth noting that the AGNES simulation system allows any functional agents to exchange messages through a "yellow pages" service [8], which supports any thread interaction topologies.

The scalability of the resulting algorithms has been tested using the AGNES simulation system. We have considered utilizing GPU-equipped supercomputers for solving the elastodynamic problem, GPU and MIC architectures for solving the astrophysical problem, and MPP and GPU architectures for the plasma physics problem.

The results of the research into the numerical simulation scalability of the elastic wave propagation problem in complex media are presented in Fig. 5. Two algorithms that solve this problem in two ways are studied. From the figures it is clear that the scalability of the two approaches slightly differs, and both algorithms are suitable for execution on a large number of cores.

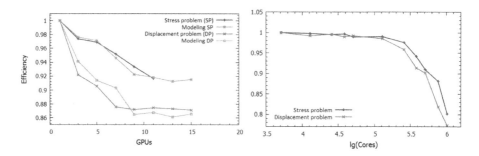

Fig. 5. The scalability of numerical modeling of the elastic wave propagation problem in complex media based on the AGNES simulation. The real calculations for verification were performed with the Siberian Supercomputer Center.

The results of the research into the numerical simulation scalability of 3D gas objects in a self-consistent gravitational field are shown in Fig. 6. The solution of this problem is studied on various types of computing nodes. As can be seen from the graphs, the algorithm shows a fairly good scalability. A better scalability is attained in calculations on nodes with accelerators from NVIDIA Kepler K40 and Intel Xeon Phi.

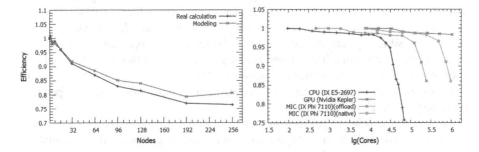

Fig. 6. The scalability of numerical modeling of 3D gas objects in a self-consistent gravitational field based on the AGNES simulation. The real calculations for verification were performed with "Polytechnic RSC PetaStream" supercomputer in St. Peterrsburg State Polytechnical University.

The results of the research into scalability of the numerical simulation of the interaction of an electron beam with plasma are shown in Fig. 7. As a step of forming the full matrix current density charge and the necessary exchange of values "all-with-all", it is noticeable that these exchanges significantly reduce the efficiency of the algorithm. The solution to this problem is investigated on different types of computing nodes. As can be seen a fairly good scalability is attained in calculations on nodes with accelerators from NVIDIA Kepler K40.

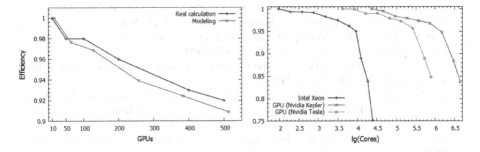

Fig. 7. The scalability of numerical modeling of the interaction of an electron beam with plasma based on the AGNES simulation. The real calculations for verification were performed with "Lomonosov" supercomputer in Moscow State University.

6   Conclusion

In this paper we propose the integrated approach to the development of algorithms and software for solving physical problems demanding a large amount of calculations. For the purpose of the co-design, we have made a comparison of the developed parallel implementations of solutions with the elastodynamic problem written in different terms for the hybrid clusters, equipped with graphics cards. The Godunov method with a modification of the Roe averaging, and a piecewise-parabolic method on a local stencil, has been chosen from several different approaches to solving the astrophysical problem. A similar approach has been taken for the plasma physics problem. The scalability of the resulting algorithms has been tested using the AGNES simulation system. In our case, it is possible to determine carrying out the simulation the optimal number of cores for a specific architecture. This allows investigating the algorithm scalability without resorting to direct time-consuming computations. The energy efficiency of the algorithm for the elastodynamic problem has been investigated on supercomputers equipped with Tesla 2090M and K40M GPUs.

As a result, a suite of parallel programs for solving physics problems has been developed based of the described approach. It is capable of carrying out 3D simulations in acceptable time, provided the resources are sufficient.

Acknowledgments. This work was supported by the Russian Foundation for Basic Research, grants 15-01-00508, 16-29-15120, 16-07-00434, 16-01-00455 and the Grants of the President of the Russian Federation for the support of young scientists MK – 1445.2017.9, MK – 152.2017.5. The plasma code development was supported by the Russian Science Foundation under grant 16-11-10028.

References

1. Reed, D.A., Dongarra, J.: Exascale computing and big data. Commun. ACM **58**(7), 56–68 (2015)
2. Keyes, D.E.: Exaflop/s: the why and the how. C.R. Mechanique **339**, 70–77 (2011)
3. Asanovic, K., Bodik, R., Demmel, J., Keaveny, T., Keutzer, K., Kubiatowicz, J., Morgan, N., Patterson, D., Sen, K., Wawrzynek, J., Wessel, D., Yelick, K.: A view of the parallel computing landscape. Commun. ACM **52**, 56–67 (2009)
4. Sterling, T.: Achieving scalability in the presence of asynchrony for exascale computing. Adv. Parall. Comput. **24**, 104–117 (2013)
5. Glinskiy, B.M., Kulikov, I.M., Snytnikov, A.V., Chernykh, I.G., Weins, D.: A multilevel approach to algorithm and software design for exaflops supercomputers (in Russian). Vychisl. Metody Programm. **16**, 543–556 (2015)
6. Wooldridge, M.: Introduction to MultiAgent Systems. John Wiley & Sons, Ltd., England (2002)
7. Bellifemine, F.L., Caire, G., Greenwood, D.: Developing Multi-Agent Systems with JADE. Wiley, Chichester (2007)
8. Podkorytov, D., Rodionov, A., Choo, H.: Agent-based simulation system AGNES for networks modeling: review and researching. In: Proceedings of the 6th International Conference on Ubiquitous Information Management and Communication (ACM ICUIMC 2012), p. 115. ACM (2012). ISBN: 978-1-4503-1172-4
9. Glinsky, B.M., Marchenko, M.A., Mikhailenko, B.G., Rodionov, A.S., Chernykh, I.G., Karavaev, D.A., Podkorytov, D.I., Vins, D.V.: Simulation modeling of parallel algorithms for Exaflop supercomputers (in Russian). Inf. Technol. Comput. Syst. **4**, 3–14 (2013)
10. Bihn, M., Weiland, T.: A stable discretization scheme for the simulation of elastic waves. In: Proceedings of the 15th IMACS World Congress on Scientific Computation, Modelling and Applied Mathematics (IMACS 1997), Berlin, vol. 2, pp. 75–80 (1997)

11. Sapetina, A.F.: Supercomputer-aided comparison of the efficiency of using different mathematical statements of the 3D geophysical problem. Bull. NCC Ser. Numer. Anal. **18**, 1–9 (2016)
12. Glinskii, B.M., Martynov, V.N., Sapetina, A.F.: 3D modeling of seismic wave fields in a medium specific to volcanic structures. Yakutian Math. J. **22**(3), 84–98 (2015)
13. Vshivkov, V.A., Lazareva, G.G., Snytnikov, A.V., Kulikov, I.M., Tutukov, A.V.: ApJS **194**, 47 (2011)
14. Kulikov, I.M.: ApJS **214**, 12 (2014)
15. Mitchell, N., Vorobyov, E., Hensler, G.: MNRAS **428**, 2674–2687 (2013)
16. Vorobyov, E., Recchi, S., Hensler, G.: A&A **579**, A9 (2015)
17. González, M., Audit, E., Huynh, P.: A&A **464**, 429–435 (2007)
18. Popov, M., Ustyugov, S.: Comput. Math. Math. Phys. **48**, 477–499 (2008)
19. Kulikov, I., Vorobyov, E.: J. Comput. Phys. **317**, 318–346 (2016)
20. Lowenthal, D., Supinski, B., Schulz, M.: Adagio: making DVS practical for complex HPC Barry Rountree. In: The 23rd International Conference on Supercomputing, ICS, New York (2009)
21. Ravi, S., Raghunathan, A., Chakradhar, S.T.: Efficient RTL power estimation for large designs. In: Proceedings of the 16th International Conference on VLSI Design, New Delhi, India, pp. 431–439, January 2003
22. Lively, C., et al.: E-AMOM: an energy-aware modeling and optimization methodology for scientific applications on multicore systems. Comput. Sci. Res. Dev. **29**(3), 197–210 (2014)
23. Ren, D.Q.: Algorithm level power efficiency optimization for CPU–GPU processing element in data intensive SIMD/SPMD computing. J. Parall. Distrib. Comput. **71**, 245–253 (2011)

Further Development of the Parallel Program Complex of SL-AV Atmosphere Model

Mikhail Tolstykh[1,2,3(✉)], Rostislav Fadeev[1,3], Gordey Goyman[1,3], and Vladimir Shashkin[1,2,3]

[1] Institute of Numerical Mathematics, Russian Academy of Sciences, Moscow, Russia
mtolstykh@mail.ru, rost.fadeev@gmail.com,
gordeygoyman@gmail.com, vvshashkin@gmail.com
[2] Hydrometcentre of Russia, Moscow, Russia
[3] Moscow Institute of Physics and Technology, Dolgorpudny, Russia

Abstract. The SL-AV global semi-Lagrangian atmosphere model is applied to the operational medium-range weather forecast at Hydrometeorological center of Russia. The works on increasing the code scalability and using future computer architectures are described. The scalable parallel multigrid algorithm for solving the linear algebraic equations systems is implemented. It is expected that the multigrid algorithm will be used instead of direct algorithm based on fast Fourier transforms requiring global communications. The results for convergence and strong scalability of the multigrid method are given.

The parallel scalability of the low-resolution versions of the SL-AV model for both seasonal and climate simulation has been evaluated at computer systems based on Intel Xeon Phi 2 (Knights Landing) processors. The results show a practical possibility to use these processors for the global atmosphere modelling with the efficiency comparable to the classical cluster systems.

Keywords: Global atmosphere model · Numerical weather prediction · Climate change modeling · Scalable algorithms for solving elliptic equations · Massively-parallel implementation of the atmosphere model

1 Introduction

Numerical weather prediction (NWP) atmosphere models require huge computer resources. The modern global NWP model has the resolution of about 10 km and about 100 vertical levels, thus the problem dimension is approaching 10^9. Operational application of an NWP model requires the 24-hour forecast be computed in 5–20 min. This means that atmosphere model should use efficiently up to 10^5 processor cores.

SL-AV is the global atmosphere model developed at the Institute of Numerical Mathematics, Russian Academy of Sciences (INM RAS) in cooperation with the Hydrometeorological centre of Russia (HMCR) [1]. SL-AV is the model acronym (semi-Lagrangian, based on Absolute-Vorticity equation). The SL-AV model is applied to the operational medium-range weather forecast at Hydrometeorological center of Russia. It is also used as a component of the long-range probabilistic forecast system. The

V. Voevodin and S. Sobolev (Eds.): RuSCDays 2017, CCIS 793, pp. 290–298, 2017.
https://doi.org/10.1007/978-3-319-71255-0_23

dynamical core of this model applies the semi-implicit semi-Lagrangian approach allowing time steps several times larger than in Eulerian methods [2]. The most part of subgrid-scale parameterizations is adopted from ALADIN/LACE model [3, 4].

SL-AV model uses a combination of MPI and OpenMP technologies. Description of the dynamical core and parallel implementation is given in [2]. Briefly, each MPI process performs computations in the band of grid latitudes during the first phase of the time-step. OpenMP threads are used to parallelize loops along longitude dividing the latitude belt into a number of parts. The second phase of SL-AV time-step consists in solving linear systems of equations using direct solvers in the space of Fourier coefficients obtained after Fast Fourier Transforms in longitude. To apply these direct solvers, the set of Fourier coefficients from all grid latitudes are gathered in the memory of specific MPI-process using data transposition. Each MPI-process performs computations for set of longitude Fourier coefficients from pole to pole. OpenMP parallelization of loops in vertical is applied.

Currently, the SL-AV code runs at 3024 cores with 70% efficiency, at 4536 cores with 63% efficiency, and at 9072 cores with 45% efficiency (while comparing with 512-cores run). This is achieved for the grid of 3024 by 1513 points in longitude and latitude respectively. This grid corresponds to 13 km resolution at the equator and has 51 levels in vertical. The data transpositions before and after the solution of elliptic problems require global communications between the processors. This will become a problem on future massively parallel computers. Therefore, the work has started to implement scalable iterative grid-point solvers. It is known that iterative solvers for elliptic problems can scale up to tens of thousands processors [5]. The second part of the SL-AV dynamical core is the semi-Lagrangian advection that is also known to scale up to 10^4 processors [6]. The replacement of the direct solver for elliptic equations on the sphere based on Fast Fourier Transforms with the multigrid solver is presented in Sect. 2 of this paper.

Recently, the application of relatively cheap massively parallel accelerators such as GPU or Intel Xeon Phi in different areas of mathematical modelling gained increased popularity in the world, especially in molecular dynamics, chemistry, electrodynamics, astrophysics etc. This was due to growing demand for computer power, from one side, and known limitations of growth for traditional cluster systems, from the other side. However, the applications of such systems in atmosphere modelling so far have been limited. The part of the problem is that the most part of the atmosphere models is written in different dialects of Fortran language and have very complex code (typically, hundreds of thousand lines) so they are not suited for GPUs. Contrary to computer systems with GPU that do not reasonably support Fortran codes, Intel Xeon Phi systems allow using Fortran. The previous generation of Intel Xeon Phi (codename Knights Corner) was designed as a coprocessor so had certain limitations, for example, on memory exchange between host and coprocessor. The recent introduction of cluster systems based on standalone many-core Intel Xeon Phi 2 processors opened a possibility to implement the existing parallel atmosphere model Fortran code directly, without any change. So the second problem considered in this paper is a study on possibility to run the existing SL-AV code at the Intel Xeon Phi processor systems. The first results available today are presented in Sect. 3.

2 Implementation of the Multigrid Solver in SL-AV Model

2.1 Problem Statement

The SLAV model uses semi-implicit semi-Lagrangian time integration scheme [7] applied to dynamics equations formulated in terms of (vertical component of the) vorticity- (horizontal) divergence [2]. After discretization in space, this approach leads to the set of 2D elliptic equations for each vertical level to be solved at each time step. First, Helmholtz equations are solved to obtain divergence field at the new time level:

$$\left(K^2 - \nabla^2\right)S = H \tag{1}$$

K is the constant depending on vertical level, ∇^2 is the horizontal Laplace operator on the sphere, S is a vector variable related to the divergence at the new time level by the linear equation $D = VS$, V being known matrix. Then the relative vorticity ω at the new time level is calculated and Poisson equations are solved to obtain streamfunction and velocity potential ψ, χ.

$$\nabla^2 \chi = D, \tag{2}$$

$$\nabla^2 \psi = \omega. \tag{3}$$

The horizontal wind velocities are then restored using relations

$$u = -\frac{1}{a}\frac{\partial \psi}{\partial \varphi} + \frac{1}{a \cos \varphi}\frac{\partial \chi}{\partial \lambda}, \tag{4}$$

$$v = \frac{1}{a \cos \varphi}\frac{\partial \psi}{\partial \lambda} + \frac{1}{a}\frac{\partial \chi}{\partial \varphi}. \tag{5}$$

Currently, the abovementioned equations are solved in the space of longitudinal Fourier components in the SL-AV model. This means that all the derivatives in longitude are replaced by multiplication by the corresponding coefficients. Compact fourth-order differences are used to approximate the derivatives in latitude. Thus, a 2D equation on the sphere is replaced by the set of 1D linear systems of equations solved by block-tridiagonal version of Thomas algorithm. The parallel implementation of this approach requires data transpositions (hence global communications between MPI processes) before and after these solvers and is a principal obstacle in implementing 2D MPI domain decomposition in this part of the model. The 1D domain decomposition currently used in the SL-AV model code limits the number of the MPI processes by the number of grid points in latitude.

We present here the results of implementing 2D finite-difference approximations for Helmholtz and Poisson equations described above. The arising linear systems of equations are solved with previously implemented multigrid parallel algorithm [8].

2.2   Discretized Equations and the Algorithm

Geometric multigrid with V-cycle [9] is chosen as a base algorithm for the new solver. Intergrid operators are bilinear interpolation and 8-point full weighting. Gauss-Seidel method with red-black ordering is used as a smoother. At the bottom level of the V-cycle, the matrix is inverted with BICGstab solver [10]. We use conditional semi-coarsening approach [11] to account for the anisotropy of the regular latitude-longitude grid near the poles.

Compact finite differences used in solvers in the current version of SL-AV code are not well suited for parallel solver because they imply global dependence of the derivative on values of the function. Thus they are replaced with the local second-order approximation in the solvers. When applying conditional semicoarsening procedure [11], the resolution in latitude becomes irregular. Let us define grid latitudes $\varphi_j, j \in [0, N_\varphi]$ with arbitrary spacing and a constant mesh size in longitude $h_\lambda = \dfrac{2\pi}{N_\lambda}$ where N_λ, N_φ are grid dimensions in longitude and latitude respectively. Let us denote

$$V_j = a^2 \left(\sin \frac{\varphi_{j+1} + \varphi_j}{2} - \sin \frac{\varphi_j + \varphi_{j-1}}{2} \right) h_\lambda, \, dS_{\varphi,j} = \frac{\varphi_{j+1} - \varphi_{j-1}}{2}, \, dS \Big|_{\lambda, j \pm \frac{1}{2}} = h_\lambda \cos \frac{\varphi_{j\pm1} + \varphi_j}{2}. \tag{6}$$

The second-order finite-volume approximation from [12] is used to discretize Laplace operator. Except for the pole points, it is written as

$$(\nabla^2 \psi)_{i,j} = \frac{1}{V_j} \left(dS_{\varphi,j} \frac{\psi_{i+1,j} - 2\psi_{i,j} + \psi_{i-1,j}}{h_\lambda \cos \varphi_j} + \left(\frac{\psi_{i,j+1} - \psi_{i,j}}{h_\varphi} dS_{\lambda,j+1/2} - \frac{\psi_{i,j} - \psi_{i,j-1}}{h_\varphi} dS_{\lambda,j-1/2} \right) \right). \tag{7}$$

For the pole grid points, the following formulae are used

$$(\nabla^2 \psi)_S = \frac{1}{a^2 \left(1 + \sin \dfrac{\varphi_1 + \varphi_0}{2} \right) h_\lambda} \sum_i \frac{\psi_{i,1} - \psi_S}{h_\varphi} dS_{\lambda, \frac{1}{2}} \tag{8}$$

$$(\nabla^2 \psi)_N = -\frac{1}{a^2 \left(1 - \sin \dfrac{\varphi_{N_\varphi} + \varphi_{N_\varphi-1}}{2} \right) h_\lambda} \sum_i \frac{\psi_N - \psi_{i,N_\varphi-1}}{h_\varphi} dS_{\lambda, N_\varphi-1/2} \tag{9}$$

In order to reconstruct the horizontal velocity field, the standard fourth-order finite difference formulae are used to approximate derivatives in longitude and in latitude, except for latitudinal derivatives near the poles where third-order formulae are used.

2.3   Convergence and Scalability

Convergence of the implemented multigrid solver for equations described in Sect. 2.1 is studied for different grid resolutions between 128×64 and 2048×1024, and different number of iterations for smoothing operator. There is practically no convergence dependence on the problem size when using two iterations of pre- and post- smoothing.

The algorithm scalability is studied at MVS10P cluster system installed at Joint Supercomputer Center (Moscow, Russia). This system is based on two-processor nodes with Intel Xeon E5-2690 8-core processors. The strong scalability for problems with grid sizes $512 \times 256 \times 28$ and $2048 \times 1024 \times 51$ is studied. These grids approximately correspond for different version of the SL-AV model. Parallel speedup with respect to 16 processor cores is presented in Fig. 1. One can see that the problem with the size of $512 \times 256 \times 28$ scales up to 256 cores with the efficiency more than 50%. The problem with the size of $2048 \times 1024 \times 51$ scales efficiently up to at least 1024 processor cores.

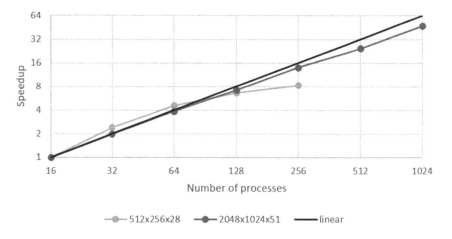

Fig. 1. Strong scalability of the multigrid solver.

2.4   Implementation in the SL-AV Model

The discretization and the algorithm described above are implemented in the SL-AV model. To test the accuracy of the new algorithm, a series of 31 numerical weather 72-hour forecasts starting with initial data of each day of January 2014 at 12 h UTC is calculated. The model version with the resolution 0.9° in longitude, 0.72° in latitude and 28 vertical levels is used. The grid dimension is $400 \times 251 \times 28$. The averaged over series root mean squared errors for forecast of geopotential heights at 850, 500 and 250 hPa are depicted in Fig. 2 for forecast lead times of 24, 48 and 72 h. The errors are averaged over Northern extratropics (20–90 N). One can see that the new solver presented above slightly reduces forecast errors of the SL-AV model as compared with the 'standard' direct solver. Similar results are obtained for other regions. Also, the tests have revealed that it is sufficient to reduce the norm of the residual by 10^4 times that corresponds to 1 to 3 V-cycle iterations.

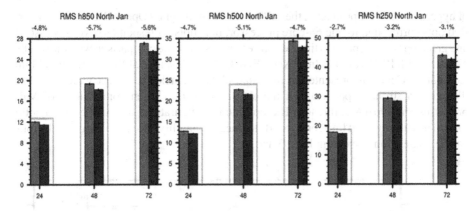

Fig. 2. Root-mean squared geopotential errors averaged over 31 forecasts and Nortern extratropics for the standard SL-AV version (red) and the SL-AV model with multigrid solvers (blue). Green boxes mean that the results are statistically significant. (Color figure online)

3   Testing SL-AV Model Code at Intel Xeon Phi 2 Systems

3.1   Model Configurations and System Description

Based on growing requirements for computer resources related to the ongoing development of the climate version of the SL-AV model, we have studied possibility to run the existing SL-AV code at the Intel Xeon Phi many-core processor systems. It is essential that no changes were made to the parallel program complex working at 'traditional' x86-based clusters. The two model versions having different resolution are tested. The first version has the horizontal resolution 0.9° in longitude, 0.72° in latitude and 28 vertical levels. The horizontal resolution of the second version is 0.56° in longitude, 0.45° in latitude, 50 vertical levels. The problem dimensions are thus $400 \times 251 \times 28$ and $640 \times 401 \times 50$ respectively.

The cluster system based on Intel Xeon Phi 2 processors is used. Each node contains processor 7250 (codename Knights Landing or KNL) with 68 cores allowing up to 272 hyperthreads. There are 16 Gbytes of fast MCDRAM memory and 48 Gbytes of DDR4 memory, Intel Omnipath interconnect allowing up to 100 Gbytes per second transfer rate. The peak node performance is about 3.04 Tflops. At the time of writing this paper, only three-node cluster was available for tests. Currently, more nodes with such processors are being installed at Joint Supercomputer Center RAS in Moscow, so such tests allow preparing for a proper use of these resources.

3.2   Results and Conclusions

The elapsed times for the SL-AV model time step as a function of hyperthreads number are depicted in Fig. 3 for $400 \times 251 \times 28$ version and in Fig. 4 for $640 \times 401 \times 50$ version. The model time step here comprises one time step calling solar and longwave radiation computations and three time steps without radiation computations. The number of hyperthreads is the product of the MPI processes number by OpenMP threads number. Red dots correspond to one 68-core processor, blue dots correspond to two processors, and green ones correspond to the use of three processors. The comparison with the results obtained with RSC Tornado Intel Xeon E2690 cluster installed at Roshydromet's Main Computing Center is presented at the same plots in black lines. Vast variety of dots illustrates a strong dependence of the elapsed time on chosen combination of MPI processes and OpenMP threads. The optimum configurations are marked with solid lines. The black dashed line corresponds to linear scalability. The following conclusions can be made upon inspecting Figs. 3 and 4.

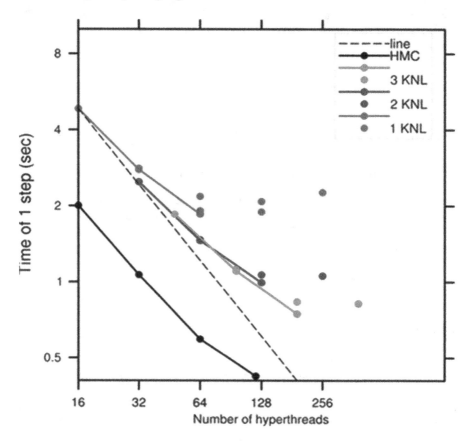

Fig. 3. Scalability of SL-AV model code with $400 \times 251 \times 28$ grid for different combinations of MPI processes and OpenMP threads. See text for details.

The SL-AV model code for tested versions scales at KNL processor systems similar to classical x86-based systems if an optimum combination of MPI processes and OpenMP threads is used. The absence of significant jumps in the scalability curve when increasing the number of KNL processors shows sufficient interconnect exchange rate.

The efficient use KNL requires good code vectorization. One can see from Figs. 3 and 4 that increasing the number of hyperthreads at single KNL processor slows down

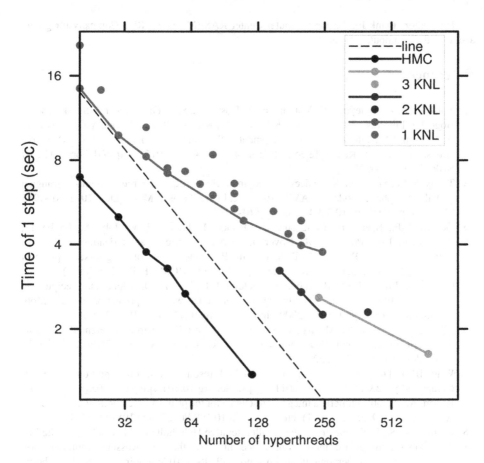

Fig. 4. Scalability of SL-AV model code with $640 \times 401 \times 50$ grid for different combinations of MPI processes and OpenMP threads. See text for details.

the execution of 28-level version while 50-level version continues to accelerate. The last result can be explained by the specifics of SL-AV parallel code. Indeed, 1D MPI decomposition in latitude is supplemented with OpenMP parallelization along longitude. The innermost vectorizable loops in many computationally demanding model blocks are in longitude or in the vertical and increasing both dimensions improves vectorization.

The results demonstrate the possibility to use Intel Xeon Phi 2 processors and cluster systems based on these processors for SL-AV model computations. Important points needed to achieve good performance are the choice of proper combination of numbers of MPI processes and OpenMP threads and good code vectorization.

Acknowledgements. This study was carried out at the Institute of Numerical Mathematics, Russian Academy of Sciences. The study presented in Sect. 2 was supported with the Russian Science Foundation grant No. 14-27-00126, the work described in Sect. 3 was supported with the Russian Academy of Sciences Program for Basic Researches I.33P.

The authors thank Joint Supercomputer Center RAS (Moscow), RSC Company for giving access to their computer resources.

References

1. Tolstykh, M.A., Geleyn, J.-F., Volodin, E.M., Kostrykin, S.V., Fadeev, R.Y., Shashkin, V.V., Bogoslovskii, N.N., Vilfand, R.M., Kiktev, D.B., Krasjuk, T.V., Mizyak, V.G., Shlyaeva, A.V., Ezau, I.N., Yurova, A.Y.: Development of the multiscale version of the SL-AV global atmosphere model. Russ. Meteor. Hydrol. **40**, 374–382 (2015). https://doi.org/10.3103/S1068373915060035
2. Tolstykh, M., Shashkin, V., Fadeev, R., Goyman, G.: Vorticity-divergence Semi-Lagrangian global atmospheric model SL-AV20: dynamical core. Geosci. Model Dev. **10**, 1961–1983 (2017). https://doi.org/10.5194/gmd-10-1961-2017
3. Geleyn, J.-F., Bazile, E., Bougeault, P., Deque, M., Ivanovici, V., Joly, A., Labbe, L., Piedelievre, J.-P., Piriou, J.-M., Royer, J.-F.: Atmospheric parameterization schemes in Meteo-France's ARPEGE N.W.P. model. In: Parameterization of subgrid-scale physical processes, ECMWF Seminar Proceedings, pp. 385–402. RCMWF, Reading (1994)
4. Gerard, L., Piriou, J.-M., Brožková, R., Geleyn, J.-F., Banciu, D.: Cloud and precipitation parameterization in a meso-gamma-scale operational weather prediction model. Mon. Weather Rev. **137**, 3960–3977 (2009). https://doi.org/10.1175/2009MWR2750
5. Müller, E., Scheichl, R.: Massively parallel solvers for elliptic partial differential equations in numerical weather and climate prediction. Q. J. R. Meteorol. Soc. **140**, 2608–2624 (2014). https://doi.org/10.1002/qj.2327
6. White III, J., Dongarra, J.: High-performance high-resolution tracer transport on a sphere. J. Comput. Phys. **230**, 6778–6799 (2011). https://doi.org/10.1016/j.jcp.2011.05.008
7. Robert, A.: A stable numerical integration scheme for the primitive meteorological equations. Atmos. Ocean **19**, 35–46 (1981). https://doi.org/10.1080/07055900.1981.9649098
8. Goyman, G.S.: Development of the parallel iterative Helmholtz problem solver for the SL-AV global atmospheric model. In: Proceedings of the 2nd Russian Conference on Supercomputing - Supercomputing Days, pp. 959–966 (2016). (Goiman, G.S.: Razrabotka parallelnogo iterativnogo bloka reshenia tipa Gelmgoltca dlia globalnoi modeli atmosferi PLAV - Trudi mezhdunarodnoi konferencii 'Superkomp'iuternie dni Rossii'. Izdatelstvo MGU im. M.V. Lomonosova, pp. 959–966 (2016))
9. Trottenberg, U., Oosterlee, C.W., Schuller, A.: Multigrid, 631 p. Academic press, London (2000)
10. Van der Vorst, H.A.: Bi-CGSTAB: a fast and smoothly converging variant of Bi-CG for the solution of nonsymmetric linear systems. SIAM J. Sci. Stat. Comput. **13**, 631–644 (1992). https://doi.org/10.1137/0913035
11. Larsson, J., Lien, F.S., Yee, E.: Conditional semicoarsening multigrid algorithm for the Poisson equation on anisotropic grids. J. Comput. Phys. **208**, 368–383 (2005). https://doi.org/10.1016/j.jcp.2005.02.020
12. Barros, S.R.M.: Multigrid methods for two-and three-dimensional Poisson-type equations on the sphere. J. Comput. Phys. **92**, 313–348 (1991). https://doi.org/10.1016/0021-9991(91)90213-5

The Supercomputer Simulation of Nanocomposite Components and Transport Processes in the Li-ion Power Sources of New Types

V.M. Volokhov, D.A. Varlamov[✉], T.S. Zyubina, A.S. Zyubin, A.V. Volokhov, and E.S. Amosova

Institute of Problems of Chemical Physics of the Russian Academy of Sciences, Academician Semenov Avenue 1, Chernogolovka, Moscow Region 142432, Russian Federation
{vvm,dima,zyubin,vav,aes}@icp.ac.ru

Abstract. As a result of a large amount of computational experiments on a number of supercomputer resources quantum-chemical and molecular dynamic modeling of various nanocomposite components of Li-ion power sources was performed. Various aspects of transport, structural and energy processes inside LPS during numerous cycles of charging and discharge were simulated. By tools of molecular dynamics estimated influence of various external conditions on structure of nanocomposites and characteristics of above-stated processes.

Keywords: Computer simulation · Silicon-carbon nanocomposites · Solid electrolytes · Li-ion power sources · VASP applied package · Quantum chemistry · Molecular dynamic

1 Introduction

In this article the main results of works on the project "Computer simulation of absorption and transport properties of solid electrolytes and nanostructured electrodes based on carbon and silicon in Li-ion power sources" are summed up. The aim of this project is the supercomputer simulation of quantum-chemistry and molecular dynamics of new nanocomposite materials (based on silicon and carbon) and solid electrolytes with high ionic conductivity, as well as non-reactive electrode materials during operation of a current source. Also, transports, structural and energetic processes occurring in the modeled nanostructures and at the "interface" between them have been simulated.

Li–ion power sources (LPS) are currently the most promising and common types of power sources and batteries. LPS are based on the transport of Li-ions through a liquid or solid electrolyte from cathode to anode (and back when charging). The design of new types of LPS is needed to improve their efficiency parameters, such as energy capacity, number of charge-discharge cycles, resistance to external conditions (temperatures), safety of their production and utilization from an environmental point of view, and cost (prime cost of materials in main components).

Here is a brief description of the operating principle of Li-ion power sources (Fig. 1).

© Springer International Publishing AG 2017
V. Voevodin and S. Sobolev (Eds.): RuSCDays 2017, CCIS 793, pp. 299–312, 2017.
https://doi.org/10.1007/978-3-319-71255-0_24

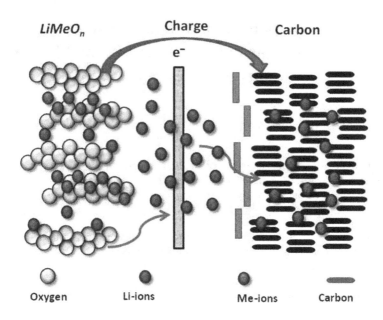

Fig. 1. Schematic diagram of a Li-ion power source

The following reactions occur in a Li-ion power source during charge:

on the positive plates: on the negative plates:

$$LiMeO_2 \rightarrow Li_{1-x}MeO_2 + xLi^+ + xe^-$$ $$C + xLi^+ + xe^- \rightarrow CLi_x$$

The reverse reactions occur when discharging. Therefore, the modeling of lithiation processes (saturation of the anode by lithium == discharge process) and delithiation (lithium ions return to electrolyte and cathode == charging process) is basic for a comprehension of processes of functioning of LIA in general, estimation of the limiting factors and prediction of the most perspective nanocomposite materials.

Simulated materials should be the basis for the design and creation of new types of electrochemical and ecologically safe Li-ion power sources (LPS). These power sources will be able to operate at low and medium temperatures, provide significantly higher energy densities, and improve operational and cost characteristics.

The synthesis of new nanocomposite materials, the study of their properties and predictable applications are only possible as a result of a detailed computer modeling of crystalline composite structures, elementary processes and different mechanisms of chemical reactions and transport processes at molecular level.

Experimental studies of factors having a major influence on the solution of the issues listed above are complex, expensive, not always possible, and in most cases, do not give clear answers to the following questions: mechanisms of ongoing physical and chemical processes, reasons for their differences depending on the composition of the system and external conditions, possible directions of reactions, etc.

The experimental (analogue) simulation of the influence of various factors on the properties of the constituent components of Li-ion power sources and the processes occurring in them poses labor-intensive and costly tasks.

Since experiments give only initial and final information about processes, it is quite difficult to build a genuine analytical model. Such tasks can be solved partially in laboratory conditions, where analytic experiments give incomplete or indirect information about mechanisms and structures of experiment components. However, modern numerical methods of quantum-chemistry and molecular-dynamics simulation can provide substantial assistance in determining the characteristics of processes and assessing the impact of individual factors with a high degree of accuracy. These methods allow obtaining new theoretical data on the structure and properties of both nanostructured cathode-anode systems and ion-conducting solid electrolytes, making it possible to subsequently develop new highly effective materials for electrochemical devices.

A detailed simulation of elementary processes as well as mechanisms of lithiation/delithiation and ion-transport processes in Li-ion power sources at the micro level leads to a better control over chemical reactions occurring in them, allowing to design the most appropriate anode materials in terms of efficiency of electricity generation, lithiation processes, stability of materials during numerous charge-discharge cycles, cost of LPS constructive materials and environmental recycling processes.

Also, the created models can be reviewed for adequacy by comparing them (and the properties of materials modeled on their basis) against observable analytical, experimental and theoretical data published in specialized literature references.

For this task, the authors carried out a detailed quantum-chemical and molecular dynamic simulation of various nanosystems based on carbon and silicon, as well as solid electrolytes with high ionic conductivity, both in cluster approximation and for periodic boundary conditions with projector-augmented wave (PAW), using VASP, CPMD and Gaussian application packages on a number of high-performance computing resources [1–3].

The objects of the computer simulation are composites based on carbon and silicon, which have the ability to repeatedly absorb Li without damage and are promising materials for Li-ion power sources (nanoclusters, nanotubes, nanowires, nanopapers and active crystal surfaces). Also objects of this computer simulation are solid electrolytes with high ion conductivity based on glasses, salts and polymer composites that do not react with the electrode material during operation of a current source.

Some simulation experiments were conducted using authors' computer system based on up-to-date software packages for quantum-chemistry and molecular dynamics, "hybrid" computing technologies, web services, data storage, visualization of results, etc. Using high-performance resources (supercomputers, problem-oriented clusters and hybrid systems) would greatly improve the details and the quality of the created models of nano-objects and those of the processes accompanying them, and would also allow to solve tasks previously inaccessible due to their computational complexity.

2 Simulation Methods

The models of nanocomposite materials and processes occurring in them were constructed by methods of quantum chemistry computer simulation on the clusters of the Computation Center at the Institute of Problems of Chemical Physics (IPCP) and at the Supercomputer Center of Moscow State University "Lomonosov" [4], using the applied software packages VASP (Vienna Ab initio Simulation Package, https://www.vasp.at) and CPMD (Car-Parinello Molecular Dynamics, http://www.cpmd.org) for the calculation of complex nanostructures and the dynamics of their behavior depending on time and temperature.

The VASP applied package has been used by the authors during a long time for modeling materials and components of complex electrochemical objects. This package is applied to the simulation of various processes both in the volume and on the surface of solids (first of all, catalysis and ionic conductivity) within the non-empirical approaches based on the use of density functional theory with periodic boundary conditions and a plane wave basis set. VASP allows to optimize the structures and to model processes within a molecular dynamics framework.

VASP implements effective schemes of iterative matrix diagonalization and the highly efficient Broyden–Pulay electronic charge density mixing. In addition, the MSP processes convergence procedure (self-consistent field) and optimization are significantly improved, which greatly increases the efficiency of calculations. This package provides a good accuracy of description for structural and energy characteristics of systems containing up to several hundred atoms. First of all, we conducted a full optimization of the geometric and energy parameters of molecules under consideration within the established basis and method of calculation.

In this paper, we applied an approach based on density functional theory (DFT) with periodic boundary conditions to simulate learning systems. We applied the projector-augmented wave (PAW) with the corresponding PAW pseudopotentials and PBE functional (Perdew–Burke–Ernzerhof). The limit of energy (E_c) defining the completeness of the basis set was established at 400 eV. When simulating two-dimensional plates, the vacuum layer between them was not less than 10 Å. To simulate the $Li_{10}GeP_2S_{12}$ electrolyte volume, we used a canned double cell $Li_{20}Ge_2P_4S_{24}$ involving 50 atoms; for the simulation of the surface, four such cells (200 atoms) were used.

To solve the problem of interaction of the surfaces between electrodes and electrolytes, we modeled (with full optimization of geometric parameters) a structure continuously propagating in two directions, solid electrolyte fragments (propagated $Li_{80}Ge_8P_{16}S_{96}$-fragment), a silicon-carbon paper (propagated $Si_{32}C_{38}$-fragment) and the result of their interaction (propagated $[Si_{32}C_{38}]*[Li_{80}Ge_8P_{16}S_{96}]$-fragment).

In the case of polymer electrolytes, we modeled (with full optimization of geometric parameters) the structure of infinite nanowires of Li*Nafion**nDMSO (n = 0,1,8,16), and also spatially propagated fragments of $Li(C_{15}O_5F_{29}S)*n(H_6C_2OS)$ and $[Li(C_{15}O_5F_{29}S)*n(H_6C_2OS)]_2$ of 51 to 262 atoms.

For the optimization, we applied the Methfessel–Paxton method of electronic state (with blur parameter (σ) 0.2 and energy approximation of the value $\sigma = 0$). This approach allows for the automatic detection of system's multiplicity. The estimate of energy stability of combined systems was determined according to De/n(Li), computed as the difference between the calculated energy of the system and the total energy of isolated lithium atoms divided by the number of atoms of adsorbed lithium, for example, De/n(Li) = -[E((SiC)$_k$Si$_m$Li$_n$) - E((SiC)$_k$Si$_m$) - nE(Li)]/n.

We used two approaches for the simulation of transport processes in the framework of an ab initio non-empirical molecular dynamics with periodic boundary conditions: CPMD (Car-Parrinello approximation), in which the calculated wave function for the starting configuration is approximated by a set of classically-moving low-mass particles, and a more accurate but slower approximation, namely MD-VASP (MD/PBE/PAW), which uses the same algorithms as normal optimization structures, but with rougher calculation criteria.

Generally, the use of MD-VASP allows a substantially faster simulation than CPMD. MD-VASP requires about 6 to 8 times less computation steps to achieve the same penetration depth.

3 Computational Complexity and Efficiency of Calculations

In earlier times, similar computer simulations were hindered by a catastrophic lack of computing resources, since calculating the behavior of small atomic clusters of the Si$_{7-126}$ type, even in a simplified form, required months, and modeling systems as a whole (containing thousands of atoms) required approximately n·10^6 CPU-hours per year.

Only in recent times, the same simulation became feasible using high-performance supercomputing centers and grid polygons. Currently, the use of computing resources with speeds of the order of teraflops and petaflops allows to make sufficiently detailed simulations of geometrical and energy characteristics of modeled nanostructures. It is also possible to study the effects of various factors and processes occurring in these nanostructures for a variety of conditions determining the efficiency of the created LPS.

Let us summarize the computational complexity and use efficiency of computing resources in the process of quantum-chemical simulation of learnt structures. We used the IPCP cluster (15 teraflops: 176 dual-node HP Proliant, making a total of 1472 cores based on 4- and 6-core Intel Xeon processors 5450 and 5670 3 GHz, 8 and 12 GB of RAM per node; InfiniBand DDR communication network, transport and network management – Gigabit Ethernet; hard drives – no less than 36 GB per node), and the SCC of MSU supercomputing installations "Lomonosov-1,2" having various pools of processors (8 to 128 CPU) with obligatory presence of local drives and no less than 2 GB of RAM per core.

A sufficient effective acceleration of the VASP package for this type of tasks was observed for 40 to 48 CPU. The further growth of the efficiency of task parallelization is limited (or even reduced) by the rate of data exchange due to a significant increase in the amount of data being transferred between nodes. Thus, increasing the number of CPU over 48 (at least for this task variant) is meaningless for the moment. If the number of processors is more than 64, the dependence of the acceleration on the number of processors is practically absent or even falls [1].

The average effective time for calculation of Si_n clusters (n = 2÷350) and C_nSi_m nanofibers increases as the dimension of the silicon-carbon fragment increases, taking up to 4 days (78 h on a pool based on 4-core Intel Xeon 5450 3 GHz processors) and even more (due to complications of the structure). The calculation time of lithiated large mesostructures of silicon and aggregates reinforced with nanotubes or nanowires took tens of days to complete.

The most critical calculation parameter is the amount of RAM per core, with an effect of acceleration of calculations with a decrease in the number of allocated cores by increasing the amount of RAM per core. For MD calculations, we used 14 000 steps per calculation (for example, heating up to 400 K for 2000 steps, holding at 400 K for 10 000 steps, cooling down to 10 K for 2000 steps, and optimizing the structure in standard mode; the time step model was 1 femtosecond). The calculation of complex structures, such as those described in Sect. 4.6, requires up to 80 000 CPU-hours.

In the latest versions of VASP, starting with version 5.4.1 (February 2015), the application package supports CUDA technology for the calculation method of standard and hybrid DFT (Hartree–Fock equation). For most tasks, using DFT on Tesla C2075 accelerators at the IPCP, we achieved (comparing VASP versions with support and without support of GPU acceleration on hybrid workstations with combination – 1 GPU with 2x6 cores CPU) 1.6- to 6-fold accelerations depending on the dimension of the problem and its type. This gives the prospect of a significant acceleration for VASP calculations on "hybrid" computing nodes (following VASP upgrade to versions above 5.4), including existing CPU-GPU pools on the SC "Lomonosov-1.2" and hybrid IPCP stations (in experiments we use combination 1 GPU Nvidia Tesla C2075 plus 2x6 CPU 3.46 GHz Intel® Xeon® X5675, 48 Gb RAM). In addition to upgrading VASP, it is necessary to do a further reconfiguration of VASP settings files, and update CUDA library to version 7.5 (current version: 8.0).

The total number of computing experiments performed at all stages of the work reached more than 2000. If to speak about use of computing resources, then it is estimated as follows: SC "Lomonosov-1" (+"Chebyshev") – about 40-45% of experiments, IPCP cluster – 50%, IPCP and IEM workstations with GPU Nvidia support – 2-3%, MVS-100 – single experiments.

4   Simulation Results

The results of the multi-step simulation have been described in detail in a number of publications by the authors of the present work [5–10]. Here is a brief description of the most representative results of the computer simulation of nanostructures and processes occurring in them.

4.1   Computer Simulation of Various Types of Porous Nanocomposite Materials Based on Carbon and Silicon

Computer models of the following types of Si–C nanocomposites have been constructed by the authors [1, 5, 6]:

1. pure silicon aggregates with different morphologies (clusters of "snowballs", "core/shell", etc., size up to 3 nm), and a number of silicon atoms ranging from 2 to 350;
2. silicon clusters with silicon carbide core (rod-shaped), 1.2 to 2.8 nm in diameter, and nanofibers of Si_nC_m type, n/m = 1÷3;
3. carbon nanotubes (CNT) with dimension (6,6) and 0.8 nm in diameter, surrounded by a layer of silicon clusters of various dimensions;
4. silicon nanowires with a rod on the basis of silicon carbide and silicon shell;
5. infinite carbon nanofibers coated with silicon nanoclusters;
6. silicon-carbon "nanopapers".

A conclusion following from our research is that the use of different types of simulated nanocomposites may be a promising opportunity in the construction of new types of electrodes for LPS. Examples of the simulated nanostructures are shown in Fig. 2.

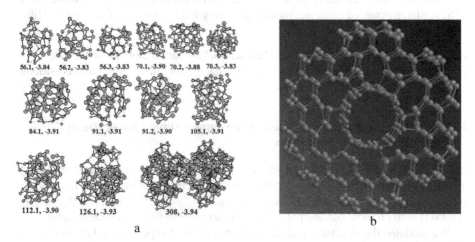

Fig. 2. Examples of nanocomposites based on Si–C: a) mesoclusters Si_n (n = 56÷308); b) carbon nanotub with silicon atoms around Si_{288}/C_{120}. The models have been derived from the authors' computer simulation using the VASP package (PBE/PAW level of calculation).

4.2 Quantum-Chemical Simulation of Transport Processes of Lithium Ions in Nanocomposite Materials Based on Carbon and Silicon

On the basis of the constructed models of nanocomposites (see above), we made [7–9] a quantum-chemical simulation of various processes occurring during charge-discharge cycles of LPS (i.e. processes of lithiation and delithiation on electrodes based on the above-described nanostructures). A majority of characteristics of these processes have been established, including:

1. Li-ion transport processes and processes of lithium consistent implementation in Si–C nanostructures of various types and dimensions;
2. structural and energetic changes of nano-objects in processes of absorption of lithium atoms;

3. possible paths and migration barriers for lithium atoms in the process of nanoparticle saturation;
4. construction of models of sequential removal of lithium atoms from lithiated nanoparticles and determination of structural and energetic changes identified in this process;
5. determination of the limits of resistance to fracture for nanoparticles during delithiation processes.

4.3 Computer Models of Aggregation Processes of Initial and Lithiated Nanoparticles

In LPS operation, there are aggregation processes of small nanostructures into larger ones and vice versa, affecting greatly the characteristics of components and the whole LPS. We have obtained computer models of aggregation processes of original and lithiated nanoparticles [8, 9], including:

1. formation of a mesostructure based on original silicon-carbon nanoparticles;
2. formation of a mesostructure based on lithiated (saturated lithium atoms) silicon-carbon nanocomposite structures.

4.4 Quantum-Chemical and Molecular Dynamics Simulation of Highly Conductive Solid Electrolytes

This work presents the results of the computer quantum-chemistry and molecular dynamics simulation [10] of highly conductive solid electrolytes based on $Li_{10}GeP_2S_{12}$ systems and polymer electrolytes based on Li*Nafion*™ * dimethylsulfoxide (Li*Nafion* * 8DMSO) with an ionic conductivity that is higher than that of liquid electrolytes.

We modeled the structures and the contact surface of superionic solid electrolytes. In this connection, the ionic conductivity mechanism was determined during simulation. We also defined the types of surface structure and the nature of the $Li_{10}GeP_2S_{12}$ electrolyte contacts with anode nanocomposite materials (carbon fibers coated with Si_nC_m silicon nanoclusters and silicon-carbon "nanopaper").

We modeled the contact surfaces of superionic solid electrolytes with different Si-C nanocomposites. It was shown that a layer of liquid or plastic polymer electrolyte, such as dimethylsulfoxide (hereinafter DMSO), can be used to enhance the contact between solid surfaces. The simulated structure and its surface are shown in Fig. 3.

We constructed a computer model of interaction of solid and polymer lithium-based electrolytes with composites based on carbon fibers and silicon nanoclusters.

Lithium transition across the interface "electrode–electrolyte", as well as the determination of the migration channels and the potential barrier were modeled taking as example the interaction of solid and polymer lithium electrolytes with composites based on carbon fibers and silicon nanoclusters.

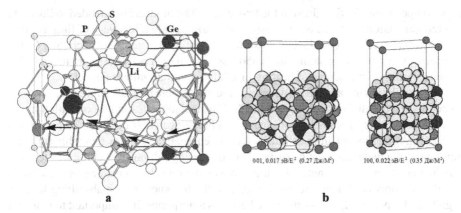

Fig. 3. Simulated structure (a) and (b) two types of modeled surfaces (001) and (100) $Li_{10}GeP_2S_{12}$ solid electrolyte crystal

4.5 Simulation of Lithium Ions Migration in Non-aqueous Polymer Electrolytes

Using methods of quantum chemistry and molecular dynamics, we modeled [2, 3, 10] various aspects of the migration of lithium ions in a complex electrolyte (Li*Nafion* * nDMSO, n = 0÷18) as well as the structure, stability and electronic properties of membranes based on the electrolyte, including the effect of various parameters, such as the degree of swelling of the electrolyte and a number of physico-chemical properties of the plasticizer containing the molar volume, the viscosity and the coordination number.

On the basis of the transport models, we made some conclusions on the possible paths of lithium migration and energy parameters: (1) four-coordinated lithium transition through three-coordinated state to the next position surrounded by four DMSO molecules, and (2) movement of the $Li(DMSO)^{4+}$ tetrasolvate complex.

4.6 Computer Simulations of Repeated Lithiation/Delithiation Cycles Depending on the Degree of Lithium Saturation and Temperature Conditions

Along with the simulation of individual LPS components (electrodes, electrolytes) and processes occurring at the interfaces between them during operation of the battery, a great importance for the creation of new types of power sources is ascribed to the stability of these elements over time in case of multiple "charge-discharge" cycles, depending on the power capacity of the system (amount of lithium) and temperature conditions.

The molecular dynamics simulation is performed to assess the feasibility of the composite mesostructure return to its original state after repeated lithiation/delithiation cycles depending on the degree of lithium saturation and temperature conditions. MD-VASP (MD/PBE/PAW) approximation is used. For MD calculations, we used 14 000 steps per calculation (for example, heating up to 400 K for 2000 steps, holding at 400 K for 10 000 steps, cooling down to 10 K for 2000 steps, and optimizing the structure in standard mode; the time step model was 1 femtosecond). We took as initial model the nanocomposite models obtained in the first stages. The models are intended to illustrate the reorganization of the Li/Si layer structure during gradual recovery of lithium from the surface (i.e. discharge): the effect exerted by heating and subsequent cooling on the structure of silicon-carbon delithiated nanosystems and their possible return to the initial state according to the degree of heating of lithium and the saturation level.

The simulation has allowed to determine the most stable mesostructures for electrode materials, the optimal ratio of Li:Si in the Si–C nanocomposites saturation with lithium, and the best energy parameters of charge-discharge cycles. It has been demonstrated that the introduction of lithium into silicon is energetically more favorable than the formation of a metal layer on its surface, but increasing lithium concentrations leads to a reduction of energy difference, i.e. the implementation is less advantageous, the mesh of silicon atoms is broken into smaller pieces, the thickness of the absorbing layer is significantly increased, and its structure becomes amorphous. It is important to note that the energy in the modeled systems does not lead after cooling to stabilization to the substantial structural rearrangement that makes LPS components more resistant.

4.7 Computer Model of Ionic Transport in Lithium-ion Batteries

We constructed computer models of ion transport using different combinations of the three main components of LPS, namely anode, cathode and membrane (electrolyte). These models allowed defining the basic characteristics of the energy system and evaluating the properties of the target battery based on calculations of the structure and transport properties of the electrode and electrolyte at molecular level. Comparing the results of different simulation options, we could identify the most promising areas of construction of lithium-ion batteries of new type and their characteristics during lithiation/delithiation.

Examples of the simulated complicate complexes "electrolyte–Si_nC_m" of different dimensions are shown in Fig. 4.

4.8 Estimation of Adequacy of Created Models

The assessment of adequacy, reliability and accuracy of the constructed computer models of nanocomposite materials and processes with their participation requires carrying out a number of tests of models by various methods.

In the model there should be no obvious contradictions to the observed physical and chemical effects in the evolution of real simulated systems (for example, during heating-cooling cycles) and the absence in the model of inconsistencies in the physical and chemical state of the simulated substances (for example, the formation of metallic lithium or the decomposition of electrolytes).

The program of tests of model includes the following methods:

1. comparison of the data obtained in the simulation with independent external data (analytical, experimental, theoretical, reference) by comparison of the received

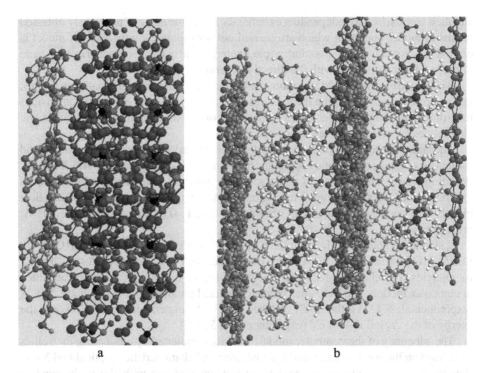

Fig. 4. Structure models of solid electrolyte complexes $Li_{80}Ge_8P_{16}S_{96}$ (a) and polymer electrolyte [LiNafion*8DMSO (b) with layers of $Si_{32}C_{38}$ silicon-carbon "nanopaper"

   parameters, for example, the average bond energy of atoms in a crystal, the parameters of a crystal cell, photoelectronic absorption spectra, etc.) with earlier known from literature or reference sources;
2. comparison of simulation data of model operation received during the work of model in general to the data obtained earlier at model operation of separate components or processes with their participation;
3. check of a correctness and stability of work of model when using a wide range of various combinations of the simulated substances – nanocomposite electrodes and solid/polymeric electrolytes at various external parameters and in the conditions of multiple cycles of a lithiation/delithiation;
4. assessment of a correctness and independence of work of model when carrying out computing experiments on various high-performance resources with different versions of the computer equipment (random access and disk memory, quantity of computed nodes, versions of the applied application packages)

The main test method is carrying out mass computing experiments on various computing resources on the basis of the created computer model with use of a wide range of input parameters with the subsequent analysis of the received results and selection of data for comparison. Further, the analysis of consistency of the received results from the point of view of physical and chemical criteria, correctness of the received results and their comparison to independent external data or to the array of previously obtained results of model operation which are carried out for separate components of model is carried out. For calculated quantitative parameters, a fairly accurate numerical estimate of the level of compliance with the directly observable data is possible.

It allows to estimate both the common adequacy of model to the simulated processes, and a correctness of use of the received results for analytical and expected conclusions regarding nanocomposite electrodes on the basis of carbon-silicon and solid/polymeric electrolytes for creation of new types of LPS.

The Gaussian software package (http://gaussian.com) was used for comparison and estimation of the accuracy of the simulation of some nano-objects at DFT/B3LYP level. By comparing different levels of calculation, we noted that the calculated values used in VASP and Gaussian software for average bond energies and distances of identical objects give consistent results with accuracy of 0.02 to 0.04 eV and 0.005 to 0.01 Å, respectively.

It should be noted that the difference of calculation results at B3LYP/6-31G (d, p), PBE/6-31G (d, p) and PBE/PAW levels does not exceed 0–2% for distances and 1–13% for energies. The chosen calculation level provides the following calculation accuracy in computer models: the Si crystal lattice calculated parameters a = b = c are 5.48 Å (experimental: 5.43 Å), the Si–Si distance is 2.37 Å (experimental: 2.34 Å), and the energy of the crystal is 4.44 eV (experimental: 4.52 eV).

The adequacy of the computer models was also evaluated by comparing the values calculated on the basis of their physical and chemical characteristics (optical and X-ray spectra, thermodynamic measurements, energy parameters) with those observed in physical experiments. For example, the calculated structural parameters for crystal electrolytes (a = b = 8.79 Å and c = 12.80) are in good agreement with X-ray experiments (a = b = 8.72 Å and c = 12.63 Å).

5   Conclusion

Thus, on the basis of a large number (more 2000) of numerical experiments regarding computer quantum chemistry and molecular dynamics simulation, we calculated the structures and surfaces of solid and polymeric electrolytes of a new type for LPS, their interaction with various nano-objects based on carbon and silicon with different morphologies, spatial rigidity, power characteristics, saturation potential with lithium ions. We also calculated transport processes of lithium ions (delithiation-lithiation) in nanocomposites, including structural energy characteristics and structures evolving over time (depending on the number of cycles of lithiation).

The model structures calculated, as well as the characteristics of electrolyte and anode materials for LPS and their interaction during charge and discharge were used to simulate a whole picture of lithiation and delithiation processes in Li-ion cells, the interaction of lithium ions with the surfaces of carbon and silicon nanomaterials, the determination of the "container" received by the anode materials, and also to model both components and new LPS types in general.

The simulation results will be used to determine the optimal conditions for the synthesis and production of the most energetically favorable and industrially suitable electrolyte and anode materials for new types of Li-ion power sources.

During the project are received the Certificate on the state registration of the database "Database on Structures and Physical and Chemical Properties Silicon and Silicon-Carbon Anodes for Lithium-Ion Accumulators" for No. 2016620100 and the Certificate on the state registration of the computer program No. 2017610081 "System of visualization of results of quantum and chemical modeling".

For the development of the "Database on the structures and physical and chemical properties of silicon and silicon-carbon anodes for lithium-ion batteries," received a diploma and a silver medal at the Moscow International Salon of Inventions and Innovative Technologies "Archimedes-2016".

For development "The database on structures and physical and chemical properties silicon and silicon-carbon anodes for lithium-ion accumulators" the author's team gained the diploma and a silver medal on the 19th Moscow International Salon of inventions and innovative technologies "Arkhimed-2016".

Acknowledgement. The activity is part of the work «Creation of an effective environment of computer modeling of quantum-chemical processes and nanostructures on the basis of a program complex of newest computing services and high-level web- and grid-interfaces to them», supported by the Russian Foundation of Basic Research (project no. 15-07-07867-a)

References

1. Volokhov, V.M., Varlamov, D.A., Zyubina, T.S., Zyubin, A.S., Volokhov, A.V., Pokatovich, G.A.: The supercomputer simulations of nanocomposites based on carbon and silicon in the new type of Li-ion power sources. In: Proceedings of the International Scientific Conference on Russian Supercomputing Days 2015, 28-29 September 2015, pp. 453–464. Publishing of the Moscow State University named Lomonosov, Moscow (2015) (in Russian)

2. Volokhov, V.M., Varlamov, D.A., Zyubina, T.S., Zyubin, A.S., Volokhov, A.V., Pokatovich, G.A.: Supercomputer simulation of transport and energetic processes in nanocomposite materials on the basis of carbon and silicon. In: Proceedings of the International Scientific Conference on Parallel Computational Technologies (PCT 2016), 28 March – 1 April 2015, pp. 105–117. Publishing of the South Ural State University, Arkhangelsk, Chelyabinsk (2016) (in Russian)

3. Volokhov, V.M., Varlamov, D.A., Zyubina, T.S., Zyubin, A.S., Volokhov, A.V.: The supercomputer simulation of processes of interaction silicon-carbonic nanostructured electrodes and solid electrolytes in new types of Li-ion power sources. In: Proceedings of the International Scientific Conference on Russian Supercomputing Days 2016, 26–27 September 2016, pp. 690–699. Publishing of the Moscow State University named Lomonosov, Moscow (2016) (in Russian)

4. Voevodin Vl, V., Zhumatij, S.A., Sobolev, S.I., Antonov, A.S., Bryzgalov, P.A., Nikitenko, D.A., Stefanov, K.S., Voevodin Vad, V.: Practice of a supercomputer "Lomonosov". In: Open Systems, vol. 7, pp. 36–39 (2012) (in Russian)

5. Zyubin, A.S., Zyubina, T.S., Dobrovol'skii, Y.A., Volokhov, V.M.: Silicon- and Carbon-based anode materials: a quantum-chemical modeling. Russ. J. Inorg. Chem. **61**(1), 48–54 (2016)

6. Zyubina, T.S., Zyubin, A.S., Dobrovolsky, Y., Volokhov, V.M.: Quantum chemical modeling of nanostructured silicon Si_n (n = 2-308). The snowball-type structures. Russ. Chem. Bull. **V65**(3), 621–630 (2016)

7. Zyubin, A.S., Zyubina, T.S., Dobrovol'skii, Y.A., Volokhov, V.M.: Quantum-chemical modeling of lithiation of a silicon–silicon carbide composite. Russ. J. Inorg. Chem. **61**(11), 1423–1429 (2016)

8. Zyubina, T.S., Zyubin, A.S., Dobrovol'skii, Y.A., Volokhov, V.M.: Lithiation-delithiation of infinite nanofibers of the Si_nC_m type – the possible promising anodic materials for lithium-ion batteries. quantum-chemical modeling. Russ. J. Electrochem. **52**(10), 988–991 (2016)

9. Zyubina, T.S., Zyubina, A.S., Dobrovol'skii, Y.A., Volokhov, V.M.: Quantum-chemical modeling of lithiation-delithiation of infinite fibers [SinCm] k (k = ∞) for n = 12–16 and m = 8–19 and small silicon clusters. Russ. J. Inorg. Chem. **61**(13), 1677–1687 (2016)

10. Zyubina, T.S., Zyubin, A.S., Dobrovol'skii, Y.A., Volokhov, V.M.: Migration of lithium ions in a nonaqueous nafion-based polymeric electrolyte: quantum-chemical modeling. Russ. J. Inorg. Chem. **61**(12), 1545–1553 (2016)

Possibility of Physical Detonation in the Flow of Vibrationally Preexcited Hydrogen in a Shock Tube

Sergey V. Kulikov[✉], Nadezda A. Chervonnaya,
and Olga N. Ternovaya

Institute of Problems of Chemical Physics, RAS, Chernogolovka, Russia
{kuls,nadan,olg}@icp.ac.ru

Abstract. The direct simulation Monte Carlo method was used to numerically simulate processes in the shock wave front in vibrationally excited hydrogen flowing in the shock tube. The cases of partially and completely excited hydrogen were considered. Equilibrium hydrogen was applied as a pusher gas, but its concentration was 50 times higher than the hydrogen concentration in the low-pressure channel. In addition, the strength of the shock wave was varied by heating the pusher gas. Number of employed processor was equal to 274. The modeling domain was split into 274 sub-domains, in each of which the evolution of the system was simulated with a single processor. The parameters of the wave in the case of physical detonation become dependent on the vibrational-to-thermal energy conversion and independent of the way of its initiation. This served as a criterion for the appearance of the physical detonation in the numerical experiment. It turned out that this phenomenon occurs until the degree of pre-excited hydrogen is not less than 85% in the low pressure channel. And the vibrational temperature is not less than 2800 K.

Keywords: Supercomputer · Simulation · Block decomposition · Vibrational excitation · Shock tube

1 Introduction

The detonation of gas mixtures is an interesting and complex phenomenon that has been systematically studied for more than a century [1]. The question arises as to whether physical, rather than chemical, energy concentrated on internal degrees of freedom of, say, vibrationally excited molecules may cause detonation under certain conditions. If so, this detonation can be called physical detonation [2]. Vibrationally preexcited hydrogen is the most appropriate gas for checking the feasibility of physical detonation [3].

Gas detonation and flow in a shock tube are very special gasdynamic processes. Parameters of the flow change drastically in a very narrow zone with the local characteristic size, L, comparable with the local mean free molecular path, λ, i.e. $\lambda \sim L$. At the same time the local mean free molecular path is much less than the local characteristic size of other parts of the flow. However, the processes that take place in the narrow zone influence the whole flowfield. And they should be considered at the molecular level. Generally speaking, in other parts of the flow one can restrict

© Springer International Publishing AG 2017
V. Voevodin and S. Sobolev (Eds.): RuSCDays 2017, CCIS 793, pp. 313–324, 2017.
https://doi.org/10.1007/978-3-319-71255-0_25

consideration by the hydrodynamic approach, but the problem of "cross-linking" different solution methods appears in this case. Of course, it is more correct to use a method that makes it possible to consider all problems at the molecular level. At the modern state of computation techniques, this can be made by the Monte Carlo non-stationary method of statistical simulation, also known as direct simulation Monte Carlo (DSMC). The idea of the method was proposed by Bird [4]. The method gives a result without solving the Boltzmann equation and automatically takes into account all details of mass and heat transfer.

Conditions of realization of the physical detonation in a shock tube have been numerically obtained in [5]. It has been shown that the phenomenon took place in a fully vibrationally preexcited hydrogen placed in the low-pressure channel (LPC) of a shock tube. The cases of completely and partially excited hydrogen were considered. The initial vibrational temperature (T_V) was equal to 3000 K. It has been shown in [5] that, if the prestored vibrational energy is weakly converted to translational energy, the shock wave slows down over time. If the energy conversion is sufficiently intense, when the pusher gas is warm and only completely vibrationally excited hydrogen is in the low-pressure channel, the wave gains velocity over time (its velocity increases roughly by a factor of 1.5). This causes physical detonation, in which case the parameters of the wave become dependent on the vibrational-to-thermal energy conversion and independent of the way of its initiation. The latter has been shown in [5] by heating the hydrogen in the high-pressure chamber (HPC) to different temperatures (T_H) (439 and 585 K) that resulted in waves of varying intensity.

Below we present results which complement the results of work [5]. Namely, cases of partially preexcited hydrogen and of lower temperatures of vibrational preexcitation were considered.

2 Statement of the Problem

At the beginning, the LPC of a shock tube is filled with two portions of hydrogen. The first portion is vibrationally preexcited hydrogen The rotational and translational temperatures were assumed to be equal to room temperature $T_1 = 292$ K. The second portion consisted of totally equilibrium hydrogen with T_1. In other words, it was assumed that only part the hydrogen was excited (for example, by electrical discharge). A HPC was initially filled with hydrogen as pusher gas at a much higher pressure. The strength of the shock wave (SV) was varied by heating the pusher gas. Then the numerical simulation of the process in the shock tube started.

3 Simulation Technique

Simulation was performed in a 1D coordinate space and a 3D velocity space using DSMC method. The basics of shock and detonation wave simulation in a shock tube with taking into account rotational and vibrational degrees of freedom of molecules were presented elsewhere [6–8]. Below, we cite the simulation algorithm for the reader's convenience.

A real medium (a medium to be simulated) was replaced by a set of model particles. According to initial conditions, the particles at zero time have the given velocities and are distributed over cells into which the coordinate space is split. It is assumed that particle collisions are binary and may occur between the particles occupying the same cell with a certain probability.

The evolution of the system over time interval Δt is divided into two stages: (i) the movement of particles with constant velocities (stage A) and (ii) variation of the particle velocities due to collisions. (stage B).

The ith model particle of species l denoted by $A_l^{(i)}$ was characterized by mass m_l, velocity $c_l^{(i)}(u_l^{(i)}, v_l^{(i)}, w_l^{(i)})$ coordinate in the flow $x_l^{(i)}$, and weighting factor η_l. This weighting factor indicates the number of actual molecules represented by the given model particle. Thus, the concentration of actual molecules in the jth cell of volume V_j is defined as

$$n_l^{(j)} = \sum \eta_l / V_j$$

Simulation of stage A is very simple. As a result of this simulation, the new position of model particle $A_l^{(i)}$ in the flow is given by

$$x_l^{(i)*} = x_l^{(i)} + u_l^{(i)} \Delta t.$$

The impact parameter used here is the integrated cross section σ_{lm}^{ik} of elastic scattering between molecules $A_l^{(i)}$ and $A_m^{(k)}$. Molecules were considered as a perfectly rigid spheres.

We will henceforth use the following notation:

$$\vartheta_{lm} = max\{\eta_l, \eta_m\}, \quad \theta_{lm} = min\{\eta_l, \eta_m\}.$$

Here and below, the indices of numbers of particles and cells will be omitted for simplicity whenever possible.

At stage B, the evolution of the system was simulated in several (k) steps. At each of these steps, the interaction of pairs of particles in the cell under investigation occurred during time period $\Delta t_* = \Delta t / k$. Simulation of each such step was performed in accordance with the ballot box scheme of testing. To this end, all pairs of particles in a cell were divided into aggregates characterized by species of particles forming the pairs (e.g., pairs of particles of species 1, pairs of particles of species 1 and 2, and so on). In each aggregate, only one pair of particles (e.g., A_l and A_m) is chosen equiprobably. The evolution of the state of the chosen pair of particles was simulated in accordance with the scheme described below.

Step 1. The interaction of particles A_l and A_m was simulated with probability

$$Q_{lm} = K_{lm} \vartheta_{lm} \sigma_{lm} g_{lm} \Delta t_* / V$$

Here, K_{lm} is the number of pairs of particles in the aggregate under investigation. Number k was chosen so that Q_{lm} was slightly less than unity. If the result of the test is negative, the next step for the given pair of particles was not made.

Step 2. Velocities of particles A_l and A_m were replaced by velocities after the collision c_l^* and c_m^* with probabilities θ_{lm}/η_l and θ_{lm}/η_m, respectively. This is the so called improved ballot box scheme of tests for simulating the collision stage.

This method was used to numerically simulate the problem of the shock wave front in vibrationally excited hydrogen flowing in the shock tube. The cases of completely and partially excited hydrogen were considered. It was supposed that unexcited H_2 was in full thermodynamic equilibrium. In the simulation, the rotational and vibrational degrees of freedom of molecules were taken into account in the simplest terms [5, 7, 8]. An energy sink model (see Sect. 11.3 in [4]) was employed. It should be emphasized that Monte Carlo (statistical) simulation considers the kinetic temperature as the mean energy associated with the respective degrees of freedom of a molecule. In the case of equilibrium over degrees of freedom, this temperature is the thermodynamic temperature. Each time, at the beginning of the collision stage, the total (over all translational degrees of freedom) kinetic translational temperature (T) was determined for the given type of molecules in the cell. Then, at each collision adopted, the difference Δ_i between temperature T and the temperature of a given inner degree of freedom was determined. Parameters R_{ij} were set (i - is the component number, and $j = 1$ or 2 for rotational and vibrational degrees of freedom, respectively). Then the internal temperature (and the corresponding internal energy) changed in the product $R_{ij}\Delta_i$ in the direction of approximation to T. When the equilibrium internal energy was determined, it was assumed that the specific heat of a molecule for the rotational and vibrational energies equals k (k is the Boltzmann constant; in units used in the simulation, $k = 0.5$) and $R_{i1} = 0.01$. However, R_{i2} depended on relative velocity g of colliding particles. At $g < 3.726$ m/s, $R_{i2} = 0$; at $3.726 \leq g \leq 9749$ m/s, $R_{i2} = 0.00005$; and, at $g \geq 9748$ m/s, $R_{i2} = 0.01$. The parameters of this model were selected so as to provide real values of vibrational relaxation time for H_2 at different temperature. When the post-collision velocities of a pair of particles were determined, it was assumed that the change in the translational energy of this par of particles is equal in magnitude and opposite in sigh to the change in the internal energy of these particles. It should be noted that vibrational degrees of freedom of H_2 are not excited when 292 K. Therefore, the initial vibrational temperature and energy of the unexcited hydrogen in the LPC were supposed equal to zero.

Equilibrium hydrogen was applied as a pusher gas in HPC, but its concentration was 50 times higher than the hydrogen concentration in the LPC.

For linear size Δx of a spatial cell to be shorter than free path λ of molecules in a gas, the modeling domain in the HPC was first split into cells 20 times smaller than those in the LPC. In the process of evolution of the system, size Δx in that part of the LPC to which the gas is delivered from the HPC was also decreased by 20 times.

In addition, the strength of the shock wave was varied by heating the pusher gas in HPC. Velocity D can be increased by raising the pressure drop between the HPC and LPC. This can be done by increasing the concentration of molecules in the HPC. In this case, however, Δx in the HPC decreases further. Correspondingly, the counting time increases and the number of processors exceeds a reasonable value. A more simpler and more effective way is to raise the gas temperature in the HPC (with the number of processors remaining the same) and upgrade the computational program to a minor extent. The counting time will be change slightly. However, it should be noted that, if D is too high, the temperature in the post shock flow may also rise to excess and the change of vibrational energy will become minor again.

With the aim of using a reasonable number of model particles, the weight factor for all particles of H_2 in the HPC was taken to be equal to 5 and that for excited particles of H_2 in the LPC to 1. The weight factor for of the equilibrium unexcited hydrogen in the LPC (η_{uH}) was determined from the equation:

$$\eta_{uH} = 1/\alpha - 1.$$

Here, α is the fraction of vibrationally excited hydrogen in the LPC.

The value of Δt was set equal to 0.04. Henceforth, t is normalized to λ_1/u, where λ_1 − λ in the LPC at zero time and $u = (2kT/m)^{0.5}$ is the most probable thermal velocity of the particles in the gas medium ahead of the SW (m is the molecular mass of hydrogen). The distance is normalized to λ_1. In the LPC, initially, $\Delta x = 0.15$. The linear dimensions of the HPC and LPC were 4795.2 and 10873.2, respectively. The boundary between LPC and HPC was placed at the point $x = 0$. At the beginning, the average number of model particles of each sort in the cell equaled 90.

Particles are elastically reflected from the boundaries of the modeling domain.

A multiprocessor computer MBC100 K installed in the Joint Supercomputer Center of RAS was used. Its peak performance is equal to 227.94 Teraflops. The supercomputer consists of 1275 computing module (10572 cores).

The block decomposition of the modeling domain was applied [6, 7]. The modeling domain was split into 274 sub-domains, in each of which the evolution of the system was simulated with a single processor (core). After each step of movements, information about particles that leave sub-domains occupied by them at the beginning of the given step and pass to neighboring sub-domains was transferred to the latter sub-domains using the SEND and RECV procedures from the MPI library [9]. Thus, by increasing the number of processors up to several thousand, one can extend the modeling domain virtually without increasing the counting time (at a fixed time of the system's evolution), since data exchange takes place between neighboring sub-domains. The transfer time for this information is almost independent of the number of processors. Experience has shown [6, 7] that this organization of parallel computation is the most reasonable.

4   Results of the Simulation

First, the simulations were carried out for cases when the fractions of vibrationally preexcited hydrogen were equal to 0.85 and 0.75 at the vibrational temperature of 3000 K. It was shown that only the initial excitation of 85% H_2 led to the physical detonation. And it was absent for lesser degree of excitation. An increase in wave velocity with time had occurred for $\alpha = 0.85$ at T_H 584, 657 and 730 K (see Fig. 1)

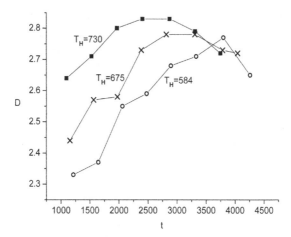

Fig. 1. Wave velocity D vs. simulation time t for $T_V = 3000$ at different T_H, $\alpha = 0.85$.

Henceforth, D is normalized to u. The accuracy of determining of D values was within ±0.05.

Figures 2 and 3 demonstrate simulation results obtained for the case when the HPC was initially filled with completely equilibrium hydrogen heated to $T_H = 584$ K, $\alpha = 0.85$. Simulation time t was equal to 4257.775. This time corresponds to the moment when the front came very close to the left end of the simulation region. Profiles of parameters in the post-shock flow (a region of a shock-heated gas in the LPC) are shown in Fig. 2 for vibrationally preexcited H_2. Figure 3 demonstrate profiles of parameters of initially completely equilibrium hydrogen. Henceforth, concentration n is normalized to initial concentration of vibrationally preexcited H_2 in the LPC. Total

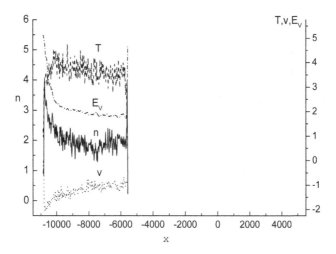

Fig. 2. Profiles of concentration (n), temperature (T), velocity of the flow (v) and internal vibrational energy of initially excited H_2 (E_V), $T_H = 584$ K.

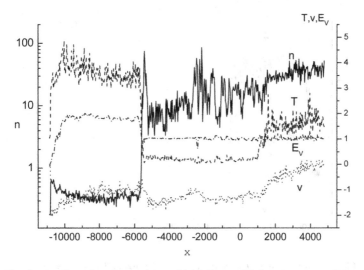

Fig. 3. Profiles of n, T, v and E_V of initially equilibrium H_2, $T_H = 584$ K.

kinetic translational temperature T is normalized to T_1. Velocity of the flow v is normalized to u. Normalized energy E_V is given by equation

$$E_V = k(T_V/T_1),$$

where T_V is the running vibrational temperature. The dimensionless initial vibrational energy E_V of the hydrogen equaled 5.14 for initial $T_V = 3000$ K. It is necessary to remind that k is the Boltzmann constant. In units used in the simulation, k is equal to 0.5.

Figures 4 and 5 show results obtained for the case when the HPC was initially filled with completely equilibrium hydrogen heated to $T_H = 730$ K, $\alpha = 0.85$. Simulation time t was equal to 3742.85. This time corresponds to the moment when the front came very close to the left end of region of the simulation. As above, profiles of parameters in the post-shock flow are presented in Fig. 4 for vibrationally preexcited H_2. Figure 5 demonstrates profiles of parameters of initially completely equilibrium hydrogen.

Figures 3, 4 and 5 illustrate the flow picture when high wave velocity was already installed. The profiles of flow parameters in the post-shock flow coincide with each other with accuracy of statistical dispersion at T_H 584 and 730, despite the varying intensity of the initial shock waves. The profiles of parameters obtained at T_H 675 K give that kind of same picture. And they are not shown in the article. All these results convincingly testifie to the achievement of the physical detonation.

Next, simulations were carried out for initial excitation of 85% H_2 to the vibrational temperatures 2900 and 2800 K. Figures 6 and 7 demonstrate the growth of wave velocities over simulation time and achievement their stationary values. Figure 6 shows the dependence of wave velocity on the simulation time at T_H 584 and 675 K for $T_V = 2900$ K.

These results and the corresponding profiles of flow parameters in the post shock flow at high D (for example at t more 4000) testifies to the achievement of the physical

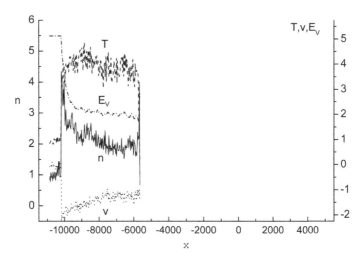

Fig. 4. Profiles of n, T, v and E_V of initially excited H_2, T_H = 730 K.

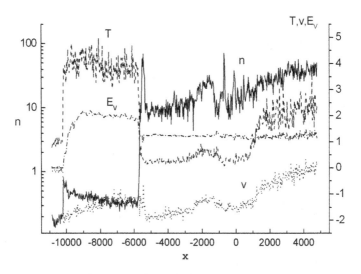

Fig. 5. Profiles of n, T, v and E_V of initially equilibrium H_2, T_H = 730 K.

detonation. The correspondent profiles at T_H 584 and 657 K are very similar to each other as above (see Figs. 3, 4, 5 and 6). They are not given in the article due to lack of space.

Figure 7 shows the dependence of wave velocity on the simulation time at T_H 657 and 730 K for T_V = 2800 K.

Figures 8 and 9 show results obtained for the case of T_V 2800 K when the HPC was initially filled with completely equilibrium hydrogen heated to T_H = 730 K, α = 0.85. Simulation time t was equal to 4095.425. This time corresponds to the moment when the

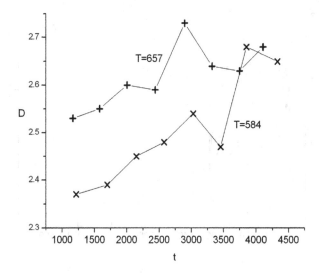

Fig. 6. Wave velocity D vs. simulation time t for $T_V = 2900$ K at different T_H, $\alpha = 0.85$.

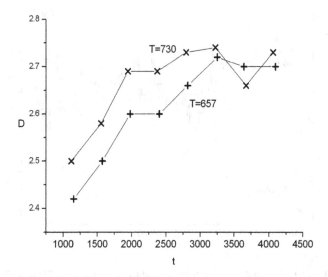

Fig. 7. Wave velocity D vs. simulation time t for $T_V = 2800$ K at different T_H, $\alpha = 0.85$.

front came very close to the left end of region of the simulation. Profiles of parameters in the post-shock flow are presented in Fig. 8 for vibrationally preexcited H_2.

Figure 9 demonstrates profiles of parameters of initially completely equilibrium hydrogen.

Figures 10 and 11 show results obtained for the case when the HPC was initially filled with completely equilibrium hydrogen heated to $T_H = 730$ K, $\alpha = 0.85$. Simulation time t was equal to 4058.875. This time corresponds to the moment when the

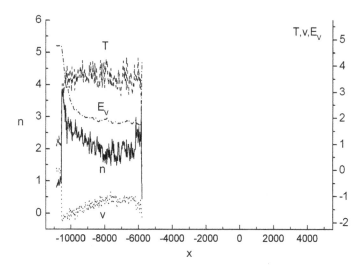

Fig. 8. Profiles of n, T, v and E_V of initially excited H_2 for $T_V = 2800$ K, $T_H = 657$ K.

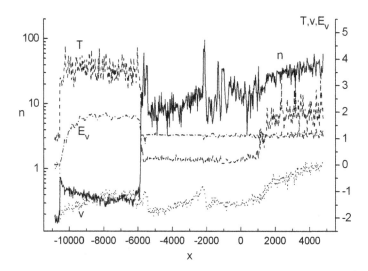

Fig. 9. Profiles of n, T, v and E_V of initially equilibrium H_2 for $T_V = 2800$ K, $T_H = 657$ K

front came very close to the left end of region of the simulation. Profiles of parameters in the post-shock flow are presented in Fig. 10 for vibrationally preexcited H_2.

Figure 11 demonstrates profiles of parameters of initially completely equilibrium hydrogen.

Figures 8, 9, 10 and 11 illustrate the flow picture when high wave velocity was already installed. The profiles of flow parameters in the post-shock flow coincide with each other with accuracy of statistical dispersion at T_H 657 and 730, despite the varying

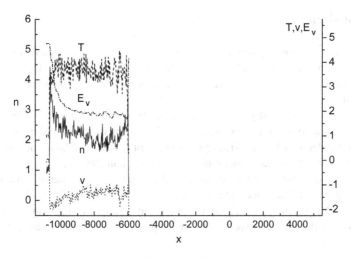

Fig. 10. Profiles of n, T, v and E_V of initially excited H_2 for $T_V = 2800$ K, $T_H = 730$ K.

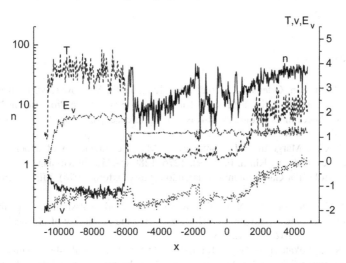

Fig. 11. Profiles of n, T, v and E_V of initially equilibrium H_2 for $T_V = 2800$ K, $T_H = 730$ K.

intensity of the initial shock waves. All these results convincingly testifie the achievement of the physical detonation.

The profiles of parameters obtained at T_V 2900 K give that kind of same picture. And they are not shown in the article.

The normalized initial vibrational energy of hydrogen is consistent with 4.97 and 4.79 for the initial vibrational temperature 2800 and 2900 K. This information is presented for the reader's convenience.

5   Conclusions

A powerful computational program was created for the supercomputer. Simulation of the flow of hydrogen in a shock tube was performed. The cases of the vibrationally preexcited hydrogen were considered. The number of used processors (cores) was equal to 274.

It was shown numerically that the physical detonation may take place at milder conditions compared with the predicted in [5]. It turned out that this phenomenon occurs until the degree of preexcited hydrogen is not less than 85% in the low pressure channel. And the vibrational temperature is not less than 2800 K.

The simulation results obtained for a wider region of parameters confirmed the conclusions given in the paper [5]. Indeed, as for a classical detonation wave, the parameters of the physical detonation wave are independent of the way of detonation initiation. When the shock wave evolves to a physical detonation wave, its velocity markedly grows.

The obtained results can be useful for researchers in their experimental realization of the physics detonation.

The authors are grateful to the Joint Supercomputer Center of RAS for providing computing resources.

References

1. Physics of Explosion, Orlenko, L.P. (ed.), vol. 1. Fizmatlit, Moscow (2002). (in Russian)
2. Drozdov, M.S., Kulikov, S.V.: Physical detonation. In: Irreversible Processes in Nature and Technology. Proceedings of the 7th all-Russian Conference, part 1, pp. 10–14. Bauman, LPI, Moscow (2013). (in Russian)
3. Evtyukhin, N.V., Margolin, A.D., Shmelev, V.M.: The interaction of shock waves with vibrationally excited gas. Khim. Fiz. 4(9), 1276–1280 (1985). (in Russian)
4. Bird, G.A.: Molecular Gas Dynamics. Clarendon Press, Oxford (1976). Mir, Moscow (1981). (in Russian)
5. Kulikov, S.V., Chervonnaya, N.A., Ternovaya, O.N.: Statistical simulation of the flow of vibrationally preexcited hydrogen in a shock tube and the possibility of physical detonation. Tech. Phys. 61(8), 1162–1167 (2016)
6. Kulikov, S.V., Pokatovich, G.A., Ternovaya, O.N.: A molecular-level simulation of a detonation wave in a gas initiated by an instantaneous heating of an endplate of a tube. Russ. J. Phys. Chem. 2(3), 371–374 (2008)
7. Kulikov, S.V.: Modeling of processes in gases in shock tubes with consideration for internal degrees of freedom by using the direct simulation Monte Carlo Method. Russian J. Phys. Chem. B 2(6), 894–899 (2008)
8. Genich, A.P., Kulikov, S.V., Manelis, G.B., Chereshnev, S.L.: Thermophysics of translational relaxation in shock waves in gases. Sov. Tech. Rev. B Therm. Phys. 4, part 1, 1–69 (1992)
9. Snir, M., Otto, S., Huss-Lederman, S., Walker, D., Dongarra, J.: MPI: The Complete Reference, vol. 1. The MPI Core, MIT, Boston (1998)

Supercomputer Modelling of Electromagnetic Wave Scattering with Boundary Integral Equation Method

Andrey Aparinov[1], Alexey Setukha[2(✉)], and Stanislav Stavtsev[3]

[1] Central Aerohydrodynamic Institute, Moscow Region,
Zhukovsky Str. 1, Zhukovsky 140180, Russia
andrey.aparinov@gmail.com
[2] Lomonosov Moscow State University, LeninskieGory,
GSP-1, Moscow 119991, Russia
setuhaav@rambler.ru
[3] Institute of Numerical Mathematics Russian Academy of Sciences,
Gubkin str. 8, Moscow 119333, Russia
sstass2000@mail.ru

Abstract. Authors consider the approaches to increase efficiency of calculations in the problem of numerical modelling of electromagnetic wave scattering. To solve such kind of problems authors develop innovative variation of boundary integral equations method based on utilisation of integral equation with hyper singular integrals which can be solved with methods of piece-wise approximations and collocations. From numerical point of view the problem reduces to the solution of the system of linear equations which coefficients present the influence of cells of mesh on collocation points. The specialities of parallel algorithms for diffraction problems are described as for the straight solutions of the appearing linear systems so as for the approach utilising mosaic-skeleton approximation method which allows to solve linear equation system calculating only small part of matrix elements.

Keywords: Supercomputer modelling · Electromagnetic scattering · Integral equations · Fast matrix methods

1 Introduction

Mathematical modelling of electromagnetic wave scattering by surfaces with complex shapes is actual problem for which solution there are various approaches. If wave length is much less than typical sizes of reflecting objects methods of physical optics and asymptotic methods work well. But in case of wave length comparable to objects sizes it is critical to formulate and numerically solve exterior boundary value problem for electromagnetic field in space outside the bodies. Modern grid and finite-element methods based on discretization of electromagnetic field in space allow to consider complex, including non-homogeneous, structure of the environment and different physical effects [1].

© Springer International Publishing AG 2017
V. Voevodin and S. Sobolev (Eds.): RuSCDays 2017, CCIS 793, pp. 325–336, 2017.
https://doi.org/10.1007/978-3-319-71255-0_26

However, the significant problem here is that to fulfil boundary conditions on infinity one have to use calculating domain many times exceeding size of bodies. This leads to the great calculating difficulty of such methods. Herewith the requirement of smallness of space discretization step in comparison to wave length puts the limitation on utilisation of non-uniform meshes.

In case of modelling monochromatic wave processes in homogeneous environment the approach based on the method of boundary integral equations is very efficient [2,3]. Here the solution of boundary problem is found from the integral representation with the integrals written for the boundary of the problem solution domain (body surfaces). The whole problem reduces to integral equations written on this boundary. With this the boundary conditions on infinity are fulfilled automatically and obtained solutions exactly satisfy the equations in the problem solution domain. For the numerical solution the grid is needed only on body surfaces. The problem of calculation efficiency remains actual here and raises from the necessity to model diffraction on bodies with complex shapes and the requirements to wide the diapason of investigated wave lengths.

Specific of methods of boundary integral equations is that in their discretization appear systems of linear equations with filled matrices which rank is defined by the number of cells in mesh. Here both problems of computation time reduction and element storage in operating memory are of great importance and practice shows that in many cases the memory problem is leading.

In this article two approaches to the solution of high complexity problems are described. First is the application of mosaic-skeleton approximations which allows to approximately solve the linear system calculating only comparably small number of its matrix elements [4]. This leads both to significant gain in computation time and in memory utilisation. Second is parallel computing. The implemented algorithms are based on a variation of numerical method for boundary integral equations developed in [5,6]. This approach utilize integral equations with strongly singular integrals which can be solved by methods of piece-wise constant approximations and collocations. Authors describe the special aspects of parallel algorithms as for the straight solution of linear equation system so for approximate solution with mosaic-skeleton approximations.

2   Reduction of Problem to Integral Equation

Authors consider a 3D problem of scattering of a monochromatic electromagnetic field by a body or a system of bodies. Each body can be either a solid object bounded by a closed surface or a thin surface (a screen). The surfaces of the bodies are assumed to be ideally conducting, and the ambient medium is assumed to be homogeneous. The described below problem statement is classical [2,7].

Let Σ — total surface of bodies and screens which can be closed (surface of ideally conducting body), opened (ideally conducting screen) or consists of several components of such kind. Let us call Ω — space domain outside considered bodies. The problem is to find the electric and magnetic field intensities, which will be sought in the form $\mathbf{E}_{full}(x)e^{-i\omega t}$, $\mathbf{H}_{full}(x)e^{-i\omega t}$, $x \in \Omega$, where ω —

circular frequency of electromagnetic field, t — time, $x = (x_1, x_2, x_3) \subset R^3$ — points in space. It is assumed than full electromagnetic field is inducted by primary electromagnetic emission where electric and magnetic field intensities can be represented as $\mathbf{E}_{ent}(x)e^{-i\omega t}$ and $\mathbf{H}_{ent}(x)e^{-i\omega t}$, respectively. With this full electric and magnetic field intensities we will find in form

$$\mathbf{E}_{full}(x) = \mathbf{E}_{ent}(x) + \mathbf{E}(x), \mathbf{H}_{full}(x) = \mathbf{H}_{ent}(x) + \mathbf{H}(x), \tag{1}$$

$\mathbf{E}, \mathbf{H}$ — unknown intensities of electric and magnetic fields which have to satisfy Maxwell equations ([2], p.109):

$$rot\,\mathbf{E} = i\omega\mu\mathbf{H}, \quad rot\,\mathbf{H} = -i\omega\varepsilon\mathbf{E}. \tag{2}$$

Here ε and μ — dielectric and magnetic conductivity of environment. Either must be fulfilled Sommerfeld radiation conditions at infinity ([2], p.69, 116):

$$\left\{ \frac{\partial\mathbf{E}}{\partial\tau} + ik\mathbf{E} = o\left(|x|^{-1}\right), \frac{\partial\mathbf{H}}{\partial\tau} + ik\mathbf{H} = o\left(|x|^{-1}\right) |x| \to \infty, \tag{3}\right.$$

where $\partial/\partial\tau$ — derivative in the direction of vector $\tau = x/|x|$, and the condition $|\nabla\mathbf{E}| \subset L_2^{loc}, |\nabla\mathbf{H}| \subset L_2^{loc}$ ([7], subsection 22).

On the surfaces of irradiated objects Σ the condition of equality to zero of tangential component of full electric field must be fulfilled and it may be written in form

$$\mathbf{n} \times \mathbf{E} = \mathbf{f}, \tag{4}$$

where $\mathbf{f} = -\mathbf{n} \times \mathbf{E}_{ent}$, where $\mathbf{n}$ — unit normal vector to the surface.

From now on we consider that on closed components of surface Σ vector $\mathbf{n}$ has the direction outside the body, on each opened component it has violent direction but to one side on all surface.

Unknown tension of secondary electric field we'll find using known integral representation ([2], p.110):

$$\mathbf{E}(x) = \int_{\Sigma} \mathbf{e}(\mathbf{j}(y), x, y)d\sigma_y, x \in \Sigma, \tag{5}$$

where $\mathbf{j} = \mathbf{j}(x)$, $x \in \Sigma$ — unknown tangential vector field on surface Σ (surface currents),

$$\mathbf{e}(\mathbf{j}, x, y) = \{grad_x div_x[\mathbf{j}\Phi(x - y)] + k^2\mathbf{j}\Phi(x - y)\} \text{ where } \mathbf{j} \in C^3, x, y \in \Sigma, x \neq y, \tag{6}$$

$k^2 = \omega^2\varepsilon\mu$,

$$\Phi(x) = \frac{e^{ikr}}{4\pi r}, \quad r = |x|.$$

Herewith Maxwell equations (2) and conditions on infinity (3) are fulfilled automatically.

As it shown in [5], for predefined surface field $\mathbf{j} = \mathbf{j}(x), x \in \Sigma$, when special smoothness requirements of this field are completed, vector field $\mathbf{E}$, defined by (5), has boundary values on each side of surface on surface Σ and

$$\mathbf{n} \times \mathbf{E}^+ = \mathbf{n} \times \mathbf{E}^- = \mathbf{n} \times \mathbf{E},$$

where $\mathbf{E}$ — straight value, got from the expression (5) when placing in it point $x \in \Sigma$. Herewith under integral expression, defined by formula (6) has singularity of order $\mathcal{O}\left(|x - y|^{-3}\right)$ and the integral should be understood as hyper singular in the sense of the Hadamard finite value. Placing unknown field $\mathbf{E}(x)$ in boundary value (4), we get boundary integral equation with hyper singular integral:

$$\mathbf{n}(x) \times \int_{\Sigma} \{grad_x div_x [\mathbf{j}(y)\Phi(x - y)] + k^2 \mathbf{j}(y)\Phi(x - y)\} d\sigma_y = \mathbf{f}(x), \quad x \in \Sigma. \quad (7)$$

3   Numerical Scheme

For the numerical solution of integral Eq. (7) authors use the collocation method with utilisation of rectangle type quadratures basing on values of unknown function in nodes coinciding with collocation points, developed in [5]. Total surface Σ is approximated by set of cells σ_i, $i = 1, ..., n$. Authors use surface mesh which is constructed by following method. Surface Σ is divided into modules, each of which is approximated by spline surface and comes a mapping of a plane rectangle to 3D space. Then this rectangle is divided to rectangular cells and this partition arises on module of surface Σ some set of surface cells, where each has 4 vertices (surface may have poles near which cells have triangle form but considered as quadrangles with 2 coincided vertices).

In work [5] was developed the numerical method for approximation of integral Eq. (7), which uses only information about cell vertices and doesn't need any other information about surface parametrization. On each cell the collocation point is chosen x_i as the weight center of cell vertices (in the assumption that all vertices have equal masses), and normal ort is constructed $\mathbf{n}_i$ as a vector orthogonal to the diagonals of the cell. After that on each cell local orthonormal coordinate system is constructed with vectors $\mathbf{e}_{i1}$ and $\mathbf{e}_{i2} = \mathbf{n}_i \times \mathbf{e}_{i1}$ in plane, orthogonal to vector $\mathbf{n}_i$). Vector directions $\mathbf{e}_{i1}$ can be chosen violent in specified plane.

Let $\mathbf{j}_i$ — approximate value of function $\mathbf{j}(y)$ in point $x_i \in \sigma_i$, $i = 1, ..., n$,

$$\mathbf{j}_i^*(y) = (\mathbf{j}_i \times \mathbf{n}_i) \times \mathbf{n}(y), \quad (8)$$

$y \in \sigma_i$ — tangential vector field in cell σ_i, approximating on this cell function $\mathbf{j}(x)$. Replacing in Eq. (7) on each cell σ_j function $\mathbf{j}(y)$ to function $\mathbf{j}_j^*$ and writing this equation in nodes x_i, we get system of operator equations:

$$\sum_{j=1}^{n} A_{ij} \mathbf{j}_j = \mathbf{f}(x_i), \quad i = 1, ..., n, \quad (9)$$

$$A_{ij}\mathbf{j}_j = \mathbf{n}_i \times \int_{\sigma_j} \mathbf{e}(\mathbf{j}_j^*(y), x, y)d\sigma_y. \tag{10}$$

Equation system (9) we can rewrite in the form of system of linear algebraic equations in respect to the vector coordinates $\mathbf{j}_i$, $i = 1, ..., n$, in local bases constructed in grid cells:

$$\mathbf{j}_j = j_j^1\mathbf{e}_j^1 + j_j^2\mathbf{e}_j^2. \tag{11}$$

Placing vector $\mathbf{j}_j$ in form (11) to Eq. (9) and multiplying each equation on vectors $\mathbf{e}_i^l$, $l = 1, 2$, we get system of linear algebraic equations:

$$\sum_{\substack{j=1,...,n \\ l=1,2}} a_{ij}^{ml} j_j^l = f_i^m, i = 1, ..., n, m = 1, 2, \tag{12}$$

where $f_i^m = (\mathbf{f}_i, \mathbf{e}_i^m)$, $i = 1, n$, $m = 1, 2$, $a_{ij}^{ml} = (A_{ij}\mathbf{e}_j^l, \mathbf{e}_i^m), m, l = 1, 2, i, j = 1, ..., n$.

While calculating the coefficients of equation system (12) the integrals (10) are calculated by formulas from work [5] based on extraction of main singularity in explicit form. Herewith the integrals from dominant terms are calculated analytically. The remaining weakly singular integrals can be calculated numerically by method of additional partition of each cell and utilisation of rectangular type formulas with smoothing of singularity by multiplying on smoothing function. The details of calculation of weakly singular integrals are described in works [6].

In the examples below problem of plane wave scattering by ideally conducted bodies is considered. In this case primary field is written as:

$$\mathbf{E}_{ent}(x) = \mathbf{E}_0 e^{i\mathbf{k}\mathbf{r}}, \mathbf{H}_{ent}(M) = \frac{e^{i\mathbf{k}\mathbf{r}}}{\omega\mu}\mathbf{k} \times \mathbf{E}_0, \tag{13}$$

where $\mathbf{k}$ — wave vector (herewith $k = |\mathbf{k}|$), $\mathbf{r}$ — radius vector of the point x, $\mathbf{E}_0$ — defined vector orthogonal to vector $\mathbf{k}$ (vector $\mathbf{E}_0$ defines wave polarization).

One of the purposes of solving scattering problem is to find directional pattern of secondary field. There patterns characterize dependence of radar-cross section σ (RCS) in the direction of pre-set unit vector τ defined by formula:

$$\sigma(\tau) = \lim_{R \to \infty} 4\pi R^2 \frac{|\mathbf{E}(R\tau)|^2}{|\mathbf{E}_{ent}|^2}$$

from vector direction τ. Directional patterns usually made in form of dependence of values σ from some angle, which defines this vector.

If the tension of electric field is represented in form (5), then for value $\sigma(\tau)$ the following formula is true:

$$\sigma(\tau) = \frac{4\pi}{|\mathbf{E}_{ent}|^2} \left| \int_\Sigma \frac{k^2}{4\pi} e^{-ik(\tau,y)} (\mathbf{j}(y) - \tau(\mathbf{j}(y), \tau)) d\sigma_y \right|^2.$$

In numerical solution the last integral is calculated numerically with rectangle quadrature formula on the base of calculated approximate values $\mathbf{j}_i$, $i = 1, ..., n$ of function $\mathbf{j}(y)$ on the cells of surface mesh.

4   Numerical Complexity of the Algorithm, Parallel Algorithm

Main calculation costs in numerical implementation of the algorithm are related to the solution of the system of linear equations (12), which consists of $2n$ complex equations, where n — number of cells. As it was mentioned above the number of cells is defined by the requirement of utilising small cells in comparison to body sizes and to wave length. So the calculation difficulty grows with increase of frequency of falling field.

In practice it is possible to save out following 2 classes of problems. First is to investigate characteristics of electromagnetic field in space and to construct directional pattern of secondary field for known primary field. Second class is to calculate inverse RCS which characterize intensity of secondary field in the direction back to the direction of falling field of predefined frequency, (with condition that $\tau = -\mathbf{k}$), depending on the direction of vector $\mathbf{k}$. In first case the system (12) is solved single-fold and afterwards the result processing is done. In second case the system (12) is solved many times with the same matrix (matrix depends only on parameter k which is constant if frequency doesn't change) and different right-hand sides.

Authors have implemented three different algorithms of linear system solution: with LU decomposition, with GMRES iteration algorithm [8] and with mosaic-skeleton matrix approximation [4] and GMRES algorithm.

In the first and second variants of algorithms parallel calculation of matrix elements is done and all elements are stored in RAM memory (own block for each processor) and afterwards the solution itself for one or several right-hand sides is done. The solution is implemented using standard Scalapack procedures for LU factorization or with GMRES iterative method. It is notable that GMRES does not give any advantages in time or in memory in this case because of very slow convergence (more than 1000 iterations) and implementations with restart do not converge at all. Herewith authors didn't find any preconditioners that are able to achieve better convergence.

It was pointed out from practical calculations that problems of outstanding interest usually require meshes with 50000 and more cells. So it becomes clear that the main deficit resource is RAM and the algorithms of first interest are those that allow to calculate and store only small part of matrix elements.

The software was tested on supercomputer "Chebyshev" in Lomonosov Moscow State University supercomputer center and on INM RAS computer cluster. In first case 150 processors and 225 Gb of RAM were used in second case 16 processors and 180 Gb of RAM were used. In both cases different problems with grids up to 50 000 cells were successfully calculated. Significant increase of cell number required the increase in operational memory (as predicted). So successful calculations for the problem with 100000 cells were made on "Lomonosov" supercomputer in Lomonosov Moscow State University supercomputer center. About 400 Gb of RAM was used in calculations.

Further increase of cell number is limited by the lack of RAM memory to store matrix elements. So authors used mosaic-skeleton method in combination with GMRES method to solve the linear system. Mosaic-skeleton method allows to calculate and store about one percent of matrix elements and to use GMRES for system solution. In spite of slow convergence of iteration method taking into account significant memory economy for matrix storage this approach allows to reduce significantly the required RAM size and to increase the size of initial problem. Its implementation in described below.

5   Utilisation of Mosaic-Skeleton Approximations to the Solution of Diffraction Problems

To increase the calculation efficiency of the algorithm for linear system (12) solution the method of mosaic-skeleton approximations described in [4,9–12] was implemented. The overview of recent methods of matrix compression can be found in works [13,14]. The advantage of this method in comparison to others is generality and automatic precision control (the iteration nature of mosaic-skeleton approximations algorithm lead to precision increase on each step). The lack is implementation difficulty comparing to multipole methods, for example. Implementing this algorithm authors used great groundwork made in INM RAS.

Mosaic-skeleton approximation method is based on hierarchical decomposition of mesh cells into clusters basing on binary tree. Possessing a pair of cluster trees corresponding to pair of meshes representing our discretization we decompose matrix to a list of blocks of different sizes where each block represents interaction of a group of points-emitters x_j (in our case center of each cell) with group of collocation points x_i. Blocks that representing interaction of geometrically distant clusters can be approximated with low-rank matrix (Fig. 1). On Fig. 1 grey color marks dense blocks, those for which all their elements should be calculated. Other blocks assumed to be low-rank. Using incomplete crest approximation algorithm [4] such blocks can be presented in form $B = U \cdot V^T$, where for block B of size $m \times n$ matrices U and V have sizes $m \times r$ and $n \times r$ respectively, $r \ll \min(m,n)$ — rank of block B. In such a way instead of storing $\mathcal{O}(mn)$ block elements it is possible to store only $\mathcal{O}((m+n)r)$ complex numbers.

Mosaic-skeleton approximations allow to compress matrix. To solve the system authors use GMRES [8] adapted to work with compressed matrices. This algorithm is based on parallel procedure of matrix-vector multiplication, where matrix is presented in skeleton format. No other operations are needed to solve the system.

To use GMRES for linear systems with multiple right-hand sizes authors made some modifications. Among right-hand sides let's choose vector with residual with maximal norm. Let's solve the system with GMRES method for this right-hand side and construct bases of subspace where the residual minimizes. Then we calculate the residual of remaining right-hand sides and again choose vector with maximal residual and repeat GMRES for this right-hand side with this widening the set of basic vectors. Using in such manner bases from previous iterations for new right-hand side we do much less steps for this right-hand side. So we solve the system for vectors from right-hand side until on the total base we got the maximal residual of remaining vectors is not achieved, which gives us the solution of desired accuracy. Finally the number of sorted out right-hand sides decreases in times.

The main time-consuming operation in matrix approximation is calculation of block approximations. The advantage of mosaic-skeleton method in that each block-approximation is independent from others and so can be paralleled into multiple processors by distributing blocks on different processors. Each processor has to calculate, construct and store approximations only for its own blocks. Processor intercommunication occurs only during the system solution step while matrix-vector multiplication is done. In iteration algorithm of system solution each processor multiplies only its blocks on vector and the results from different processors are summed.

Mosaic-skeleton method requires $\mathcal{O}((m + n)r)$ to approximate block $m \times n$ of rank r. If r is known, then block number can be distributed to processors before the block approximation itself. Calculating experiments made for different integrands show that value r is equal to $\mathcal{O}(\log^\gamma(m + n))$, where γ depends on integrand. For integral equation used in the described problem the calculation shows that γ is approximately equal to $3/2$. Notable that $3/2$ does not depend on wave number.

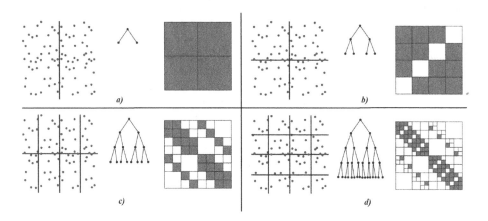

Fig. 1. Domain partition on clusters, partition tree and matrix partition on blocks: (a) 1 level; (b) 2 levels; (c) 3 levels; (d) 4 levels.

6   Numerical Results and Discussion

As an example calculation results and calculation costs are shown for the problem of plane wave scattering by a circular cylinder of finite length. Calculations were made on processors Intel Xeon E5-2670v3 2.30 GHz of INM RAS cluster (http://cluster2.inm.ras.ru/). Intel Fortran Compiler 9.0 for Linux (9.0.033) and OpenMPI Scalapack 2.0.2–4.3 was used.

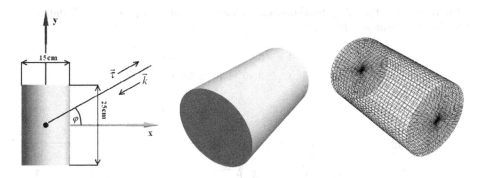

Fig. 2. Geometry and grid of the object.

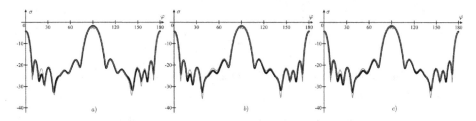

Fig. 3. 4 GHz (*a*) 13482 cells; (*b*) 21760 cells; (*c*) 45784 cells.

Figure 2 shows the cylinder form and an example of grid on its surface. Calculations were made for frequencies 4 GHz (wave length $\lambda = 7.5$ cm $= H/3.3$), 8 GHz ($\lambda = 3.75$ cm $= H/6.6$) and 16 GHz ($\lambda = 1.875$ cm $= H/13.2$), where H — cylinder height.

Table 1 shows required storage for matrix of linear equation system compressed with accuracy 10^{-3} from the number of cells in mesh and emission frequency. Last column shows memory required to store full matrices.

Table 1. Required storage for matrix.

n	4 GHz	8 GHz	16 GHz	Full matrix
21760	1.962 Gb (7.0%)	2.340 Gb (8.3%)	2.999 Gb (11%)	28.223 Gb
45784	4.457 Gb (3.6%)	5.369 Gb (4.3%)	6.927 Gb (5.6%)	124.920 Gb
273600	—	—	48.998 Gb (4.39%)	1115.456 Gb

Table 2. Acceleration of matrix calculations for various number of processors. Number of cells 273600. Frequency 16 GHz.

n_p	1	2	4	8	16	32	64	128
	1.00	1.91	3.39	6.22	11.07	19.95	30.52	42.10

Table 3. Acceleration of system solution for various number of processors. Number of cells 273600. Frequency 16 GHz.

n_p	1	2	4	8	16	32	64	128
	1.00	1.83	3.17	3.40	4.46	6.39	6.81	6.29

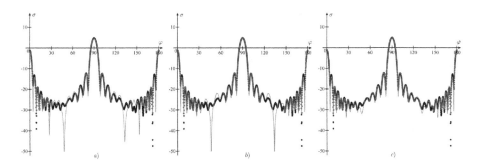

Fig. 4. 8 GHz (a) 13482 cells; (b) 21760 cells; (c) 45784 cells.

In brackets there is compression coefficient which is calculated as relation of memory required to store compressed matrix to memory required to store full matrix. Easy to mention that compression coefficient decreases with matrix size growth and increases with frequency growth.

Tables 2 and 3 show parallel acceleration rate for matrix compress operation and for linear system solution with GMRES method on multiple processors. Parallel acceleration rate shows the relation of times required to make the same calculations on one and n_p processors. Note that computation time on 64 processors needed for matrix compression was 5 min 55 s and for solving system of linear equations was about 26 h.

Finally Figs. 3, 4 and 5 show RCS diagrams obtained from calculations. The correspondence of RCS represented in decibels $\tilde{\sigma} = 10 \log \sigma$ in the direction of vector $\tau = -\mathbf{k}$ from angle φ defining the direction of vector $\mathbf{k}$ – see Fig. 2. Herewith considered vertical polarization of falling wave so vector E_0 in the expression (13) is orthogonal to plane Oxy. Calculation results (gray line) are compared to experimental data (black line) received from ITAE RAS. It can be seen that for frequency 4 GHz ($\lambda = 7.5$ cm) the mesh of 13482 cells (maximal size of cell side is $h = 0.375$ cm) is enough for good agreement of calculation with experiment. It was shown that for frequency 8 GHz ($\lambda = 3.75$ cm) the same mesh is also more or less enough for calculations. Local runs on graphs with calculation results disappear on mesh with 45784 cells ($h = 0.2$ cm). For the frequency 16 GHz ($\lambda = 1.875$ cm) good agreement with experiment (without parasite local runs on the curve) was achieved on the mesh with 273600 cells ($h = 0.1$ cm). Note that all data were presented for vertical polarization because in horizontal polarization (vector $\mathbf{E_0}$ in the expression (13) lies in plane Oxy)

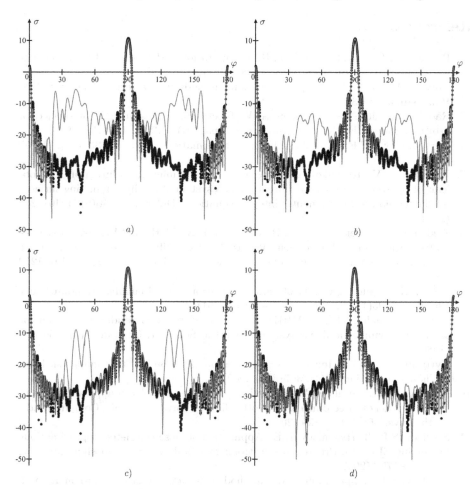

Fig. 5. 16 GHz (*a*) 13482 cells; (*b*) 21760 cells; (*c*) 45784 cells; (*d*) 273600 cells.

for all frequencies coarser meshes were enough to achieve good agreement with experiment.

Hence, the calculation difficulty of the scattering problems grows with primary field frequency increase. This is caused by several reasons: the need to cut the mesh in correspondence to wave length; wave number k growth leads to increase of compress coefficient of matrix of linear equations system (12). Besides that iteration method convergence speed falls with growth of wave number k. So even for bodies with simple geometries parallel technologies are required when wave length is less than body size by an order or more.

Acknowledgments. Authors appreciate much SCC MSU for provided opportunity to use supercomputers "Chebychev" and "Lomonosov". The work was supported by the Russian Science Foundation, grant 14-11-00806.

References

1. Peterson, A.F., Ray, S.L., Mittra, R.: Computational Methods for Electromagnetics. Wiley-IEEE Press, New York (1998). 592 p
2. Colton, D., Kress, R.: Integral Equation Methods in Scattering Theory. SIAM, Philadelphia (2013). 286 p
3. Rao, S.M., Wilton, D.R., Glisson, A.W.: Electromagnetic scattering by surfaces of arbitrary shape. IEEE Trans. Antennas Propag. **AP–30**(3), 409–418 (1982)
4. Tyrtyshnikov, E.E.: Mosaic-skeleton approximations. Calcolo **33**(1–2), 47–57 (1996)
5. Zakharov, E.V., Ryzhakov, G.V., Setukha, A.V.: Numerical solution of 3D problems of electromagnetic wave diffraction on a system of ideally conducting surfaces by the method of hypersingular integral equations. Differ. Eqn. **50**(9), 1240–1251 (2014)
6. Zakharov, E.V., Setukha, A.V., Bezobrazova, E.N.: Method of hypersingular integral equations in a three-dimensional problem of diffraction of electromagnetic waves on a piecewise homogeneous dielectric body. Differ. Eqn. **51**(9), 1197–1210 (2015)
7. Hoenl, H., Maue, A.W., Westpfahl, K.: Theorie der Beugung. Handbuch der Physik, vol. 25(1), pp. 218–573. Springer (1961)
8. Saad, Y., Schultz, M.H.: GMRES: a generalized minimal residual algorithm for solving nonsymmetric linear systems. SIAM J. Sci. Stat. Comput. **7**(3), 856–869 (1986)
9. Tyrtyshnikov, E.E.: Incomplete cross approximation in the mosaic-skeleton method. Computing **64**(4), 367–380 (2000)
10. Aparinov, A.A., Setukha, A.V.: Application of mosaic-skeleton approximations in the simulation of three-dimensional vortex flow by vortex segments. Comput. Math. Math. Phys. **50**(5), 890–899 (2010)
11. Stavtsev, S.L., Tyrtyshnikov, E.E.: Application of mosaic-skeleton approximations for solving EFIE. In: Progress in Electromagnetics Research Symposium, Moscow, pp. 1752–1755 (2009)
12. Stavtsev, S.L.: Application of the method of incomplete cross approximation to a nonstationary problem of vortex rings dynamics. Russ. J. Numer. Anal. Math. Model. **27**(3), 303–320 (2012)
13. Yokota, R., Ibeid, H., Keyes, D.: Fast multipole methods as a matrix-free hierarchical low-rank approximation, p. 19 (2016). https://arxiv.org/abs/1602.02244
14. Mikhalev, AYu., Oseledets, I.V.: Iterative representing set selection for nested cross approximation. Numer. Linear Algebra Appl. **23**(2), 230–248 (2016)

Parallel FDTD Solver with Optimal Topology and Dynamic Balancing

Gleb Balykov[✉]

Moscow, Russia
balykov.gleb@yandex.ru

Abstract. Finite-difference time-domain method (FDTD) is widely used for modeling of computational electrodynamics by numerically solving Maxwell's equations and finding approximate solution at each time step. The FDTD method was originally developed by K. Yee in 1966 and is still improving to fulfill the needs of researchers. Highly parallel Maxwell's equations solvers based on the FDTD method allow to model sophisticated structures on large grids with acceptable performance and required accuracy. This article describes parallel FDTD solver for different dimensions with comparison method for virtual topologies of computational nodes' grid, which allows to choose the best virtual topology for target architecture. Developed solver also incorporates dynamic balancing of computations between computational nodes. Measurements for presented algorithms are provided for IBM Blue Gene/P supercomputer. Further directions for optimizations are also discussed.

Keywords: Computational electrodynamics · FDTD · Parallel FDTD · MPI

1 Overview

The FDTD method is widely used in electrodynamics solvers as well as its different parallelization techniques. After it had originated in 1966 [1], it had a long road from sequential algorithm to implementations of high-performance parallel versions. This happened along with development of new hardware and architectures, giving engineers opportunities to develop and evolve FDTD algorithm.

Three commonly used parallelization technologies for the FDTD method are MPI, OpenMP and Cuda. Each of them serves its own purpose: MPI is a standard for high-performance parallel computations on architectures with distributed memory, OpenMP is a standard for high-performance parallel computations on architectures with shared memory, Cuda is a parallel computing platform and API for parallel computations on Nvidia GPUs.

The most common trend in parallelization of FDTD algorithm is still a combination of MPI and OpenMP, however, interest in massive parallel computations rapidly shifts towards GPUs and computations on them and FDTD algorithm is no exception. Cuda FDTD solvers give engineers opportunities to perform

© Springer International Publishing AG 2017
V. Voevodin and S. Sobolev (Eds.): RuSCDays 2017, CCIS 793, pp. 337–348, 2017.
https://doi.org/10.1007/978-3-319-71255-0_27

electrodynamic modeling on systems varying from personal computers equipped with Nvidia GPUs to GPU clusters. But FDTD solvers developed for heterogeneous architectures with support of MPI, OpenMP and Cuda present the most interest [2–4].

Load balancing in parallel FDTD algorithms allows to achieve the best performance possible for current parameters of computation and characteristics of the computational system. In general, load balancing couldn't be done without characteristics of the system, on which computations are performed [5]. However, for homogeneous architectures load balancing could be performed statically before computations in some cases.

In this article, parallel FDTD solver with optimal topology and dynamic balancing is introduced [6], which incorporates algorithm of choosing of optimal virtual topologies for computational nodes' grid for homogeneous systems and dynamic balancing of computations between computational nodes. Solver has UPML and TF/SF support and supports both complex and real values with different precision. Besides, solver supports Cuda and could perform computations on GPUs. Combination of MPI and Cuda, which could be enabled separately, allows to achieve significant speed up on a wide range of target architectures and high portability of developed solver. In case of heterogeneous architectures, dynamic balancing could also be applied. Two other possible solutions for heterogeneous architectures are also discussed.

2   Parallel Algorithm Description

Electrodynamics modeling could be performed in different dimensions, i.e. one-dimensional modeling (1D), two-dimensional (2D) and three-dimensional (3D). For all dimensions Cartesian computational grid is introduced: to be specific, Ox axis is defined in case of 1D mode, Ox and Oy axes in case of 2D mode, Ox,Oy and Oz axes in case of 3D mode. Yee grid [7] for field components is then set, and all points of Yee grid are spread between all computational nodes. In case of sequential solver, all points of Yee grid remain on the one and only computational node. Thus, each point of Yee grid is assigned to one or another computational node.

In parallel FDTD algorithm, described here, points of Yee grid are spread between computational nodes in a very natural way: Yee grid is divided in rectangular chunks and each chunk is assigned to computational node. Besides, each computational node has buffer points on its borders in order to store data from neighboring computational nodes. This computational nodes' grid maps directly on MPI virtual topology, where each MPI process is launched on different computational node and virtual topologies are simply MPI virtual topologies.

Share operations between computational nodes are performed at each time step, so overall computational time is sum of computational time and share time for each time step. Note that only maximum sum of computational time and share time for each time step is taken into account, i.e. if one computational node performs its computations much slower than other nodes, all other nodes would have to wait for it to finish.

Each computational node performs computations on chunk of Yee grid points assigned to it and then performs share operations with all its neighboring computational nodes. Computations are the same as for sequential algorithm and could be performed either on CPU, or on GPU if computational node has one.

Share operations consist of the next steps. All directions in which share operations could be performed are considered one after another, and each computational node sends data in this directions, and also receives data from the opposite directions at the same time. In case of send operation, data in border points of node's chunk is send, in case of receive, received data is stored in buffers.

For example, for 2D mode there are 8 directions — same as number of neighbors of computational node: 1 direction — positive by Ox axis, 2 — positive by Ox and Oy axes (diagonal), 3 — positive by Oy axis, 4 — negative by Ox and positive by Oy axis (diagonal), 5 — negative by Ox axis, 6 — negative by Ox and Oy axes (diagonal), 7 — negative by Oy axis, 8 — positive by Ox and negative by Oy axis (diagonal). Figure 1 on the left shows for computational node marked with number 0 all 8 possible send directions, Fig. 1 on the right shows send procedure in direction 1 for 9 computational nodes. Arrows show direction in which data is sent, so each computational node receives data from the opposite direction (in case it has such neighbor).

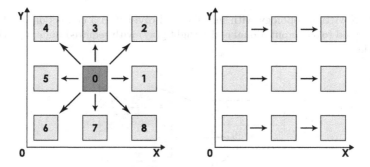

Fig. 1. 8 possible send directions for computational node marked 0 for 2D mode (on the left) and send procedure for 9 computational nodes for 2D mode (on the right). For a single computational node arrows show direction where data is sent, data is received from the opposite direction.

Division of Yee grid in rectangular chunks could be done in different ways. For 1D mode there is only one way — to divide Ox axis in chunks. In this case computational nodes would perform share operation only along Ox axis. Let's call this division of Yee grid 1D-X virtual topology. Oy and Oz axes could be divided the same way. Combining different axes divisions one can yield that for 2D mode there are 3 options: 2D-X, 2D-Y, 2D-XY virtual topologies, and for 3D mode there are 7 options: 3D-X, 3D-Y, 3D-Z, 3D-XY, 3D-YZ, 3D-XZ, 3D-XYZ virtual topologies. Figures 2, 3 and 4 show three kinds of virtual topologies for 2D mode. On the left full Yee grid is shown divided in chunks with data shown

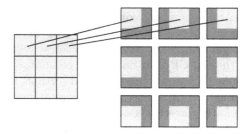

Fig. 2. 2D-XY virtual topology with full Yee grid divided in 9 chunks on the left and chunks assigned to 9 computational nodes (light gray) with required buffers (dark gray) on the right.

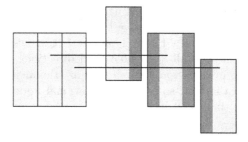

Fig. 3. 2D-X virtual topology with full Yee grid divided in 3 chunks on the left and chunks assigned to 3 computational nodes (light gray) with required buffers (dark gray) on the right.

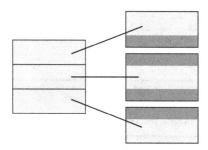

Fig. 4. 2D-Y virtual topology with full Yee grid divided in 3 chunks on the left and chunks assigned to 3 computational nodes (light gray) with required buffers (dark gray) on the right.

in light gray and on the right data assigned to each computational node is shown separately with required buffers shown in dark gray.

There are two main cases for computational nodes and communicational network: all computational nodes are the same by performance and share time for all nodes is the same (homogeneous computing system), computational nodes are not the same by performance or share time for nodes is not the same (heterogeneous computing system).

Let N be the number of computational nodes used in computations. Let's consider it being determined somehow for now (e.g. by user of the solver).

First, let's consider case of homogeneous computing system for 2D and 3D modes (for 1D mode there is only one kind of virtual topology).

2.1   2D Mode for Homogeneous Computing System

Let $a > 0$ be the size of Yee grid by Ox axis and $b > 0$ be the size of Yee grid by Oy axis. Goal is to identify virtual topology to use for computations so that the overall computational time is minimal.

Let virtual topology have size of $n > 0$ computational nodes by Ox axis and $m > 0$ computational nodes by Oy axis, $N = n * m$. 2D-X virtual topology will be used in case $n = N$ and $m = 1$, 2D-Y topology in case $n = 1$ and $m = N$, otherwise 2D-XY virtual topology will be used, where $n \neq 1$ and $m \neq 1$.

Then, single computational node will have $a_1 = \lfloor a/n \rfloor$ grid points by Ox axis in the chunk assigned to it and $b_1 = \lfloor b/m \rfloor$ grid points by Oy axis. Total size of chunk assigned to computational node is $a_1 * b_1 = \lfloor a/n \rfloor * \lfloor b/m \rfloor$ grid points. Let's consider only cases where $a \bmod n = 0$ and $b \bmod m = 0$, which leads to the size of chunk being equal to

$$a_1 * b_1 = \frac{a}{n} * \frac{b}{m} = \frac{a * b}{N} \qquad (1)$$

Computational time on single time step is proportional to number of Yee grid points in chunk of computational node $a_1 * b_1$ and share time on a single time step is proportional to the surface area of chunk $a_1 + b_1$. Number of Yee grid points in chunk is the same for all computational nodes, which means that computational time should also be the same. This is accurate in case each computational node performs same amount of computations, for example, this is not accurate if there is a point wave source with sophisticated wave function calculated only on one computational node and not calculated on others. In cases when nodes perform different amounts of computations on the same amount of grid points, dynamic information could be used and methods of solving such tasks are the same as for heterogeneous systems and are discussed later.

Thus, the minimal overall computational time could be achieved by minimizing share time on a single time step:

$$a_1 + b_1 = \frac{a}{n} + \frac{b}{m} = \frac{a}{n} + \frac{b * n}{N} = f(n) \qquad (2)$$

Function $f(n)$ has only one extremum for $n > 0$ — global minimum:

$$n_0 = \sqrt{\frac{a * N}{b}} \qquad (3)$$

$$m_0 = \frac{N}{n_0} = \frac{N}{\sqrt{\frac{a*N}{b}}} = \sqrt{\frac{b * N}{a}} \qquad (4)$$

However, obtained values n_0 and m_0 might not be integer or might not be dividers of N, and a and b correspondingly. Values of n and m, which satisfy this conditions, have to be found.

First, all pairs of n and m which satisfy next conditions have to be found: n and m are dividers of N, and a and b correspondingly, and $N = n * m$. These pairs define set, which contains possible optimal values of n and m. In order to find these pairs, n has to be set equal to all dividers of $GCD(a, N)$, including 1 and $GCD(a, N)$ itself, and only those pairs have to be chosen for which corresponding $m = N/n$ is divider of b.

After that, two pairs (n_0', m_0') and (n_0'', m_0'') have to be found, for which for n the next conditions are satisfied: $n_0' < n_0$ and $n_0'' \geq n_0$, and there are no pairs in the range $(n_0'; n_0'')$ (i.e., values n_0' and n_0'' are the closest possible to n_0 from different directions). From two pairs (n_0', m_0') and (n_0'', m_0'') one has to be chosen, for which the value of $f(n)$ is smaller. Chosen pair describes the optimal virtual topology.

So, algorithm for choosing optimal virtual topology for homogeneous computational system for defined a, b, N consists of the next steps:

- Identify all pairs (n, m), for which n is divider of a, m is divider of b, $n*m = N$.
- Find n_0 using relation (3).
- Choose from the pairs found on the first step two (n_0', m_0') and (n_0'', m_0''), for which $n_0' < n_0$ and $n_0'' \geq n_0$, and there are no pairs in the range $(n_0'; n_0'')$. From this two pairs one has to be chosen as optimal, for which the value of $f(n)$ is smaller.

There could be a case, when no appropriate pair is found (e.g., Yee grid could not be divided in chunks for N computational nodes). In this case N should be increased or decreased. Cases when $a \bmod n \neq 0$ or $b \bmod m \neq 0$ are not discussed here because they require dynamic information about computational system.

2.2 3D Mode for Homogeneous Computing System

Let $a > 0$ be the size of Yee grid by Ox axis, $b > 0$ be the size of Yee grid by Oy axis and $c > 0$ be the size of Yee grid by Oz axis.

Let virtual topology have size of $n > 0$ computational nodes by Ox axis, $m > 0$ computational nodes by Oy axis, $k > 0$ computational nodes by Oz axis, $N = n * m * k$. For example, 3D-X virtual topology will be used in case $n = N$, $m = 1$, $k = 1$.

Similarly to 2D mode, number of grid points in chunk assigned to computational node $a_1 * b_1 * c_1$ is constant (in case $a \bmod n = 0$, $b \bmod m = 0$ and $c \bmod k = 0$). In order to minimize overall computational time, the next function $f(n, m)$ has to be minimized (share time on a single time step is proportional to $f(n, m)$):

$$f(n, m) = \frac{a * b}{n * m} + \frac{b * c * n}{N} + \frac{a * c * m}{N} + 4 * \left(\frac{a}{n} + \frac{b}{m} + \frac{c * n * m}{N} \right) \quad (5)$$

Again, all triples of (n, m, k) have to be identified, for which n is divider of a, m is divider of b, k is divider of c, $n*m*k = N$. One of these triples describes the optimal virtual topology. In order to find these triples n has to be set equal to all dividers of $GCD(a, N)$, including 1 and $GCD(a, N)$ itself, then m has to be set equal to all dividers of $GCD(b, N)$, including 1 and $GCD(b, N)$ itself, and only those triples have to be chosen for which corresponding $k = N/(n*m)$ will be a divider of c.

Then for all m from found triples the minimum of function $f(n, m)$ has to be found. This leads to the next formulas for chosen m:

$$n_0(m) = \sqrt{\frac{a * N}{m * c}} \tag{6}$$

$$k_0(m) = \frac{N}{n_0(m) * m} \tag{7}$$

Similarly to 2D mode, obtained values n_0 and k_0 might not be integer or might not be dividers of N, and a and c correspondingly. n and k, which satisfy this conditions, have to be found. Two triples (n_0', m, k_0') and (n_0'', m, k_0'') have to be found, for which for n the next conditions are satisfied: $n_0' < n_0$ and $n_0'' \geq n_0$, and there are no triples in the range $(n_0'; n_0'')$ (i.e., values n_0' and n_0'' are the closest possible to n_0 from different directions). From two triples (n_0', m, k_0') and (n_0'', m, k_0'') one has to be chosen, for which the value of $f(n, m)$ is smaller. Chosen triple describes the optimal virtual topology for specified m. After all values of m are handled, triples, found for each m, have to be compared and one triple with smallest $f(n, m)$ has to be chosen as the optimal virtual topology.

So, algorithm for choosing optimal virtual topology for homogeneous computational system for defined a, b, c, N consists of the next steps:

- Identify all triples (n, m, k), for which n is divider of a, m is divider of b, k is divider of c, $n*m*k = N$.
- For all allowed values for m find $n_0(m)$ using (6).
- Choose from the triples found on the first step two (n_0', m_0', k_0') and (n_0'', m_0'', k_0''), for which $n_0' < n_0$ and $n_0'' \geq n_0$, and there are no triples in the range $(n_0'; n_0'')$. From two triples one has to be chosen as optimal for specified m, for which the value of $f(n, m)$ is smaller.
- Virtual topologies found for each m have to be compared and one triple with smallest $f(n, m)$ has to be chosen as the optimal virtual topology.

2.3 Non-specified Number of Computational Nodes for Homogeneous Computing System

In case when number of computational nodes N is not specified, there are two possible directions for optimization. First is to optimize by memory, i.e. choose the smallest number of computational nodes, memory of which is capable to store Yee grid. Second is to optimize by performance, i.e. choose the number of computational nodes in such a way that time of computations is minimal.

The second problem requires dynamic information and methods of solving such tasks are the same as for heterogeneous systems and are discussed later.

Let's consider the first problem with optimization by memory for 2D mode. Problem is to choose N so that for specified a and b the number of Yee grid points per node $\frac{ab}{N}$ is maximal.

First the size of memory of a single computational node S has to be identified. The amount of memory occupied on a single computational node is $F(a, b, N)$:

$$F(a, b, n, m) = k_1 * \left(\frac{a * b}{n * m}\right) + k_2 * \left(\frac{a}{n} + \frac{b}{m}\right) + k_3 \tag{8}$$

where s is number of Yee grid points and k_1, k_2, k_3 — constants, which are defined by parameters of computation statically (during compilation of solver). Let's perform transformation in order to remove dependency on virtual topology.

$$k_1 * \left(\frac{a * b}{n * m}\right) + k_2 * \left(\frac{a}{n} + \frac{b}{m}\right) + k_3 < k_1 * \left(\frac{a * b}{n * m}\right) + k_2 * (a + b) + k_3 \tag{9}$$

Let's consider $F_1(a, b, N)$:

$$F_1(a, b, N) = k_1 * \left(\frac{a * b}{N}\right) + k_2 * (a + b) + k_3 \tag{10}$$

N_s has to be found, for which $F_1(a, b, N)$ is equal to S for specified a and b:

$$N_s = \frac{k_1 * a * b}{S - k_3 - k_2 * (a + b)} \tag{11}$$

An answer N_0 is the first divider of $a * b$ in ascending order, which will be greater or equal to N_s: $N_0 \geq N_s$ and $F_1(a, b, N_0) \leq F_1(a, b, N_s)$. Different choices of virtual topology will not affect the found answer, because it was taken into account. For 3D mode N_0 could be obtained in a similar way.

2.4   Heterogeneous Computing System

In case of heterogeneous systems, the algorithms described above couldn't be applied because either computational time on different computational nodes is proportional to number of grid points in chunk with different proportional coefficient, or share time has different proportional coefficients for different computational nodes. In such cases more sophisticated methods should be applied. Each one of the methods described below could give results, but the best option would be to use their combination.

Dynamic Balancing (Dynamic Redistribution). Dynamic redistribution of Yee grid points between computational nodes will allow to assign chunks of different sizes for a single computational node during computations (opposed to previously described distribution before computations). This process could be triggered based on some dynamic information, i.e. computation time and share time of each computational node. Thus, computational nodes, which have higher performance, could be dynamically assigned larger chunks and computational nodes with less performance — smaller chunks. One drawback of this method is that redistribution process takes time and will affect computational time. However, benefit of this method is the ability to distribute Yee grid points between computational nodes more efficiently without the need to identify optimal virtual topology before computations.

In developed solver dynamic balancing was implemented for 1D-X, 2D-X, 2D-Y, 3D-X, 3D-Y and 3D-Z virtual topologies. Before start of computations some virtual topology has to be chosen (machine learning or saved dynamic profile could be used to choose initial virtual topology in future). Then after M time steps, during which computational time T was gathered, redistribution is performed. Let's consider 2D-X case. Total size of grid is $a * b$, chunk of grid points, assigned to i ($i \geq 0, i < N$) computational node has size $S_i = a_i * b$ and

$$\sum_i a_i = a \tag{12}$$

Performance of i computational node is calculated like this

$$perf_i = \frac{a_i * b * M}{T_i} \tag{13}$$

where T_i is computational time of i computational node for M time steps. Then, new a_i' is calculated

$$a_i' = \left[\frac{a * perf_i}{\sum_i perf_i}\right] \tag{14}$$

In case $a' = a - \sum_i a_i' > 0$, a' is spread between all computational nodes. After this procedure computational nodes will have chunks with new sizes, and distribution of computations between computational nodes will be better in terms of reduction of total computational time. This procedure could be repeated later to further improve distribution of computations.

Saving the Profiling Data. During first computation on the computational system profiling data could be saved for each computational node as a recommendation for virtual topology chooser. On the second launch, this saved data could be used to identify optimal virtual topology. Drawback of this method is the need to perform first computation with gathering of dynamic information in order to save it later. This process takes time and will affect computational time. However, starting from the second computation launch virtual topology will be chosen more optimally. Besides, even the second computation launch could also gather dynamic information and update the saved one. So, the benefit of this method is that after K computation launches with profiling $K + 1$ computation launch will use the most optimal virtual topology from all, which could be chosen according to saved dynamic information. Besides, $K + 1$ computation launch could be performed without profiling, thus, without performance degradation.

Some modification of this method is to perform benchmarking of computational system before computations, when profiling data is gathered not on some random computations but on one which is optimized to save more relevant dynamic data. Such a benchmark then has to be found. This method is to be discussed in detail in further work.

Machine Learning. Neural network could be used in this method, trained on characteristics of different computational systems and optimal virtual topologies. Then, before computation on some computational system optimal virtual topology will be found using trained neural network. Benefit of this method is the lack of need to do additional activities during computations because optimal virtual topology is identified before first computation on the system. However, drawback is that identified virtual topology could be not the most optimal even

Table 1. Measurements for 2D mode for 4 computational nodes for Yee grid with size $a = 256$ and $b = 256$ and 10000 time steps.

Virtual topology	Value of $f(n)$	Execution time, seconds
2D-X with $n = 4, m = 1$	320	3015.16
2D-XY with $n = 2, m = 2$	256	3001.69

Table 2. Measurements for 2D mode for 4 computational nodes for Yee grid with size $a = 8192$ and $b = 8$ and 10000 time steps.

Virtual topology	Value of $f(n)$	Execution time, seconds
2D-X with $n = 4, m = 1$	2056	2605.45
2D-XY with $n = 2, m = 2$	4100	3192.83

Table 3. Measurements for 3D mode for 8 computational nodes for Yee grid with size $a = 64$, $b = 64$ and $c = 64$ and 1000 time steps.

Virtual topology	Value of $f(n, m)$	Execution time, seconds
3D-X with $n = 8, m = 1, k = 1$	5664	2106.54
3D-XY with $n = 4, m = 2, k = 1$	4032	2090.29
3D-XYZ with $n = 8, m = 1, k = 1$	3456	2069.23

Table 4. Measurements for 3D mode for 8 computational nodes for Yee grid with size $a = 4096$, $b = 8$ and $c = 8$ and 1000 time steps.

Virtual topology	Value of $f(n, m)$	Execution time, seconds
3D-X with $n = 8, m = 1, k = 1$	10368	1587.31
3D-XY with $n = 4, m = 2, k = 1$	16464	1815.29
3D-XYZ with $n = 2, m = 2, k = 2$	24624	2069.98

for well trained neural network. This method is to be discussed in detail in further work.

3   Measurements

All measurements were performed on IBM Blue Gene/P supercomputer for different virtual topologies. IBM Blue Gene/P is a massively parallel computational system. It contains 8192 calculation cores (2048 calculation nodes, 4 core each) with peak performance at 27.9 tflops. It supports both MPI and OpenMP technologies. Single calculation core is a PowerPC 450 with frequency at 850 MHz having 4 GB of RAM. Communicational network is a three-dimensional torus and unites all the nodes. Single node has 6 bidirectional connections with 6 neighbors and throughput of each of these 12 connections is 425 MB/s. Blue Gene/P has GCC 4.2 compiler.

Basic FDTD computation was chosen as a benchmark (no PML, no TF/SF, point wave source for each computational node). In each computation virtual topology was mapped on computational nodes of Blue Gene/P in such a way that virtual topology matches physical topology, so, computational nodes, which are neighbors in virtual topology, will be neighbors in physical topology too, and no additional share expenses arise.

As Tables 1, 2, 3 and 4 show, the smallest computational time is achieved with the virtual topology that is optimal for current grid size, and variation of computational times for different virtual topologies could be significant, varying from 0.5% to 18.4% for 2D mode and from 1.8% to 23.3% for 3D mode for different Yee grid sizes, which could be significant for long running tasks. These results depend on the size of Yee grid and variation could be even higher for larger grids. Besides, obtained results depend heavily on the target architecture, and for architectures where share operations are heavy in terms of time, results could be even more significant.

Measurements for dynamic balancing for two computational nodes for 2D mode and 2D-X virtual topology for Yee grid with size $a = 1000$ and $b = 1000$ were performed for two cases: 0 computational node has point wave source, nodes don't have point wave sources, i.e. the difference between two cases is the calculation of wave function for 0 node. After some execution time Yee grid appeared to be divided in chunks in the next way. In the first case, 0 node had chunk with size of 47% of total Yee grid size and 1 node had chunk with size of 53% of total Yee grid size. In the second case computational nodes had chunks of the same size. This proves that dynamic balancing allows to spread computations optimally even for homogeneous architectures, if each node has to perform different amount of computations.

4   Conclusion

Developed FDTD solver provides features for optimal computations distribution between computational nodes. Measurements prove described algorithm of choosing of optimal virtual topology for homogeneous architectures and that there is no one "silver bullet" virtual topology to choose for different Yee grid sizes. This allows solver to be more efficient in terms of computational time. Besides, dynamic balancing was shown to spread computations optimally through all computational nodes for homogeneous target architectures in case of different amount of computations on each computational node. In further work dynamic balancing would be improved for both homogeneous and heterogeneous target architectures and other dynamic methods would be described in detail.

References

1. Yee, K.S.: Numerical solution of initial boundary value problems involving Maxwell's equations in isotropic media. IEEE Trans. Antennas Propag. **14**(3), 303–307 (1966)
2. Zunoubi, M.R., Payne, J., Roach, W.P.: CUDA-MPI-FDTD implementation of Maxwell's equations in general dispersive media. In: Proceedings of the SPIE, vol. 8221, id. 822115 (2012)
3. Zakirov, A.V., Levchenko, V.D., Perepelkina, A.Yu., Zempo, Y.: High performance FDTD code implementation for GPGPU supercomputers. Keldysh Institute Preprints, No. 44 (2016)
4. He, B., Tang, L., Xie, J., Wang, X., Song, A.: Parallel numerical simulations of three-dimensional electromagnetic radiation with MPI-CUDA paradigms. Math. Prob. Eng. **2015**, 1–9 (2015). Article ID 823426
5. Shams, R., Sadeghi, P.: On optimization of finite-difference time-domain (FDTD) computation on heterogeneous and GPU clusters. J. Parallel Distrib. Comput. **71**(4), 584–593 (2011)
6. Parallel FDTD solver. https://github.com/zer011b/fdtd3d
7. Taflove, A., Hagness, S.C.: Computational Electrodynamics: The Finite-Difference Timedomain Method, 3rd edn. Artech House (2000)

High Performance Architectures, Tools and Technologies

Retrospective Satellite Data in the Cloud: An Array DBMS Approach

Ramon Antonio Rodriges Zalipynis[(✉)], Anton Bryukhov, and Evgeniy Pozdeev

National Research University Higher School of Economics, Moscow, Russia
rodriges@wikience.org, asbryukhov@gmail.com, jonnypozdeev@gmail.com

Abstract. Earth remote sensing has always been a source of "big" data. Satellite data have inspired the development of *"array"* DBMS. An array DBMS processes N-dimensional (N-d) arrays utilizing a declarative query style to simplify raster data management and processing. However, raster data are traditionally stored in files, not in databases. Respective command line tools have long been developed to process these files. Most tools are feature-rich and free but optimized for a single machine. The approach of partially delegating in situ raster data processing to such tools has been recently proposed. The approach includes a new formal N-d array data model to abstract from the files and the tools as well as new distributed algorithms based on the model. This paper extends the approach with a new algorithm for the reshaping (tiling) of N-d arrays. The algorithm physically reorganizes the storage layout of N-d arrays to obtain an order of magnitude speedup. The extended approach outperforms SciDB up to 28× on retrospective Landsat data – one of the most typical and popular kind of satellite imagery. SciDB is the only freely available distributed array DBMS to date. Experiments were carried out on an 8-node cluster in Microsoft Azure Cloud.

Keywords: ChronosServer · SciDB · Raster Data · Cloud computing · Remote sensing · Array DBMS · Command Line Tools · Landsat

1 Introduction

Earth remote sensing is increasingly becoming a data-rich, practically important and commercially attractive domain. The most prominent example is the Landsat Program – the longest continuous space-based record of Earth's land in existence. The Program lasts from 1972 onwards and has accumulated over 6.8×10^6 scenes mostly in GeoTIFF files ($\approx$ 6 PB in total) [8]. Landsat data are so popular that Amazon and Google provide Landsat scenes via commercial clouds [5]. The number of practical Landsat applications is rapidly growing [7]. A retrospective time series of Landsat scenes for a particular area is of great importance since it makes it possible to track area changes that were happening over the past decades.

The file-centric model of raster data storage resulted in a broad set of highly optimized raster file formats. For example, GeoTIFF represents an effort by over

© Springer International Publishing AG 2017
V. Voevodin and S. Sobolev (Eds.): RuSCDays 2017, CCIS 793, pp. 351–362, 2017.
https://doi.org/10.1007/978-3-319-71255-0_28

160 different companies and organizations to establish interchange format for georeferenced raster imagery [6]. Decades of development and feedback resulted in numerous feature-rich, elaborate, free and quality-assured tools for processing raster files. For example, NCO (NetCDF Operators) are are under development since about 1995 [10], GDAL (Geospatial Data Abstraction Library) has over one million lines of code made by hundreds contributors [4].

An array DBMS is one of the tools to streamline raster data processing. The idea of partially delegating raster data processing to existing command line tools was first presented and proved to outperform SciDB on NetCDF data 3× to 193× on a single machine [16] and up to 1000× running both SciDB and ChronosServer on a computer cluster (Microsoft Azure Cloud) [17]. ChronosServer is the system into which the delegation ability is being integrated [15].

The formal array model and formal distributed algorithms are given in [17]. The new two-level data model was designed to uniformly represent diverse raster data types and formats, take into account the distributed context, and be independent of the underlying raster file formats at the same time [17].

The main goal of this paper is to advance the proposed delegation approach and to show its exceptional suitability for satellite data processing. ChronosServer outperforms SciDB on raw Landsat scenes. To obtain an order of magnitude speedup, a physical reorganization of the storage layout of 2-d Landsat scenes is carried out by cutting and joining them into 3-d arrays. This case is generalized and a generic reshaping algorithm is proposed to transform a set of N-d arrays with arbitrary shapes to a set of M-d arrays with a fixed shape, where $N - M \in \mathbb{Z}$. The new algorithm is useful on a "data cooking" stage to spend some time to prepare data and make further algorithms to run much faster.

In summary, the major contributions of this paper are (i) the generic N-d reshaping algorithm and (ii) an experimental evaluation of ChronosServer and SciDB on retrospective Landsat 8 data in the Cloud.

The rest of the paper is organized as follows. For the sake of completeness, Sect. 2 describes the array model, dataset model, and ChronosServer architecture [17]. Section 3 gives generic distributed algorithms for in situ processing of arbitrary N-d arrays. The algorithms are refined in order to treat NetCDF and GeoTIFF formats and delegate portions of work to NCO and GDAL tools. Section 4 presents the N-d reshaping algorithm. Section 5 gives the performance evaluation. Section 6 reviews the related work. Section 7 concludes the paper.

2   ChronosServer

2.1   ChronosServer Multidimensional Array Model

In this paper, an N-dimensional array (N-d array) is the mapping $A : D_1 \times D_2 \times \cdots \times D_N \mapsto \mathbb{T}$, where $N > 0$, $D_i = [0, l_i) \subset \mathbb{Z}$, $0 < l_i$ is a finite integer, and $\mathbb{T}$ is a numeric type (to be specific about value ranges, size in bytes, precision, etc., a C++ type according to ISO/IEC 14882 can be taken). l_i is said to be the *size* or *length* of ith dimension (in this paper, $i \in [1, N] \subset \mathbb{Z}$).

Let us denote the N-d array by

$$A\langle l_1, l_2, \ldots, l_N \rangle : \mathbb{T} \tag{1}$$

By $l_1 \times l_2 \times \cdots \times l_N$ denote the *shape* of A, by $|A|$ denote the *size* of A such that $|A| = \prod_i l_i$. A *cell* or *element* value of A with integer indexes $(x_1, x_2, \ldots, x_N)$ is referred to as $A[x_1, x_2, \ldots, x_N]$, where $x_i \in D_i$. Each cell value of A is of type $\mathbb{T}$. The array may be initialized after its definition by enumerating the values of the cells. For example, the following defines and initializes a 2-d array of integers: $A\langle 2, 2 \rangle : int = \{\{1, 2\}, \{3, 4\}\}$. In this example, $A[0, 0] = 1$, $A[1, 0] = 3$, $|A| = 4$, and the shape of A is 2×2.

Indexes x_i are optionally mapped to specific values of ith dimension by *coordinate* arrays $A.d_i\langle l_i \rangle : \mathbb{T}_i$, where $\mathbb{T}_i$ is a totally ordered set, and $d_i[j] < d_i[j+1]$ for all $j \in D_i$. In this case, A is defined as

$$A(d_1, d_2, \ldots, d_N) : \mathbb{T} \tag{2}$$

A *hyperslab* $A' \sqsubseteq A$ is an N-d subarray of A. The hyperslab A' is defined by the notation

$$A[b_1 : e_1, \ldots, b_N : e_N] = A'(d'_1, \ldots, d'_N) \tag{3}$$

where $b_i, e_i \in \mathbb{Z}$, $0 \leqslant b_i \leqslant e_i < l_i$, $d'_i = d_i[b_i : e_i]$, $|d'_i| = e_i - b_i + 1$, and for all $y_i \in [0, e_i - b_i]$ the following holds

$$A'[y_1, \ldots, y_N] = A[y_1 + b_1, \ldots, y_N + b_N] \tag{4a}$$
$$d'_i[y_i] = d_i[y_i + b_i] \tag{4b}$$

Equations (4a) and (4b) state that A and A' have a common coordinate subspace over which cell values of A and A' coincide.

2.2   ChronosServer Datasets

A *dataset* $\mathbb{D} = (A, M, P)$ contains a *user-* or *higher-level* array $A(d_1, \ldots, d_N) : \mathbb{T}$ and the set of *system-* or *lower-level* arrays $P = \{(A_k, B_k, E_k, M_k, node_k)\}$, where $A_k \sqsubseteq A$, $k \in \mathbb{N}$, $node_k$ is an identifier of the cluster node storing array A_k, M_k is metadata for A_k, $B\langle N \rangle : int = \{b_1, b_2, \ldots, b_N\}$, $E\langle N \rangle : int = \{e_1, e_2, \ldots, e_N\}$ such that $A_k = A[b_1 : e_1, \ldots, b_N : e_N]$. A user-level array is never materialized and stored explicitly: an operation with A is mapped to a sequence of operations with respective arrays A_k. Let us call a user-level array and a system-level array an array and a subarray respectively for short. A dataset also contains metadata $M = \{(key, val)\}$, where key is a string and val is a string or a number. Dataset metadata includes two types of information: general dataset properties (name, description, contacts, etc.) and metadata valid for all $p \in P$ (array data type $\mathbb{T}$, storage format, etc.). For example, $M = \{(name = $ "Landsat 8 Band 1"$), (type = int16), (format = GeoTIFF)\}$. Let us refer to an element in a tuple $p = (A_k, B_k, \ldots) \in P$ as $p.A$ for A_k, $p.B$ for B_k, etc. Example of a subarray metadata $p.M = \{(key, val)\}$ is $p.M = \{(date = $ "2016-Aug-08"$, bounding_box = $ "WKT$(\ldots)$"$, projection = $ "EPSG:32637"$)\}$.

2.3   ChronosServer Architecture

ChronosServer runs on a computer cluster of commodity hardware. Files are distributed among cluster nodes without changing their names and formats. A file is always stored entirely on a node in contrast to parallel or distributed file systems. Workers run on each node and are responsible for data processing. One Gate at a dedicated node receives client queries and coordinates workers. A file may be replicated on several nodes for fault tolerance and load balancing.

Gate stores metadata for all datasets and subarrays. Consider a dataset $\mathbb{D} = (A, M, P)$. Arrays $A.d_i$ and elements of $\forall p \in P$ except $p.A$ are stored on Gate. In practice, array axes usually have coordinates such that $A.d_i[j] = start + j \times step$, where $j \in [0, |A.d_i|) \subset \mathbb{N}$, $start, step \in \mathbb{R}$. Only $|A.d_i|$, $start$, and $step$ values have to be usually stored. ChronosServer array model merit is that it has been designed to be generic as much as possible but allowing to establish 1:1 mapping of a $p \in P$ to a physical dataset file at the same time.

Upon startup workers connect to Gate and receive a list of all available datasets and file naming rules. Workers scan their local filesystems to discover datasets and create $p.M$, $p.B$, $p.E$ by parsing file names or reading file metadata. Workers transmit to Gate the described information.

A user-level array may have a *virtual* dimension. Values for virtual dimensions are taken from the subarrays metadata. For example, Landsat files are 2-d arrays $A(lat, lon)$ without a temporal axis. Virtual axis "time" in $A(time, lat, lon)$ may contain scenes acquisition dates extracted from the file names. This makes it possible to treat a set of Landsat scenes as a 3-d array.

3   Array Operations

3.1   Aggregation

The aggregate of an N-d array $A(d_1, d_2, \ldots, d_N):\mathbb{T}$ over axis d_1 is the $(N-1)$-d array $A_{aggr}(d_2, \ldots, d_N) : \mathbb{T}$ such that $A_{aggr}[x_2, \ldots, x_N] = f_{aggr}(cells(A[0 : |d_1| - 1, x_2, \ldots, x_N]))$, where $x_2, \ldots, x_N$ are valid integer indexes, $f_{aggr} : T \mapsto w$ is an aggregation function, T is a multiset of values from $\mathbb{T}$, $w \in \mathbb{T}$, $cells : A' \mapsto T$ is the multiset of all cell values of an array $A' \sqsubseteq A$.

Algorithm 1 performs aggregation of system-level arrays.

Algorithm 1. Distributed in situ array aggregation with delegation to an external command line tool (procedure AGGREGATE is executed on workers).

Input: wid is the identifier of the worker performing final aggregation

```
 1: procedure AGGREGATE(𝔻, f_aggr, wid)              ▷ 𝔻 is a dataset, see section 2.2
 2:     aggregate all p ∈ P residing on this worker into p'        ▷ DELEGATION
 3:     if the id of this worker equals to wid then
 4:         accept subarrays from other workers: P_aggr
 5:         aggregate p' and all p ∈ P_aggr into p_aggr
 6:         report success to Gate
 7:     else send p' to worker with id = wid
```

The generic aggregation Algorithm 1 is based on the dataset model from Sect. 2.2 and takes into account that system-level arrays may overlap and cover the N-d space irregularly (e.g., scenes following a riverbed). Also, Landsat scenes for the same path and row may be shifted relatively to each other (it is hard to capture precisely the same area each time).

Algorithm 1 is designed for $f_{aggr} \in \{max, min, sum\}$. The basic idea is that workers aggregate in parallel all subarrays residing locally into one subarray and send it to a worker calculating the final result. Calculating $mean$ is reduced to calculating the sum and dividing the result onto the number of participating subarrays. The Gate is responsible for calculating this number and sending it to the worker performing final aggregation (not shown in the algorithm).

In Sect. 5, we always set wid to the largest possible. Array p' on Line 2 may grow large in volume and require splitting. We leave this case for future work. Line 2 is highlighted with light gray to accent the work being delegated to an external tool: gdal_calc.py for GeoTIFF format and an NCO tool (ncra, ncwa, or ncap2 depending on file structure) for NetCDF format.

3.2 Chunking

Chunking is the process of partitioning original array onto a set of smaller subarrays called chunks. Chunks are autonomous, possibly compressed subarrays (hyperslabs) with contiguous storage layout. Given chunk shape $c_1 \times \cdots \times c_N$ and an N-d array $A(d_1, \ldots, d_N) : \mathbb{T}$, the *exact chunking* operation reorganizes cells in array A such that all cells of A with coordinates $(x_1, \ldots, x_N)$ and $(y_1, \ldots, y_N)$ belong to the same chunk if x_i div $c_i = y_i$ div c_i for all i. Due to space constraints, please, find the illustration and benefits of chunking in [16].

The exact chunking of an array may lead to data movement between files and cluster nodes. However, in practice the condition $c_i \ll |A.d_i|$ usually holds. This translates to $c_i \ll |p.A.d_i|$ for $\forall p \in P$ (in practice, raster data are already shipped in wisely cut files satisfying this condition). For example, in climate modeling it is common to split a time series with hourly time step onto files storing yearly or monthly data.

A good practical approach is to do inexact user-level array chunking and exact independent chunking of its subarrays. More chunks will smaller shapes than the given one will appear. However, the fraction of such small chunks will be negligible and they will not influence significantly the I/O performance. Note that if $|A.d_i|$ mod $c_i \neq 0$, then even the exact chunking of a user-level array is not possible leading to a certain amount of chunks with smaller shapes.

In practice, inexact chunking is even more desirable in many cases: it is much faster and more consistent than the exact chunking. Recall that files under ChronosServer control are directly accessible by a user and any other software. Consider the climate modeling example given above. In this case, it is inconsistent to have a perfectly chunked file named "2015.nc" and supposed to store data for year 2015 but with extra grids from the next and/or previous years.

Chunking is delegated to gdal_translate (GDAL) and ncks (NCO) for GeoTIFF and NetCDF file formats respectively.

4   Generic Reshaping Algorithm

This section presents an algorithm to reshape system-level arrays. The algorithm is useful on a "data cooking" stage to spend some time to reshape the subarrays to speedup further raster operations.

It is costly to aggregate and/or hyperslab large number of files. For example, aggregating a time series of 2-d scenes makes the hidden asymptotic constants quite noticeable. Changing position of a virtual axis during reshaping requires complex data moves between files (reshaping operation is defined in [17]). Chunking along a virtual axis $A.v_i$ could be implemented as co-locating $p_1, p_2 \in P$ on a single machine such that $p_1.v_i[x]$ and $p_2.v_i[y]$ are in the same chunk. This "virtual" chunking will not speedup the I/O as chunking of a single physical file.

The reshaping algorithm overcomes these limitations. For example, initial shape of raw Landsat files (system-level arrays) is $1 \times L_1 \times L_2$ ($time \times lat \times lon$), where $L_1 \approx L_2 \approx 8000$. Reshaping subarrays, say, to $5 \times L_1/4 \times L_2/4$ will accelerate aggregation and hyperslabbing (hyperslabbing is defined in [17]). In this case, virtual $time$ dimension will become a regular physical dimension and will explicitly present in dataset files. This will make it possible to delegate dimension permutation of Landsat 8 scenes (please, refer to [17] for details on dimension permutation) as well as chunking (Sect. 3.2) to an external tool.

Algorithm 2 takes as input a set of N-d subarrays P with arbitrary shapes and produces a set of subarrays P' such that $\forall p' \in P'$ has shape $s_1 \times s_2 \times \cdots \times s_N$ (except border cases) and sides of p' are parallel to the coordinate axes.

Given $M \neq N$, Algorithm 2 reshapes subarrays from N-d to M-d form using virtual axes supported by ChronosServer data model. A virtual axis can be made a physical one to get subarrays with greater physical dimensionality $M > N$ (the case with Landsat scenes described above). An axis can be made virtual and deleted from the files if the axis has a unit length in all resulting subarrays.

This will produce subarrays with lower physical dimensionality, i.e. $M < N$.

The basic idea is to cut each $p \in P$ onto smaller pieces $P' = \{p' : p' \sqsubseteq p\}$, assign each piece a key, and merge all pieces with the same key into a single, new system-level array. For $x \in \mathbb{N}$, $lag(x)$ is defined below.

$$lag(x) = \begin{cases} 0, & \text{if } x = 0 \\ x - 1, & \text{if } x \geqslant 1 \end{cases} \tag{5}$$

The RESHAPE function of Algorithm 2 implements the idea outlined above. Each subarray is cut independently by CUT-ONE procedure, line 3. Set $\mathbb{K}$ collects N-tuples which are N-d keys of cut pieces collected in $\mathbb{C}$. Pieces with the same key are merged into a single subarray on lines 4–7. Line 6 is highlighted with light gray since merging of files is possible to delegate to an external tool.

Algorithm 2 is best illustrated on a 2-d case. Consider a 2-d array $A(lat, lon)$, Fig. 1a. Array A has shape 10×15 and consists of 6 subarrays separated by thick blue lines. The reshaping produces 2-d subarrays with shape 3×3, $s_1 = s_2 = 3$.

Resulting subarrays P' are separated with dashed red lines. The hatched area marks one of the resulting subarrays $A[3\!:\!5, 3\!:\!5]$.

Algorithm 2. Generic Reshaping.

Input: $\mathbb{D} = (A, M, P)$ ▷ Dataset, section 2.2
   $S = (s_1, s_2, \dots, s_N)$ ▷ Target shape for arrays from P is $s_1 \times s_2 \times \cdots \times s_N$
   $\mu = (\mu_1, \mu_2, \dots, \mu_N)$ ▷ Shift, fig. 1b
Output: $\mathbb{D}' = (A, M, P')$ ▷ A is the same, $\forall p' \in P'$ shape is $s_1 \times s_2 \times \cdots \times s_N$
                                                    ▷ (except border cases)
Require: $s_i \in [1, \Theta_{axis}] \subset \mathbb{N}$,   ▷ Resulting pieces are not too large
   $\prod_{i=1}^{N} s_i \leqslant \Theta_{shape}$, $\mu_i \in [0, s_i - 1] \subset \mathbb{N}$
1: **function** RESHAPE$(\mathbb{D}, S, \mu)$
2:    $\mathbb{C} \leftarrow \{\}$ and $\mathbb{K} \leftarrow \{\}$ ▷ $\mathbb{C}$: line 22, $\mathbb{K}$: line 23 of **procedure** CUT-ONE
3:    **for each** $p \in P$ **do** CUT-ONE$(\mathbb{C}, \mathbb{K}, p, S, \mu)$
4:    **for each** $key \in \mathbb{K}$ **do**
5:      $C \leftarrow \{a \in \mathbb{C} : a.key = key\}$
6:      $p_{new} \leftarrow$ merge all $a \in C$ given $a.B_{new}$ and $a.E_{new}$ ▷ DELEGATION
7:      $P' \leftarrow P' \cup \{p_{new}\}$
8:    **return** $\mathbb{D}' = (A, M, P')$

1: **procedure** CUT-ONE$(\mathbb{C}, \mathbb{K}, p, S, \mu)$
2:    **if** $b_i \geqslant \mu_i$ **then**   ▷ $b_i = p.B[i], e_i = p.E[i]$
3:      $\eta_i \leftarrow s_i - ((b_i - \mu_i) \bmod s_i)$
4:    **else**
5:      $\eta_i \leftarrow \mu_i - b_i$
6:    **if** $\eta_i = 0$ **then** $\eta_i \leftarrow s_i$
7:    **if** $\eta_i > e_i - b_i$ **then**
8:      $x_i \leftarrow 1$
9:    **else**
10:      $x_i \leftarrow (e_i - b_i + 1 - \eta_i) \text{ div } s_i + sgn((e_i - b_i + 1 - \eta_i) \bmod s_i) + 1$
11:    **for each** $y_i \in [0, x_i - 1] \subset \mathbb{N}$ **do**
12:      $b'_i \leftarrow s_i \times lag(y_i) + \eta_i \times sgn(y_i)$ ▷ $0 \leqslant b'_i \leqslant e'_i < |p.d_i|$
13:      $e'_i \leftarrow min(s_i \times y_i + \eta_i - 1, e_i - b_i)$ ▷ b'_i, e'_i are local indexes within p
14:      $p' \leftarrow p[b'_1 : e'_1, b'_2 : e'_2, \dots, b'_N : e'_N]$ ▷ DELEGATION
15:      **if** $b'_i + b_i > \mu_i$ **then**
16:        $k_i \leftarrow (b'_i + b_i - \mu_i) \text{ div } s_i + sgn(\mu_i)$
17:      **else**
18:        $k_i \leftarrow 0$ ▷ $k_i \geqslant 0$
19:      $key \leftarrow (k_1, k_2, \dots, k_N)$ ▷ $key \subset \mathbb{Z}_{\geqslant 0}^N$
20:      $B_{new}\langle N \rangle : int = \{b_1 + b'_1, b_2 + b'_2, \dots, b_N + b'_N\}$ ▷ global indexes for p'
21:      $E_{new}\langle N \rangle : int = \{e_1 + e'_1, e_2 + e'_2, \dots, e_N + e'_N\}$ ▷ within $A.d_i$
22:      $\mathbb{C} \leftarrow \mathbb{C} \cup \{(p', key, B_{new}, E_{new})\}$ ▷ $\mathbb{C}$ is the set of cuts from all $p \in P$
23:      $\mathbb{K} \leftarrow \mathbb{K} \cup \{key\}$ ▷ $\mathbb{K}$ is the set of all generated merge keys

---

Subarray $A[5\!:\!9, 5\!:\!9]$ will be cut on 9 pieces separated by the red lines: $A[5,5]$, $A[5, 6\!:\!8]$, $A[5, 9]$ and so on. Each of them will be assigned a 2-d key. Resulting subarray $A[3 : 5, 3 : 5]$ will be assembled from 4 pieces cut from $A[5 : 9, 0 : 4]$, $A[5\!:\!9, 5\!:\!9]$, $A[0\!:\!4, 0\!:\!4]$, and $A[0\!:\!4, 5\!:\!9]$. These 4 pieces are $A[5, 5]$, $A[5, 3\!:\!4]$, $A[3\!:\!4, 3\!:\!4]$, and $A[3\!:\!4, 5]$. All 4 pieces will have key $(1, 1)$.

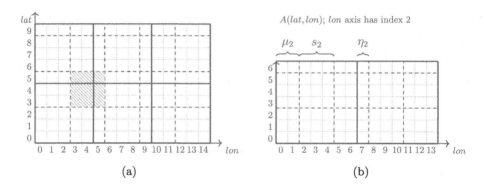

Fig. 1. Reshaping system-level arrays. (Color figure online)

Algorithm 2 accepts a shift μ_i for the ith axis in order to start cutting with an "indent", Fig. 1b. Resulting subarrays containing a cell with a zero coordinate will have shape $s'_1 \times s'_2 \times \cdots \times s'_N$, where $s'_i = \mu_i$ if $\mu_i \neq 0$; $s'_i = s_i$ otherwise.

Lines 2–6 of the CUT-ONE procedure calculate η_i which is a "local indent" within the current subarray p along the ith axis: η_i cells from p along the ith axis must go to the resulting subarray contained in p and other source subarrays bordering with p. Then, x_i is found which is the number of pieces to be cut along the ith axis. Thus, $\prod_i x_i$ is the total number of pieces to be cut from p. Loop on lines 11–23 cuts one piece at a time. Lines 12–13 find indexes within p to cut a piece on line 14 by the delegation to an external tool. The piece is assigned the key which is an N-d 0-based index of the resulting subarray to which the piece belongs. The ith tuple element of an N-d key is the index along the ith axis.

5   Performance Evaluation

Microsoft Azure Cloud was taken for the experiments. Azure cluster creation, scaling up and down with given network parameters, number of virtual machines, etc. was fully automated using Java Azure SDK [17]. The latest version of Ubuntu Linux on which SciDB 16.9 runs is 14.04 LTS. We rented standard D2 v2 machines with 2 CPU cores (Intel Xeon E5-2673 v3 (Haswell) 2.4 GHz), 7 GB RAM, 100 GB local SSD drive (4 virtual data disks), max 4×500 IOPS. Although Azure states the disk to be SSD, after the creation of such a disk Azure displays the disk to be a standard HDD disk backed by a magnetic drive.

We selected band 1 from nine Landsat 8 scenes for path 190 and row 31 ($\approx$585 MB in total) such that the cloud cover percent for the majority of scenes is less than 20%. We evaluated the latest SciDB version 16.9 released in November, 2016. We have written a Java program that converts GeoTIFF files to CSV files to feed the latter to SciDB. To date, this is the only way to import an external file into SciDB 16.9. We aligned all scenes in UTM coordinates since they are slightly shifted relatively to each other and imported the scenes into a SciDB array with shape $9 \times 7971 \times 7941$. We filled the cells with NULL values for areas in some scenes that appeared in the result of extension of that scenes during their alignment. Import time of one Landsat 8 scene into SciDB takes about **40 min** on a cloud node, not local machine. Therefore, we imported all 9 scenes on a local computer, exported the resulting SciDB array into a file of proprietary SciDB binary format, and copied that file in the Cloud when needed (SciDB imports data from its proprietary format much faster).

Cluster in order to deploy it on a cluster. SciDB is mostly written in C++, parameters used: 0 redundancy, 2 instances per machine, 5 execution and prefetch threads, 1 prefetch queue size, 1 operator threads, 1024 MB array cache, etc. ChronosServer has 100% Java code, ran one worker per node, OracleJDK 1.8.0_111 64 bit, max heap size 978 MB (-Xmx). We used NCO and GDAL tools available from the standard Ubuntu 14.04 repository. NCO v4.4.2, last modified 2014/02/17. GDAL v1.10.1, released 2013/08/26.

We have evaluated cold query runs (a query is executed for the first time). Every runtime reported is the average of 3 runtimes of the same query. Respective OS commands were issued to free `pagecache`, `dentries` and `inodes` each time before executing a cold query to prevent data caching at various OS levels. ChronosServer benefits from native OS caching and is much faster during hot runs when the same query is executed for the second time on the same data. There is no significant runtime difference between cold and hot SciDB runs.

To increase the data volume and to avoid waiting for loading more scenes, we attached SciDB array to itself to get the time dimension of size 18. We could not attach the resulting array to itself again. We tried in many ways including array import with different chunk shapes but SciDB had been always failing with **not enough memory** error. As of 29-May-2017, we did not receive any feedback from SciDB developers on this issue [11]. The same errors prevented us to measure SciDB chunking performance (Sect. 3.2). Chunking is one of the slowest SciDB queries even on small arrays [16]. We replicated 9 scenes to get 18 scenes and placed them by 2–3 on each node for ChronosServer.

Table 1 summarizes ChronosServer performance on raw and preprocessed Landsat scenes as well as SciDB performance with automatically chosen chunk shape for the SciDB array. Given array $A(time, lat, lon)$, "cut $m \times n$" means extracting a hyperslab $A[0 : |time| - 1, x_1 : x_1 + m, x_2 : x_2 + m]$, where x_1, x_2 are random indexes for array A. Line "Time series" reports hyperslabbing a time series for a single point $A[0 : |time| - 1, x_1, x_2]$. Hyperslabbing is an extraction of a hyperslab from an array. "Chunk" lines report chunking of A (for raw data, $time$ is a virtual axis and its chunk size equals to 1).

Table 2 shows the time for "cooking" Landsat 8 scenes for further speedup of the queries. Algorithm from Sect. 4 is implemented in a serial mode: all 18 scenes were processed on a single node. Future work includes assigning a set of keys to a node which will merge all cut subarrays having the given keys.

Table 1. Performance for 18 scenes, 8 cluster nodes

Operation	Time, sec.			Ratio,
	ChronosServer (raw data)	ChronosServer ("cooked" data)	SciDB	SciDB/Chronos
Average	38.36	8.12	230.74	6.02	28.42
Maximum	38.83	4.56	127.71	3.29	28.00
Minimum	38.98	4.63	125.70	3.22	27.15
Cut 512×512	1.79	1.01	1.98	1.11	1.96
Cut 1024×1024	3.34	2.14	3.41	1.02	1.59
Time series	0.53	0.31	0.84	1.58	2.71
Chunk $1 \times 64 \times 64$	22.37	—	—	—
Chunk $1 \times 128 \times 128$	22.49	—	—	—

Table 2. Preprocessing Landsat data (Sect. 4): 18 scenes, 1 cluster node

Target shape	Time, sec.	Target shape	Time, sec.
$4 \times 512 \times 512$	410.25	$9 \times 1024 \times 1024$	216.62
$9 \times 512 \times 512$	376.32	$4 \times 4096 \times 4096$	56.87
$4 \times 1024 \times 1024$	187.41	$9 \times 4096 \times 4096$	55.45

6   Related Work

Numerous techniques exist for remote sensing data processing. This work is novel because it is in the context of array DBMS research field. Four modern raster data management trends are relevant to this paper: industrial raster data models, formal array models and algebras, in situ data processing algorithms, and raster (array) DBMS. Good survey of the algorithms is in [3]. A recent survey of array DBMS and similar systems is in [16]. It is worth mentioning SciDB [18], Oracle Spatial [12], ArcGIS IS [1], RasDaMan [14], Intel TileDB [19], and PostGIS [13].

A recent survey on the array models and algebras as well as industry standard data models is in [17]. Work [17] outlines the peculiar features and merits of ChronosServer data model. It is shown that the most popular array models and algebras can be mapped to Array Algebra [2]. Industry data models are also mappable to each other [9]. SciDB does not have a formal description of its data model. SciDB neither allows array dimensions to be of temporal or spatial types making it difficult or sometimes impossible to process many real-world datasets.

7   Conclusions

ChronosServer delegates portions of raster data processing to feature-rich and highly optimized command line tools. This makes ChronosServer run much faster than SciDB. ChronosServer is up to 6× faster on raw Landsat 8 scenes than SciDB on its native storage (the same Landsat 8 scenes imported into SciDB). ChronosServer is up to 28× faster than SciDB after preprocessing the scenes which takes 105× to 780× less time than SciDB import.

Future work includes designing a distributed version of the reshaping algorithm proposed in this paper. It could be also beneficial to incorporate fault-tolerance during the reshaping once it will be parallelized.

Acknowledgments. This work was partially supported by Russian Foundation for Basic Research (grant №16-37-00416). We also thank anonymous reviewers for their helpful and inspiring comments.
Contributions. Rodriges: all text, figures, algorithms, ChronosServer, its data model, Azure management code, SciDB import code, experimental setup. Pozdeev: SciDB cluster deployment. Bryukhov: partial implementation of the reshaping algorithm for one machine, adapted SciDB import code to Landsat data. All authors: experiments.

References

1. ArcGIS for server—Image Extension. http://www.esri.com/software/arcgis/arcgisserver/extensions/image-extension
2. Baumann, P., Holsten, S.: A comparative analysis of array models for databases. Int. J. Database Theory Appl. **5**(1), 89–120 (2012)
3. Blanas, S., Wu, K., Byna, S., Dong, B., Shoshani, A.: Parallel data analysis directly on scientific file formats. In: ACM SIGMOD 2014, pp. 385–396 (2014)
4. Coverity scan: GDAL. https://scan.coverity.com/projects/gdal
5. Earth on AWS. https://aws.amazon.com/earth/
6. GeoTIFF. http://trac.osgeo.org/geotiff/
7. Landsat apps. https://aws.amazon.com/blogs/aws/start-using-landsat-on-aws/
8. Landsat project statistics. https://landsat.usgs.gov/landsat-project-statistics
9. Nativi, S., Caron, J., Domenico, B., Bigagli, L.: Unidata's common data model mapping to the ISO 19123 data model. Earth Sci. Inform. **1**, 59–78 (2008)
10. NCO homepage. http://nco.sourceforge.net/
11. Not enough memory error - SciDB forum. http://forum.paradigm4.com/t/problem-with-memory-while-stacking-array/1838
12. Oracle spatial and graph. http://www.oracle.com/technetwork/database/options/spatialandgraph/overview/index.html
13. PostGIS raster data management. http://postgis.net/docs/manual-2.2/using_raster_dataman.html
14. RasDaMan homepage. http://rasdaman.org/
15. Rodriges Zalipynis, R.A.: Chronosserver: real-time access to "native" multi-terabyte retrospective data warehouse by thousands of concurrent clients. Inform. Cybern. Comput. Eng. **14**(188), 151–161 (2011)
16. Rodriges Zalipynis, R.A.: ChronosServer: fast in situ processing of large multidimensional arrays with command line tools. In: Voevodin, V., Sobolev, S. (eds.) RuSCDays 2016. CCIS, vol. 687, pp. 27–40. Springer, Cham (2016). https://doi.org/10.1007/978-3-319-55669-7_3

17. Rodriges Zalipynis, R.A.: Distributed in situ processing of big raster data in the cloud. In: Perspectives of System Informatics - 11th International Andrei Ershov Informatics Conference, PSI 2017, Moscow, Russia, June 27–29, 2017, Revised Selected Papers. Lecture Notes in Computer Science, LNCS. Springer (2017, in press)
18. SciDB homepage. http://www.paradigm4.com/
19. TileDB. http://istc-bigdata.org/tiledb/index.html

The Architecture of Specialized GPU Clusters Used for Solving the Inverse Problems of 3D Low-Frequency Ultrasonic Tomography

Alexander Goncharsky and Sergey Seryozhnikov[✉]

Lomonosov Moscow State University, Moscow, Russia
gonchar@srcc.msu.ru, s2110sj@gmail.com

Abstract. This paper is dedicated to the development of the architecture of specialized GPU clusters that can be used as computing systems in medical ultrasound tomographic facilities that are currently being developed. The inverse problem of ultrasonic tomography is formulated as a coefficient inverse problem for a hyperbolic equation. An approximate solution is constructed using an iterative process of minimizing the residual functional between the measured and simulated wave fields. The algorithms used to solve the inverse problem are optimized for a GPU. The requirements for the architecture of a GPU cluster are formulated. The proposed architecture accelerates the reconstruction of ultrasonic tomographic images by 1000 times compared to what is achieved by a personal computer.

Keywords: Ultrasonic tomography · Coefficient inverse problems · Finite-difference time-domain (FDTD) method · GPU clusters · Medical imaging

1 Introduction

This paper focuses on using specialized supercomputers for medical ultrasonic tomography imaging. The primary application is the differential diagnosis of breast cancer. The development of ultrasonic tomography devices is currently at the prototype stage [1–3]. One of the most difficult problems in designing ultrasonic tomographic scanners is that the inverse problems of high-resolution wave tomography are nonlinear and have a very large number of unknowns — up to 10^8. The experimental data gathered in one examination amounts to approximately 5 GB. Solving such problems using precise mathematical models that take into account the diffraction, refraction and absorption of ultrasonic waves can be carried out only with the help of powerful modern supercomputers.

However, a general-purpose supercomputer cannot be included as a part of a tomographic setup. The aim of this study is to develop the architecture of specialized supercomputers for medical ultrasonic tomography. A specialized supercomputer should have an energy consumption not exceeding 10–20 kW and should fit into a single rack. This specialized supercomputer should be optimized for the most effective implementation of the iterative gradient method developed in the authors' previous works [4–8]. Preliminary studies have shown that the optimal choice for this task is a GPU cluster. Using

© Springer International Publishing AG 2017
V. Voevodin and S. Sobolev (Eds.): RuSCDays 2017, CCIS 793, pp. 363–375, 2017.
https://doi.org/10.1007/978-3-319-71255-0_29

modern hardware, it is possible to design a specialized GPU cluster, which can be included in an ultrasonic tomography facility.

2 Formulation of the Inverse Problem of Ultrasonic Tomography

A simple mathematical model that takes into account the ultrasound diffraction and absorption effects is a scalar wave model based on a second-order hyperbolic equation. In this model, the acoustic pressure $u(r, t)$ satisfies the equation:

$$c(r)u_{tt}(r, t) + a(r)u_t(r, t) - \Delta u(r, t) = \delta(r - q) \cdot f(t), \qquad (1)$$

$$u(r, t = 0) = 0, \quad u_t(r, t = 0) = 0, \partial_n u(r, t)|_{ST} = p(r, t). \qquad (2)$$

Here, $c(r) = 1/v^2(r)$, where $v(r)$ is the speed of sound in the medium, $r \in \mathbf{R}^3$ is the point in space, $a(r)$ is the absorption coefficient, and Δ is the Laplace operator with respect to r. The sounding pulse generated by the point source at q is described by the function $f(t)$; $\partial_n u(r, t)|_{ST}$ is the derivative along the normal to the surface S of the domain Ω, where $(r, t) \in S \times (0, T)$; the function $p(r, t)$ is known. The conditions (2) represent the boundary and initial conditions. It is assumed that $v(r) = v_0 = const, a(r) = 0$ outside of the studied object. This simple model of wave propagation (1) can be used to describe ultrasonic waves in soft tissues.

Figure 1 shows a typical scheme of the tomographic experiment. The studied object G is located inside the region Ω. For simplicity, we assume that the domain Ω is a cube of height H. The free space L is filled with water with a known sound speed v_0. The sources are located on the boundary S of the domain Ω in several planes: h_1, h_2, h_3. The detectors can be located on the side and bottom faces of the cube Ω. In ultrasound mammography applications, sources and detectors cannot be located on the upper side; thus, this is an incomplete-data tomography problem [6].

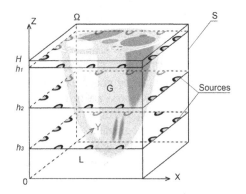
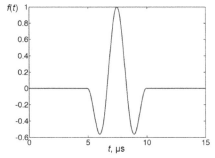

Fig. 1. The scheme of the experiment.   **Fig. 2.** Waveform of the sounding pulse.

Let us consider the inverse problem of reconstructing the unknown coefficients $c(r)$ and $a(r)$ in Eq. (1), given that the acoustic pressure $U(s, t)$ is measured at the points s of the boundary S for the time interval $(0; T)$. The value of T is chosen to be large enough ($\sim$250 µs) so that all the waves passing through and reflected from the object are registered by the detectors. The measurements are performed for source positions q. Figure 2 shows a typical waveform of sounding pulses $f(t)$ emitted by the sources. For low-frequency acoustic tomography in the 500 kHz band, the duration of the pulses is 3–10 µs. The low-frequency approach allows to use a much smaller number of sources but requires precise measurements and a precise mathematical model [5].

The exact solution of the inverse problem includes the coefficients $c(r)$, $a(r)$, which, when substituted into Eqs. (1)–(2), produce the wave field $u(r, t)$ equal to the measured wave field $U(s, t)$ at the detector points s. Because the inverse problem is ill-posed, we formulate it as a problem of minimizing the residual functional with respect to its argument (c, a):

$$\Phi\left(u(c, a)\right) = \frac{1}{2} \int_0^T \int_S \left(u(s, t) - U(s, t)\right)^2 ds\,dt. \tag{3}$$

Here, $U(s, t)$ is the acoustic pressure measured on the boundary S for the time interval $(0, T)$; $u(r, t)$ is the solution of the direct problem (1)–(2) for the given $c(r)$ and $a(r)$.

We use the gradient method to minimize the residual functional (3). Representations of the gradient $\Phi'(u(c,a))$ in various formulations were obtained in [7, 8]. In [9] and [10], expressions for the gradient in the time-domain formulation were derived. The gradient $\Phi'(u(c, a)) = \{\Phi'_c(u), \Phi'_a(u)\}$, representing the linear part of the increment of the functional $\Phi(u(c,a))$ (3) with respect to the variation of the sound speed and the absorption coefficient $\{dc, da\}$, has the form:

$$\Phi'_c(u(c)) = \int_0^T w_t(r, t)u_t(r, t)dt, \quad \Phi'_a(u(a)) = \int_0^T w_t(r, t)u(r, t)dt. \tag{4}$$

Here, $u(r, t)$ is the solution of the main problem (1)–(2), and $w(r, t)$ is the solution of some "conjugate" problem for the given $c(r)$, $a(r)$ and $u(r, t)$:

$$c(r)w_{tt}(r, t) - a(r)w_t(r, t) - \Delta w(r, t) = 0, \tag{5}$$

$$w(r, t = T) = 0, \quad w_t(r, t = T) = 0, \quad \partial_n w|_{ST} = u|_{ST} - U. \tag{6}$$

At the points of the boundary S where no measured data are present, the boundary condition $\partial_n w|_{ST} = 0$ is applied. To calculate the gradient $\Phi' = \{\Phi'_c(u), \Phi'_a(u)\}$ using formula (4), it is necessary to solve the direct problem (1)–(2) and the "conjugate" problem (5)–(6). With the calculated gradient, we can use various iterative algorithms to minimize the residual functional (4).

3 Numerical Algorithms for Solving Inverse Problems of Low-Frequency Ultrasonic Tomography

We use the finite-difference time-domain (FDTD) method to solve problems (1)–(2) and (5)–(6) numerically. Let us introduce a uniform discrete grid for the spatial coordinates (x, y, z) at the time t: $x_i = ih$, $y_j = jh$, $z_l = lh$, $t_k = k\tau$; $i, j, l = 1,...,N$, $k = 1,...,M$, where h is the grid step, and τ is the time step. To approximate Eq. (1), we use the following second-order finite-difference scheme:

$$c_{ijl} \frac{u_{ijl}^{k+1} - 2u_{ijl}^k + u_{ijl}^{k-1}}{\tau^2} + a_{ijl} \frac{u_{ijl}^{k+1} - u_{ijl}^{k-1}}{\tau} - \frac{\Delta u_{ijl}^k}{h^2} = 0. \tag{7}$$

Here, u_{ijl}^k are the values of $u(\mathbf{r}, t)$ at the point (i,j,l) at the time step k; c_{ijl} and a_{ijl} are the values of $c(\mathbf{r})$ and $a(\mathbf{r})$ at the point (i,j,l). The first term in (7) approximates $c(\mathbf{r})u_{tt}(\mathbf{r}, t)$, and the second term approximates $a(\mathbf{r})u_t(\mathbf{r}, t)$. The discrete Laplacian is denoted by Δu_{ijl}^k. It is computed using the formula:

$$\Delta u_{i_0 j_0 l_0}^k = \sum_{i=i_0-1}^{i_0+1} \sum_{j=j_0-1}^{j_0+1} \sum_{l=l_0-1}^{l_0+1} b_{ijl} u_{ijl}^k. \tag{8}$$

The b_{ijl} coefficients were presented, for example, in [11]. The parameters h and τ for the three-dimensional problem are connected by the Courant-Friedrichs-Lewy (CFL) stability condition: $\tau < h/\sqrt{3c}$. Collecting the terms with u_{ijl}^{k+1} in (7), we obtain an explicit finite-difference scheme for the wave Eq. (1). A similar scheme is used for Eq. (5). To solve the direct problem (1)–(2), non-reflecting boundary conditions [12] are applied at the boundary of the computational domain.

The components of the gradient of the residual functional are computed using the formulas:

$$(\Phi'_c)_{ijl} = \sum_{k=2}^{M-2} \frac{u_{ijl}^{k+1} - u_{ijl}^k}{\tau} \frac{w_{ijl}^{k+1} - w_{ijl}^k}{\tau} \tau, \quad (\Phi'_a)_{ijl} = \sum_{k=2}^{M-2} u_{ijl}^k \frac{w_{ijl}^{k+1} - w_{ijl}^{k-1}}{\tau} \tau, \tag{9}$$

where M is the number of time steps.

The iterative gradient descent algorithm is used to minimize the residual functional. As initial approximations for $c(\mathbf{r})$ and $a(\mathbf{r})$, we use $c^{(0)} = c_0 = \text{const}$, $a^{(0)} = 0$. These values correspond to the parameters of the environment. For water, $c_0 = 1500$ m·s^{-1}. The following actions are performed at each iteration (m):

1. The initial pulse is computed.
2. The direct problem (1)–(2) is solved, given that $c(\mathbf{r}) = c^{(m)}$, $a(\mathbf{r}) = a^{(m)}$. The acoustic pressure $u^{(m)}(\mathbf{r}, t)$ is calculated using formula (7). The values of $u(s, t)$ at the points s, where the detectors are located, are stored in memory.
3. The residual functional $\Phi^{(m)} = \Phi(u^{(m)}(\mathbf{r}))$ is computed using formula (3).
4. The "conjugate" problem (5)–(6) is solved to compute the wave field $w^{(m)}(\mathbf{r}, t)$.

5. The gradient $\Phi'(u^{(m)})$ is computed using formula (9). The stages 1–5 are repeated for all the sources, and the values of $\Phi^{(m)}$ и $\Phi'(u^{(m)})$ are summed for all the sources.

6. The current approximation is updated: $c^{(m+1)} = c^{(m)} + \lambda^{(m)}\Phi'_c(u^{(m)}(r))$, $a^{(m+1)} = a^{(m)} + \lambda^{(m)}\Phi'_a(u^{(m)}(r))$. The process returns to stage 2.

The step of the gradient descent $\lambda^{(m)}$ is chosen using a priori considerations. During the iterative process, the step is automatically corrected: $\lambda^{(m)}$ is decreased by 1.5 times if $\Phi^{(m)} > \Phi^{(m-1)}$; otherwise, it is increased by 25%. These rates are tuned for typical sound speed variations in soft tissues (1400–1600 m·s^{-1}).

4 GPU Implementation of Computations for the Direct and Inverse Problems of 3D Ultrasound Tomography

4.1 Specific Features of Graphics Processors

Graphics processors have become the first widely available parallel architecture and are already used in ultrasonic tomography applications [3].

The specific feature of the GPU memory hierarchy is a very high memory performance combined with a slow communication channel. As in most modern systems, the performance of arithmetic units is much higher than the memory performance.

Graphics processors are designed for data-parallel tasks, where each of the thread blocks processes its own data area, for example, a part of the image that does not overlap with other parts. Thus, an algorithm optimized for a GPU cluster must first divide the problem into processes that require a relatively small amount of fast memory; second, it must subdivide the task for each GPU into completely independent thread blocks.

4.2 Specific Features of the Problem and Optimization of the Algorithm

The inverse problem of low-frequency ultrasound tomography has some specific features that allow optimizing the algorithm and reducing the computational complexity.

1. The main computational complexity of the algorithm lies in computing the gradient of the residual functional using the formulas presented in (9), which includes computing the wave fields $u(r, t)$ and $w(r, t)$. Functional (3) is the sum of the squared differences $\|U(s, t)-u(s, t)\|^2$ for all the sources and detectors. It can be computed for each source separately, and the results can be added together. The gradient (4) can also be computed as the sum of partial gradients for each source.

In the proposed architecture of the GPU cluster, each computing node calculates the gradient for one of the sources. A scheme for parallelizing the computations for sources S_1–S_6 is shown in Fig. 3. The input data for all computing nodes are the same: $c^{(m)}$, $a^{(m)}$ at the m-th iteration of the gradient descent method. The data transfers between the nodes occur only when an iteration of the gradient method is completed and when the current approximation is updated.

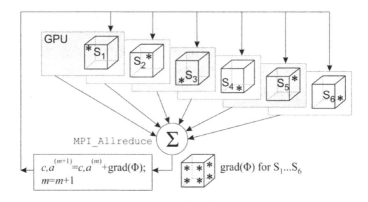

Fig. 3. Parallelizing the computations for multiple sources.

The total value of the gradient is computed and broadcast to all the nodes using the MPI_Allreduce operation. Then the new iterative approximation $c^{(m+1)}$, $a^{(m+1)}$ is computed. Various parameters of the algorithm are synchronzed via MPI_Bcast operations.

The amount of transferred data for a problem of size 400^3 is approximately 256 MB ($400 \times 400 \times 400 \times 4$ bytes) for each node in each direction. Thus, the parallelization of computations by sources proves to be very effective. The total overhead is a few seconds per minute.

2. In medical diagnostic applications, the variance of the sound speed in soft tissues does not exceed 10–15%. This allows us to estimate the volume V, in which the wave propagates after being emitted from the source, in advance and to perform the computations within only this volume. The volume V is a sphere of radius $v_{\max}t$, where $v_{\max}$ is the maximum permissible sound speed in the model, and t is the current simulation time. This optimization is easily accomplished using the GPU by executing only the blocks for which $r \leq v_{\max}t$. This optimization reduces the computation time by 25%.

3. The gradient of residual functional (4) is an integral over time. This means that to calculate the gradient, it is not necessary to store all the values of $u(r, t)$ in the $X \times Y \times Z \times T$ region, which would require a huge amount of memory. It is sufficient to compute $u(r, t_k)$ sequentially at time steps t_k. Thus, the required amount of memory is proportional to N^3, and the number of operations is proportional to N^4, where N is the number of grid points along one dimension.

Taking into account these features of the ultrasonic tomography problem, we propose a two-stage method according to the scheme presented in Fig. 4. In the first stage ("Forward-time computation"), the wave field $u(r, t)$ generated by the source S in the volume V is computed sequentially in time. An absorbing layer of width d and simple non-reflecting boundary conditions [12] are used to cancel the reflected waves. A typical value of d is 32 grid points.

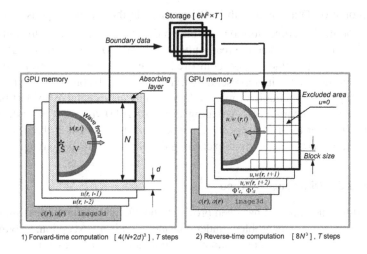

Fig. 4. The scheme of GPU computations performed in two stages.

To use the FDTD scheme (7), it is necessary to store $u(r, t)$, $u(r, t-1)$, $u(r, t-2)$, which amounts to $3 \cdot (N + 2d)^3$ 32-bit words, and the coefficients $c(r)$ and $a(r)$ in the GPU memory. The coefficients do not change over time; thus, we can use the `image3d` structure, which enables texture caching, improving the performance by 10%.

Since the coefficients are unknown and determined only approximately, the 16-bit `half` data type is sufficient for their representation. The error between the exact and approximate solutions is ~2 m·s^{-1} for $c(r)$, which is ~2% of $|c(r)-c_0|$, and even worse for $a(r)$ [4]. This means that such variations of the coefficients do not noticeably affect the simulated ultrasound wave. Thus, a rounding error not exceeding 0.2% is acceptable.

The values of $u|_{ST}$ at the boundary of the computational domain are stored for use in the second stage. Fast access to these data is not needed; therefore, we place them in the system RAM. The required RAM capacity is $6\,N^2 \cdot T$, where T is the number of time steps. It follows from the FDTD stability conditions that $T \approx 3\,N$, and the amount of memory needed for the boundary values is $18\,N^3$ 32-bit words.

To compute the gradient using formula (4), we need the values of $u(r, t)$ and $w(r, t)$ at the same points. To start computing $w(r, t)$ from the last time step $t = T$, it is necessary to compute $u(r, t)$ for all $t \leq T$ first. In the second stage ("Reverse-time computation"), we solve Eq. (1) for $u(r, t)$ and problem (5)–(6) for $w(r, t)$ simultaneously in reverse time. The values of $u(r, t)$ and $w(r, t)$ are computed based on $u(r, t + 1)$, $w(r, t + 1)$, $u(r, t + 2)$ and $w(r, t + 2)$. The boundary conditions for $w(r, t)$ are determined from the experimental data $U(s, t)$ by using formula (6).

To fill in the missing values of $u(r, t)$ at the boundary, we use the values of $u|_{ST}$ stored in the memory in the first stage. The wave field $u(r, t)$ obtained in this way is equal to $u(r, t)$ computed in the first stage. The recalculation of $u(r, t)$ allows us to use a data array of size X × Y × Z, not X × Y × Z × T. The numerical error introduced by the recalculation does not exceed 10^{-5}, making it negligible for practical measurements.

The amount of data stored in the GPU memory in the second stage is $8\ N^3$: $u(r, t)$ and $w(r, t)$ for the three time steps (six 32-bit words), the gradient values Φ'_c and Φ'_a (one word), which are updated, and the read-only coefficients $c(r)$, $a(r)$ (one word).

Since the gradient value is small compared to the coefficients $c(r)$, $a(r)$, the short data type is sufficient for its representation. The use of a 16-bit integer type reduces memory usage and computation time, while providing an acceptable level of precision to accumulate the integrals (4). A scaling factor of $24000/\text{max}|\Phi'|$ is applied to the gradient Φ' in order to limit the possible increase of the gradient at the next iteration and to prevent the coefficients from exceeding the physically realistic range.

4.3   The Finite Difference Method

The finite difference method has been proven to be efficient for numerical simulations of physical processes. The problem under consideration is no exception. Solving problems (1)–(2) and (5)–(6) requires numerical simulation of the wave field with given parameters. An explicit FDTD scheme is a naturally data-parallel algorithm because the values at all the grid points are computed in the same way and do not depend on each other. Such algorithms fit well in SIMD/SPMD-architectures.

To compute the 3D wave fields at each time step, the volume V is divided into blocks that are processed by each thread block of the GPU. The blocks that are far from the ultrasound source, where $u(r, t) = 0$ for the current simulation time t, are excluded. The "Z-marching" method is used inside each block because the optimal number of parallel threads in a typical GPU is several hundreds per multiprocessor (MP), and the dimension of the problem is approximately 400^3. The typical number of 3D blocks to be processed is approximately 10000. The thread blocks are two-dimensional (x, y), and each thread computes the data sequentially along the Z-axis.

The discrete Laplacian (8) can be reduced to scalar products of three-component vectors by collecting the terms, because only four of the b_{ijl} coefficients have non-repeating values:

$$\Delta u(i, j, z_0) = \boldsymbol{b}_0 \cdot \boldsymbol{u}(z_0) + \boldsymbol{b}_1 \cdot (\boldsymbol{u}(z_0 - 1) + (\boldsymbol{u}(z_0 + 1)),$$
$$\boldsymbol{b}_0 = \{b_{000}, b_{100}, b_{110}\}, \boldsymbol{b}_1 = \{b_{100}, b_{110}, b_{111}\},$$
$$\boldsymbol{u}(z) = \{u_{ij}, u_{ij+1} + u_{i+1j} + u_{ij-1} + u_{i-1j},\ u_{i+1j+1} + u_{i+1j-1} + u_{i-1j+1} + u_{i-1j-1}\}.$$

To compute Δu sequentially along the Z-axis, we need to keep three vectors per thread in the registers: $\boldsymbol{u}(z_0)$, $\boldsymbol{u}(z_0-1)$, and $\boldsymbol{u}(z_0 + 1)$. At each step of the Z-marching method, $\boldsymbol{u}(z_0 + 1)$ becomes $\boldsymbol{u}(z_0)$, the new $\boldsymbol{u}(z_0 + 1)$ is computed, and the results for the $z = z_0$ plane are saved in the global memory. Small amount of data per thread and mostly sequential memory access pattern allow for an efficient GPU implementation.

4.4   Profiling the Algorithm

For profiling, a test run of 6 iterations of the gradient method is carried out for a $320 \times 320 \times 320$ problem using the NVidia GeForce GTX Titan graphics card. The

OpenCL interface was used for GPU programming. Program profiling statistics are shown in Table 1. The computational kernels that require less than 0.01% of the GPU time were left out. The start and end times of each kernel run were obtained using OpenCL profiling events, then the execution times and overlap times were computed. The total time of all GPU operations corresponds to 100%.

Table 1. Execution time for computational kernels and data transfers.

Runs	avg, µs	min, µs	max, µs	Overlap, s	Total time s	% GPU	Function
293216	271	252	499	0	79.47	0.94	LoadBound
306172	11184	112	15842	3423.1	3424.39	0.02	SaveBound
1375	911	385	3049	0	1.25	0.01	Initialize
306172	1215	1139	2041	0	372.11	4.41	FwdBoundCond
306172	11645	275	16146	0	3565.55	42.21	ForwardWave
293216	14292	712	17981	0	4190.81	49.62	BackwardWave
12331	11076	38	16143	0	136.59	1.62	SaveExData
293216	28	23	251	0	8.37	0.10	LoadExData
14671	2017	1247	3615	0	29.60	0.35	DisplayGL
8512	1100	1063	1245	0	9.37	0.11	ScalarMax
DATA TRANSFERS							
306172	251	26	1506	76.19	77.08	0.01	_LoadFromGPU
293216	32	26	51	0	9.51	0.11	_SaveToGPU
293216	33	32	65	0	9.91	0.12	_SaveExData
39767	76520	0	47662	0	30.43	0.36	_BufferOp

The test shows that, as expected, almost all the time is spent on the calculation of 3D problems (1)–(2) and (5)–(6) ("ForwardWave" and "BackwardWave" functions). The boundary conditions require approximately 6% of the time because the memory access pattern is mostly random. The performance impact decreases as the problem size N increases, because the boundary contains $\sim N^2$ elements, while the volume contains $\sim N^3$ elements.

The data transfers between the GPU and the system memory require less than 1% of the time. Some data transfers are performed in parallel with the calculations. The total program execution time exceeded the total GPU time by 7.5%. This value shows the overhead costs that are not parallel with the GPU computations, like summation and data distribution via the MPI interface.

Further optimization of the algorithm includes choosing the size of the thread blocks, which determines the optimal use of the GPU register files and memory access circuits. Figure 5a shows the execution time for different block sizes on a GTX Titan device. The optimal block size choice can provide up to 15% performance boost, and the size of 32×4 was found to be optimal for this problem on all of the tested devices.

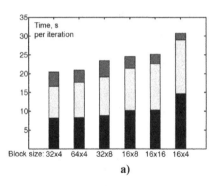
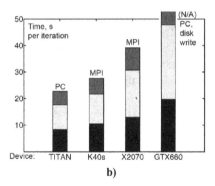

Fig. 5. Execution time: direct problem (bottom), "conjugate" problem (middle), other (top): (a) for various block sizes, GTX Titan; (b) for various devices, 32 × 4 block size.

The algorithm was tested using a number of GPU devices: NVidia Tesla K40 s on the "Lomonosov-2" supercomputer, NVidia Tesla X2070 on the "Lomonosov" super-computer of the Moscow State University Supercomputer Center [13], NVidia GeForce GTX Titan and GTX 660 on personal computers.

Figure 5b shows the execution time for various devices. On the supercomputers, the test run comprised executing 8 tasks in parallel on 8 devices and collecting the data using the MPI interface, as shown on Fig. 3; on PCs, a single task was executed and the data were saved to a disk. The `MPI_Allreduce` operation required less than 1% of the total time (280 ms for 8 parallel tasks, 340 ms for 48 parallel tasks on the "Lomonosov" supercomputer equipped with a 40 Gbit/s QDR Infiniband network).

The tests demonstrated a direct relationship between the performance and the memory bandwidth of the device. The more expensive Tesla devices showed lower performance, which means that the specific Tesla features are not relevant to this partic-ular problem. The parameters of the algorithm, such as caching in local memory vs. automatic caching, were tuned for best performance on each device.

5   The Architecture of the Computing System for Solving Inverse Problems of 3D Ultrasonic Tomography

The main problem in 3D wave tomography imaging is that in a typical problem of size 400^3, the number of unknowns reaches 10^8. The number of ultrasound sources required to collect enough data is approximately 20–40, and the total number of computed data points reaches 10^{12}. These computations have to be performed within a reasonable time.

The FDTD method is a data-intensive task, for which the memory performance is of prime importance. Therefore, graphics processors are a natural choice. GPU computing performance remains high as long as the data fits into the on-board memory of the device.

Let us formulate the requirements for a computer system that can be used in a tomo-graphic complex. To determine these requirements, we ran a series of tests. Table 2 shows the execution times and memory requirements for different problem sizes. These tests were performed on an NVidia GeForce GTX Titan device.

Table 2. Memory requirements and execution times for various problem sizes.

3D problem size (N)	256	288	320	384	416	448	512
GPU memory used, GB	0.8	1.2	1.5	2.2	2.8	3.5	5.0
System RAM used, GB	1.5	2	3	5	6	8	12
Time per one iteration, s	11	16	23	48	63	92	160

The performance is a primary limiting factor here because the execution time is proportional to N^4, whereas the amount of memory is proportional to N^3. This means that there is an optimal amount of on-board GPU memory. Thus, using expensive devices with large amounts of memory is impractical because of the greatly increased time needed to process such amount of data.

Let us assume that a practically acceptable computation time is 1 h for 100 gradient method iterations (36 s per iteration). Then, we can use one device with 3 GB of on-board memory for each ultrasound source when the problem size is limited to 360^3. To tackle problems of sizes up to 400^3, we can use two such devices per source, or a single higher-end device. This setup requires 6 GB of system RAM and 3.5 GB of GPU memory for each source. For example, the NVidia GeForce GTX 690 graphics card, which contains two 2-GB GPU devices on a single board, can be used to this end. The proposed algorithm theoretically allows splitting the processed volume between two GPUs across the Z-axis. To balance the load, the partitions should include the same number of blocks, which is known a priori. This is a standard approach to parallelizing 3D FDTD schemes.

Recently announced devices with High Bandwidth Memory architecture (HBM) have approximately 3 times the performance compared to devices with GDDR5 memory. Using one such device per source, the problem size can be increased to 480^3. In this setup, the device should have 6 GB of on-board memory, and 10 GB of system RAM per source is required. A problem size of 480^3 is close to the practical requirements for ultrasonic mammography applications. This grid size provides a resolution of 0.4 mm over a 20 cm range.

Let us formulate the essential characteristics of the specialized GPU computer.

Graphics processors and RAM storage. Each computing node must contain at least a sufficient number of GPU devices to compute the residual functional gradient for a single source in a reasonable time. For problem sizes up to 360^3 — one GPU device (~250 GB/s memory bandwidth, 3 GB of on-board memory) and 4 GB of RAM storage for each ultrasound source. For larger problems — two consumer-class GPU devices (4–6 GB of total GPU memory), or one higher-end or HBM-class device, and 10–12 GB of RAM storage per source. The total number of ultrasound sources is 20–30. The number of GPU devices per node should be maximized in order to reduce the total number of hardware components. Currently available mainboards can support up to 4 devices.

Central processors. The CPUs distribute data to the GPUs and between computing nodes. The CPUs must meet the minimal requirements.

Communication network. The network is used to combine the data from all the nodes to compute the gradient and to distribute the next iterative approximation to all the nodes (Fig. 3). These actions are carried out only once per iteration. Because only the all-reduce and broadcast operations are needed, the optimal network topology is a star or a tree topology. The minimal required bandwidth is ~200 MB/s.

Disk storage. The disk storage must meet the minimal requirements. The amount of data to be stored is under 50 GB for one experiment.

These requirements can be met using common solutions that fit in a single rack and have a power consumption of 10–20 kW. Modern graphics cards require 150–300 W per unit, and this value steadily decreases as the technology improves. A mainboard with CPU and RAM requires no more than 250 W; thus, a node containing a mainboard and four GPU devices requires at most 1.5 kW.

Using widely available hardware components, we can build computing systems that provide the medical image reconstruction using the 3D wave tomography technology. The performance gain relative to a single-CPU personal computer is on the order of 1000 times. This estimation is based on a typical 30-fold difference between CPU and GPU implementations of 3D FDTD methods [14], multiplied by an estimated number of devices in the cluster of 32.

Graphics cards and GPU supercomputers continue to improve. There is no doubt that in a few years, the performance of graphics processors will increase by several times, while the energy footprint will decrease. All this speaks in favour of using GPU clusters as specialized supercomputers for the new ultrasonic tomographic systems currently being developed.

6   Conclusions and Discussion

The requirements for a GPU cluster that provides an efficient implementation of the iterative gradient methods of the reconstruction of tomographic images are formulated. The specialized GPU cluster can achieve a 1000-fold performance increase compared to that of a single-CPU personal computer. The characteristics, such as the size, power consumption, and cost, of a GPU cluster allow it to be used as a computing system in the new medical ultrasonic tomographic complexes being developed.

There are other high-performance solutions that can be used for solving the inverse problems of ultrasound imaging. Modern multicore systems, such as the Intel Xeon Phi, have performances comparable to that of a GPU, but these systems are much more expensive because they are much more complex devices designed for a wide range of applications. CPU-based systems require a large number of memory access channels and a large cache to be efficient for 3D imaging. According to the authors, GPU clusters have the most promising architecture for high-performance 3D image reconstruction.

Acknowledgements. This research was supported by Russian Science Foundation (project No. 17–11-01065). The study was carried out at the Lomonosov Moscow State University.

References

1. Duric, N., Littrup, P., Li, C., Roy, O., et al.: Breast ultrasound tomography: bridging the gap to clinical practice. Proc. SPIE. Med. Imaging **8320**, 832000 (2012)
2. Wiskin, J., Borup, D., Andre, M., Klock, J., et al.: Three-dimensional nonlinear inverse scattering: quantitative transmission algorithms, refraction corrected reflection, scanner design, and clinical results. J. Acoust. Soc. Am. **133**, 3229 (2013)
3. Birk, M., Dapp, R., Ruiter, N.V., Becker, J.: GPU-based iterative transmission reconstruction in 3D ultrasound computer tomography. J. Parallel Distrib. Comput. **74**, 1730–1743 (2014)
4. Goncharsky, A.V., Romanov, S.Y.: Inverse problems of ultrasound tomography in models with attenuation. Phys. Med. Biol. **59**, 1979–2004 (2014)
5. Goncharsky, A.V., Romanov, S.Y., Seryozhnikov, S.Y.: A computer simulation study of soft tissue characterization using low-frequency ultrasonic tomography. Ultrasonics **67**, 136–150 (2016)
6. Goncharsky, A.V., Romanov, S.Y., Seryozhnikov, S.Y.: Inverse problems of 3D ultrasonic tomography with complete and incomplete range data. Wave Motion **51**, 389–404 (2014)
7. Goncharskii, A.V., Romanov, S.Y.: Two approaches to the solution of coefficient inverse problems for wave equations. Comput. Math. Math. Phys. **52**, 245–251 (2012)
8. Goncharsky, A.V., Romanov, S.Y.: Iterative methods for solving coefficient inverse problems of wave tomography in models with attenuation. Inverse Prob. **33**(2), 025003 (2017)
9. Natterer, F.: Possibilities and limitations of time domain wave equation imaging. In: Contemporary Mathematics, vol. 559, pp. 151–162. American Mathematical Society, Providence (2011)
10. Beilina, L., Klibanov, M.V., Kokurin, M.Y.: Adaptivity with relaxation for ill-posed problems and global convergence for a coefficient inverse problem. J. Math. Sci. **167**, 279–325 (2010)
11. Mu, S.-Y., Chang, H.-W.: Dispersion and local-error analysis of compact LFE-27 formula for obtaining sixth-order accurate numerical solutions of 3D Helmholtz equation. Prog. Electromagnet. Res. **143**, 285–314 (2013)
12. Engquist, B., Majda, A.: Absorbing boundary conditions for the numerical simulation of waves. Math. Comput. **31**, 629–651 (1977)
13. Voevodin, V.V., Zhumatiy, S.A., Sobolev, S.I., Antonov, A.S., Bryzgalov, P.A., Nikitenko, D.A., Stefanov, K.S., Voevodin, V.V.: Practice of "Lomonosov" supercomputer. Open Syst. J. **7**, 36–39 (2012)
14. Zhang, L., Du, Y., Wu, D.: GPU-Accelerated FDTD simulation of human tissue using C++ AMP. In: 31st International Review of Progress in Applied Computational Electromagnetics (ACES), Williamsburg, VA, pp. 1–2 (2015)

The Energy Consumption Analysis for the Multispectral Infrared Satellite Images Processing Algorithm

Ekaterina Tyutlyaeva[(✉)], Sergey Konyukhov, Igor Odintsov,
and Alexander Moskovsky

ZAO RSC Technologies, Kutuzovskiy avenue, 36, building 23, 121170 Moscow, Russia
{xgl,s.konyuhov,igor_odintsov,moskov}@rsc-tech.ru

Abstract. This paper includes the energy consumption analysis of the testing mini-application that implements night time infrared remote sensing algorithm `Nightfire`. On this stage of our project computational nodes with Intel Xeon E5 and Intel Xeon Phi processors were tested.

The correlation analysis between the number of used MPI ranks - OMP threads and total energy consumptions was performed for each of tested computational nodes. The optimized benchmarking parameters were used to compare energy efficiency of tested nodes.

Moreover, the analysis of mini-application statements blocks was carried out for the following computation phases: I/O with HDF5 and ENVI data; the data processing using Nelder-Mead method. The impact of each computation phase to the total energy consumptions was determined so it gives new insights to possible ways of further optimization.

Based on obtained results, the effectiveness of tested computational architectures for multispectral satellite images processing was evaluated.

Keywords: Energy consumption · Energy efficiency · Satellite image processing · Power-aware execution · Hybrid parallel programming

1 Introduction

Nowadays power-aware program executions are one of the most immediate and pressing challenges facing software developers. The software energy-efficient design and execution tuning are especially relevant to the reusable applications that are executed regularly because, among other things, energy consumption depends on the algorithmic structure and computation phases. Thereby, the reasonable strategy of hardware usage by program application can result in substantial reduction of supercomputer operational costs. In that context, the power consumption during the application runtime should be explored, and the most power efficient execution configuration should be suggested before the productive use of software.

© Springer International Publishing AG 2017
V. Voevodin and S. Sobolev (Eds.): RuSCDays 2017, CCIS 793, pp. 376–387, 2017.
https://doi.org/10.1007/978-3-319-71255-0_30

In this research we conducted the energy-efficiency analyses of `Nightfire` algorithm for multispectral infrared satellite images (MISI) [1]. The multispectral data for this algorithm are collected globally, each night, by the Visible Infrared Imaging Radiometer Suite (VIIRS) [2] operated by Suomi National Polar Partnership (NPP) [3]. Thus, software implementing `Nightfire` algorithm is planned to be used on a daily basis, and so it should be tuned with power concerns.

In our tests we used own mini-application implementing `Nightfire` algorithm by so-called hybrid programming model that combines shared-memory (OpenMP) and distributed-memory (MPI) programming models. The hybrid programming model allows to use many- and multi-core processors in more efficient way and provides opportunities for flexible application tuning against target architectures. Besides, the hybrid programming model can utilize all available computing cores and RAM resources more efficiently than pure MPI model what may cause a program to run faster [4].

On the other hand, due to the flexible nature of hybrid programming model it is possible to optimize program runtime parameters - first of all, the numbers of MPI processes and OMP threads - regarding not only to the execution time, but also to the power consumption.

Finally, drawing up energy profiles along the critical path of program execution allows to identify those blocks of the studied algorithm that incur the most energy consumption, which helps to determine further directions of code optimization.

2 Related Work

Power-aware execution of parallel programs is now a primary concern in large-scale HPC environments. Dong Li et al. [5] presented and evaluated solutions for power-efficient execution of programs written in the hybrid program model targeting large-scale distributed systems with multicore nodes. The authors used a new power-aware performance prediction model of hybrid MPI/OpenMP applications to derive a novel algorithm for power-efficient execution of realistic applications.

Power-aware policies were the focus of many researchers. Typically, the studies proposes to expand the time-saving strategies with power-saving features. As an example, Wenlei Bao et al. [6] proposed a novel power-aware WCET (Worst Case Execution Time) analysis technique to improve system energy efficiency and simultaneously validate real-time tasks. The Power-Aware Linear Programming based Affinity Scheduling Policy was described in [7].

In this paper we studied energy efficiency of MISI processing on the various Intel Xeon E5 and Intel Xeon Phi architectures. We carried out cross-architectural research that includes computation phase analysis and CPU/DRAM power consumption distribution. Our methodology agrees with common profiling approach and can be used for wide range of software to give recommendations for power-aware executions.

3   Algorithm

For our energy consumption analysis we used own mini-application implementing Nightfire algorithm for the multispectral infrared satellite images processing. This algorithm detects and characterizes sub-pixel hot sources using multispectral data collected for different infrared spectral ranges. Nightfire algorithm relies on the approximation of direct composition of theoretical Planck curves describing radiation of subpixel heat source(s) and sea/land surface temperature background. The Nelder-Mead simplex algorithm is used for the unconstrained nonlinear fit of a weighted mixture of the Planck curves to observed multispectral radiances.

The method can be used to process the data from the last generation of infrared sensors at the environmental satellites: VIIRS (solar synchronous Suomi NPP and JPSS-1, NOAA/NASA), ABI (geostationary GOES-R, NOAA) [8], Landsat 8 (low orbiting, USGS) [9], and AHI (geostationary Himawari 8-9, JMA) [10].

The basic flowchart illustrating the structure of Nightfire algorithm is shown on the Fig. 1.

Nightfire algorithm has the high scalability potential because it processes images by applying Nelder-Mead optimization method [11] to individual image pixels independently from each other. Moreover, the most computational intensive part of Nelder-Mead optimization method, namely, the evaluation of target

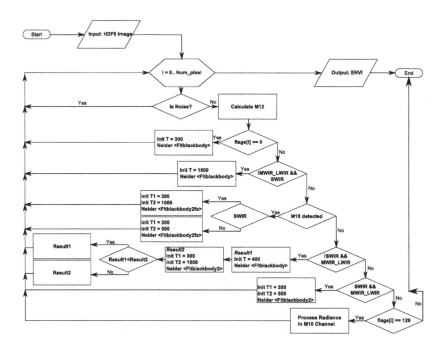

Fig. 1. Nightfire Algorithm Flowchart

function values, is also well suited for parallelization. Nevertheless, huge computational demands of this task stem from the necessity to process large number of images with large number pixels during short time period.

We used the heterogeneous MPI + OMP parallelization scheme to effectively utilize the multicore architectures. The manual and the compiler-supported code vectorization for Intel architectures also were applied before the analysis stage.

4   Hardware

In the Table 1 codenames and specifications of studied testbeds are listed.

Table 1. Testbeds Specifications

Codename	CPU	# Cores	Memory	GB per Core
Haswell	Intel Xeon E5-2697 v3	2x 14	8x DRAM Samsung 16 GB DDR4/2133MHz	4.57
Broadwell	Intel Xeon E5-2697A v4	2x 16	8x DRAM Samsung 16 GB DDR4/2133MHz	4
KNL	Intel Xeon Phi 7250	68	MCDRAM Intel 16 GB + 6x DRAM Micron 32GB DDR4/2133MHz	2.8

5   Measurements

The energy consumption of the tested mini-application was studied using the Intel Running Average Power Limit (RAPL) counters [13]. RAPL provides a way to measure the power consumption on processor packages and DRAM. According to the recent studies this software power model matches the actual power measurements [14]. Further all results are given as averaged results of the multiple executions.

PAPI performance application programming interface [15] was used to gather RAPL data statistics. The distribution of total energy consumption for different MPI/OMP numbers are shown on the Fig. 2.

We used the following execution parameters: 2/4/6/8/10 MPI processes and 1xN/2xN/3xN/4xN OMP threads (in total), where N is the number of cores per one CPU. Consequently, due to the usage of Intel Hyper-Threading Technology for Intel processors [12] the maximal number of OMP threads per CPU core was 56 for Haswell (2 (CPUs per node) × 14 (Cores per CPU) × 2 (Hyper-threading)); 64 for Broadwell (1 (CPU per node) × 14 (Cores per CPU) × 2

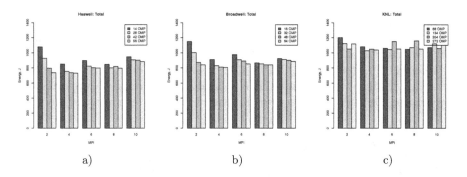

Fig. 2. Estimated Energy Consumption for DRAM and CPU Packages on (a) Haswell, (b) Broadwell, (c) KNL architectures

(Hyperthreading)) architectures, and 272 (2 (CPUs per node× 68 (Cores per CPU) × 4 (Hyperthreading) for KNL architecture.

The bar graph on the Fig. 3 expresses the total execution time in seconds for the tested mini-application.

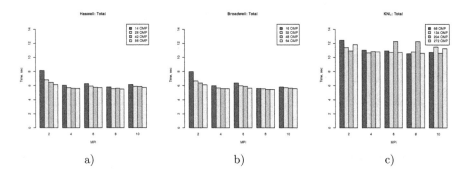

Fig. 3. Total Execution Time on (a) Haswell, (b) Broadwell, (c) KNL architectures

The minimal energy consumption results for the tested mini-application are presented in Table 2.

Obtained results show that for all testbeds the most optimal processes/ threads configuration was 4 MPI/2 × Num_{cores} OMP threads per node, where Num_{cores} is the number of physical cores in a computational node (Fig. 3).

While there is some correlation between the execution time and the consumed energy, it has complex noninear character. The values of energy (ETS) and time (TTS) to solve test problem for all testbeds are presented in the Tables 3, 4 and 5.

For example, the most power-aware MPI/OMP execution configuration for Haswell processor took 5.596 sec and 730 J. The most time-efficient (fast) result

Table 2. Minimal energy consumption for the tested architectures, J

Application	Haswell	Broadwell	KNL
DRAM	38.9043 (4 MPI 56 OMP)	184.749 (8 MPI 64 OMP)	27.071333 (8 MPI 272 OMP)
Processors	691.584 (4 MPI 56 OMP)	618.51 (4 MPI 64 OMP)	995.386667 (4 MPI 136 OMP)
Total	730.488 (4 MPI 56 OMP)	807.076 (4 MPI 64 OMP)	1027.73 (4 MPI 136 OMP)

Table 3. Energy (ETS, J) and time (TTS, sec) obtained for the test on Haswell testbed with different MPI and OMP configurations

MPI	Type	14 OMP	28 OMP	42 OMP	56 OMP
2	TTS	8.139	6.829	6.466	6.149
	ETS	1077.12	927.882	796.54	737.235
4	TTS	6.055	5.726	5.619	5.596
	ETS	850.526	753.846	737.405	730.488
6	TTS	6.276	5.935	5.762	5.706
	ETS	898.315	823.753	802.762	797.944
8	TTS	5.814	5.558	5.624	5.514
	ETS	847.626	800.677	818.487	796.684
10	TTS	6.151	5.904	5.879	5.738
	ETS	945.801	906.363	901.844	882.429

Table 4. Energy (ETS, J) and time (TTS, sec) obtained for the test on Broadwell testbed with different MPI and OMP configurations

MPI	Type	16 OMP	32 OMP	48 OMP	64 OMP
2	TTS	7.993	6.674	6.375	6.1236
	ETS	1150.2	1003.82	871.343	840.002
4	TTS	5.979	5.664	5.581	5.571
	ETS	910.977	830.855	811.053	807.076
6	TTS	6.373	5.987	5.897	5.654
	ETS	975.988	908.189	892.213	852.781
8	TTS	5.611	5.549	5.462	5.465
	ETS	866.233	853.055	839.569	839.433
10	TTS	5.805	5.719	5.653	5.582
	ETS	922.358	912.005	901.284	887.115

Table 5. Energy (ETS, J) and time (TTS, sec) obtained for the test on KNL testbed with different MPI and OMP configurations

MPI	Type	68 OMP	136 OMP	204 OMP	272 OMP
2	TTS	12.441	11.417	10.935	11.838
	ETS	1199.53	1121.73	1049.95	1118.04
4	TTS	11.045	10.626	10.841	10.788
	ETS	1078.75	1027.73	1047.4	1037.99
6	TTS	10.921	10.716	12.300	10.732
	ETS	1057.9	1044.61	1149.79	1050.69
8	TTS	10.518	10.794	12.257	10.598
	ETS	1045.54	1069.55	1158.77	1049.24
10	TTS	10.724	11.491	10.600	11.281
	ETS	1068.12	1118.9	1058.22	1100.22

was 5.514 sec, but the energy consumption for this configuration was 796 J, that is 66 J more then the energy efficient one.

For Broadwell processor, the most energy efficient result was 5.571 sec for runtime and 807 J for energy consumption.

It is worth noting that, while Haswell architecture demonstrated the best total energy efficient result, the best results for CPU energy consumption were yielded on Broadwell testbed (Fig. 4).

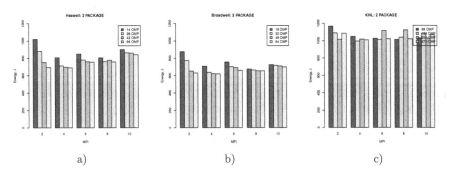

a)   b)   c)

Fig. 4. CPU Energy Consumption for (a) Haswell, (b) Broadwell, (c) KNL architectures

However, the energy-efficiency of Broadwell processors was neglected by DRAM consumption (Fig. 5). Broadwell architecture has demonstrated the highest rates of DRAM energy consumption, while KNL DRAM energy consumption is the lowest one.

The difference in DRAM energy consumption between Haswell and Broadwell architectures was even more remarkable because DRAM models for these nodes were absolutely identical.

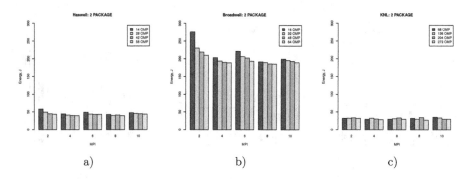

Fig. 5. DRAM Energy Consumption for (a) Haswell, (b) Broadwell, (c) KNL architectures

6 Computation Stages

6.1 Input/Output

For this work, the mini-example of satellite image was prepared in HDF5 data format [16]. Input data contained 9830400 pixels, file size was 236 Mb. The HDF5 I/O library supports parallel I/O, so we have used parallel data reading operations.

However, the processors energy consumption increases steadily with the number of used MPI processes for all architectures (see Fig. 6):

- Haswell testbed – 179,5 J in average for 2 MPI executions, 265,75 J for 10 MPI executions;
- Broadwell testbed – 175 J in average for 2 MPI executions, 238 J for 10 MPI executions;
- KNL testbed – 500,25 J in average for 2 MPI executions, 536 J for 10 MPI executions.

DRAM energy consumption also increased with the growing number of MPI ranks used (see Fig. 7).

The output data were prepared in ENVI format and took about 300 Mb of disk space. While ENVI output weren't parallelized, KNL energy consumption was comparable to Haswell and Broadwell energy consumption for this phase of algorithm and took about 250 J (vs 500 J for HDF5 input) (see Fig. 8). Consequently, for Intel Xeon Phi architecture special attention should be drawn to HDF5 input phase.

6.2 Image Processing

On the contrary to the input phase, for the image processing stage the test run with 2 MPI ranks was the least successful case in energy consumption terms for processors (see Fig. 9) as well as DRAM (Fig. 10).

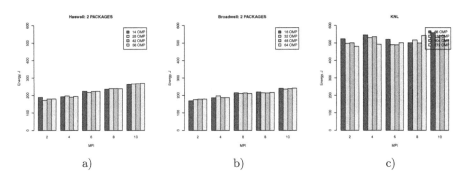

Fig. 6. CPU Energy Consumption during the Input Stage on (a) Haswell, (b) Broadwell, (c) KNL architectures

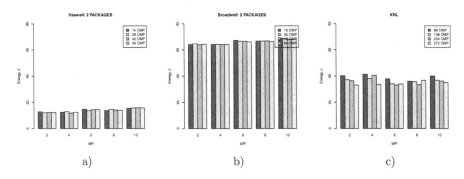

Fig. 7. DRAM Energy Consumption during the Input Stage on (a) Haswell, (b) Broadwell, (c) KNL architectures

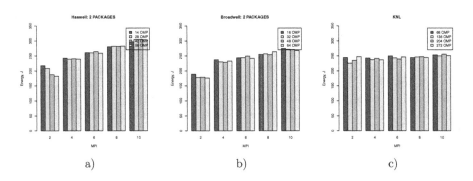

Fig. 8. CPU Energy Consumption during the Output Stage on (a) Haswell, (b) Broadwell, (c) KNL architectures

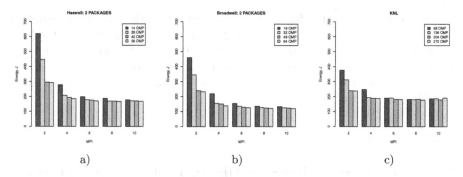

Fig. 9. CPU Energy Consumption during the Image Processing Stage on (a) Haswell, (b) Broadwell, (c) KNL architectures

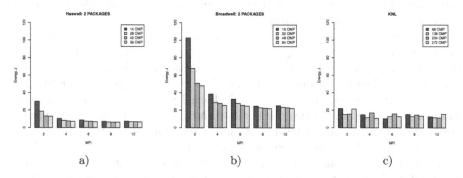

Fig. 10. DRAM Energy Consumption during the Image Processing Stage on (a) Haswell, (b) Broadwell, (c) KNL architectures

It is also worth noting that for this stage of algorithm the energy consumption difference between test runs with different MPI ranks was significant, so the less efficient configurations should be avoided.

The most power efficient MPI/OMP cases on the most computation demanding stage (2 MPI 14 OMP for Haswell and 2 MPI 16 OMP for Broadwell respectively) didn't utilize all computation resources[1], that leads to the execution time and the total energy consumption increasing.

The energy effective MPI/OMP combinations can be proposed as follows:

- The best configurations for Haswell testbed include 8/10 MPI, 28/42/56 OMP cases (The results varies in ranges 165-170 J per CPUs, 6.1-6.7 J per DRAM).
- The best configurations for Broadwell testbed include 8/10 MPI, 32/48/64 OMP cases (The results varies in ranges 120-126 J per CPUs, 22-23 J per DRAM).
- The best possible options for KNL cover 6/8/10 MPI, 68/136/204/272 OMP cases (The results varies in ranges 180-190 J per CPU, 11-15 J per DRAM).

---

[1] If it be taken into account the hyperthreading technology.

It is worth mention that for the input/output stages the energy consumption was higher than for the image processing stage. This difference is particularly notable for KNL testbed, so the major optimization efforts should be aimed to the input/output phases before the processing of big amounts of data. We plan in the future work to study the energy-efficiency of possible models of the usage of HBM^2 on KNL testbeds.

7   Conclusions

The results of our test runs have showed that the isolated CPU energy consumption didn't reflect a full picture. For example, for Broadwell testbed processors have showed the most power-aware execution results, but DRAM energy consumption was the substantial and the largest among the considered testbeds share of the total energy consumption.

Our experimental studies have demonstrated the efficiency of simple energy measurements to reveal energy consumption flaws. As given test results illustrate, the execution time could be slightly different for different runtime configurations, but the energy consumption for them might change considerably.

So it seems to be possible to tailor suitable TTS/ETS configurations for each satellite processing algorithm in order to reduce the total energy consumption. For Nightfire algorithm of MISI processing we can propose 4 MPI/$2 \times Num_{cores}$ configurations as the most power-aware configurations.

In the future, we plan to continue these experiments with larger amounts of real satellite data (up to 20–30 TB) and proof the scalability of our assumptions.

Currently, the VIIRS boat detector algorithm is developed for subsequent study. It is an innovative method to robustly identify fishing boats at night using visible and infrared images from the SNPP satellite. We are planning to study the inter-node communications and larger amount of data using this algorithm in the nearest future.

Acknowledgments. This research was supported by the Russian Federal Science and Technology Program grant 14.607.21.0165 Efficient co-design of massively parallel computer for multispectral night-time remote sensing.

References

1. Elvidge, C.D., Zhizhin, M., Hsu, F.-C., Baugh, K.E.: VIIRS nightfire: satellite pyrometry at night. Remote Sens. **5**, 4423–4449 (2013)
2. Cao, C., Shao, X., Xiong, X., Blonski, S., Liu, Q., Uprety, S., Shao, X., Bai, Y., Weng, F.: Suomi NPP VIIRS sensor data record verification, validation, and long-term performance monitoring. J. Geophys. Res. Atmos. (2013). https://doi.org/10.1002/2013JD020418

---

^{2} On-package high-bandwidth memory based on the multi-channel dynamic random access memory, MCDRAM.

3. Suomi NPP (National Polar-orbiting Partnership), Home Page, http://rammb. cira.colostate.edu/projects/npp/. Accessed 13 Apr 2017
4. Wright, N.J., Fuerlinger, K., Shan, H., Drummond, T., Canning, A., Shalf, J.: Best Practices for Hybrid OpenMP/MPI Programming on Hopper. The Cray Center of Excellence: Performance Optimization for the Multicore Era //NERSC/LBNL, Princeton Plasma Physics Lab; slides, October 2010, https://www.nersc.gov/ assets/NUG-Meetings/NUG-Oct2010-Wright.pdf
5. Li, D., de Supinski, B.R., Schulz, M., Cameron, K., Nikolopoulos, D.S.: Hybrid MPI, OpenMP power-aware computing. In: IEEE International Symposium on Parallel & Distributed Processing (IPDPS), Atlanta, GA, pp. 1–12 (2010). https:// doi.org/10.1109/IPDPS.2010.5470463
6. Bao, W., Tavarageri, S., Ozguner, F., Sadayappan, P.: PWCET: power-aware worst case execution time analysis. In: 2014 43rd International Conference on Parallel Processing Workshops, Minneapolis, MN, pp. 439–447 (2014). https://doi.org/10. 1109/ICPPW.2014.64
7. Al-Daoud, H., Al-Azzoni, I., Down, D.G.: Power-aware linear programming based scheduling for heterogeneous computer clusters. Future Gener. Comput. Syst. **28**(5), 745–754 (2012)
8. NOAA GOES-R Web Site, A collaborative NOAA & NASA program, http://www. goes-r.gov/. Accessed 13 Apr 2017
9. LANDSAT 8 (L8) DATA USERS HANDBOOK, LSDS-1574, Sioux Falls, South Dakota, March 29 (2016), https://landsat.usgs.gov/sites/default/files/documents/ Landsat8DataUsersHandbook.pdf. Accessed 13 Apr 2017
10. Himawari User's Guide, Home Pagev, Japan Meteorological Agency (JMA), http://www.jma-net.go.jp/msc/en/support/index.html. Accessed 13 Apr 2017
11. Lagarias, J.C., Reeds, J.A., Wright, M.H., Wright, P.E.: Convergence properties of the Nelder-Mead simplex method in low directions. SIAM J. Optim. **9**, 112–147 (1998)
12. Intel Hyper-Threading Technology, http://www.intel.com/content/www/us/en/ architecture-and-technology/hyper-threading/hyper-threading-technology.html. Accessed 10 Apr 2017
13. Intel 64 and IA-32 Architectures Software Developers Manual, Volume 3B: System Programming Guide, Part 2, http://www.intel.com/content/dam/www/public/ us/en/documents/manuals/64-ia-32-architectures-software-developer-vol-3b-part-2-manual.pdf
14. Rotem, E., Naveh, A., Ananthakrishnan, A., Weissmann, E., Rajwan, D.: Power-management architecture of the intel microarchitecture code-named sandy bridge. IEEE Micro **32**(2), 20–27 (2012). https://doi.org/10.1109/MM.2012.12
15. Jagode, H., YarKhan, A., Danalis, A., Dongarra, J.: Power management and event verification in PAPI. In: 9th Parallel Tools Workshop, Dresden, Germany, 2–3 September (2015)
16. HDF5 Tutorial, Parallel Topics, https://support.hdfgroup.org/HDF5/Tutor/ parallel.html

Automatic SIMD Vectorization of Loops: Issues, Energy Efficiency and Performance on Intel Processors

Olga Moldovanova[1,2(✉)] and Mikhail Kurnosov[1,2]

[1] Siberian State University of Telecommunications and Information Sciences,
Novosibirsk, Russia
{ovm,mkurnosov}@isp.nsc.ru
[2] Rzhanov Institute of Semiconductor Physics,
Siberian Branch of Russian Academy of Sciences, Novosibirsk, Russia

Abstract. In this paper we analyse how well compilers vectorize a well-known benchmark ETSVC consisting of 151 loops. The compilers we evaluated were Intel C/C++ 17.0, GCC C/C++ 6.3.0, LLVM/Clang 3.9.1 and PGI C/C++ 16.10. In our experiments we use dual CPU system (NUMA server, 2 x Intel Xeon E5-2620 v4, Intel Broadwell microarchitecture) with the Intel Xeon Phi 3120A co-processor. We estimate time, energy and speedup by running the loops in scalar and vector modes for different data types (double, float, int, short int) and determine loop classes which the compilers fail to vectorize. The Running Average Power Limit (RAPL) subsystem is used to obtain the energy measurements. We analyzed and proposed transformations for the loops that compilers failed to vectorize. After applying proposed transformations loops were successfully auto-vectorized by all compilers. The most part of the transformations based on loop interchange, fission by name and distribution.

Keywords: Loops · Compilers · Automatic vectorization · CPU energy consumption · Intel Xeon · Intel Xeon Phi

1 Introduction

Modern high-performance computer systems are multiarchitectural systems and implement several levels of parallelism: process level parallelism (PLP, message passing), thread level parallelism (TLP), instruction level parallelism (ILP), and data level parallelism (data processing by several vector arithmetic logic units). Processor vendors pay great attention to the development of vector extensions (Intel AVX, IBM AltiVec, ARM NEON SIMD). In particular, Fujitsu announced in its future version of the exascale K Computer system a transition to processors with the ARMv8.2-A architecture, which implements scalable vector extensions. And Intel extensively develops AVX-512 vector extension. That is why problem definitions and works on automatic vectorizing compilers have given

© Springer International Publishing AG 2017
V. Voevodin and S. Sobolev (Eds.): RuSCDays 2017, CCIS 793, pp. 388–399, 2017.
https://doi.org/10.1007/978-3-319-71255-0_31

the new stage in development in recent decades: OpenMP and Cilk Plus SIMD directives; Intel ISPC and Sierra language extensions; libraries: C++17 SIMD Types, Boost.SIMD, gSIMD, Cyme.

In this work we studied time, energy and speedup by running the loops in scalar and vector modes for different data types (double, float, int, short int) and compilers (Intel C/C++ Compiler, GCC C/C++, LLVM/Clang, PGI C/C++). The main goal is to determine loop classes which the compilers fail to vectorize. The Running Average Power Limit (RAPL) subsystem is used to obtain the energy measurements.

Since there was no information about vectorizing methods implemented in the commercial compilers, the evaluation was implemented by the "black box" method. We used the Extended Test Suite for Vectorizing Compilers [1–4] as a benchmark for our experiments to estimate an evolution of vectorizers in modern compilers comparing to an evaluation made in [1]. We determined classes of typical loops that the compilers used in this study failed to vectorize and evaluated them.

The rest of this paper is organized as follows: Sect. 2 discusses the main issues that explain effectiveness of vectorization; Sect. 3 describes the benchmark we used; Sect. 4 presents results of our experiments; and finally Sect. 5 concludes.

2 Vector Instruction Sets

Instruction sets of almost all modern processor architectures include vector extensions: MMX/SSE/AVX in the IA-32 and Intel 64 architectures, AltiVec in the Power architecture, NEON SIMD in the ARM architecture family, MSA in the MIPS. Processors implementing vector extensions contain one or several vector arithmetic logic units (ALU) functioning in parallel and several vector registers. Unlike vector systems of the 1990s, modern processors support execution of instructions with relatively short vectors (64–512 bits), loaded in advance from the RAM to the vector registers ("register-register" vector systems).

The main application of the vector extensions consists in decreasing of time of one-dimensional arrays processing. As a rule, a speedup achieved using the vector extensions is primarily determined by the number of array elements that can be loaded into a vector register. For example, each of 16 AVX vector registers is 256-bit wide. This allows loading into them 16 elements of the short int type (16 bits), 8 elements of the int or float type (32 bits) and 4 double elements (64 bits). Thus, when using AVX the expected speedup is 16 times for operations with short int elements, 8 times for int and float, and 4 for double.

The Intel Xeon Phi processors support AVX-512 vector extension and contain 32 512-bit wide vector registers. Each processor core with the Knights Corner microarchitecture contains one 512-bit wide vector ALU, and processor cores with the Knights Landing microarchitecture have two ALUs.

To achieve a maximum speedup during vector processing it is necessary to consider the microarchitectural system parameters. One of the most important of them is an alignment of array initial addresses (32-byte alignment for AVX and

64-byte alignment for AVX-512). Reading from and writing to unaligned memory addresses is executed slower. Effectiveness decreasing can also be caused by a mixed usage of SSE and AVX vector extensions. In such a case during transition from execution of one vector extension instructions to another one a processor stores (during transition from AVX to SSE) or restores (in another case) highest 128 bits of YMM vector registers (AVX-SSE transition penalties) [5].

When vector instructions are used, the achieved speedup can exceed the expected one. For example, after vectorization of the loop, which calculates an elementwise sum of two arrays, the processor overhead decreases due to reducing the number of add instruction loads from the memory and its decoding by the processor; the number of memory accesses for operands of the add instruction; the amount of calculations of loop end condition (the number of accesses to the branch prediction unit of the processor).

Besides that, a parallel execution of vector instructions by several vector ALUs can be a reason of additional speedup. Thus, an efficiently vectorized program overloads subsystems of a superscalar pipelined processor in a less degree. This is the reason of less processor energy consumption during execution of a vectorized program as compared to its scalar version [6].

Application developers have different opportunities to use vector instructions:

- inline assembler – full control of vectorization usage, least portable approach;
- intrinsics – set of data types and internal compiler functions, directly mapping to processor instructions (vector registers are allocated by compiler);
- SIMD directives of compilers, OpenMP and OpenACC standards;
- language extensions, such as Intel Array Notation, Intel ISPC, Apple Swift SIMD and libraries: C++17 SIMD Types, Boost.SIMD, SIMD.js;
- automatic vectorizing compiler – ease of use, high code portability.

In this work, we study the last approach. Such vectorizing technique does not require large code modification and provides its portability between different processor architectures.

3   Related Works and Benchmarks

We used the Extended Test Suite for Vectorizing Compilers (ETSVC) [2] as a benchmark containing main loop classes, typical for scientific applications in C language. The original package version was developed in the late 1980s by the J. Dongarra's group and contained 122 loops in Fortran to test the analysis capabilities of automatic vectorizing compilers for vector computer systems: Cray, NEC, IBM, DEC, Fujitsu and Hitachi [3,4]. In 2011 the D. Padua's group translated the TSVC suite into C and added to it new loops [1]. The extended version of the package contains 151 loops. The loops are divided into categories: dependence analysis (36 loops), vectorization (52 loops), idiom recognition (reductions, recurrences, etc., 27 loops), language completeness (23 loops). Besides that, the test suite contains 13 "control" loops, trivial loops that are expected to be vectorized by every vectorizing compiler.

The loops operate on one- and two-dimensional 16-byte aligned global arrays. The one-dimensional arrays contain $125 \cdot 1024/$`sizeof(TYPE)` elements of the given type `TYPE`, and the two-dimensional ones contain 256 elements by each dimension.

Each loop is contained in a separate function (see Listing 1). In the `init` function (line 5) an array is initialized by individual for this test values before loop execution. The outer loop (line 7) is used to increase the test execution time (for statistics issues). A call to an empty `dummy` function (line 10) is used in each iteration of the outer loop so that, in case where the inner loop is invariant with respect to the outer loop, the compiler is still required to execute each iteration rather than just recognizing that the calculation needs to be done only once [4].

After execution of the loop is complete, a checksum is computed by using elements of the resulting array (`check` function, line 16).

4 Results of Experiments

4.1 Test Environment

We used two systems for our experiments. The first system was a server based on two Intel Xeon E5-2620 v4 CPUs (Intel 64 architecture, Broadwell microarchitecture, 8 cores, Hyper-Threading was on, AVX 2.0 support), 64 GB RAM DDR4, GNU/Linux CentOS 7.3 x86-64 operating system (linux 3.10.0-514.2.2.el7 kernel). The second system was Intel Xeon Phi 3120 A co-processor (Knights Corner microarchitecture, 57 cores, AVX-512 support, 6 GB RAM, MPSS 3.8) installed in the server.

The compilers evaluated in these experiments were Intel C/C++ Compiler 17.0; GCC C/C++ 6.3.0; LLVM/Clang 3.9.1; and PGI C/C++ 16.10. The vectorized version of the ETSVC benchmark was compiled with the command line options shown in Table 1 (column 2). To generate the scalar version of the test suite the optimization options were used with the disabled compilers vectorizer (column 3, Table 1).

32-byte aligned global arrays were used for the Intel Xeon processor, and 64-byte aligned global arrays were used for the Intel Xeon Phi processor. We used arrays with elements of `double`, `float`, `int` and `short` data types for our evaluation.

4.2 Results for Intel 64 Architecture

The following results were obtained for the `double` data type on the Intel 64 architecture (Intel Xeon Broadwell processor). The Intel C/C++ Compiler vectorized 95 loops in total, 7 from which were vectorized by it alone. For GCC C/C++ the total amount of vectorized loops was 79. But herewith there was no loop that was vectorized only by this compiler. The PGI C/C++ vectorized the largest number of loops, 100, 13 from them were vectorized by it alone. The minimum number of loops was vectorized by the LLVM/Clang compiler,

Table 1. Compilers options

Compiler	Compilers options	Disabling vectorizer

Intel C/C++ 17.0	`-O3 -xHost -qopt-report3`
	`-qopt-report-phase=vec,loop`
	`-qopt-report-embed`
		`-no-vec`

GCC C/C++ 6.3.0	`-O3 -ffast-math -fivopts`
	`-march=native -fopt-info-vec`
	`-fopt-info-vec-missed`
		`-fno-tree-vectorize`

LLVM/Clang 3.9.1	`-O3 -ffast-math -fvectorize`
	`-Rpass=loop-vectorize`
	`-Rpass-missed=loop-vectorize`
	`-Rpass-analysis=loop-vectorize`
		`-fno-vectorize`

PGI C/C++ 16.10	`-O3 -Mvect -Minfo=loop,vect` `-Mneginfo=loop,vect`	`-Mnovect`

52, 4 from which were vectorized only by it. The number of loops unvectorized by any compiler was equal to 28.

We compared the obtained results with the evaluation done in [1]. The comparison shows that the vectorizer of the GCC C/C++ compiler has been significantly improved: 52.3% of vectorized loops from ETSVC in 2017 versus 32% in 2011 (see Table 2).

Table 2. Comparison of results with previous evaluations of compilers

2011 (Padua et al. [1])		2017 (our work)	
Intel C/C++ 12.0	90 loops (59.6 %)	Intel C/C++ 17.0	95 loops (62.9 %)
GCC C/C++ 4.7.0	59 loops (39 %)	GCC C/C++ 6.3.0	79 loops (52.3 %)

The similar results were obtained for arrays with elements of the `float` and `int` types by all compilers. The consistent results were obtained for the `short` type when Intel C/C++ Compiler, GCC C/C++ and LLVM/Clang were used. The exception to this rule was the PGI C/C++ compiler that vectorized no loops processing data of this type.

Figure 1 shows the results of loop vectorization for the `double` data type on the Intel 64 architecture. Abbreviated notations of the vectorization results are shown in the table cells. They were obtained from vectorization reports of compilers for all 151 loops. The full form of these notations is shown in Table 3. The similar results were obtained for other data types.

In the "Dependence analysis" category 9 loops were not vectorized by any compiler for the `double` data type. The compilers used in this study failed to vectorize loops with linear dependences (1st order recurrences), induction variables together with conditional and unconditional (`goto`) branches, loop nesting

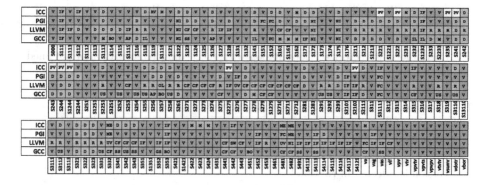

Fig. 1. Results of loops vectorization (Intel 64 architecture, `double` data type)

and variable values of lower and/or upper loop bounds and/or iteration step. In the last case, no compiler could determine whether a data dependence was present and took a pessimistic decision that the dependence existed.

In the "Vectorization" category the compilers failed to vectorize 11 loops. These loops required transformations as follows: loop fission, loop interchange, node splitting (to avoid cycles in data dependence graphs and output and anti-dependences [7]) and array expansions. Among causes of problems were interdependence of iteration counts of nested loops; linear dependencies in a loop body (1st order recurrences); conditional and unconditional branches in a loop body.

The following idioms (6 loops) from the "Idiom recognition" category were not vectorized by the compilers used: 1st and 2nd order recurrences, array searching, loop rerolling and reduction with function calls. The loops with recurrences were not vectorized because of linear data dependence. In a loop with array searching for the first element meeting a condition the unconditional branch `goto` prevented vectorization.

Compilers execute rerolling for loops that were unrolled by hand before vectorization [8]. The compilers in this study decided that vectorization of such loops was possible but inefficient. The reason was an indirect addressing in array elements access: $X[Y[i]]$, where X is a one-dimensional array of the `float` type, Y is a pointer to a one-dimensional array of integers, i is a loop iteration count.

The next challenging idiom was a reduction, namely sum of elements of a one-dimensional array. In this case the idiom was not vectorized because of test function calls. This function calculated sum of 4 array elements beginning from the one passed as the function argument. The Intel C/C++ Compiler reported that vectorization was possible but inefficient. Other compilers reported a function call as a reason of vectorization failing.

The "Language completeness" category contain 2 loops unvectorized by any compiler. The problem of both loops consisted in breaking loop computations (exit in the first case and break in the second case). Compiler vectorizers could not analyze control flow in these loops.

Total execution time of the benchmark (all loops) for each data type and compiler is shown in Fig. 2. A median value and maximum speedups of vectorized

Table 3. Abbreviated notations of vectorization results

V	Loop is vectorized
PV	Partial loop is vectorized (loop fission with succeeding vectorization of obtained loops)
RV	Remainder is not vectorized
IF	Vectorization is possible but seems inefficient
D	Vector dependence prevents vectorization (supposed linear or non-linear data dependence in a loop)
M	Loop is multiversioned (multiple loop versions are generated, unvectorized version is selected in runtime)
BO	Bad operation or unsupported loop bound (e.g., `sinf` or `cosf` function is used)
AP	Complicated access pattern (e.g., value of iteration count is more than 1)
R	Value that could not be identified as function is used outside the loop (induction variables are present in a loop)
IL	Inner-loop count not invariant (e.g., iteration count of inner loop depends on iteration count of outer loop)
NI	Number of iterations cannot be computed (lower and/or upper loop bounds are set by function's arguments)
CF	Control flow cannot be substituted for a select (conditional branches inside loop)
SS	Loop is not suitable for scatter store (e.g., in case of packing a two-dimensional array into a one-dimensional array)
ME	Loop with multiple exits cannot be vectorized (`break` or `exit` are present inside a loop)
FC	Loop contains function calls or data references that cannot be analyzed
OL	Value cannot be used outside the loop (scalar expansion or mixed usage of one- and two-dimensional arrays in one loop)
UV	Loop control flow is not understood by vectorizer (conditional branches inside a loop)
SW	Loop contains a `switch` statement
US	Unsupported use in statement (scalar expansion, wraparound variables recognition)
GS	No grouped stores in basic block (unrolled scalar product)

loops are shown in Figs. 3 and 4. The maximum speedup obtained on the Intel 64 architecture by the Intel C/C++ was 6.96 for the `double` data type, 13.89 for the `float` data type, 12.39 for `int` and 25.21 for `short int`. The maximum speedup obtained by GCC C/C++ was equal to 4.06, 8.1, 12.01 and 24.48 for types `double`, `float`, `int` and `short int`, correspondingly. The LLVM/Clang obtained results as follows: 5.12 (`double`), 10.22 (`float`), 4.55 (`int`) and 14.57 (`short int`). For PGI C/C++ these values were 14.6, 22.74, 34.0 and 68.0,

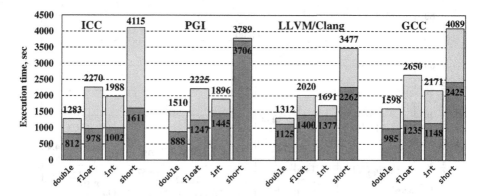

Fig. 2. Execution time of the benchmark (all loops) on the Intel Xeon E5-2620 v4 CPU: *outer columns* – scalar version of the benchmark; *inner columns* – vectorized version of the benchmark

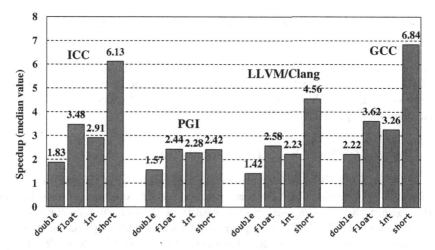

Fig. 3. Median value of speedup for vectorized loops on the Intel Xeon E5-2620 v4 CPU (only speedups above 1.15 were considered)

correspondingly. The speedup is the ratio of the running time of the scalar code over the running time of the vectorized code.

As our evaluation showed maximum speedups for Intel C/C++ Compiler, GCC C/C++ and LLVM/Clang correspond to the loops executing reduction operations (sum, product, minimum and maximum) with elements of one-dimensional arrays of all data types. These loops belong to the "Idiom recognition" category in the ETSVC. For PGI C/C++ maximum speedup was achieved for the loop calculating an identity matrix ("Vectorization" category) for the **double** and **float** data types. And for **int** and **short** this value was obtained in the loop calculating product reduction ("Idiom recognition" category).

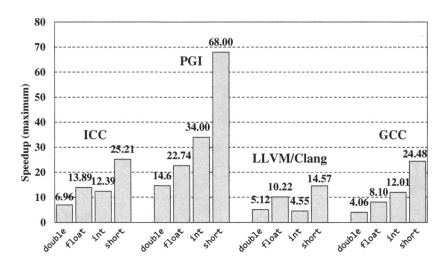

Fig. 4. Maximum speedup for vectorized loops on the Intel Xeon E5-2620 v4 CPU

However, the obtained speedup is not always a result of vectorization. For the PGI C/C++ compiler the speedup value 68.0 for the short data type can be explained by the fact that calculations in a loop are not executed at all because of the compiler optimization.

4.3 Results for Intel Xeon Phi Architecture

On the Intel Xeon Phi architecture we studied vectorizing capabilities of the Intel C/C++ Compiler 17.0. The -mmic command line option was used instead of the -xHost during compilation. The results of the experiments for two data types are shown in Fig. 5. The compiler could vectorize 99 loops processing data of the double type and 102 of the float type. Supposed data dependencies (28 loops for the double type and 27 for the float type) were the main reason of loop vectorization failing. 12 loops were partially vectorized for both data types. Similar results were obtained for the int and short types.

In this case the maximum speedup for the double type was 13.7, for float – 19.43, int – 30.84, and short – 46.3. For float and short maximum speedups were obtained for loops executing reduction operations for elements of one-dimensional arrays. For the double data type sinf and cosf functions were used in a loop. In the case with int it was a "control" loop vbor calculating a scalar product of six one-dimensional arrays.

4.4 Effect of Vectorization on CPU Energy Consumption

We modified the ETSVC benchmark to measure the CPU (Intel Xeon E5-2620v4) energy for each loop. The measurements were accomplished by using the Intel RAPL (Running Average Power Limit) subsystem before and after each

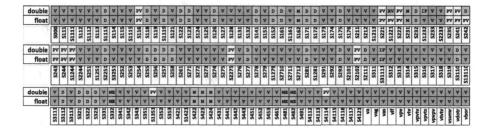

Fig. 5. Results of loops vectorization (Intel Xeon Phi architecture)

Table 4. Statistical characteristics for the decrease E of CPU energy consumption (for successfully vectorized loops, execution time of which is less than execution time of their scalar versions at least on 15%)

Compiler	Data type	Min, %	Max, %	Avg, %	Median, %
Intel C/C++ 17.0	double	13	85	42	41
	float	14	91	64	70
	int	13	92	62	65
	short	70	99	94	96
GCC C/C++ 6.3.0	double	17	75	52	60
	float	16	99	71	73
	int	13	91	67	70
	short	13	96	80	85
LLVM/Clang 3.9.1	double	17	79	37	28
	float	21	99	62	59
	int	26	77	50	52
	short	46	99	92	96
PGI C/C++ 16.10	double	11	93	48	39
	float	15	96	60	58
	int	10	96	52	54
	short	52	99	92	96

loop execution. We requested information about total CPU energy consumption (RAPL PKG domain) and DRAM controller energy consumption (RAPL DRAM domain) from the RAPL subsystem.

For every loop we determined the decrease E of CPU energy consumption (RAPL PKG domain) for vectorized loop execution against its scalar version execution:

$$E = (E_{novec} - E_{vec})/E_{novec} \cdot 100\%, \qquad (1)$$

where E_{novec} is CPU energy for scalar loop execution ($[E_{novec}] = $ J), E_{vec} is CPU energy for vectorized loop execution ($[E_{vec}] = $ J).

In Table 4 we show the results for the decrease E of CPU energy consumption only for successfully vectorized loops, execution time of which is less than execution time of their scalar versions at least on 15%.

For arrays with elements of `double` type vectorized loops decreased the CPU energy consumption by a mean of 45% as compared to their scalar versions. For `float`, `int` and `short int` the CPU energy consumption decrease was 64%, 58% and 90%, correspondingly.

It is apparent that for the ETSVC benchmark increasing the number of array elements which can be loaded into a vector register (due to decreasing the size of data type) results in decreasing the CPU energy consumption.

5   Conclusion

In this work we studied auto-vectorizing capabilities of modern optimizing compilers Intel C/C++ Compiler, GCC C/C++, LLVM/Clang, PGI C/C++ on the Intel 64 and Intel Xeon Phi architectures. Our study shows that the compilers evaluated could vectorize 39–77 % of the total number of loops in the ETSVC package. The best results were shown by the Intel C/C++ Compiler, and the worst ones – by the LLVM/Clang compiler. The compilers failed to vectorize loops containing conditional and unconditional branches, function calls, induction variables, variable loop bounds and iteration count, as well as such idioms as 1st or 2nd order recurrences, search loops and loop rerolling. We analyzed and proposed transformations for the loops that compilers failed to vectorize. After applying proposed transformations loops was successfully auto-vectorized by all compilers. The most part of the transformations based on loop interchange, fission and distribution.

We estimated the CPU energy consumption for execution of vectorized loops against their scalar versions. The experiments show that increasing the number of array elements which can be loaded into a vector register (due to decreasing the size of data type) results in decreasing the CPU energy consumption.

The future work will consist of evaluation and development of vectorizing methods (polyhedral model) for the obtained class of challenging loops, applicability analysis of JIT compilation [9] and profile-guided optimization.

Acknowledgement. This work is supported by Russian Foundation for Basic Research (projects 16-07-00992, 15-07-00653).

References

1. Maleki, S., Gao, Y., Garzarán, M.J., Wong, T., Padua, D.A.: An evaluation of vectorizing compilers. In: Proceedings of the International Conference on Parallel Architectures and Compilation Techniques, pp. 372–382 (2011)
2. Extended Test Suite for Vectorizing Compilers. http://polaris.cs.uiuc.edu/~maleki1/TSVC.tar.gz

3. Callahan, D., Dongarra, J., Levine, D.: Vectorizing compilers: a test suite and results. In: Proceedings of the ACM/IEEE Conference on Supercomputing, pp. 98–105 (1988)
4. Levine, D., Callahan, D., Dongarra, J.: A comparative study of automatic vectorizing compilers. J. Parallel Comput. **17**, 1223–1244 (1991)
5. Konsor, P.: Avoiding AVX-SSE Transition Penalties. https://software.intel.com/en-us/articles/avoiding-avx-sse-transition-penalties
6. Jibaja, I., Jensen, P., Hu, N., Haghighat, M., McCutchan, J., Gohman, D., Blackburn, S., McKinley, K.: Vector parallelism in JavaScript: language and compiler support for SIMD. In: Proceedings of the International Conference on Parallel Architecture and Compilation Techniques, pp. 407–418 (2015)
7. Program Vectorization: Theory, Methods, Implementation (1991)
8. Metzger, R.C., Wen, Z.: Automatic Algorithm Recognition and Replacement: A New Approach to Program Optimization. MIT Press, Cambridge (2000)
9. Rohou, E., Williams, K., Yuste, D.: Vectorization technology to improve interpreter performance. ACM Trans. Archit. Code Optim. **9**(4), 26:1–26:22 (2013)

Improving the Performance of an AstroPhi Code for Massively Parallel Supercomputers Using Roofline Analysis

Boris Glinskiy, Igor Kulikov, and Igor Chernykh[✉]

Institute of Computational Mathematics and Mathematical Geophysics SB RAS,
Lavrentjeva Ave. 6, 630090 Novosibirsk, Russia
gbm@sscc.ru, {kulikov,chernykh}@ssd.sscc.ru

Abstract. Astrophysics is the branch of astronomy that employs the principles of physics and chemistry "to ascertain the nature of the heavenly bodies, rather than their positions or motions in space". Numerical modeling plays a key role in modern astrophysics. It is the main tool for the research of nonlinear processes and provides communication between the theory and observational data. New massive parallel supercomputers provide an opportunity to simulate these kinds of problems in high details. Our astrophysics code AstroPhi was written for new massive parallel supercomputers based Intel Xeon Phi architecture. The original numerical method based on the combination of the Godunov method, operator splitting approach and piecewise-parabolic method on local stencil was used for numerical solution of the hyperbolic equations. The piecewise-parabolic method on local stencil provides the high-precision order. After the transition of AstroPhi to KNL architecture, we obtained abnormally low performance of solver on KNL cores. In this paper, we will show the roofline analysis using Intel Advisor application and the results of the AstroPhi optimizations.

Keywords: Massively parallel supercomputers · Astrophysics · Roofline analysis

1   Introduction

Numerical simulation in astrophysics allows to research many important problems such as the collision and evolution of galaxies, chemical evolution of stars, identification of dark matter and more. Modern supercomputers have given us the possibility of detailed astrophysics modeling that considers different physical effects such as magnetohydrodynamics, chemical kinetics, cooling/heating, and more. One of the most interesting developments in supercomputer technology at this moment is massively parallel supercomputers. The main concept of this technology is based on the possibility of massive usage of computational cores of CPUs or GPUs. Recently the scientific community is widely discussing the transition to exascale supercomputers. The main global challenge is the development of algorithms that can consider the massive exascale level parallelism. One of the problems is the difficulty of debugging and optimization of massively-parallel codes. The difficulties of debugging of parallel code connect with the problem of a greater propensity for race conditions, asynchronous events, and the general difficulty of trying to understand N processes simultaneously executing. Modern parallel debugging tools such

© Springer International Publishing AG 2017
V. Voevodin and S. Sobolev (Eds.): RuSCDays 2017, CCIS 793, pp. 400–406, 2017.
https://doi.org/10.1007/978-3-319-71255-0_32

as Eclipse Parallel Tools Platform [1], Intel Debugger [2], Nvidia nsight [3] helps in this problem. The difficulties of massively-parallel code optimizations are based on array dependence analysis, pointer alias analysis, loop transformations, adaptive profile-directed optimizations, and dynamic compilation [4]. Last years the roofline model [5] became a very popular tool for application performance analysis and optimization. The Roofline model is an intuitive visual performance model used to provide performance estimates of a given compute kernel or application running on multicore, many-core, or accelerator processor architectures, by showing inherent hardware limitations, and potential benefit and priority of optimizations. By combining locality, bandwidth, and different parallelization paradigms into a single performance figure, the model can be an effective alternative to assess the quality of attained performance instead of using simple percent-of-peak esti-mates, as it provides insights on both the implementation and inherent performance limi-tations [6]. The main idea of roofline model based on visualizing of application perform-ance as a function of arithmetic intensity, where the application performance is the number of floating point operations per second (FLOPS) and the arithmetic intensity is the ratio of application performance to the memory traffic created by the application. Modern roofline analysis tools such as Intel Advisor [7] shows this data for each loop in the application and visualize the machine peak performance and machine peak memory bandwidth for target CPU architecture. Figure 1 showing the typical roofline chart [8]. We can see in this figure that some of the application's loops using cache because the arithmetic intensity is higher than DRAM peak bandwidth. Figure 2 helps to identify which kind of your application is: memory-bound, memory/compute-bound or compute-bound application. After classifica-tion of application, this chart helps to build the strategy for optimization. Vectorization of application will increase the performance of the compute-bound code. L1/L2 cache opti-mizations of code will increase the performance of the memory-bound code.

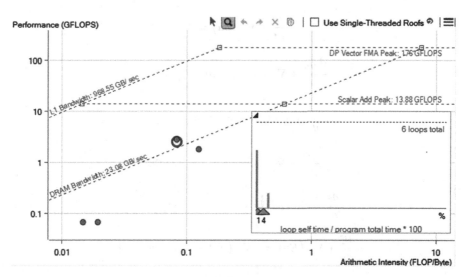

Fig. 1. Typical roofline chart with memory bandwidth and peak performance data of the target architecture [8].

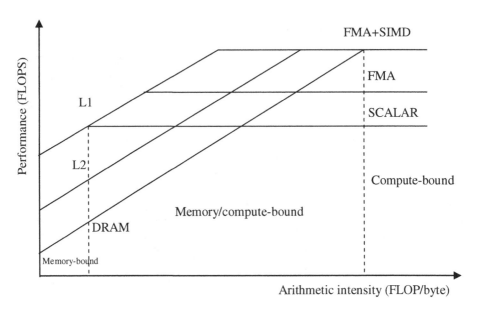

Fig. 2. Classification of application with roofline analysis.

2   Mathematical Model and Numerical Method

In our work, we use a multicomponent hydrodynamic model of galaxies considering the chemodynamics of molecular hydrogen and cooling in the following form:

$$\frac{\partial \rho}{\partial t} + \nabla \cdot \left(\rho \vec{u}\right) = 0,$$

$$\frac{\partial \rho_{H_2}}{\partial t} + \nabla \cdot \left(\rho_{H_2} \vec{u}\right) = S\left(\rho, \rho_H, \rho_{H_2}\right),$$

$$\frac{\partial \rho_H}{\partial t} + \nabla \cdot \left(\rho_H \vec{u}\right) = -S\left(\rho, \rho_H, \rho_{H_2}\right),$$

$$\frac{\partial \rho \vec{u}}{\partial t} + \nabla \cdot \left(\rho \vec{u} \vec{u}\right) = -\nabla p - \rho \nabla \Phi,$$

$$\frac{\partial \varepsilon}{\partial t} + \nabla \cdot \left(\varepsilon \vec{u}\right) = -(\gamma - 1)\varepsilon \nabla \cdot \left(\vec{u}\right) - Q,$$

$$\frac{\partial E}{\partial t} + \nabla \cdot \left(E\vec{u}\right) = -\nabla \cdot \left(p\vec{u}\right) - \left(\rho \nabla \Phi, \vec{u}\right) - Q,$$

$$\Delta \Phi = 4\pi G \rho,$$

$$E = \varepsilon + \frac{\rho \vec{u}}{2},$$

$$p = (\gamma - 1)\varepsilon,$$

where ρ is density, ρ_H is atomic hydrogen density, ρ_{H_2} is molecular hydrogen density, $\vec{u}$ is the velocity vector, ε is internal energy, p is pressure, E is total energy, γ is the ratio of specific heats, Φ is gravity, G is the gravitational constant, S is the formation rate of molecular hydrogen, and Q is a cooling function. A detailed description of this model can be found in [9].

The formation of molecular hydrogen is described by an ordinary differential equation [10]:

$$\frac{dn_{H_2}}{dt} = R_{gr}(T)n_H\left(n_H + 2n_{H_2}\right) - \left(\xi_H + \xi_{diss}\right)n_{H_2},$$

where n_H is the concentration of atomic hydrogen, n_{H_2} is the concentration of molecular hydrogen, and T is temperature. Detailed descriptions of the H_2 formation rate R_{gr} and the photodissociation ξ_H, ξ_{diss} of molecular hydrogen, can be found in [11, 12]. Chemical kinetics was don with using of CHEMPAK tool [13, 14].

The original numerical method based on the combination of the Godunov method, operator splitting approach and piecewise-parabolic method on local stencil was used for numerical solution of the hyperbolic equations [15]. The piecewise-parabolic method on local stencil provides the high-precision order. The equation system is solved in two stages: at the Eulerian stage, the equations are solved without advective terms and at the Lagrangian stage, the advection transport is being performed. At the Eulerian stage, the hydrodynamic equations for both components are written in the non-conservative form and the advection terms are excluded. As the result, such a system has an analytical solution on the two-cell interface. This analytical solution is used to evaluate the flux through the two-cell interface. In order to improve the precision order, the piecewise-parabolic method on the local stencil (PPML) is used. The method is the construction of local parabolas inside the cells for each hydrodynamic quantity. The main difference of the PPML from the classical PPM method is the use of the local stencil for computation. It facilitates the parallel implementation by using only one layer for subdomain overlapping. It simplifies the implementation of the boundary conditions and decreases the number of communications thus improving the scalability. The detailed description of this method can be found in [16]. The same approach is used for the Lagrangian stage. Now the Poisson equation solution is based on Fast Fourier Transform method. This is because the Poisson equation solution takes several percents of the total computation time. After the Poisson equation solution, the hydrodynamic equation system solution is corrected. It should be noticed here that the system is over defined. The correction is performed by means of the original procedure for the full energy conservation and the guaranteed entropy nondecrease. The procedure includes the renormalization of the velocity vector length, its direction remaining the same (on boundary gas-vacuum) and the entropy (or internal energy) and dispersion velocity tensor correction. Such a modification of the method keeps the detailed energy balance and guaranteed non-decrease of entropy.

3   Roofline Analysis

Roofline analysis with using of Intel Advisor consists of 3 steps: survey collection, trip count collection, visualization and/or extraction of the collected data into a report. Before analysis, the application should be compiled in debug mode.

1. Survey collection by command line with advisor:
   ```
   mpirun  -n  <number  of  KNL  nodes>  advixe-cl  -collect
   survey --trace-mpi - ./<app_name>
   ```
2. Trip count collection by command line with advisor:
   ```
   mpirun -n <number of KNL nodes> advixe-cl -collect trip-
   counts -flops-and-masks --trace-mpi - ./<app_name>
   ```
3. Extraction of the data in a report:
   ```
   advixe-cl   -report   survey   -show-all-columns   --
   format=text -- report-output report.txt
   ```

In our research, we used RSC Tornado-F [17] experimental node with Intel Xeon Phi 7250 (16 GB MCDRAM) processor. We used all 68 cores for the tests. Figure 3 shows roofline chart for AstroPhi application before optimizations. We can see that main loop of the application has very low arithmetic intensity (less than 0.1 FLOP/byte) and very low performance (less than 1 GFLOPS).

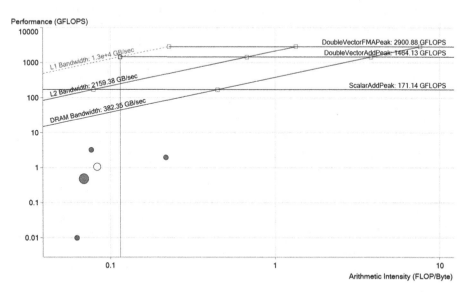

Fig. 3. Roofline chart for AstroPhi code before optimizations. The red dot is the main loop of the application. (Color figure online)

Intel Advisor proposed some optimizations for gaining performance. The main of these optimizations are to remove the vector dependencies, to optimize memory access patterns, to move source loop iterations from peeled/remainder loops to the loop body.

After the optimization roofline analysis was repeated on the same hardware with the same analysis steps. Figure 4 shows roofline chart for AstroPhi application after optimizations. After all improvements in AstroPhi application, we achieved 190GFLOPS performance and 0.3 FLOP/byte arithmetic intensity with 100% mask utilization and 573 GB/s memory bandwidth.

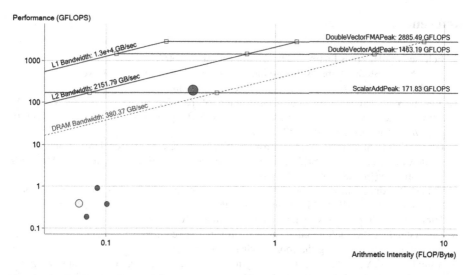

Fig. 4. Roofline chart for AstroPhi code after optimizations. The red dot is the main loop of the application. (Color figure online)

4   Conclusion

Numerical modeling plays a key role in modern astrophysics. It is the main tool for the research of nonlinear processes and provides communication between the theory and observational data. Numerical simulation in astrophysics allows detailed investigation of the collision and evolution of galaxies. Author's astrophysics code was written for new massive parallel supercomputers based Intel Xeon Phi architecture. The original numerical method based on the combination of the Godunov method, operator splitting approach and piecewise-parabolic method on local stencil was used for numerical solution of the hyperbolic equations. The piecewise-parabolic method on local stencil provides the high-precision order. After the transition of AstroPhi to Intel Xeon Phi KNL architecture, we obtained abnormally low usage of KNL's cores. The roofline analysis of our code with using of Intel Advisor showed that main loop has very low arithmetic intensity (less than 0.1 FLOP/byte) and very low performance (less than 1 GFLOPS). Due to recommendations of Intel Advisor, vector dependencies were removed, memory operations were optimized, and arrays sizes were adapted for KNL architecture. After these improvements, we achieved 190GFLOPS performance and 0.3 FLOP/byte arithmetic intensity with 100% mask utilization and 573 GB/s memory bandwidth. This arithmetic intensity is standard for this kind of algorithms.

Acknowledgments. This work was partially supported by the Grant of the President of Russian Federation for the support of young scientists number MK – 1445.2017.9, RFBR grant 15-01-00508, 16-29-15120 and 16-07-00434.

References

1. Eclipse Parallel Tools Platform. http://www.eclipse.org/ptp/
2. Intel Parallel Studio. https://software.intel.com/en-us/intel-parallel-studio-xe
3. Nvidia Nsight. http://www.nvidia.com/object/nsight.html
4. Sarkar, V.: Challenges in code optimization of parallel programs. In: de Moor, O., Schwartzbach, M.I. (eds.) CC 2009. LNCS, vol. 5501, p. 1. Springer, Heidelberg (2009). https://doi.org/10.1007/978-3-642-00722-4_1
5. Ofenbeck, G., Steinmann, R., Caparros, V., Spampinato, D.G., Püschel, M.: Applying the roofline model. In: 2014 IEEE International Symposium on Performance Analysis of Systems and Software (ISPASS), pp. 76–85 (2014)
6. Ren, D.Q.: Algorithm level power efficiency optimization for CPU–GPU processing element in data intensive SIMD/SPMD computing. J. Parallel Distrib. Comput. **71**, 245–253 (2011)
7. Intel Advisor. https://software.intel.com/en-us/intel-advisor-xe
8. Understanding the roofline chart. https://software.intel.com/en-us/intel-advisor-2017-user-guide-linux-understanding-the-roofline-chart?language=fr
9. Vshivkov, V.A., Lazareva, G.G., Snytnikov, A.V., Kulikov, I.M., Tutukov, A.V.: Hydrodynamical code for numerical simulation of the gas components of colliding galaxies. Astrophys. J. Suppl. Ser. **194**(47), 1–12 (2011)
10. Bergin, E.A., Hartmann, L.W., Raymond, J.C., Ballesteros-Paredes, J.: Molecular cloud formation behind shock waves. Astrophys. J. **612**, 921–939 (2004)
11. Khoperskov, S.A., Vasiliev, E.O., Sobolev, A.M., Khoperskov, A.V.: The simulation of molecular clouds formation in the Milky Way. Mon. Not. R. Astron. Soc. **428**(3), 2311–2320 (2013)
12. Glover, S., Mac Low, M.: Simulating the formation of molecular clouds. I. Slow formation by gravitational collapse from static initial conditions. Astrophys. J. Suppl. Ser. **169**, 239–268 (2006)
13. Chernykh, I., Stoyanovskaya, O., Zasypkina, O.: ChemPAK software package as an environment for kinetics scheme evaluation. Chem. Prod. Process Model. **4**(4), 1934–2659 (2009)
14. Snytnikov, V.N., Mischenko, T.I., Snytnikov, V., Chernykh, I.G.: Physicochemical processes in a flow reactor using laser radiation energy for heating reactants. Chem. Eng. Res. Des. **90**(11), 1918–1922 (2012)
15. Godunov, S.K., Kulikov, I.M.: Computation of discontinuous solutions of fluid dynamics equations with entropy nondecrease guarantee. Comput. Math. Math. Phys. **54**, 1012–1024 (2014)
16. Kulikov, I., Vorobyov, E.: Using the PPML approach for constructing a low-dissipation, operator-splitting scheme for numerical simulations of hydrodynamic flows. J. Comput. Phys. **317**, 316–346 (2016)
17. RSC Tornado. http://www.rscgroup.ru/en/our-technologies/267-rsc-tornado-cluster-architecture

Using Simulation to Improve Workflow Scheduling in Heterogeneous Computing Systems

Alexey Nazarenko and Oleg Sukhoroslov[✉]

Institute for Information Transmission Problems
of the Russian Academy of Sciences, Moscow, Russia
nazar@phystech.edu, sukhoroslov@iitp.ru

Abstract. Workflows is an important class of parallel applications that consist of many tasks with logical or data dependencies. A multitude of scheduling algorithms have been proposed to optimize the workflow execution in heterogeneous computing systems. However, in order to be efficiently applied in practice, these algorithms require accurate estimates of task execution and communication times. In this paper two modifications of the well-known HEFT algorithm are investigated that use simulation instead of simple analytical models in order to better estimate data transfer times. The results of experimental study show that the proposed approach can improve makespan for data-intensive workflows with high parallelism and communication-to-computation ratio.

Keywords: Workflow · Scheduling · Simulation · Heterogeneous systems · Distributed computing

1 Introduction

Heterogeneous computing systems (HCSs) composed of different computational units or standalone resources, which can be local or geographically distributed, are widely used nowadays for executing parallel applications. Workflows [14] is an important class of such applications that consist of many tasks with logical or data dependencies which can be modeled as directed acyclic graphs (DAGs).

The efficiency of executing workflows in HCS critically depends on the methods used to schedule the workflow tasks, i.e. decide when and which resource must execute the tasks of the workflow. The main objective is to minimize the overall completion time or makespan subject to possible additional constraints such as meeting a deadline or using a fixed budget. In comparison to homogeneous systems, the task scheduling problem in HCS is more complicated because of the different execution rates of individual resources and different communication rates of links between these resources.

The DAG scheduling problem has been shown to be NP-complete [9], even for the homogeneous case. This makes it practically impossible to obtain the

© Springer International Publishing AG 2017
V. Voevodin and S. Sobolev (Eds.): RuSCDays 2017, CCIS 793, pp. 407–417, 2017.
https://doi.org/10.1007/978-3-319-71255-0_33

optimal schedule even for the simplest formulations of practical interest. Therefore the research effort in this field has been mainly to obtain low complexity heuristics that produce good schedules. Since the late 1990s and until now, a multitude of workflow scheduling algorithms [18] based on different heuristics and metaheuristics have been proposed. However, in order to be efficiently applied in practice, these algorithms require accurate estimates of task execution and communication times.

In this paper we focus on the accuracy of models used for estimation of data transfer times. The presented experimental results provide a strong evidence against the widely used approach based on simple Hockney's model [11] that disregard network topology and bandwidth allocation. The schedules produced by static algorithms using this model clearly demonstrate that even for the modestly parallel workloads with sufficiently large data items the effect of competing data transfers may lead to the drastic underestimation of the communication time and the makespan degradation.

To address this issue we propose to incorporate simulation inside a workflow scheduling algorithm in order to improve the data transfer time estimates. Simulation, involving computer modeling of the process of application execution in HCS, has been actively used in scheduling algorithm research. The main advantage of simulation in comparison to the real-world experiments is the ability to perform a statistically significant number of experiments in a reasonable amount of time while ensuring the reproducibility and having moderate hardware resource requirements. However, while being widely used to evaluate the scheduling algorithms, the simulation has been rarely used inside the algorithms.

In this paper we investigate the use of more accurate simulation models instead of simple analytical models inside a workflow scheduling algorithm. Two modifications of the well-known HEFT algorithm [15] are proposed that use simulation in order to estimate data transfer times. The proposed modifications are compared with original HEFT and other scheduling algorithms using the developed simulation framework. The obtained experimental results show that the proposed approach can improve the makespan for workflows with high parallelism and communication-to-computation ratio.

The paper is structured as follows. Section 2 describes the used system and application models along with the used simulation framework. Section 3 provides an overview of HEFT algorithm and presents the proposed algorithm modifications. Section 4 presents and discusses the results of simulation experiments. Section 5 concludes and discusses future work.

2   Simulation Framework

To study the workflow scheduling algorithms in this paper we use simulation by modeling the process of application execution in a distributed computing system. In comparison with the full-scale experiments on real systems, simulation allows to significantly reduce the time needed to run an experiment and to ensure the reproducibility of produced results, while having moderate requirements to the used hardware resources. However, when using simulation it is important to ensure the accuracy, i.e. minimal deviation from the results of real-world experiments, and the scalability, i.e. the ability to conduct large-scale experiments, of the used simulation model.

The simulation model used in this paper is implemented on the base of Sim-Grid[1] [6], a simulation toolkit for studying the behaviour of large-scale distributed systems. The toolkit provides the required fundamental abstractions for the discrete-event simulation of parallel applications in distributed environments. The choice of SimGrid was motivated by the maturity of the toolkit, the soundness and high level of verification of embedded models, and the active support of developers. An important factor is also the versatility of the toolkit that allows one to simulate grids, cloud infrastructures, peer-to-peer systems and MPI applications.

Many studies also used WorkflowSim [7], an open source toolkit for simulating scientific workflows based on CloudSim simulator. We avoided the use of WorkflowSim as it has been shown that CloudSim among other simulators has flaws in its network model [17].

The heterogeneous computing system is modeled as a set of hosts and network links between them as depicted on Fig. 1. Each host is characterized by its performance expressed in FLOPS. In this study it is assumed that each host can process a single task at a time. The execution of any task is considered non-preemptive. Network links are characterized by their bandwidth and latency.

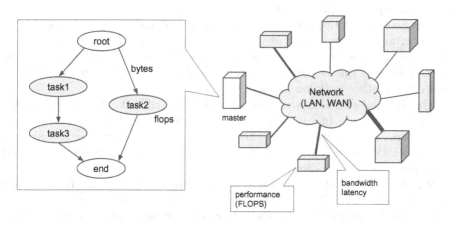

Fig. 1. Workflow and heterogeneous computing system models.

While the simulation is widely used to assess scheduling algorithms, the researchers often neglect the accuracy of the used models, especially network ones. In particular, in many papers authors assume a contention-free network model in which a network host can simultaneously send to or receive data from

---

[1] http://simgrid.gforge.inria.fr/.

as many hosts as possible without experiencing any performance degradation. However, this model is not representative of real world networks. In this study we use the bounded multiport model provided by SimGrid. In this model, a host can communicate with several other hosts simultaneously, but each communication flow is limited by the bandwidth of the traversed route, and communications using a common network link have to share bandwidth. This scheme corresponds well to the behavior of TCP connections on a LAN. The validity of this network model has been demonstrated in [16].

SimGrid supports simulation of various network topologies including hierarchies and combinations of autonomous systems with different internal routing strategies. In this study we consider systems with a simple topology where each host is connected to a central backbone via a dedicated link as depicted on Fig. 1, and a route between any two hosts contains the two respective links. The backbone, which can correspond to the LAN switch or the WAN, doesn't impose additional latency or bandwidth constraints in this model. Therefore the rate of communication between any pair of hosts is determined only by characteristics of the corresponding pair of links.

The workflow application is modeled as a directed acyclic graph (DAG), whose vertices correspond to individual tasks and directed edges represent the data dependencies between tasks as depicted on Fig. 1. Each vertex is characterized by its size, i.e. the amount of computations in flops associated with the corresponding task. Similarly, each edge is characterized by the amount of communication in bytes between the corresponding pair of tasks. The size of task input data equals to the sum of sizes of incoming edges.

The two special tasks with zero size are introduced in order to model the staging of workflow input and output data. The *root* task passes the input data to the initial tasks, i.e. those that do not depend on other workflow tasks. The *end* task receives the output data from the final tasks, i.e. those that do not pass their data to other workflow tasks.

The *root* and *end* tasks are executed on a dedicated host called *master*, which does not participate in computations. This host corresponds to the machine, which stores the input data and where the output data should be placed after the application execution. In practice, this host often performs submission and management of the workflow.

While the SimGrid toolkit has been used previously for studying workflow scheduling algorithms [2,12], to the best of our knowledge there are no published open source implementations of such algorithms for SimGrid. Therefore we have implemented a number of well-known static and dynamic algorithms, such as HEFT [15], HCPT [10], Lookahead [5], PEFT [1], OLB [3], MCT [13], MinMin [8,13], MaxMin [8,13] and Sufferage [13], following their original papers.

To simplify the implementation of scheduling algorithms for our experiments we have developed a *pysimgrid* library[2]. This library implements a thin wrapper around the native SimGrid API and provides a convenient interface for development of scheduling algorithms in Python language. The library also includes

---

[2] https://github.com/alexmnazarenko/pysimgrid.

auxiliary tools for generation of synthetic systems and workflows, batch execution of simulation experiments and analysis of simulation results.

3 Modifications of HEFT Algorithm Using Simulation

3.1 HEFT Overview

Heterogeneous Earliest Finish Time (HEFT) [15] is probably the most cited workflow scheduling algorithm. Being relatively simple and proved to be consistently more efficient than other algorithms, HEFT is commonly used as a reference for evaluation of new algorithms.

HEFT can be described as a variant of static list scheduling algorithms that prioritize tasks having the most influence on the total workflow execution time (makespan). Such algorithms operate in two phases. During the first phase the algorithm computes the rank of each task according to some criterion that takes into account the position of the task in the DAG, its dependencies, etc. The output of the ranking phase is a list of tasks sorted by their rank. During the second phase the algorithm iterates over the list and assigns each task to a host that minimizes some criterion, for example task completion time.

The rank of a task T_i in HEFT is recursively defined by

$$rank(T_i) = \overline{EET}(T_i) + \max_{T_j \in succ(T_i)} \left(\overline{ECOMT}(data_{ij}) + rank(T_j) \right), \qquad (1)$$

where $\overline{EET}(T_i)$ is the average execution time of the task across all hosts, $succ(T_i)$ is the set of immediate successors of the task, $ECOMT(c_{ij})$ is the average communication time corresponding to the transfer of $data_{ij}$ bytes via edge (i, j).

The $\overline{EET}(T_i)$ is computed by averaging the estimated execution time $EET(T_a, H_i)$ of a task on each host H_i which is assumed to be known beforehand. In our model we compute accurate estimates using the task size and host performance.

The $\overline{ECOMT}(data_{ij})$ is computed in HEFT using the Hockney's model [11] as

$$\overline{ECOMT}(data_{ij}) = \overline{L} + \frac{data_{ij}}{\overline{B}}, \qquad (2)$$

where $\overline{L}$ is the average latency and $\overline{B}$ is the average bandwidth of communication links between the hosts in the system.

The tasks in HEFT are scheduled in decreasing order of their rank. Each task is scheduled to a host with a minimum estimated completion time

$$ECT(T_i, H_j) = EST(T_i, H_j) + EET(T_i, H_i), \qquad (3)$$

where $EST(T_i, H_j)$ is the earliest start time of the task on a given host

$$EST(T_i, H_j)$$
$$= \max \{ avail(H_j), \max_{T_k \in pred(T_i)} (ECT(T_k, H_k) + ECOMT(data_{ki}, H_k, H_i)) \},$$
$$(4)$$

where $avail(H_j)$ is the earliest time the host is ready for task execution, $pred(T_i)$ is the set of immediate predecessors of the task.

Note the important feature of the rank function — it defines a valid topological order for the tasks. All tasks are scheduled after their parents, so it is possible to compute the required estimates of parent tasks' completion and communication times.

The communication time between tasks T_i and T_j running on hosts H_i and H_j respectively is computed as

$$ECOMT(data_{ij}, H_i, H_j) = L_{ij} + \frac{data_{ij}}{B_{ij}}, \tag{5}$$

where L_{ij} and B_{ij} are the latency and the bandwidth of the link between the given hosts.

3.2   Modified HEFT Versions

The simple linear model used in HEFT to estimate communication times doesn't take into account network topology and bandwidth allocation. A shown in Sect. 4, even for the modestly parallel workflows with sufficiently large data dependencies the effect of competing data transfers may lead to the drastic underestimation of the communication time and the degradation of HEFT performance. This is due to the fact that the inaccurate estimates of $ECOMT(data_{ij}, H_i, H_j)$ lead to inaccurate estimates of $ECT(T_i, H_j)$, and these inaccuracies accumulate during the scheduling of subsequent tasks. The ranking function also doesn't take into account the bandwidth contention by using a simple $\overline{ECOMT}(data_{ij})$ estimate.

To address this problem we modified HEFT to use simulation instead of analytical models to improve the used estimates. The proposed HEFT modification, hereinafter referred as SimHEFT, uses the same ranking phase as HEFT. However, during the task assignment phase SimHEFT uses simulation instead of analytical models to compute $ECT(T_i, H_j)$. For each host H_j, the execution of the workflow subgraph including already scheduled tasks and the current task T_i assigned to H_j is simulated. Note that these simulations are independent and, therefore, can be run in parallel. Then the task is scheduled to a host that corresponds to a minimum task completion time observed in simulations.

We also have tried to change the criterion used for selection of hosts during the task scheduling. Indeed, by optimizing the completion time of individual task it is possible to significantly degrade the completion times of already scheduled tasks due to the communication interference. The SimHEFT* variant schedules each task on a host that minimizes the overall makespan of currently scheduled subgraph instead of the task completion time. The intuition behind this variant is to minimally degrade the overall makespan during the scheduling of individual tasks.

The main advantage of the proposed approach is the minimal modification of the original algorithm. However, it is not clear without the experimental evaluation whether it is sufficient to improve only the task assignment phase while keeping the original ranking function and task scheduling order. Another concern is the additional overhead of simulation that can significantly increase the algorithm execution time. Note, however, that the proposed modifications allow running multiple simulations in parallel during the task assignment phase.

4   Experimental Evaluation

In this section we present the results of simulation experiments that compare the performance of proposed HEFT modifications with original HEFT and other workflow scheduling algorithms for a range of workflow and system configurations using the described simulation framework.

Besides HEFT we used two well-known dynamic algorithms - OLB and MCT. Opportunistic Load Balancing (OLB), which is widely used in modern HCSs, assigns available tasks to resources currently being idle without any a priori information about tasks. Minimum Completion Time (MCT) assigns each available task to a resource that is expected to finish the task the earliest.

We use the *makespan*, i.e. the measured total run time of a workflow in a given system according to a schedule produced by an algorithm, as the basis for comparison of algorithm performance. For each simulated pair system-application we run all algorithms and then normalize their makespans by the makespan achieved by the simplest algorithm - OLB. Finally, to reduce the variance, we compute the mean of normalized makespans across all simulations.

The experiments use a fixed set of workflows while varying the system characteristics. The used workflows are based on real world scientific applications [4]:

- **LIGO Inspiral:** analyses and filters the time-frequency data from the Laser Interferometer Gravitational Wave Observatory experiment (LIGO);
- **Epigenomics:** automates various genome sequencing operations (USC Epigenome Center);
- **Montage:** stitches together multiple images of the sky to create large-scale custom mosaics (NASA/IPAC);
- **CyberShake:** characterizes earthquake hazards in a region (SCEC).

The simulated systems have 5, 10 or 20 hosts with performance varying in a range of 1 to 4 GFlops. The network links have identical characteristics selected to be close to the Gigabit Ethernet network (bandwidth: 100 MBytes/sec, latency: 100 µs). For each host count 100 distinct systems are randomly generated. The mean normalized makespans achieved by each algorithm in the experiments are presented in Table 1.

As it can be seen, HEFT outperforms the dynamic algorithms for the LIGO, Epigenomics and Montage workflows. The maximum speedup achieved in comparison to OLB varies among the workflows due to the different amount of inherent parallelism. However, for the CyberShake workflow both dynamic algorithms show similar results and outperform HEFT. The analysis of this workflow revealed that it has two distinguishing properties — high parallelism and high

Table 1. Mean normalized makespan

Hosts count	OLB	MCT	HEFT	SimHEFT	SimHEFT*
LIGO Inspiral, 100 tasks					
5	1.0000	0.9839	0.9651 (0.9608)	0.9652	1.0229
10	1.0000	0.9182	0.8792 (0.8602)	0.8791	1.0338
20	1.0000	0.7885	0.6898 (0.6865)	0.6898	0.9384
Epigenomics, 100 tasks					
5	1.0000	0.9753	0.9376 (0.9311)	0.9376	0.9368
10	1.0000	0.9014	0.8459 (0.8405)	0.8458	0.8437
20	1.0000	0.7942	0.7093 (0.6740)	0.7099	0.7067
Montage, 100 tasks					
5	1.0000	0.9791	0.9769 (0.9683)	0.9766	0.9766
10	1.0000	0.9639	0.9635 (0.9478)	0.9629	0.9636
20	1.0000	0.9109	0.9165 (0.9023)	0.9156	0.9172
CyberShake, 100 tasks					
5	1.0000	1.0104	1.0616 (0.5074)	1.0760	1.0395
10	1.0000	0.9972	1.0846 (0.3789)	1.1354	1.0325
20	1.0000	0.9845	1.1038 (0.2958)	1.3803	1.0244

communication-to-computation ratio (CCR). This could lead to a network contention resulting in a significant mismatch between the simple network model used in HEFT for estimation of $ECOMT$ and the accurately modeled network in the simulator.

To confirm this hypothesis, we obtained the estimated makespan from the internal state of HEFT. These values, normalized to the simulated OLB makespan, are presented in brackets after the simulated HEFT makespan in Table 1. As it can be seen for the CyberShake workflow, HEFT expects to achieve a drastically different makespan than the one produced after the simulation. Ignoring the network contention effect resulted in more than 200% error in the makespan estimation. This result emphasizes the importance of accurate estimations of communication times during the workflow scheduling.

As for SimHEFT, it fails to improve the HEFT makespan for the Cybershake workflow while having a similar performance for other workflows. Contrary to expectations, SimHEFT behaves even worse than HEFT on Cybershake by showing up to 25% makespan degradation. We hypothesize that by optimizing the completion time of individual task it is possible to significantly degrade the completion times of already scheduled tasks due to the communication interference.

The SimHEFT* results confirm the above hypothesis. The results from Table 1 show that SimHEFT* was able to improve the Cybershake makespan by 2–7% in comparison to HEFT while having a similar performance for Epigenomics and Montage. However, SimHEFT* behaves significantly worse than

HEFT and SimHEFT on the LIGO workflow. This can be explained by the fact that this workflow has the lowest CCR ratio and therefore is less sensitive to errors in estimated communication time. In this case the modified scheduling criterion doesn't bring any improvements over the original criterion and, as it can be seen, can even worsen the schedule.

While improving the Cybershake makespan, SimHEFT* is still up to 4% worse than dynamic MCT algorithm. This could indicate that it is not sufficient to improve only the task assignment phase while keeping the original HEFT ranking function and task scheduling order intact.

5   Conclusion and Future Work

In this paper we have investigated the use of simulation instead of simple analytical models inside a workflow scheduling algorithm to improve the estimation of communication times. Two extensions of the well-known HEFT algorithm that use simulation during the task assignment phase have been proposed. The experimental study of proposed modifications showed that it is not sufficient to simply plug simulation into the HEFT assignment phase (SimHEFT variant). However, by modifying the host selection criterion it is possible to improve the makespan for workflows with high parallelism and communication-to-computation ratio (SimHEFT* variant). As was demonstrated, such workflows suffer the most from inaccurate estimations of simple analytical models.

While it is demonstrated that the proposed approach have some potential, there are remaining challenges and room for improvement. The SimHEFT* variant is still behind simple dynamic algorithms for Cybershake and doesn't work well for workflows with low CCR ratio. The possible improvements here include modifications of the ranking phase and adapting the algorithm behaviour depending on the CCR ratio. The use of simulation significantly (up to two orders) increased the scheduling time. However, it is possible to decrease this time, e.g. by running the simulations in parallel during the task assignment phase. We plan to address the mentioned challenges in the future work and to perform an extended experimental study across a wide range of synthetic workflows.

Acknowledgments. This work is supported by the Russian Foundation for Basic Research (projects 15-29-07068, 15-29-07043).

References

1. Arabnejad, H., Barbosa, J.G.: List scheduling algorithm for heterogeneous systems by an optimistic cost table. IEEE Trans. Parallel Distrib. Syst. **25**(3), 682–694 (2014)
2. Arabnejad, H., Barbosa, J.G., Prodan, R.: Low-time complexity budget-deadline constrained workflow scheduling on heterogeneous resources. Future Gener. Comput. Syst. **55**, 29–40 (2016)

3. Armstrong, R., Hensgen, D., Kidd, T.: The relative performance of various mapping algorithms is independent of sizable variances in run-time predictions. In: 1998 Seventh Heterogeneous Computing Workshop, 1998, (HCW 98) Proceedings, pp. 79–87. IEEE (1998)

4. Bharathi, S., Chervenak, A., Deelman, E., Mehta, G., Su, M.H., Vahi, K.: Characterization of scientific workflows. In: 2008 Third Workshop on Workflows in Support of Large-Scale Science, pp. 1–10, November 2008

5. Bittencourt, L.F., Sakellariou, R., Madeira, E.R.M.: Dag scheduling using a lookahead variant of the heterogeneous earliest finish time algorithm. In: 2010 18th Euromicro Conference on Parallel, Distributed and Network-based Processing, pp. 27–34, February 2010

6. Casanova, H., Giersch, A., Legrand, A., Quinson, M., Suter, F.: Versatile, scalable, and accurate simulation of distributed applications and platforms. J. Parallel Distrib. Comput. **74**(10), 2899–2917 (2014)

7. Chen, W., Deelman, E.: WorkflowSim: a toolkit for simulating scientific workflows in distributed environments. In: 2012 IEEE 8th International Conference on e-Science (e-Science), pp. 1–8. IEEE (2012)

8. Freund, R.F., Gherrity, M., Ambrosius, S., Campbell, M., Halderman, M., Hensgen, D., Keith, E., Kidd, T., Kussow, M., Lima, J.D., et al.: Scheduling resources in multi-user, heterogeneous, computing environments with SmartNet. In: 1998 Seventh Heterogeneous Computing Workshop, 1998, (HCW 98) Proceedings, pp. 184–199. IEEE (1998)

9. Graham, R.L., Lawler, E.L., Lenstra, J.K., Kan, A.R.: Optimization and approximation in deterministic sequencing and scheduling: a survey. Ann. Discrete Math. **5**, 287–326 (1979)

10. Hagras, T., Janecek, J.: A simple scheduling heuristic for heterogeneous computing environments. In: 2003 Second International Symposium on Parallel and Distributed Computing, Proceedings, pp. 104–110, October 2003

11. Hockney, R.W.: The communication challenge for MPP: intel paragon and Meiko CS-2. Parallel Comput. **20**(3), 389–398 (1994)

12. Hunold, S., Rauber, T., Suter, F.: Scheduling dynamic workflows onto clusters of clusters using postponing. In: 2008 8th IEEE International Symposium on Cluster Computing and the Grid, CCGRID 2008, pp. 669–674. IEEE (2008)

13. Maheswaran, M., Ali, S., Siegal, H.J., Hensgen, D., Freund, R.F.: Dynamic matching and scheduling of a class of independent tasks onto heterogeneous computing systems. In: 1999 Proceedings of Eighth Heterogeneous Computing Workshop, (HCW 1999), pp. 30–44. IEEE (1999)

14. Taylor, I.J., Deelman, E., Gannon, D.B., Shields, M.: Workflows for e-Science: Scientific Workflows for Grids. Springer, London (2014)

15. Topcuoglu, H., Hariri, S., Wu, M.Y.: Performance-effective and low-complexity task scheduling for heterogeneous computing. IEEE Trans. Parallel Distrib. Syst. **13**(3), 260–274 (2002)

16. Velho, P., Legrand, A.: Accuracy study and improvement of network simulation in the SimGrid framework. In: Proceedings of the 2nd International Conference on Simulation Tools and Techniques, p. 13. ICST (Institute for Computer Sciences, Social-Informatics and Telecommunications Engineering) (2009)

17. Velho, P., Schnorr, L.M., Casanova, H., Legrand, A.: On the validity of flow-level TCP network models for grid and cloud simulations. ACM Trans. Model. Comput. Simul. (TOMACS) **23**(4), 23 (2013)
18. Yu, J., Buyya, R., Ramamohanarao, K.: Workflow scheduling algorithms for grid computing. In: Xhafa, F., Abraham, A. (eds.) Metaheuristics for Scheduling in Distributed Computing Environments. Studies in Computational Intelligence, vol. 146, pp. 173–214. Springer, Heidelberg (2008). https://doi.org/10.1007/978-3-540-69277-5_7

C++ Playground for Numerical Integration Method Developers

Stepan Orlov$^{(\boxtimes)}$

Computer Technologies in Engineering Department,
Peter the Great St. Petersburg Polytechnic University,
St. Petersburg, Russian Federation
`majorsteve@mail.ru`

Abstract. A C++ framework for investigating numerical integration methods for ordinary differential equations (ODE) is presented. The paper discusses the design of the software, rather than the numerical methods. The framework consists of header files defining a set of template classes. Those classes represent key abstractions to be used for constructing an ODE solver and to monitor its behavior. Several solvers are implemented and work out-of-the-box. The framework is to be used as a playground for those who need to design an appropriate numerical integration method for the problem at hand. An example of usage is provided. The source code of the framework is available on GitHub under the GNU GPL license.

Keywords: C++ · Extensible framework · Object-oriented programming · Numerical integration · Differential equations

1 Introduction

In this paper we consider the numerical solution of the initial value problem for a system of ordinary differential equation in the normal form:

$$\dot{\mathbf{x}} = \mathbf{f}(t, \mathbf{x}), \quad \mathbf{x}\big|_{t=t_0} = \mathbf{x}_0, \tag{1}$$

where $\mathbf{x} = [x_1, \ldots, x_n]^T$ is the vector of n state variables of the system, t is the time, dot denotes the time derivative, and $\mathbf{f}$ is the ODE right hand side vector. Sometimes we also consider a more general case of ODE, namely

$$\dot{\mathbf{x}} = \mathbf{f}(t, \mathbf{x}, \phi), \quad \mathbf{x}\big|_{t=t_0} = \mathbf{x}_0, \quad \phi\big|_{t=t_0} = \phi_0, \tag{2}$$

where $\phi = [\phi_1, \ldots, \phi_m]^T$ is the vector of m discrete state variables. Each of the variables ϕ_k may only change at discrete time instants identified by the relation

$$e_k(t, \mathbf{x}) = 0, \quad k = 1, \ldots, m; \tag{3}$$

functions e_k are called *event indicators*, and time instants satisfying (3) are called *events*. How the variables ϕ change is determined by a state machine that has

© Springer International Publishing AG 2017
V. Voevodin and S. Sobolev (Eds.): RuSCDays 2017, CCIS 793, pp. 418–429, 2017.
https://doi.org/10.1007/978-3-319-71255-0_34

to be defined along with the ODE system and event functions. See [1] for more information.

Of course, there is software, including open source software, implementing numerical solvers of ODE initial value problem. For example, the SUNDIALS software suite [2] is capable of solving problems (1) and (2), and also differential-algebraic equations. The `odeint` library [3] from the `boost` project [4] provides many solvers for problems (1). There are many more, written in different programming languages. Nevertheless, we chose to create another piece of software for solving ODE initial value problem, named `ode_num_int`[1]. While existing software is focused on obtaining the numerical solution, the idea behind our framework is different. Our goal is to provide user with flexible components to create new solvers and to investigate how these solvers behave.

The `ode_num_int` framework currently covers single-step numerical integration schemes, including Runge–Kutta schemes [5]. It also provides a template class for building solvers based on Richardson extrapolation [5, ch. II.9]. It contains code to solve linear and nonlinear algebraic equations, so explicit and implicit schemes are easily constructed. The framework is designed for systems of medium scale, with n up to several thousands.

While the implementation of explicit ODE solvers is typically straightforward and requires no special tuning to get them work, the implementation of implicit solvers may require a lot of effort from a developer, especially if the goal is to have an efficient solver in terms of CPU time consumed. The latter applies to fully implicit schemes such as SDIRK [5, ch. II.7] and to linearly implicit methods of Rosenbrock type [6], such as W-methods [7]. Notice that both classes of methods require the Jacobian of ODE right hand side, $\mathbf{J} = D\mathbf{f}/D\mathbf{x}$, or an approximation to it, which we denote as $\mathbf{A}$. When a researcher starts applying such a method to the problem at hand, he or she may face a number of difficulties listed below.

– The quality of solution obtained with a linearly implicit W-method may depend on how close $\mathbf{A}$ is to $\mathbf{J}$. It is therefore natural for a developer to experiment with different strategies to update the $\mathbf{A}$ matrix. It should also be noticed that keeping the same $\mathbf{A}$ for as many time steps as possible is what can make a W-method outperform any other implicit method. Each time $\mathbf{A}$ changes, and each time the step size h changes, the matrix $\mathbf{W} = \mathbf{I} - hd\mathbf{A}$ (where $\mathbf{I}$ is the identity matrix and d is a parameter) has to be factorized, which consumes CPU time.
– The ways of calculation of the Jacobian matrix $\mathbf{J}$ may be different. For some ODE systems, it is easy to provide an explicit formula for $\mathbf{J}(t, \mathbf{x})$; for more complicated systems, it may still be possible to compute $\mathbf{J}$ analytically with an automatic differentiation tool like ADOL-C [8]; for complex systems, one has to compute the Jacobian numerically using finite differences.
– The Jacobian may happen to be a sparse matrix, and taking its sparsity structure into account during its numerical calculation and the solution of linear system becomes crucial for overall performance of a solver, as soon

---

[1] The source code of the framework is available at https://github.com/deadmorous/ode_num_int.

as n is not too small. In addition, sparse structure of the Jacobian imposes certain constraints on the choice of algorithms to update its approximation **A**.

- The convergence and performance of Newton-type method [9] used to solve nonlinear algebraic system at a time step may depend on several factors. It might require too many iterations to converge if **A** is not updated frequently enough; on the other hand, it may take too long time if we enforce **A** = **J**; it also may fail to converge unless a specific regularization strategy is applied. Last but not least, the number of iterations may strongly depend on the initial guess to the solution.

The `ode_num_int` framework has been designed with the idea to equip developers with useful abstractions helping to build problem-specific solvers that best fit the ODE system at hand. The construction of such a solver is often an investigation and requires from a developer to try various combinations of components and algorithms and to observe how the resulting solver works.

2   Software Design

The framework is written in C++11 and is a set of template classes. There are also a few translation units that support the dynamic creation of instances and implement some timing utilities. Other functionality is implemented in header files. Subsections below outline framework components and actually explain how an extensible system can be designed in C++.

2.1   Common Infrastructure

In this subsection, we describe how some general design patterns are employed in the `ode_num_int` framework and how they help to build consistent easy-to-use software.

Observers. In C++, functions are not really first-class citizens, like, e.g., in JavaScript. However, C++ allows classes (so called functors) pretending to be functions; moreover, C++11 allows to easily create functors using lambda expressions. The `Observers` template class represents an array of such functors with certain signature, which is the template parameter pack. User can add to or remove from this array using corresponding methods. Besides, `Observers` itself is a functor. When it is invoked, it in turn invokes all functors from the array. This pattern exists in other programming languages and is similar to `Boost.Signals2`, but our implementation is more lightweight.

Interfaces make use of the `Observers` template class by declaring public fields where interface users can add arbitrary callbacks. Interface method implementations invoke the callbacks by "calling" those fields (they are functors).

Property Holders. To hold a member variable, a C++ class can simply declare it. However, it is a good practice to keep the field private and use getter and setter methods to access the field; it is also sometimes desirable to notify any interested party about member modification when the setter is called. Another desirable thing is being able to hold similar members in different classes. An elegant solution to this is to put member declaration, as well as getter and setter code, into a separate class, and to inherit that class. This way we also follow the single responsibility principle. Since there might be many different things to be stored like this, the type of the member and the names of getter and setter should be different in each certain case, we have come to the solution using a preprocessor macro to declare such classes. We call such classes *property holders*.

Factories. Factory is a well-known design pattern used to provide a way to create instances of classes that implement certain interface [10]. Implementations do not need to be known when the factory is designed and even when their instances are created: the exact type of the instance is identified, e.g., by a number or a string.

In our framework, there are two template classes to support the pattern. We have at most one factory per interface, therefore each interface for a dynamically creatable entity inherits the `Factory` template class and gives it itself as the template parameter. On the other hand, each creatable implementation of the interface inherits the `FactoryMixin` template class instantiated with two template parameters, class type and interface type. Finally, each creatable type has to be registered in the factory, which can be done with a macro declaring a static registrator variable, or in a number of different ways.

Optional Parameters. It is often necessary to provide parameter values for object instances. Parameter types could be numbers (e.g., a tolerance), strings (e.g., a file name), or typed objects (e.g., an ODE solver). When parameters are specified directly from C++ code, there is no problem. However, it might be necessary to read all parameters from a file and set all of them to appropriate objects, and also to create objects by type identifiers found in the file. To support this, two classes have been designed, `OptionalParameters` and `Value`. The former one is an interface declaring methods to read parameters from the object and to set parameters. The latter one is for storing a single value of arbitrary type; it is similar to the `QVariant` type from the Qt library [11], although it has a feature providing interoperability with factories: if a value is a pointer to an interface with a factory, assigning a string to it leads to the creation of appropriate instance, followed by assigning the created instance to the value. The string in this case is treated as a type identifier. As long as any type can be stored in a `Value`, it is easy to have a tree of parameters and to organize its transformation to any suitable format, e.g., XML, JSON, or plain text.

Timing Utilities. In software engineering, the standard tool used to identify performance bottlenecks is a profiler. However, since the performance of an ODE solver and its parts is always an important issue for a developer, we included a few lightweight tools to measure time intervals. The `TickCounter` class has methods to count number of CPU cycles between method invocations. Its design is similar to that of `QTimer` from Qt, but the measurement is much more precise due to the use of the `rdtsc` instruction. On multi-core AMD CPUs, however, this approach requires pinning threads to CPU cores. To convert CPU cycle counts into milliseconds, the `TimerCalibrator` class can be used.

The `TimingStats` class is convenient to manage timing statistics for multiple invocation of the same code. It counts the total time, the invocation count, and can compute average time per one invocation.

The `TickCounter` and `TimingStats` classes can be used in combination with `Observers` to easily measure the time spent in any part of solver code.

2.2   Linear Algebra

There are several implementations of linear algebra code (e.g., Intel MKL or AMD ACML libraries). In order to simplify building from source code, the `ode_num_int` framework does not depend on any of them and implements a minimal set of linear algebra operations internally. At the same time the design of template class for a vector allows interoperability with such implementations.

The `VectorTemplate` template class provides an interface to a column vector. It defines linear operations, such as addition and multiplication by a number, and provides their reference implementations. There is one template parameter, `VectorData`, that determines how to access actual data in a vector, element type, and how vector data is copied. Different instantiations of vector template can be seamlessly assigned to each other, which improves code flexibility. The `ode_num_int` framework provides an implementation of `VectorData` that stores the data in `std::vector`. Other implementations are possible. Such design helps to avoid data copying in some cases, e.g., when it is necessary to represent a part of a vector (for example, its upper half) as another vector. To achieve this, we use the `VectorProxy::Block` as an implementation of `VectorData`, thus employing the *proxy* pattern [10]. Another similar example is when we need a scaled vector. To avoid immediate copying and multiplication, the `VectorProxy::Scale` class is used in place of `VectorData`. Importantly, `VectorData` is a template parameter for most of the template classes of the framework.

The `SparseMatrixTemplate` template class provides an interface to a sparse matrix, as follows from its name. Its design is also split into the interface class with the above name and its template parameter, `MatrixData`. Rectangular blocks of sparse matrices can be represented as separate matrices without copying by using proxies similar to those for vectors. To actually store matrix data, there are two implementations of `MatrixData`. One of them (call it D to be short) allows dynamically changing the sparsity pattern and stores matrix elements in `std::map`. Matrix operations, like matrix-vector multiplication, require iteration over an associative array, and are not really fast in this case. For fixed matrix sparsity pattern, there is a faster solution (call it F), with matrix elements stored in `std::vector`. Like for vectors, different instantiations of sparse matrices can be seamlessly assigned to each other, which allows the following usage pattern, e.g., to compute the Jacobian. If sparsity pattern is not known, use `SparseMatrixTemplate<D>` to compute it, but don't store the resulting matrix; instead, store an instance of type `SparseMatrixTemplate<F>` and copy the first matrix into it. Then perform operations on the second matrix. When the Jacobian is computed next time and its sparsity pattern remains the same, take it into account to compute the Jacobian (probably much faster [12]); do not use D at all until the sparsity pattern changes.

The `LUFactorizer` template class is capable of solving sparse linear systems using the LU factorization [13]. It has a method to specify a sparse matrix, another method to update matrix with the same sparsity pattern (faster than the first one because doesn't require memory allocation and the calculation of layout of matrices **L** and **U**), and a method to solve linear system with the specified right hand side. The factorization is done when necessary. The class automatically manages timing statistics for setting the matrix, decomposition, and backward substitution using `TimingStats` fields.

2.3   Nonlinear Algebraic Newton-Type Solver

This section presents abstractions specific to the solution of systems of nonlinear equations at a time step of an implicit ODE solver, as well as several implementations. The functionality is split into interchangeable components in order to provide great flexibility in combining different algorithms. The components naturally follow the single responsibility principle. Importantly, interfaces described in this and next subsections provide relevant observers (not described here due to size limitation) that can be exploited to pull all necessary information from components and use it for understanding how they perform.

An iteration of a Newton-type solver for the equation $\mathbf{f}(\mathbf{x}) = 0$ with the initial guess $\mathbf{x}^{(k)}$ can be written as follows [9]:

$$\mathbf{x}^{(k+1)} = \mathbf{x}^{(k)} + \alpha^{(k)}\mathbf{d}^{(k)}, \quad \mathbf{A}\mathbf{d}^{(k)} = -\mathbf{f}(\mathbf{x}^{(k)}), \tag{4}$$

where **A** is an approximation to the Jacobian $\mathbf{J} = D\mathbf{f}/D\mathbf{x}$, $\mathbf{d}^{(k)}$ is the *search direction*, and $\alpha^{(k)}$ is a number determined by a line search algorithm (see below).

Vector Mapping. To formulate the problem for a nonlinear solver, we use the `VectorMapping` template class that defines an interface for a vector function of a vector argument, $\mathbf{f}(\mathbf{x})$. Implementations provide actual mappings. A vector mapping can be held in a property holder (see above).

Error Estimator. The interface declares methods to compute the norms of absolute and relative error vectors of numerical solution obtained at an iteration of a Newton-type method. It also declares iteration status codes and a method returning such a code to instruct iteration performer (see below) what to do next. The default implementation of error estimator has absolute and relative tolerances and thresholds used to detect the divergence. Those are available through the `OptionalParameters` interface.

Jacobian Provider. The interface declares methods to compute the Jacobian matrix **J** and to retrieve it as a sparse matrix of type SparseMatrixTemplate<F> (see above), and also a method to inform the instance about possible change in the sparsity pattern after an event. It is up to the implementation how **J** is computed. Currently we have an implementation that computes **J** numerically, taking its sparsity pattern into account. The latter means that for mappings $\mathbf{f} : \mathbb{R}^n \to \mathbb{R}^n$ the number of **f** evaluations could be much less than $n+1$, which is the case for dense **J**. See [12] for more information.

Jacobian Trimmer. One way to speed up the solution of the linear system in (4) is to reduce the number of nonzero elements in **A**. This can be done by applying a trimmer transforming **J** into a matrix with a smaller number of nonzero elements by throwing away certain elements (for example, by limiting the bandwidth of the matrix). On the other hand, the trimming might severely impact the convergence of Newton iterations, hence much care must be taken with the trimming.

Descent Direction. The interface defines a method to compute the search direction $\mathbf{d}^{(k)}$ by solving the second of equations (4). This procedure may also involve an update of **A** according to certain strategy. Currently, a number of implementations are available: (i) $\mathbf{A} = \mathbf{J}$; (ii) **A** is subject to Broyden's update of rank 1 [14]; (iii) **A** is subject to "fake Broyden's update", such that the sparsity pattern of **A** is preserved by ignoring all elements of the update matrix that do not have corresponding nonzero elements in **A**; (iv) Hart's update directly to the LU decomposition of **A** [15]; (v) constant **A**; some more. Importantly, an implementation of the search direction interface may have an internal state changing between iterations. Therefore, there is a method telling it to reset the internal state and recompute the Jacobian next time.

Line Search. The interface provides a method to perform the search along the direction $\mathbf{d}^{(k)}$ and ensure $|\mathbf{f}(\mathbf{x}^{(k+1)})| < |\mathbf{f}(\mathbf{x}^{(k)})|$ by picking a suitable value of $\alpha^{(k)}$. Currently we have one simple implementation that starts with $\alpha^{(k)} = 1$ and divides it by two till the above condition is satisfied; if during this process $\alpha^{(k)}$ reaches a minimum threshold, the Newton iteration is considered as divergent.

Iteration Performer. The interface provides methods to specify initial guess to the solution, to perform a single iteration and to reset components. Most important, the iteration performer is a placeholder for components implementing the vector mapping specifying $\mathbf{f}(\mathbf{x})$, the error estimator, the descent direction algorithm, and the line search algorithm.

Regularization Tools. Unfortunately, Newton iterations may diverge. This problem can be addressed by introducing some kind of regularization. For example, instead of solving $\mathbf{f}(\mathbf{x}) = 0$, one could consider equations $\mathbf{g}(\mathbf{x}, \gamma) = 0$. The regularization parameter γ could vary from 0 to 1, and $\mathbf{g}$ is such that, on the one hand, $\mathbf{g}(\mathbf{x}, 1) = \mathbf{f}(\mathbf{x})$, and, on the other hand, Newton iterations converge better as γ decreases. Then we could consider iterations with γ changing gradually from 0 to 1. There is a hope that this process converges because the leading iterations give better initial guess to the problem with $\gamma = 1$. Notice that in the case of ODE numerical integration, γ could be proportional to the step size, but other choices are possible, too. The iteration performer described above can hold the regularized mapping $\mathbf{g}$ and a *regularization strategy* object that decides how γ should vary depending on iteration status.

Newton-Type Solver. The solver interface declares methods to specify the initial guess and to run Newton iterations. It is a placeholder for iteration performer, so the logics of a single iteration is out of its responsibility. The implementation of the solver just runs a loop in which it performs an iteration, and if there is no convergence so far, suggests regularization strategy to do something. When iteration count reaches certain limit (which is a parameter), the solver makes one more try with the Jacobian computed from scratch (for this, the solver instructs the iteration performer to reset any internal state of its components).

2.4   ODE Solvers

All ODE solvers implemented in our framework are considered to be single-step methods (though, the support for multistep methods is planned). Besides, we do not currently support methods of variable order, e.g., based on the extrapolation. The `OdeSolver` interface for a solver declares methods returning its order, specifying initial state, specifying initial step size, retrieving current state, and performing a single step (the latter is called `doStep`).

ODE Right Hand Side is specified by inheriting the `OdeRhs` interface. It is similar to `VectorMapping`, but is better suited for evaluating $\mathbf{f}(t, \mathbf{x})$ rather than $\mathbf{f}(\mathbf{x})$. In addition, keeping mechanical systems in mind, there is some support for second-order equations. Namely, the vector of state variables is considered to consist of coordinates $\mathbf{u}$, speeds $\mathbf{v}$ (those are vectors of size n_2), and first order variables $\mathbf{z}$ (a vector of size n_1): $\mathbf{x} = [\mathbf{u}^T, \mathbf{v}^T, \mathbf{z}^T]^T$, so the ODE right hand side is then $\mathbf{f}(t, \mathbf{x}) = [\mathbf{v}^T, \mathbf{f}_u^T, \mathbf{f}_v^T]^T$. Therefore, the interface is augmented with a method returning the number of coordinates; the total number of state variables is $n = 2n_2 + n_1$. This kind of representation allows to reduce the size of linear and nonlinear problems to be solved by an implicit solver at a step to $n_2 + n_1$, because $\mathbf{u}$ is easily excluded.

Explicit Solvers implemented in our framework can be formulated as Runge-Kutta schemes (ERK), see [5, ch. II]. There are two general implementations taking a Butcher tableau as the constructor parameter. One implementation considers schemes without automatic step size control (our framework implements Euler and RK4). The other implementation considers embedded ERK schemes with automatic step size control (currently the framework implements the DOPRI45, DOPRI56, and DOPRI78 schemes). To implement an ERK solver, one has to inherit the appropriate implementation and provide the Butcher tableau.

The Gragg's explicit solver is attractive as the reference solver for extrapolation due to its symmetry property [5, ch. II.8]. Originally, it is a two-step method. Although it can be reformulated as an ERK scheme, there is no need for that. Our implementation treats the solver as a single-step one, but internally it takes two steps of half size in its `doStep` method implementation. The solver has an additional parameter instructing it to return smoothed current state, which is the average of two last states. This allows to build the extrapolation Gragg–Bulirsch–Stoer scheme with smoothing, as described in [5]. Our experience shows that in certain cases this solver gives better results than other explicit schemes.

Extrapolation-Based Solver takes another ODE solver as the *reference solver*. Each step boils down to splitting the step interval of size h into smaller steps of sizes h/n_k and making smaller steps with the reference solver. The sequence of whole numbers n_k may vary and is specified as extrapolator parameter (we implemented the Romberg's, the Bulirsch's, and the harmonic sequences). The index k runs from 1 to certain number of stages, N, which we currently consider fixed. Finally, the Aitken–Neville's algorithm is applied to find the extrapolated solution. The details of the method can be found in [5, ch. II.9]. The extrapolation-based solver possesses great flexibility because an arbitrary reference solver and step sequence can be specified as its parameters. Another parameter is a flag telling the extrapolator if the reference solver is symmetric. In the same time, our implementation is not the most general one because the order of the method does not adjust automatically at run time.

Step Size Controller interface declares a method that suggests the value of step size to be used at the next step. The method requires current step size, scheme order, and error norm as parameters. Therefore, the interface can be used by embedded ERK schemes and by any other schemes that are capable of computing the error norm at a step. Simpler schemes would have to estimate error norm using an extrapolation-based approach. The implementation of the interface has a number of parameters, which are the acceptable tolerance and some more. Additional parameters of the step size controller may instruct it to keep the step size constant if possible. This may be very important, e.g., for W-methods, since a change in the step size also changes the matrix of linear system to be solved at step.

Event Controller interface declares two methods, one to be called in the beginning of the step, and the other one to be called at the end. Those methods check for events occurring within a step, find first of them according to (3), and change discrete state variables ϕ in (2) according to the event occurred. Finally, the second method reports the actual step size, probably truncated, and provides some information about the event. The interpolation of ODE system state inside the step is beyond the event controller responsibility; therefore, an interpolator has to be specified as a parameter for the second method. To interpolate the state vector, one can use the linear interpolator; however, if solver supports dense output it can provide a better quality interpolator.

Linearly Implicit W-Methods. The idea behind W-methods is to replace the system Jacobian $\mathbf{J}$ in Rosenbrock methods with its arbitrary approximation $\mathbf{A}$ and still to have method order conditions satisfied. When $\mathbf{A}$ is close enough to $\mathbf{J}$, the scheme is expected to have certain stability properties. Refer to [7] for more information. In our framework, two W-methods are currently implemented, the SW2(4) method [7] and the W1 method (actually used as the reference solver for extrapolator). Implementations of higher order W-methods are planned. The implementations inherit a helper class that is capable of solving the linear system at a step and to compute the $\mathbf{A}$ matrix. The same helper class can be used as a base for new implementations.

SDIRK Solvers. Implicit solvers require the solution of a nonlinear system of algebraic equations at a step. Therefore they use an instance of Newton-type solver internally. The solver and its components can be set up through the `OptionalParameters` interface. Currently the framework implements the following method:

$$\mathbf{x}_{k+1} = \mathbf{x}_k + h[(1-\alpha)\mathbf{f}(t_k, \mathbf{x}_k) + \alpha\mathbf{f}(t_k + h, \mathbf{x}_{k+1})], \tag{5}$$

where the index k denotes the step number, h is the step size, and α is a parameter of the scheme, such that it is the explicit Euler scheme at $\alpha = 0$, the implicit Euler scheme at $\alpha = 1$, and the trapezoidal rule at $\alpha = 0.5$. The implementation handles the case $n_2 > 0$ (variables of second order) such that the nonlinear system solved at a step has size $n_2 + n_1$, and the coordinates $\mathbf{u}$ are excluded. The initial guess can be specified using a predictor, which is another solver specified as a parameter (e.g., explicit Euler). Other implementations are planned.

ODE Solver Output. For convenience, there are a number of components producing some text output as an ODE solver proceeds. The `StatisticsOutput` object reports time and count statistics for various parts of code (right hand side evaluations, LU decompositions and backward substitutions, Jacobian calculations, and more, depending on solver). The `GeneralOutput` object reports current information during the solution (time, step size, error norm at step, event information, error at a Newton iteration, when applicable). The `SolutionOutput` object outputs solution vector at each step or after user-specified time intervals.

ODE Solver Configuration. The object is merely a placeholder for parameters that specify ODE solver, ODE right hand side, initial state, time interval, and output options. It is convenient for supplying the entire description of a numerical experiment from a text file.

3   Example of Application

The `ode_num_int` framework has emerged due to an attempt to design a numerical solver for the problem of continuously variable transmission (CVT) dynamics [16]. The models of CVT components developed so far are implemented in a C++ software package. System state vector contains about 3600 variables that are generalized coordinates, generalized speeds, and a few first order variables. CVT components are modeled as deformable elastic bodies. There are many contacts with friction between the bodies; in particular, torque transmission is possible only due to the friction forces. For many years, the numerical solution of the initial value problem has been done using the explicit RK4 scheme. Although the scheme works well, it requires quite small step size due to stability requirements, such that the step size is about 10^{-7}. As a result, numerical simulations take too long time. Estimations and direct calculation show that the Jacobian of the ODE right hand side has quite large eigenvalues, and they arise due to the friction characteristic. The ODE system appears to be stiff. It is known that for stiff systems implicit methods are preferable. For some of them, Rosenbrock or W-methods work good; for others, including the model of CVT, they don't.

As already said, the framework has been used for applying various solvers to the ODE system of CVT dynamics. Since the ODE system is very complex, the performance of solver should be as good as possible, which has suggested the choice of C++ as the programming language for the framework.

The design of the `ode_num_int` framework has allowed to apply many different numerical integration methods to a real-world application in quite a short period of time. There is no final result so far because it is a work in progress, but still we have found, e.g., that the trapezoidal rule at step size 10^{-5} is as good as RK4 at step size 10^{-8}. Knowing it is important as a motivation for the development of an optimized implementation of the scheme for CVT (in particular, it can only outperform RK4 if the Jacobian is computed faster, which is possible but requires tedious programming).

4   Conclusions and Future Work

The `ode_num_int` framework has been created to help developers find and tune the appropriate solver for certain ODE system. It is very flexible in combining various components together to build an ODE solver, so it can be used as a playground. Much effort has been applied to decouple functionality in different components, so each of them is responsible for one thing. That should make it quite easy for a developer to implement any missing component rather quickly, e.g., to check if some idea will work. When a suitable solver is found, it probably could be used as is; however, in many cases an optimized implementation will give better performance.

Future plans for framework development include the implementation of more components, like solvers, line search algorithms, and more. On the other hand, it is planned to provide a number of tools working out-of-the-box, like stability region generator, multi-parametric study organizer, and others.

References

1. Blochwitz, T., et al.: The functional mockup interface for tool independent exchange of simulation models. In: Proceedings of the 8th International Modelica Conference (2011)
2. Hindmarsh, A.C., et al.: SUNDIALS: suite of nonlinear and differential/algebraic equation solvers. ACM Trans. Math. Softw. **31**, 363–396 (2005)
3. Ahnert, K., Mulansky, M.: Odeint – solving ordinary differential equations in C++. In: IP Conference Proceedings, vol. 1389, pp. 1586–1589 (2011)
4. Schling, B.: The Boost C++ Libraries. XML Press (2011)
5. Hairer, E., Nørsett, S.P., Wanner, G.: Solving Ordinary Differential Equations I. Nonstiff Problems, 2nd Revised edn. Springer, Heidelberg (1993)
6. Rosenbrock, H.H.: Some general implicit processes for the numerical solution of differential equations. Comput. J. **5**, 329–330 (1963)
7. Steihaug, T., Wolfbrandt, A.: An attempt to avoid exact Jacobian and nonlinear equations in the numerical solution of stiff differential equations. Math. Comp. **33**, 521–534 (1979)
8. Griewank, A., Juedes, D., Utke, J.: Algorithm 755: ADOL-C: a package for the automatic differentiation of algorithms written in C/C++. ACM Trans. Math. Softw. **22**(2), 131–167 (1996)
9. Brown, J., Brune, P.: Low-rank quasi-Newton updates for robust Jacobian lagging in Newton-type methods. In: International Conference on Mathematics and Computational Methods Applied to Nuclear Science and Engineering, pp. 2554–2565 (2013)
10. Gamma, E.: Design Patterns. Addison-Wesley, Boston (1994)
11. Lazar, G., Penea, R.: Mastering Qt 5. Packt Publishing, Ltd. (2017)
12. Ypma, T.J.: Efficient estimation of sparse Jacobian matrices by differences. J. Comput. Appl. Math. **18**(1), 17–28 (1987)
13. Golub, G.H., Van Loan, Ch.F.: Matrix Computations. Johns Hopkins University Press, Baltimore (1996)
14. Broyden, C.: A class of methods for solving nonlinear simultaneous equations. Math. Comput. **19**(92), 577–593 (1965)
15. Hart, W.E., Soesianto, F.: On the solution of highly structured nonlinear equations. J. Comput. Appl. Math. **40**(3), 285–296 (1992)
16. Shabrov, N., Ispolov, Yu., Orlov, S.: Simulations of continuously variable transmission dynamics. ZAMM **94**(11), 917–922 (2014). Wiley-Vch Verlag GmbH & Co. KGaA, Weinheim

Efficiency Analysis of Intel and AMD x86_64 Architectures for Ab Initio Calculations: A Case Study of VASP

Vladimir Stegailov[1,2,3(✉)] and Vyacheslav Vecher[1,2]

[1] Joint Institute for High Temperatures of RAS, Moscow, Russia
v.stegailov@hse.ru, vecher@phystech.edu
[2] Moscow Institute of Physics and Technology (State University),
Dolgoprudny, Russia
[3] National Research University Higher School of Economics, Moscow, Russia

Abstract. Nowadays, the wide spectrum of Intel Xeon processors is available. The new Zen CPU architecture developed by AMD has extended the number of options for x86_64 HPC hardware. This large number of options makes the optimal CPU choice for HPC systems not a straightforward procedure. Such a co-design procedure should follow the requests from the end-users community. Modern computational materials science studies are among the major consumers of HPC resources worldwide. The VASP code is perhaps the most popular tool for these research. In this work, we discuss the benchmark metric and results based on a VASP test model that give us the possibility to compare different CPUs and to select best options with respect to time-to-solution and energy-to-solution criteria.

Keywords: Multicore · VASP · Memory wall · Broadwell · Zen

1 Introduction

Computational materials science provides an essential part of the deployment time for high performance computing (HPC) resources worldwide. The VASP code [1–4] is among the most popular programs for electronic structure calculations that gives the possibility to calculate materials properties using the non-empirical (so called *ab initio*) methods. According to the recent estimates, VASP alone consumes up to 15–20% of the world's supercomputing power [5,6]. Such unprecedented popularity justifies the special attention to the optimization of VASP for both existing and novel computer architectures (e.g. see [7]). At the same time, one can ask a question what type of processing units would be the most efficient for VASP calculations.

A large part of HPC resources installed during the last decade is based on Intel CPUs. Novel generations of Intel CPUs present the wide spectrum of multicore processors. The number Xeon CPU types for dual-socket systems is 26 for

© Springer International Publishing AG 2017
V. Voevodin and S. Sobolev (Eds.): RuSCDays 2017, CCIS 793, pp. 430–441, 2017.
https://doi.org/10.1007/978-3-319-71255-0_35

the Sandy Bridge family, 27 for Ivy Bridge, 22 for Haswell and 23 for Broadwell families. In each family, the processors share the same core type but differ by their frequency, core count, cache sizes, network-on-chip structure etc.

In March 2017, AMD released the first processors based on the novel x86_64 architecture called Zen. It is assumed that the efficiency of this architecture for HPC applications would be comparable to the latest Intel architectures (Broadwell and Skylake).

The diversity of CPU types complicates significantly the choice of the best variant for a particular HPC system. The first criterion is certainly the time-to-solution of a given computational task or a set of different tasks that represents an envisaged workload of a system under development.

Another criterion is the energy efficiency of an HPC system. Energy efficiency becomes one of the most important concerns for the HPC development today and will remain in foreseeable future [8].

The need for clear guiding principles stimulates the development of models for HPC systems performance prediction. However, the capabilities of the idealized models are limited by the complexity of real-life applications. That is why the empirical benchmarks of the real-life examples serve as a complimentary tool for the co-design and optimization of software-hardware combinations.

In this work, we present the efficiency analysis of a limited but representative list of modern Intel and AMD x86_64 CPUs using a typical VASP workload example.

2   Related Work

HPC systems are notorious for operating at a small fraction of their peak performance and the deployment of multi-core and multi-socket compute nodes further complicates performance optimization. Many attempts have been made to develop a more or less universal framework for algorithms optimization that takes into account essential properties of the hardware (see e.g. [9,10]). The recent work of Stanisic et al. [11] emphasizes many pitfalls encountered when trying to characterize both the network and the memory performance of modern machines.

The choice of the best option among several alternative GPU-systems for running the GROMACS package is the subject of the paper of Kutzner et al. [12]. In that paper, several real life examples are considered as benchmarks of the hardware efficiency. Our paper follows a similar path but for the VASP package.

The application of *ab initio* codes requires big supercomputers and the parallel scalability of the codes becomes, therefore, an important issue. The scalability of the SIESTA code was considered in [13] for several Tier0 systems from the PRACE infrastructure (although technically quite different, SIESTA shares the same field of applications in materials science as VASP). In the previous work [14], different HPC systems were compared with respect to their performance for another electronic structure code CP2K.

The increase of power consumption and heat generation of computing platforms is a very significant problem. Measurement and presentation of the results of performance tests of parallel computer systems become more and more often evidence-based [15], including the measurement of energy consumption, which is crucial for the development of exascale supercomputers [16].

The work of Calore et al. [17] discloses some aspects of relations between power consumption and performance using small Nvidia Jetson TK1 minicomputer running the Lattice Boltzmann method algorithms. An energy-aware task management mechanism for the MPDATA algorithms on multicore CPUs was proposed by Rojek et al. [18].

Our previous results on energy consumption for minicomputers running classical MD benchmarks was published previously for Odroid C1 [19] and Nvidia Jetson TK1 and TX1 [20,21].

3   Hardware and Software

In this work, we consider several Intel CPUs and the novel AMD Ryzen processor and compare the results with the data [22] for the IBM Power 7. The features of the systems considered are summarized in Table 1. We make use of the fact that the Intel X99 chipset supports both consumer Core series and server Xeon series Intel processors that share the same LGA 2011-3 socket. The Core i7-6900K is similar to the Xeon E5-2620v4 but allows us to vary CPU and DRAM frequencies.

Table 1. The main features of the systems considered

CPU type	N_{cores}	$N_{mem.ch.}$	L3 (Mb)	CPU_{freq} (GHz)	$DRAM_{freq}$ (MHz)
Single socket, Intel X99 chipset					
Xeon E5-2620v4	8	4	20	2.1	2133
Core i7-6900K	8	4	20	2.1–3.2	2133 – 3200
Xeon E5-2660v4	14	4	35	2.0	2400
Single socket, AMD B350 chipset					
Ryzen 1800X	8	2	16	3.6	2133–2400
Dual socket, Intel C602 chipset (the MVS10P cluster)					
Xeon E5-2690	8	4	20	2.9	1600
Dual socket, Intel C612 chipset (the MVS1P5 cluster)					
Xeon E5-2697v3	14	4	35	2.6	2133
Dual socket, Intel C612 chipset (the IRUS17 cluster)					
Xeon E5-2698v4	20	4	50	2.2	2400
Quad socket, IBM Power 775 (the Boreasz cluster [22])					
Power 7	8	4	32	3.83	1600

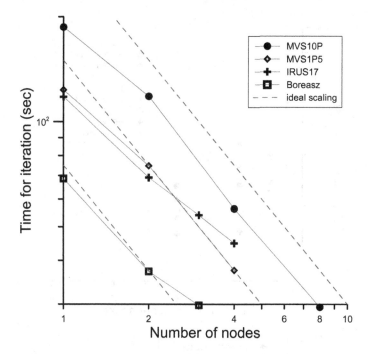

Fig. 1. Parallel scalability of the GaAs test. In all cases, 8 cores per socket are used that corresponds to 16 MPI ranks per a dual-socket node of the Xeon-based MVS10P, MVS1P5 and IRUS17 clusters and 64 MPI ranks on a quad socket node of the Power-based Boreasz cluster.

The single socket Intel Broadwell systems benchmarks are performed under Ubuntu ver. 16.04 with Linux kernel ver. 4.4.0. The single socket AMD Ryzen system is benchmarked under Ubuntu ver. 17.04 with Linux kernel ver. 4.10.0.

3.1 Test Model in VASP

VASP 5.4.1 is compiled for Intel systems using Intel Fortran, Intel MPI and linked with Intel MKL for BLAS, LAPACK and FFT calls. For the AMD system, gfortran ver.6.3 is used together with OpenMPI, OpenBLAS and FFTW libraries.

Our test model in VASP is the same as used previously for the benchmarks of the IBM 775 system [22]. The model represents a GaAs crystal consisting of 80 atoms in the supercell. The Perdew Burke Ernzerhof model for xc-functional is used. The calculation protocol corresponds to the geometry optimization. We use the time for the first iteration of electron density optimization τ_{iter} as a target parameter of the performance metric. This parameter can serve as an adequate measure of time consumption for molecular dynamics calculations as well.

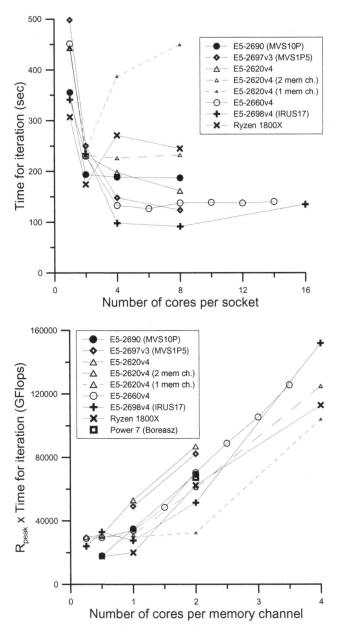

Fig. 2. Top: the dependence of the time for the first iteration of the GaAs test on the number of cores per socket. The presented time values for the single socket systems (see Table 1) is divided by 2 for comparison with the dual socket systems data. Bottom: the same data (and one point for IBM Power 7) in the reduced parameters $R_{peak}\tau_{iter}$ and $N_{cores}/N_{mem.ch.}$ (here R_{peak} is the total peak performance of the single or dual socket node).

The τ_{iter} values considered in this work are about 10–100 s and correspond mainly to a single node of an HPC cluster. At the first glance, these are not very long times to be accelerated. However, *ab initio* molecular dynamics requires usually $10^4 - 10^5$ time steps and each step consists of 3–5 such iterations. That is why the decrease of τ_{iter} by several orders of magnitude is an actual problem for modern HPC systems targeted at materials science computing.

The choice of a particular test model has a certain influence on the benchmarking results. However, our preliminary tests of other VASP models show that the main conclusions of this study do not depend significantly on a particular model. In the future, a set of regression tests would be beneficial for similar analysis.

3.2 Power Consumption Measurement

For the single socket systems considered, the power consumption measurements are performed. We use APC Back-UPS Pro BR1500G-RS and the corresponding apcupsd linux driver for digital sampling of power consumed during VASP runs. In this way, we measure the total power consumption of the CPU, the memory, the motherboard and PSU. For the evaluation of the total energy consumed for one benchmark run, we multiply the average power value during the run by the time of the first iteration τ_{iter}.

4 Results and Discussion

4.1 Where Is the Balance Between Cores, Memory Channels and L3 Cache?

VASP 5.4.1 uses MPI for parallelization. Figure 1 illustrates the acceleration of the GaAs test considered for 1–8 nodes of the MVS10P (FDR Infiniband), MSV1P5 (FDR Infiniband) and IRUS17 (Omni-path) clusters. For the modest number of nodes considered, the acceleration is very efficient. Here we do not want to analyze the limits of parallel scalability but to show that the absolute performance of the parallel code is proportional to the performance of single cluster nodes (e.g. one can mention the similarity of the strong scaling data on Fig. 1 for MVS10P and MVS1P5 that both use FDR Infiniband).

VASP is known to be both a memory-bound and a compute-bound code [7]. Modern Intel CPUs provide 4 memory channels per socket. That is why *a priori* it is not obvious how VASP performance depends on the number of cores per socket. Figure 2 shows the results of the GaAs test runs.

We see a pronounced dependence on the number of cores per socket. For majority of systems, the time per iteration saturates at 4 cores per socket and shows no significant decrease for higher core counts.

In order to understand the dependence on the number of memory channels, we perform tests with E5-2620v4 CPU with only 2 or 1 memory channels activated (when only 2 DIMMs or 1 DIMM are installed into the motherboard). The

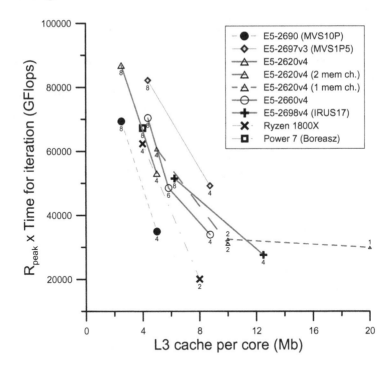

Fig. 3. The dependence of the $R_{peak}\tau_{iter}$ parameter on the L3 cache per core.

results confirm the crucial importance of the number of active memory channels for the VASP performance (this fact is a manifestation of the "memory wall" concept).

Performance comparison of different CPUs resembles usually a comparison of "apples and oranges". For comparison of CPUs with different frequencies and different peak numbers of Flops/cycle, it is better to use the reduced parameter of $R_{peak}\tau_{iter}$ [14,20]. Another reduced parameter that characterizes the memory subsystem is $N_{cores}/N_{mem.ch.}$ (for simplicity we neglect here the variation of the memory bandwidth per channel). The bottom plot of Fig. 2 presents the same data as shown on the upper plot in the reduced coordinates. In this way, we have eliminated the differences in floating point performance of different CPU core and the difference in the number of memory channels.

In these reduced coordinates, the scatter of data points is much smaller, and there is an evident common trend. The data point for the IBM Power 7 CPU is located at the same trend that suggests the low sensitivity of the results to the hardware and software differences between x86_64 and IBM Power systems.

The test model considered fixes the total number of arithmetic operations (Flops) required for its solution. The increase of $R_{peak}\tau_{iter}$ (that is proportional to the number of CPU cycles) shows the increase of the overhead due to the limited memory bandwidth. More CPU cycles are required for the CPU cores involved in computations to get data from DRAM.

The remaining scatter of data points at the bottom plot of Fig. 2 can be partially attributed to different L3 cache sizes of the CPUs considered. We select the data points from Fig. 2 that correspond to $N_{cores}/N_{mem.ch.} = 1-2$, and plot the $R_{peak}\tau_{iter}$ values as a function of the L3 cache size per core (Fig. 3). There is a visible trend: the larger is the L3 cache size per core, the smaller is the $R_{peak}\tau_{iter}$ value. The precise analysis of the data structures used by VASP and their caching is beyond the scope of this paper. However, it is evident that the main VASP computational kernel (composed of the MKL routines) accesses continuous blocks of data in DRAM, and L3 cache mechanism provides efficient acceleration.

Remarkably, the point for the IBM Power 7 benchmark corresponds to this trend very well. The rightmost point (that corresponds to the benchmark with 1 core of E5-2620v4 with 1 active memory channel) is not located at the main trend, because in this combination, presumably, all available L3 cache can not be utilized effectively.

4.2 Optimization of the Energy-to-Solution

For the single socket systems considered (see Table 1) the power consumption measurements are performed together with the VASP model test runs. The results are summarized in Fig. 4 that shows the average power and the total consumed energy as functions of τ_{iter}.

The experiments with Core i7-6900K shows that

- increasing DRAM frequency from 2133 to 3200 MHz results in 10% higher power draw but gives about 10% smaller times for iteration for 4 and 8 cores;
- decreasing CPU frequency from 3.2 to 2.1 GHz results in 20% smaller power draw but gives only about 4% larger times for iteration for 8 cores.

Comparing E5-2620v4 (with 8 cores in total) and E5-2660v4 (with 14 cores in total), we conclude that non-active cores do not contribute significantly to the power draw during VASP test runs.

AMD Ryzen shows a competitive level of power consumption. However, the increase of average power consumption after the transition from 1 to 2 cores for AMD Ryzen is more pronounced than for Intel Broadwell CPUs considered. The probable reason is the activation of both quad-core CPU-Complexes (CCX) of the Ryzen 1800X CPU.

In most cases, there is a minimum in energy consumption for a given CPU. This minimum is mainly connected with the reduction of τ_{iter}. Beyond this minimum when more cores come into play, further acceleration is connected with essentially higher power draw, or there is no acceleration at all.

The most power-efficient and energy-efficient case among the variants considered is the use of 4 cores of E5-2660v4, especially in the turbo boost mode.

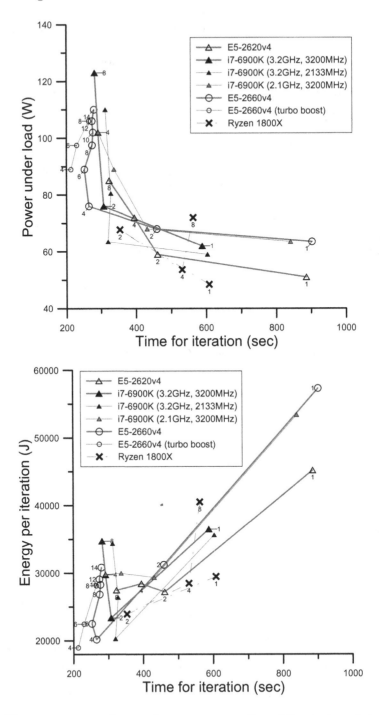

Fig. 4. The average power draw and energy consumption of the single socket systems under the VASP test model load. The number of active cores is shown near each data point.

5   Conclusions

In this work, we have considered several Intel CPUs (from Sandy Bridge, Haswell and Broadwell families), the novel AMD Ryzen CPU and used the data on IBM Power 7 for comparison. In all the cases, we have used the test VASP model of GaAs crystal as a benchmark tool. Complimentary power consumption measurements have been carried out as well.

Additionally to the variation of the CPU types, we have considered the variations in the number of active memory channels, the CPU and DRAM frequencies.

For comparison of different cases, we have used three reduced parameters: (1) the time for iteration normalized by the floating point peak performance $R_{peak}\tau_{iter}$, (2) the number of CPU cores per memory channel $N_{cores}/N_{mem.ch.}$ and (3) the L3 cache size per core.

The benchmark results correlate with these reduced parameters quite well. This fact allows us to make several conclusions on optimal VASP performance.

For VASP, the optimal number of cores per memory channel is 1–2. Using more that 2 cores per channel provides no acceleration.

For $N_{cores}/N_{mem.ch.} = 1 - 2$, VASP performance increases significantly with the increase of the L3 cache per core. Each additional Mb of L3 cache per core reduces the time-to-solution by 30–50%.

The increase of CPU frequency gives diminishing returns but increases significantly the power draw. The increase of DRAM frequency results in the proportional rise of the power draw and in the proportional acceleration.

Comparing different CPUs at the same level of performance, we conclude that CPUs with larger L3 cache size need less power and consume less energy.

Acknowledgments. The authors are grateful to Dr. Maciej Cytowski and Dr. Jacek Peichota (ICM, University of Warsaw) for the data on the VASP benchmark [22].

The authors acknowledge Joint Supercomputer Centre of Russian Academy of Sciences (http://www.jscc.ru) and Shared Resource Center "Far Eastern Computing Resource" IACP FEB RAS (http://cc.dvo.ru) for the access to the supercomputers MVS10P, MVS1P5 and IRUS17.

The work was supported by the grant No. 14-50-00124 of the Russian Science Foundation. A part of the equipment used in this work was purchased with financial support of HSE and using the President of Russian Federation grant for young researchers MD-9451.2016.8.

References

1. Kresse, G., Hafner, J.: Ab initio molecular dynamics for liquid metals. Phys. Rev. B **47**, 558–561 (1993). http://link.aps.org/doi/10.1103/PhysRevB.47.558
2. Kresse, G., Hafner, J.: Ab initio molecular-dynamics simulation of the liquid-metal-amorphous-semiconductor transition in germanium. Phys. Rev. B **49**, 14251–14269 (1994). http://link.aps.org/doi/10.1103/PhysRevB.49.14251
3. Kresse, G., Furthmuller, J.: Efficiency of ab-initio total energy calculations for metals and semiconductors using a plane-wave basis set. Comput. Mater. Sci. **6**(1), 15–50 (1996). http://www.sciencedirect.com/science/article/pii/0927025696000080

4. Kresse, G., Furthmüller, J.: Efficient iterative schemes for ab initio total-energy calculations using a plane-wave basis set. Phys. Rev. B **54**, 11169–11186 (1996). http://link.aps.org/doi/10.1103/PhysRevB.54.11169

5. Bethune, I.: Ab initio molecular dynamics. Introduction to Molecular Dynamics on ARCHER (2015). https://www.archer.ac.uk/training/course-material/2015/06/MolDy_Strath/AbInitioMD.pdf

6. Hutchinson, M.: VASP on GPUs. When and how. GPU technology theater, SC 2015 (2015). http://images.nvidia.com/events/sc15/pdfs/SC5107-vasp-gpus.pdf

7. Zhao, Z., Marsman, M.: Estimating the performance impact of the MCDRAM on KNL using dual-socket Ivy Bridge nodes on Cray XC30. In: Proceedings of the Cray User Group – 2016 (2016). https://cug.org/proceedings/cug2016_proceedings/includes/files/pap111.pdf

8. Kogge, P., Shalf, J.: Exascale computing trends: adjusting to the "new normal" for computer architecture. Comput. Sci. Eng. **15**(6), 16–26 (2013). https://doi.org/10.1109/MCSE.2013.95

9. Burtscher, M., Kim, B.D., Diamond, J., McCalpin, J., Koesterke, L., Browne, J.: Perfexpert: An easy-to-use performance diagnosis tool for HPC applications. In: Proceedings of the 2010 ACM/IEEE International Conference for High Performance Computing, Networking, Storage and Analysis, SC 2010, pp. 1–11. IEEE Computer Society, Washington, DC, USA (2010). https://doi.org/10.1109/SC.2010.41

10. Rane, A., Browne, J.: Enhancing performance optimization of multicore/multichip nodes with data structure metrics. ACM Trans. Parallel Comput. **1**(1), 3:1–3:20 (2014). http://doi.acm.org/10.1145/2588788

11. Stanisic, L., Mello Schnorr, L.C., Degomme, A., Heinrich, F.C., Legrand, A., Videau, B.: Characterizing the performance of modern architectures through opaque benchmarks: pitfalls learned the hard way. In: IPDPS 2017–31st IEEE International Parallel & Distributed Processing Symposium (RepPar workshop), Orlando, USA (2017). https://hal.inria.fr/hal-01470399

12. Kutzner, C., Páll, S., Fechner, M., Esztermann, A., de Groot, B.L., Grubmüller, H.: Best bang for your buck: GPU nodes for GROMACS biomolecular simulations. J. Comput. Chem. **36**(26), 1990–2008 (2015). https://doi.org/10.1002/jcc.24030

13. Corsetti, F.: Performance analysis of electronic structure codes on HPC systems: a case study of SIESTA. PLOS One **9**(4), 1–8 (2014). https://doi.org/10.1371/journal.pone.0095390

14. Stegailov, V.V., Orekhov, N.D., Smirnov, G.S.: HPC hardware efficiency for quantum and classical molecular dynamics. In: Malyshkin, V. (ed.) PaCT 2015. LNCS, vol. 9251, pp. 469–473. Springer, Cham (2015). https://doi.org/10.1007/978-3-319-21909-7_45

15. Hoefler, T., Belli, R.: Scientific benchmarking of parallel computing systems: twelve ways to tell the masses when reporting performance results. In: Proceedings of the International Conference for High Performance Computing, Networking, Storage and Analysis, SC 2015, pp. 73:1–73:12. ACM, New York (2015). https://doi.org/10.1145/2807591.2807644

16. Scogland, T., Azose, J., Rohr, D., Rivoire, S., Bates, N., Hackenberg, D.: Node variability in large-scale power measurements: perspectives from the Green500, Top500 and EEHPCWG. In: Proceedings of the International Conference for High Performance Computing, Networking, Storage and Analysis, SC 2015, pp. 74:1–74:11. ACM, New York (2015). http://doi.acm.org/10.1145/2807591.2807653

17. Calore, E., Schifano, S.F., Tripiccione, R.: Energy-performance tradeoffs for HPC applications on low power processors. In: Hunold, S., Costan, A., Giménez, D., Iosup, A., Ricci, L., Gómez Requena, M.E., Scarano, V., Varbanescu, A.L., Scott, S.L., Lankes, S., Weidendorfer, J., Alexander, M. (eds.) Euro-Par 2015. LNCS, vol. 9523, pp. 737–748. Springer, Cham (2015). https://doi.org/10.1007/978-3-319-27308-2_59

18. Rojek, K., Ilic, A., Wyrzykowski, R., Sousa, L.: Energy-aware mechanism for stencil-based MPDATA algorithm with constraints. In: Concurrency and Computation: Practice and Experience, p. e4016-n/a (2016). https://doi.org/10.1002/cpe.4016. Cpe.4016

19. Nikolskiy, V., Stegailov, V.: Floating-point performance of ARM cores and their efficiency in classical molecular dynamics. J. Phys. Conf. Ser. **681**(1), 012049 (2016). http://stacks.iop.org/1742-6596/681/i=1/a=012049

20. Nikolskiy, V.P., Stegailov, V.V., Vecher, V.S.: Efficiency of the Tegra K1 and X1 systems-on-chip for classical molecular dynamics. In: 2016 International Conference on High Performance Computing Simulation (HPCS), pp. 682–689 (2016). https://doi.org/10.1109/HPCSim.2016.7568401

21. Vecher, V., Nikolskii, V., Stegailov, V.: GPU-accelerated molecular dynamics: energy consumption and performance. In: Voevodin, V., Sobolev, S. (eds.) RuSC-Days 2016. CCIS, vol. 687, pp. 78–90. Springer, Cham (2016). https://doi.org/10.1007/978-3-319-55669-7_7

22. Cytowski, M.: Best Practice Guide – IBM Power 775, PRACE, November 2013. http://www.prace-ri.eu/IMG/pdf/Best-Practice-Guide-IBM-Power-775.pdf

Design of Advanced Reconfigurable Computer Systems with Liquid Cooling

Ilya Levin, Alexey Dordopulo$^{(\boxtimes)}$, Alexander Fedorov, and Yuriy Doronchenko

Scientific Research Centre of Supercomputers and Neurocomputers Co. Ltd., Taganrog, Russia
{levin,dordopulo,doronchenko}@superevm.ru, ss24@mail.ru

Abstract. The paper covers problems of design of reconfigurable computer systems with a liquid cooling system. Design principles of liquid cooling systems of open and closed types are specified, and their comparative analysis is presented. It is shown that open type liquid cooling systems are the most acceptable for high-performance computer systems design. Architecture features of various immersion liquid cooling systems are analyzed; selection criteria of the main cooling system components are given. The paper presents the results of modelling, prototyping and experimental verification of the main technical solutions for our energy-efficient computational module with liquid cooling. The design of the computational module, designed on the base of the modern FPGAs of Xilinx UltraScale series, is presented. It is shown that the developed solutions have power reserve for design of advanced computer systems, based on the new UltraScale+ FPGA family. For the UltraScale+ FPGAs it is necessary to perform some modifications, concerning both the layout of the main computational circuit board and the design of the computer unit and its cooling system. The upgraded design of our advanced computer unit with liquid cooling is presented.

Keywords: Liquid cooling · Reconfigurable Computer Systems · FPGAs · High-performance computer systems · Energy efficiency

1 Introduction

One of the most effective approaches, which provide high real performance of a computer system is adaptation of its architecture to a structure of a solving task. In this case a special-purpose computer device is created. It hardwarily implements all computational operations of the information graph of the task with the minimum delays. Here, we have a contradiction between the implementation of the special-purpose device and its general-purpose use for solving tasks from various problem areas. It is possible to eliminate these contradictions, combining creation of a special-purpose computer device with a wide range of solving tasks, within a concept of reconfigurable computer systems (RCS) based on FPGAs that are used as a principal computational resource [1].

RCS, which contain FPGA computational fields of large logic capacity, are used for implementation of computationally laborious tasks from various domains of science and technique [2–4], because they have a considerable advantage in their real performance

© Springer International Publishing AG 2017
V. Voevodin and S. Sobolev (Eds.): RuSCDays 2017, CCIS 793, pp. 442–455, 2017.
https://doi.org/10.1007/978-3-319-71255-0_36

and energetic efficiency in comparison with cluster-like multiprocessor computer systems.

The leading Russian vendor of high-performance RCS is Scientific Research Centre of Supercomputers and Neurocomputers (SRC SC & NC, Taganrog, Russia), which produce a wide range of products: from completely stand-alone small-size reconfigurable computers (the Caleano product line), desktop or rack computational modules (Rigel, based on Xilinx Virtex-6 FPGAs, Taygeta, based on Xilinx Virtex-7 FPGAs) to computer systems which consist of a set of computer racks, placed in a specially equipped computer room (RCS-7).

The main distinctive feature of the RCS, produced in SRC SC & NC, is high board density and high (not less than 90%) filling of FPGAs that, as a result, provide high specific energetic efficiency of such systems [5].

Practical experience of maintenance of large computer complexes based on RCS proves that air cooling systems have reached their heat limit. Continuous increasing of the circuit complexity and the clock rate of each new FPGA family leads to considerable growth of power consumption and to growth of the maximal operating temperature on chip. So, for the XC6VLX240T-1FFG1759C FPGAs of a computational module (CM) Rigel-2 the maximum overheat of FPGAs relative to the environment temperature of 25 °C in an operating mode and with the power of 1255 W, consumed by the CM, is 33.1 °C, i.e. the maximum temperature of the FPGA chip of the CM Rigel-2 is 58.1 °C. For the XC7VX485T-1FFG1761C FPGAs of the CM Taygeta the maximum overheat of FPGAs relative to the environment temperature of 25 °C in an operating mode and with the power of 1661 W, consumed by the CM, is 47.9 °C, i.e. the maximum temperature of the FPGA of the CM Taygeta is 72.9 °C. If we take into account that the permissible temperature of FPGA functioning, which provides high reliability of the equipment during a long operation period, is 65...70 °C, then it is evident, that maintenance of the CM Taygeta requires decrease of the environment temperature.

According to the obtained experimental data, conversion from the FPGA family Virtex-6 to the next family Virtex-7 leads to growth of the FPGA maximum temperature on 11...15 °C. Therefore further development of FPGA production technologies and conversion to the next FPGA family Virtex Ultra Scale (power consumption up to 100 W for each chip) will lead to additional growth of FPGA overheat on 10...15 °C. This will shift the range of their operating temperature limit (80...85 °C), which means negative influence on their reliability when chips are filled up to 85–95% of available hardware resource. This circumstance requires a quite different cooling method which provides keeping of performance growth rates of advanced RCS.

2   Liquid Cooling Systems for Reconfigurable Computer Systems

Development of computer technologies leads to design of computer technique which provides higher performance, and hence, more heat. Dissipation of released heat is provided by a system of electronic element cooling, that transfers heat from the more heated object (the cooled object) to the less heated one (the cooling system). If the cooled object is constantly heated, then the temperature of the cooling system grows and in some period of time will be equal to the temperature of the cooled object. So, heat transfer stops and the cooled object will be overheated. The cooling system is protected from overheat with the help of cooling medium (a heat-transfer agent). Cooling efficiency of the heat-transfer agent is characterized by heat capacity and heat dissipation. As a rule, heat transfer is based on principles of heat conduction, that require a physical contact of the heat-transfer agent with the cooled object, or on principles of convective heat exchange with the heat-transfer agent, that consists in physical transfer of the freely circulating heat-transfer agent.

To organize heat transfer to the heat-transfer agent, it is necessary to provide heat contact between the cooling system and the heat-transfer agent. Various *heat-sinks* – facilities for heat dissipation in the heat-transfer agent are used for this purpose. Heat-sinks are set on the most heated components of computer systems. To increase efficiency of heat transfer from an electronic component to a heat-sink, a *heat interface* is set between them. The heat interface is a layer of heat-conducting medium (usually multi-component) between the cooled surface and the heat dissipating facility, used for reduction of heat resistance between two contacting surfaces. Modern processors and FPGAs need cooling facilities with as low as possible heat resistance, because at present even the most advanced heat-sinks and heat interfaces cannot provide necessary cooling if an air cooling system is used.

Till 2013 air cooling systems were used quite successfully for cooling supercomputers. But due to growth of performance and circuit complexity of microprocessors and FGAs, used as components of supercomputer systems, air cooling systems have practically reached their limits for designed perspective supercomputers, including hybrid computer systems. Therefore the majority of vendors of computer technique consider liquid cooling systems as an alternative decision of the cooling problem. Today liquid cooling systems are the most promising design area for cooling modern high-loaded electronic components of computer systems.

A considerable advantage of all liquid cooling systems is heat capacity of liquids which is better than air capacity (from 1500 to 4000 times), and higher heat-transfer coefficient (increasing up to 100 times). To cool one modern FPGA chip, 1 m^3 of air or 0.00025 m^3 (250 ml) of water per minute is required. Transfer of 250 ml of water requires much less of electric energy, than transfer of 1 m^3 of air. Heat flow, transferred by similar surfaces with traditional velocity of the heat-transfer agent, is in 70 times more intensive in the case of liquid cooling than in the case of air cooling. Additional advantage is use of traditional, rather reliable and cheap components such as pumps, heat exchangers, valves, control devices, etc. In fact, for corporations and companies, which deal with equipment with high packing density of components operating at high temperatures, liquid cooling is the only possible solution of the problem of cooling of modern computer systems. Additional possibilities to increase liquid cooling efficiency are improvement of the initial parameters of the heat transfer agent: increasing of velocity, decreasing of temperature, providing of turbulent flow, increasing of heat capacity, reducing of viscosity.

Heat transfer agent of liquid cooling systems of computer technique is liquid such as water or any dielectric liquid. Heated electronic components transfer heat to the permanently circulating heat transfer agent – liquid, which, after its cooling in the external heat exchanger, is used again for cooling of heated electronic components. There are several types of liquid cooling systems. Closed loop liquid cooling systems have no direct contact between liquid and electronic components of printed circuit boards [6, 7]. In open loop cooling systems (liquid immersion cooling systems) electronic components are immersed directly into the cooling liquid [8, 9]. Each type of liquid cooling systems has its own advantages and disadvantages.

In closed loop liquid cooling systems all heat-generating elements of the printed circuit board are closed by one or several flat plates with a channel for liquid pumping [10, 11]. So, for example, cooling of a supercomputer SKIF-Avrora [12] is based on a principle "one cooling plate for one printed circuit board". The plate, of course, had a complex surface relief to provide tight heat contact with each chip. Cooling of a supercomputer IBM Aquasar is based on a principle "one cooling plate for one (heated) chip". In each case the channels of the plates are united by collectors into a single loop connected to a common heat-sink (or another heat exchanger), usually placed outside the computer case and/or rack or even the computer room. With the help of the pump the heat transfer agent is pumped through the plates and dissipates heat, generated by the computational elements, by means of the heat exchanger. In such system it is necessary to provide access of the heat transfer agent to each heat-generating element of the calculator, what means a rather complex "piping system" and a large number of pressure-tight connections. Besides, if it is necessary to provide maintenance of the printed circuit boards without any serious demounting, then the cooling system must be equipped with special liquid connectors which provide pressure-tight connections and simple mounting/demounting of the system.

In closed loop liquid cooling systems it is possible to use water or glycol solutions as the heat transfer agent. However, leak of the heat transfer agent can lead to possible ingress of electrically conducting liquid to unprotected contacts of printed circuit boards of the cooled computer, and this, in its turn, can be fatal for both separate electronic components and the whole computer system. To eliminate failures the whole complex must be stopped, and the power supply system must be tested and dried up. Control and monitoring systems of such computers always contain multiple internal humidity and leak sensors. To solve the leak problem a method, based on negative pressure of liquid in the cooling system, is frequently used. According to this method, water is not pumped in under pressure, it is pumped out, and this practically excludes leak of liquid. If airtightness of the cooling systems is damaged, then air ingresses the system but no leak of liquid happens. Special sensors are used for detection of leaks, and modular design allows maintenance without stopping of the whole system. However, all these capabilities considerably complicate design of hydraulic system.

Another problem of closed loop liquid cooling systems is a dew point problem. In the section of data processing the air is in contact with the cooling plates. It means that if any sections of these plates are too cold and the air in the section of data processing is warmer and not very dry, then moisture can condense out of the air on the plates. Consequences of this process are similar to leaks. This problem can be solved ether by hot water cooling, which is not effective, or by control and keeping on the necessary level the temperature and humidity parameters of the air in the section of data processing, which is complicated and expensive.

The design becomes even more complex, when it is necessary to cool several components with a water flow proportionally to their heat generation. Besides branched pipes, it is necessary to use complex control devices (simple T-branches and four-ways are not enough). An alternative approach is use of an industrial device with flow control, but in this case the user cannot considerably change configuration of cooled computational modules.

Advantages of closed loop liquid cooling systems are:

- use water or water solutions as the heat transfer agent which are available, have perfect thermotechnical properties (heat transfer capacity, heat capacity, viscosity), simple and comparatively safe maintenance;
- the large number of unified mechanisms, nodes and details for water supply systems, which can be used;
- great experience of maintenance of water cooling systems in industry.

However, closed loop liquid cooling systems have a number of significant disadvantages, which restrict their widespread use:

- difficulties with detection of the point of water leakage;
- catastrophic consequences that are the result of leakages not detected in time;
- technological problems of leakage elimination (a required power-off of the whole computer rack, that is not always possible and suitable);
- required support of microclimate in the computer room (a dew point problem);
- a problem of cooling of all the rest components of the printed circuit board of the RCS computational module. Even slight modification of the RCS configuration requires a new heat exchanger;
- a problem of galvanic corrosion of aluminum heat exchangers or a problem of mass and dimensions restrictions for more resistant copper heat exchangers (aluminum is three times as lighter than copper);
- air removal from the cooling system that is required before starting-up and adjustment, and during maintenance;
- complex placement of the computational modules in the rack with a large number of fittings required for plug-in of every computational module;
- necessity of use of a specialized computer rack with significant mass and dimension characteristics.

In open loop liquid cooling systems the heat transfer agent is the principal component, a dielectric liquid based, as a rule, on a white mineral oil that provides much higher heat storage capacity of the heat transfer agent, than the one of the air in the same volume. According to their design, such system is a bath filled with the heat transfer liquid (also placed into a computer rack) and which contains printed circuit boards and servers of computational equipment. The heat, generated by electronic components, is dissipated by the heat transfer agent that circulates within the whole bath. Advantages of immersion liquid cooling systems are simple design and capability of adaptation to changing geometry of printed circuit boards, simplicity of collectors and liquid connectors, no problems with control of liquid flows, no dew point problem, high reliability and low cost of the product.

The main problem of open loop liquid cooling systems is chemical composition of the used heat transfer liquid which must fulfil strict requirements of heat transfer capacity, electrical conduction, viscosity, toxicity, fire safety, stability of the main parameters and reasonable cost of the liquid.

Open loop liquid cooling systems have the following advantages:

- insensibility to leakages and their consequences, capability of operating even with local leakages of the heat transfer agent;
- insensibility to climate characteristics of the computer room;
- solution of the problem of cooling of all RCS components, because the printed circuit board of the computational module is immersed into the heat transfer agent;
- capability of modification of the configuration of the printed circuit board of the computational module without modification of the cooling system;
- simplicity of hydraulic adjustment of the system owing to lack of complex system of collectors;
- possibility of use of unified mechanisms, nodes and details, produced for hydraulic systems of machine industry, and know-how of maintenance of electrical equipment that uses dielectric oils;
- increasing of the total reliability of the liquid cooling system.

Disadvantages of open loop liquid cooling systems are the following:

- necessity of an additional pump and heat exchange equipment for improvement of thermotechnical properties (heat transfer capacity, heat capacity, viscosity) of the heat transfer agent. Here special dielectric organic liquids are used as the heat transfer agent;
- necessity of training of maintenance staff and keeping increased safety precautions for work with the heat transfer agent;
- necessity of more frequent cleaning of the computer room because of high permeability of the heat transfer agent, especially in the case of leakage;
- necessity of special equipment for scheduled and emergency maintenance operations (mounting/demounting of the computational module, loading/unloading of the heat transfer liquid, etc.);
- increasing of the maintenance cost because of necessity of regular changeout of the heat transfer liquid when its service life is over and necessity of heat transfer agent management (transporting, receipt, accounting, storing, distribution, recovery of the heat transfer agent, etc.) in the corporation.

Estimating the given advantages and disadvantages of the two liquid cooling systems we can note more weighty advantages of open loop cooling systems for electronic components of computer systems. In this connection for advanced RCS it is reasonable to use direct immersion of heat-generating system components into the mineral oil based liquid heat transfer agent.

At present the technology of liquid cooling of servers and separate computational modules is developed by many vendors and some of them have achieved success in this direction [9–11]. However, these technologies are intended for cooling computational modules which contain one or two microprocessors. All attempts of its adaptation to cooling computational modules which contain a large number of heat generating components (an FPGA field of 8 chips), have proved a number of shortcomings of liquid cooling of RCS computational modules.

The main disadvantages of existing technologies of immersion liquid cooling [10–14] for computational modules which contain FPGA computational fields are:

- poor adaptation of the cooling system for placement into standard computer racks;
- inefficiency of cooling of electronic component chips with considerable (over 50 W) heat generation;
- the thermal paste between FPGA chips and heat-sinks is washed out during long-term maintenance;
- the system of cooling liquid circulation inside the module is designed for one or two chips, but not for an FPGA field, and this fact leads to considerable thermal gradients;
- In the systems, based on the IMMERS [9] technology, all cooling liquid is circulating within a closed loop though the chiller, and this fact leads to some problems;
- necessity of computer complex maintenance stoppage for withdrawal separate components and devices;
- necessity of use of a power specialized pump and hydraulic equipment adapted to the cooling liquid;
- a complex system for control of cooling liquid circulation which causes periodic failures;
- high cost of the cooling liquid which is produced by the only one manufacturer.

The presented disadvantages can be considered as an inseparable part of other existing open loop liquid cooling systems because cooling of RCS computational modules which contain not less than 8 FPGA chips has some specific features in comparison with cooling of a single microprocessor.

The special feature of the RCS produced in Scientific Research Centre of Supercomputers and Neurocomputers is the number of FPGAs, not less than 6–8 chips on one printed circuit board and high packing density. This considerably increases the number of heat generating components in comparison with microprocessor modules, complicates application of the technology of direct liquid cooling IMMERS along with other end solutions of immersion systems, and requires additional technical and design solutions for effective cooling of RCS computational modules.

Use of open liquid cooling system is efficient owing to the heat-transfer agent characteristics and the design and specification of the used FPGA heat-sinks, pump equipment, heat-exchangers.

The heat-transfer agent must have the best electric strength, high heat transfer capacity, the maximum possible heat capacity and low viscosity.

The heat-sink must provide the maximum possible surface of heat dissipation, must allow circulation of the heat-transfer agent through itself, a turbulent heat-transfer agent flow in itself, manufacturability. The specialists of SRC SC & NC have performed heat engineering research and suggested a fundamentally new design of a heat-sink with original solder pins, which create a local turbulent flow of the heat-transfer agent. The used thermal interface cannot be deteriorated or washed out by the heat-transfer agent. Its coefficient of heat conductivity must remain permanently high. The specialists of

SRC SC & NC have created an effective thermal interface which fulfills all specified requirements. Besides, the technology of its coating and removal was also perfected.

The pump equipment is also not the least of the components of the CM cooling system. The principal criteria which must be taken into account are the following:

– provision of design operating parameters;
– outline dimension and coordinated placement of the input and the output fittings;
– the pump must be suitable for interaction with oil products with a specified viscosity and chemical composition;
– continuous maintenance mode;
– minimal vibrations;
– the pump must have the minimal permissible positive suction head;
– the protection class of the pump electric motor must be not less than IP-55.

The heat-exchanger is also an important component of the cooling system. Its design must be compact and must provide efficient heat exchange. Research, performed by the scientific team of SRC SC & NC proved that the most suitable design of the heat-exchanger is a plate-type one, designed for mineral oils cooling in hydraulic systems of industrial equipment.

The liquid cooling system must have a control subsystem which contains sensors of level, flow, temperature of the heat-transfer agent, and a temperature sensor for cooling components.

3   Reconfigurable Computer System "SKAT" Based on Xilinx Ultrascale FPGAS

Since 2013 the scientific team of SRC SC and NC has actively developed the domain of creation of next-generation RCS on the base of their original liquid cooling system for computational circuit boards with high packing density and the large number of heat generating electronic components. The basis of design criteria of the computational module (CM) of next-generation RCS with an open loop liquid cooling system are the following principles:

– the RCS configuration is based on a computational module with the 3U height and the 19" width and with self-contained circulation of the cooling liquid;
– one computational module can contain 12–16 computational circuit boards (CCB) with FPGA chips;
– each CCB must contain up to 8 FPGAs with dissipating heat flow of about 100 W from each FPGA;
– a standard water cooling system, based on industrial chillers, must be used for cooling the liquid.

The principal element of modular implementation of an open loop immersion liquid cooling system for electronic components of computer systems is a reconfigurable computational module of a new generation (see the design in Fig. 1-a). The CM casing of a new generation consists of a computational section and a heat exchange section. In the casing, which is the base of the computational section, a hermetic container with dielectric cooling liquid and electronic components with elements that generate heat during operating, is placed. The electronic components can be as follows: computational modules (not less than 12–16), control boards, RAM, power supply blocks, storage devices, daughter boards, etc. The computational section is closed with a cover.

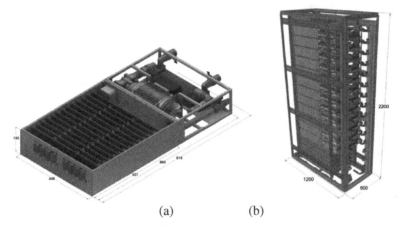

(a)   (b)

Fig. 1. The design of the computer system based on liquid cooling (a – the design of the new generation CM, b – the design of the computer rack)

The computational section adjoins to the heat exchange section, which contains a pump and a heat exchanger. The pump provides circulation of the heat transfer agent in the CM through the closed loop: from the computational module the heated heat-transfer agent passes into the heat exchanger and is cooled there. From the heat exchanger the cooled heat-transfer agent again passes into the computational module and there cools the heated electronic components. As a result of heat dissipation the agent becomes heated and again passes into the heat exchanger, and so on. The heat exchanger is connected to the external heat exchange loop via fittings and is intended for cooling the heat-transfer agent with the help of the secondary cooling liquid. As a heat exchanger it is possible to use a plate heat exchanger in which the first and the second loops are separated. So, as the secondary cooling liquid it is possible to use water, cooled by an industrial chiller. The chiller can be placed outside the server room and can be connected with the reconfigurable computational modules by means of a stationary system of engineering services. The design of the computer rack with placed CMs is shown in Fig. 1-b.

The computational and the heat exchange sections are mechanically interconnected into a single reconfigurable computational module. Maintenance of the reconfigurable computational module requires its connection to the source of the secondary cooling liquid (by means of valves), to the power supply or to the hub (by means of electrical connectors).

In the casing of the computer rack the CMs are placed one over another. Their number is limited by the dimensions of the rack, by technical capabilities of the computer room and by the engineering services.

Each CM of the computer rack is connected to the source of the secondary cooling liquid with the help of supply return collectors through fittings (or balanced valves) and flexible pipes; connection to the power supply and the hub is performed via electric connectors.

Supply of cold secondary cooling liquid and extraction of the heated one into the stationary system of engineering services connected to the rack, is performed via fittings (or balanced valves).

For testing technical and technological solutions, and for determination of expected technical and economical characteristics and service performance of the designed high-performance reconfigurable computer system with liquid cooling, we designed a number of models, experimental and technological prototypes. Figure 2-b shows the prototype of a new generation CM "Skat". For this CM a new design of a CCB with high packing density was created.

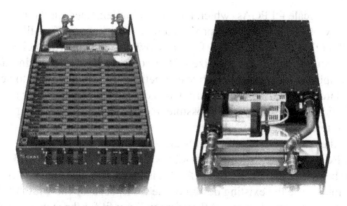

Fig. 2. The prototype of the new generation CM

The CCB of the advanced computational module contains 8 Kintex UltraScale XCKU095T FPGAs; each FPGA contains a specially designed thermal interface and a low-height heatsink for heat dissipation.

We have designed an immersible power supply unit which provides DC/DC 380/12 V transducing with the power up to 4 kWatt for 4 CCB.

The computational section of the CM "SKAT" contains 12 CCB with the power up to 800 W each, 3 power supply units. Besides, all boards are completely immersed into an electrically neutral liquid heat-transfer agent.

For creation of an effective immersion cooling system a dielectric heat-transfer agent was developed. This heat-transfer agent has the best electric strength, high heat transfer capacity, the maximum possible heat capacity and low viscosity.

The heat exchange section contains pump components and the heat exchanger, which provide the effective flow and cooling of the heat-transfer agent. The design height of the CM is 3U.

The performance of one next-generation CM "SKAT" is increased in 8.7 times in comparison with the CM "Taygeta". Such qualitative increasing of the system specific performance is provided by more than triple increasing of the system packing density owing to original design solutions, and increasing of the clock frequency and the FPGA logic capacity.

Experimental results prove that the complex of the developed solutions concerning the immersion liquid cooling system provide the temperature of the heat-transfer agent not more than 30 °C, the power of 91 W for each FPGA (8736 W for the CM) in the operating mode of the CM. At the same time, the maximum FPGA temperature during heat experiments does not exceed 55 °C. This proves that the designed immersion liquid cooling system has a reserve and can provide effective cooling for the designed RCS based on the advanced Xilinx UltraScale+ FPGA family.

4   Advanced Reconfigurable Computer System "SKAT+" Based on Xilinx Ultrascale+ FPGAS

Use of the UltraScale+ FPGAs, which have been implemented on the base of the 16-nm technology 16FinFET Plus and produced by Xilinx since 2017, will provide up to triple growth of the computational performance owing to the growth of clock frequency and FPGA circuit complexity; the size of the computer system remains unchanged. However, in spite of reduction of relative energetic consumption owing to new technological standards of FPGAs manufacturing, and owing to a certain power reserve of the designed liquid cooling system, it is possible to expect a new approach of FPGA operating temperatures to their critical values.

Besides, the new FPGAs of the UltraScale+ family have larger geometric sizes. The size of the FPGAs of the RCS "SKAT" is 42.5 × 42.5 mm. The size of the FPGAs, which are going to be placed into the RCS "SKAT+", is 45 × 45 mm. Due to this circumstance it is impossible to use the existing design of the CCB, because the width of the printed circuit board will become larger and therefore will not fit for the standard 19'' rack.

In this connection it is necessary to modify the designed open liquid cooling system and the CCB design that will lead to modification of the whole CM.

At present the scientific team of SRC of SC & NC is working on a design of an advanced RCS based on the Xilinx UltraScale+ FPGAs. Owing to these works, concerning modification of the cooling system, we are going to solve the following problems:

1. Increase of effective surface of heat-exchange between FPGAs and the heat-transfer agent.
2. Increase of the performance of the heat-transfer agent supply pump.
3. Increase of reliability of the liquid cooling system with the help of immersed pumps.
4. Experimental improvement of the heat-sink optimal design.
5. Experimental improvement of the technology of thermal interface coating.

We have designed a prototype of an advanced computational module with a modified immersed cooling system (Fig. 3). The distinctive feature of the new design is immersed pumps and the considerable reliability growth of the CM owing to reduction of the number of components and simplification of the cooling system. According to our plans, the heat exchange section will contain only the heat exchanger. We are working on experimental research of various pump equipment which can operate in the heat-exchange agent.

 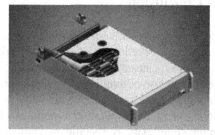

Fig. 3. A prototype of a computational module with a modified immersed cooling system

During modification of the CCB design we have created a prototype of an advanced board shown in Fig. 4. The CCB contains 8 UltraScale+ FPGAs of high circuit complexity. To provide placement of a new CCB into a 19" rack possible, it is necessary to exclude its CCB controller from its structure. The CCB controller was always implemented as a separate FPGA and provided access to FPGA computational resources of the CCB, FPGA programming, condition monitoring of the CCB resources.

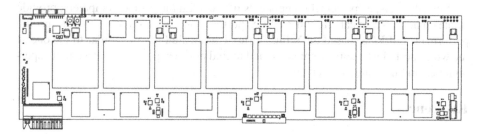

Fig. 4. The prototype of the CCB modified packing

Even if an FPGA is rather small, its resource grows permanently for each new family. At the same time, the variety of functions of the CCB controller expands slightly. As a result, at present, the resource required for implementation of all functions of the CCB controller is only some percent from the logic capacity of the used FPGAs. In this connection we assume further implementation of the CCB controller as a separate FPGA unreasonable. One of FPGAs of the computation field will perform all functions of the controller.

So, owing to breakthrough technical solutions which we have got during design of the RCS "SKAT" with the immersed liquid cooling system, we can develop this direction of high-performance RCS design, and after some design improvements we can create a computer system which provides a new level of computational performance.

5   Conclusion

Use of air cooling systems for the designed supercomputers has practically reached its limit because of reduction of cooling effectiveness with growing of consumed and dissipated power, caused by growth of circuit complexity of microprocessors and other chips. That is why use of liquid cooling in modern computer systems is a priority direction of cooling systems perfection with wide perspectives of further development. Liquid cooling of RCS computational modules which contain not less than 8 FPGAs of high circuit complexity is specific in comparison with cooling of microprocessors and requires development of a specialized immersion cooling system. The designed original liquid cooling system for a new generation RCS computational module provides high maintenance characteristics such as the maximum FPGA temperature not more than 55 °C and the temperature of the heat-transfer agent not more than 30 °C in the operating mode. Owing to the obtained breakthrough solutions of the immersion liquid cooling system it is possible to place not less than 12 CMs of the new generation with the total performance over 1 PFlops within one 47U computer rack. Power reserve of the liquid cooling system of the new generation CMs provides effective cooling of not only existing but of the developed promising FPGA families Xilinx UltraScale+ and UltraScale 2.

Since FPGAs, as principal components of reconfigurable supercomputers, provide stable, practically linear growth of RCS performance, it is possible to get specific performance of RCS, based on Xilinx Virtex UltraScale FPGAs, similar to the one of the world best cluster supercomputers, and to find new perspectives of design of super-high performance supercomputers.

References

1. Perkowski, M.: FPGA computer architectures. Northcon 1993. Conference Record, 12–14 October 1993. ISBN: 0-7803-9972-2
2. Tripiccione, R.: Reconfigurable computing for statistical physics. The weird case of JANUS. In: IEEE 23rd International Conference on Application-Specific Systems, Architectures and Processors (ASAP) (2012)
3. Baity-Jesi, M., et al.: The Janus project: boosting spin-glass simulations using FPGAs. In: IFAC Proceedings Volumes, Programmable Devices and Embedded Systems, vol. 12(1) (2013)
4. Shaw, D.E., et al.: Anton, a special-purpose machine for molecular dynamics simulation. Commun. ACM **51**(7), 91–97 (2008)
5. Kalyaev, I.A., Levin, I.I., Dordopulo, A.I., Slasten, L.M.: Reconfigurable computer systems based on Virtex-6 and Virtex-7 FPGAs. In: IFAC Proceedings Volumes, Programmable Devices and Embedded Systems, vol. 12(1), pp. 210–214 (2013)

6. http://www.coolitsystems.com/index.php/data-center/liquid-cooling-options.html. Accessed 10 June 2016
7. http://www.asetek.com/data-center/oem-data-center-coolers. Accessed 10 June 2016
8. http://www.grcooling.com/carnotjet. Accessed 10 June 2016
9. http://immers.ru/sys/immers660. Accessed 10 June 2016
10. http://www.eurotech.com/aurora. Accessed 10 June 2016
11. http://www.rscgroup.ru. Accessed 10 June 2016
12. http://www.t-platforms.ru/products/hpc/a-class/cooling.html. Accessed 10 June 2016
13. http://www.iceotope.com/product.php. Accessed 10 June 2016
14. http://www.liquidcoolsolutions.com. Accessed 10 June 2016

RAML-Based Mock Service Generator for Microservice Applications Testing

Nikita Ashikhmin[1], Gleb Radchenko[1(✉)], and Andrei Tchernykh[1,2]

[1] South Ural State University, Chelyabinsk, Russia
ashikhminna@pvc.susu.ac.ru,
gleb.radchenko@susu.ru, chernykh@cicese.mx
[2] CICESE Research Center, Ensenada, Mexico

Abstract. The automation capabilities and flexibility of computing resource scaling in cloud environments require novel approaches to application design. The microservice architectural style, which has been actively developing in recent years, is an approach to design a single application as a suite of small services. Continuous integration approach demands transition from manual testing methods to fully automated methods. The mocking is one of the methods to simplify development and testing of microservice applications. The mock service can be considered as an extension of mock object concept. It simulates the behavior of a web service based on a description of its interface. However, developers need to spend additional efforts on development and support of these mock services. We propose a method that would make it easier to generate mocks for REST services by using RAML specifications of services. Using this approach, we propose an implementation, which provides mock services generation and deployment as Docker containers.

Keywords: Microservice · Testing · Docker · REST · RAML · Mocking container

1 Introduction

The microservice model describes a cloud application as a suite of small independent services, each running in its container and communicating with other services using lightweight mechanisms. In [1], the following features of microservices are defined:

- *Open Interface* – microservice should provide an open description of interfaces and communication messages format (either API or GUI).
- *Specialization* – each microservice provides support for an independent part of application's business logic.
- *Containerization* – isolation from the execution environment and other microservices, based on a container virtualization approach. Technologies like OpenVZ, Docker and Rocket [2] became de-facto standards for implementation of such an approach.

V. Voevodin and S. Sobolev (Eds.): RuSCDays 2017, CCIS 793, pp. 456–467, 2017.
https://doi.org/10.1007/978-3-319-71255-0_37

– *Autonomy* – microservices can be developed, tested, deployed, destroyed, moved and duplicated independently and automatically. Continuous integration is the only option to deal with such a development and deployment complexity.

The complex structure of microservice applications demands that microservices should be independently deployable by fully automated machinery. Continuous integration approach demands transition from manual testing methods to fully automated methods. Newman in [3] describes following three levels of testing of microservice applications:

– *Unit tests* that typically validate a single function or method call.
– *Service tests* that are designed to test individual capabilities of isolated services.
– *End-to-End tests* verify the correctness of an entire system in its integrity.

End-to-End tests cover production codes and provide confidence that the application will behave correctly in the production environment. On the other hand, the feedback time of End-to-End Tests is significant. Finally, when such a test fails, it can be hard to determine which unit has broken.

To simplify and speed up the testing process, the developer must isolate the test of an individual service from the entire system. On the other hand, service testing will not be completed without testing its interaction with other services. To simulate the behavior of the other services in controlled ways developers use the so-called «test doubles», namely mocks.

Test Double is a generic term for any case where one replaces a production object for testing purposes. In [4], the following types of test doubles are defined:

– *Dummy objects* are objects without implemented functionality.
– *Fake objects* provide all the functionality needed by the consumer objects, but not suitable for production implementations because of some limitations in speed or effectiveness.
– *Stubs* provide canned answers to the method calls.
– *Mocks* are pre-programmed objects, which generate answers for method calls, corresponding with the interface specification.

Mock services can be used in the following cases [5]:

– *Development* – at the beginning of the development, we define protocols of the communication between services. Mock services can imitate the behavior of services that has not been implemented. This approach can provide a solution to such a «Chicken or the egg» problem when we need to develop a service which is communicating with such unimplemented services.
– *Testing* – mock services allow testing each service individually in isolation from others. It reduces time and resources required for testing. Additionally, mock services reduce the test coverage to one specific service. It helps developers to find broken functionality faster.

REST [6] is one of the most common approaches for microservice interface implementation. However, REST does not define a standard way for the interface documentation. It requires developers to provide additional information about all endpoints and call parameters using third party methods. There are two design patterns of REST interface specifications. Top-down specifications determine the behavior of the REST service independently of its implementation. On the other hand, bottom-up specifications describe the interface of the REST service based on its source code, and cannot be created independently.

We highlight three popular methods for REST interface specifications description [7].

- *SWAGGER* [8] – is a format and framework for the definition of RESTful APIs. It is used to generate server-side API code, client code, and API documentation. SWAGGER is designed as the *bottom-up specification*.
- *RAML (RESTful API Modeling Language)* [9] – is a REST-oriented non-proprietary, vendor-neutral open top-down specification language based on YAML. It focuses on the description of resources, methods, parameters, responses, media types, and other HTTP constructs. It has user-friendly syntax and is contributed by many companies, like Cisco and VMware.
- *API Blueprint* [10] – is a top-down API specification language for web APIs, based on the markdown [11] format. It requires third party server codes and specifically focuses on C++.

The aim of this work is to describe the architecture and implementation of the system, which would provide generation of mock services based on the RAML specification in the form of deploy-ready Docker containers.

We choose the RAML language for several reasons:

- this specification format is human-readable because it based on YAML language;
- it has a big community;
- it is a top-down specification so that users can generate mock service before the development of the real one.

2   Related Work

There are several systems that support the generation of mock services. Some of them allow automatic mock services generation based on an interface specification, while others use special types of "request-response" configuration to emulate service behavior.

Mountebank [12] is an open source tool, which provides cross-platform, multi-protocol test doubles for network services. An application, which is supposed to be tested, should point to the IP or URL of a Mountebank instance instead of the real dependency. Mountebank supports HTTP, HTTPS, TCP and SMTP protocols. To define the behavior of the network service, Mountebank requires a configuration, where all request and response messages for the services are specified.

SoapUI [13] is an open-source web service testing application for service-oriented architecture (SOA) and representational state transfer (REST) applications. SoapUI can generate SOAP mock service based on WSDL specification, while REST mock services must be configured by Groovy scripts.

API Designer [14] is an application that provides a web-based graphical environment for design, documentation, and testing of APIs in a web browser. API Designer creates

REST mock service using RAML specification. However, this service cannot be used by Continuous Integration systems because the generated mock services disappear when the user closes the application.

All solutions discussed above do not create deploy-ready mock services. These applications make REST mock service creation easy, but they do not make this process fully automated. Our approach generates the REST mock service based just on the RAML specification file. Furthermore, our approach would support delivery of mock services as Docker containers.

3   Mock Service Generator Requirements

The mock service generator has one functional and two nonfunctional requirements. The functional requirement is the ability to generate mock services. Nonfunctional requirements are the ability to get the file with RAML specification v0.8 as input and return created microservices as Docker images.

Use case diagram is shown in Fig. 1.

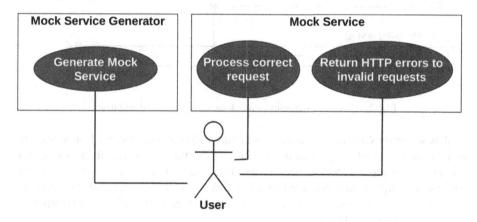

Fig. 1. The mock service generation system Use Case diagram.

Docker [15] is a lightweight mechanism, allowing to run pre-configured system images. Docker represents an implementation of container technology that is considered as an alternative to complete virtualization approach, providing a well-defined application execution environment at the operating system level. Instead of starting a complete operating system on top of a host system or a hypervisor, a container shares the kernel with the host system, which largely eliminates overheads while maintaining isolation between applications. Docker container wraps up a service inside isolated filesystem together with all required system libraries.

The mock service, created by the mock service generator, must satisfy the following functional requirements:

– process correctly GET, POST, PUT, and DELETE requests and return valid responses based on RAML specification of the service;

– return appropriate errors for all incorrect requests.

Furthermore, nonfunctional requirements for created mock services are:

– return responses based on response body examples or response body JSON schema;
– be packed inside a single Docker image.

4 System Architecture

Our system consists of two subsystems – the mock service generator and mock services. The communications between them are shown in Fig. 2.

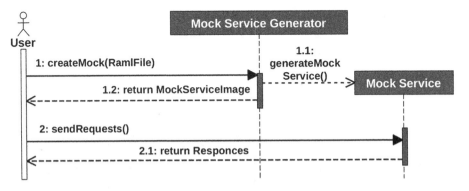

Fig. 2. Sequence diagram of mock service generation and usage.

Mock Service Generator processes user commands to generate mock services. The user can be represented by a continuous integration system that uses the mock service generator to implement testing procedures. The mock service generator processes the createMock request with one argument – a link to an RAML file that describes the interface of the service, endpoints, the format of valid requests, and expected responses for these requests (see Fig. 3).

The generator validates this file and sends the user an error message if the RAML file is incorrect. As a result, the mock service generator creates a Docker container that contains a template of mock service with the received RAML specification file and sends a link to this file to the user.

Mock Service is a service generated by the mock service generator. All mock services have the same architecture, and its behavior depends only on the RAML specification of the service. We define four components in the mock service architecture: *Gateway, Path resolver, Request validator, and Response generator* (see Fig. 4).

```
/employees:
 get:
  queryParameters:
   department:
  responses:
   200:
    body:
     application/json:
     example: |
       ...
 post:
 delete:
/{employee}:
 get:
  responses:
   200:
    body:
     application/json:
      schema: |
      { "$schema": "http://json-
        schema.org/schema",
       "type": "object",
       "properties": {
        "fullName": { "type": "string",
                     "format": "fullname"},
         "department": { "type": "string",
                        "format": "word"}
          "email": { "type": "string",
                    "format": "email"}
         },
        }
```

Fig. 3. An example of REST service specification in an RAML format.

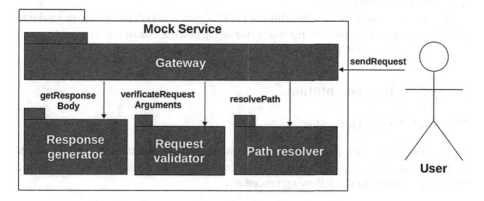

Fig. 4. The architecture of the Mock Service.

The *Gateway* implements the facade pattern and provides a single entry point for all user requests to the mock service. The Gateway receives HTTP-requests sent by a user and calls the Path resolver and Request validator to check the correctness of the received query. Further, if the results of query analysis conducted by these components indicate that the request corresponds to the RAML specification, the Gateway calls the Response generator to generate a body of the response. Finally, the Gateway forms the HTTP response and returns it to the client.

Path resolver. This component validates endpoints of requests based on the RAML-based specification of the service. The Path resolver determinates the correctness of requests endpoints.

The *Request validator* component responds for the validation of parameters of the request, based on the RAML specification of the service. The component checks the compliance of the received parameters with the limitations of the RAML specification. The example of an RAML description of parameters is shown in Fig. 5.

```
/{documentId}
  uriParameters:
    id:
      description: document identification number
      type: string
      minLength: 20
      pattern: ^[a-zA-Z]{2}\-[0-9a-zA-Z]{3}\-\d{2}$
```

Fig. 5. An example of the RAML definition of parameters of a request.

The *Response generator* component generates a body of the response for the request. The RAML language provides two ways to describe the response body. The first way is to declare an example of valid response in JSON format. The second way is to specify a JSON Schema [16], a special format that allows defining the structure of JSON documents. The body generation component to generate the response body by JSON Schema if JSON Schema exists. Otherwise, the component returns an example of response specified in the RAML file.

This component uses Elizabeth library [17] for generation dummy human-readable data. Users can define "format" parameter for string and use one of following integrated formats*: ipv4/ipv6, email, URI, date, time, name, username, surname, word* (Fig. 3).

5 System Implementation

5.1 Mock Service Generator

Mock service generator is a standalone command line application that creates Docker images of mock services. It consists of Python script that generates containers, and operates according to the following procedure:

- the user runs the generator with the following parameters: the link to RAML file, and the name of resulting Docker container;
- mock service generator creates a temporary folder and copies the template mock services files;

– generator downloads the RAML specification file by the user's link, and adds it to the folder;
– the app generates a Docker file (see Fig. 6);

```
FROM python:3-onbuild
EXPOSE 5000
ENV PYTHONPATH .
CMD ["python", "./server/main.py"]
```

Fig. 6. The Docker file for mock service.

– the app runs Docker build command that creates the Docker container. This container includes the Python interpreter, all required libraries and isolates the mock service from the other system;
– mock service generator returns a link to the container to the user.

5.2 Mock Service

The mock service is a web service based on the Flask framework [18]. The processing of user requests by the mock service is shown in Fig. 7.

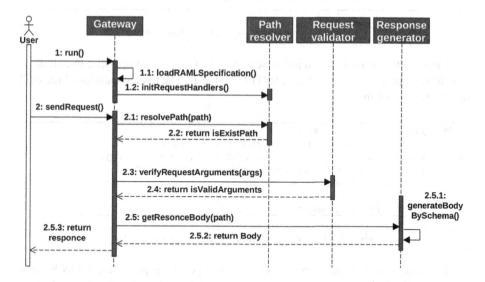

Fig. 7. Mock service user interaction sequence diagram.

The source code of all mock services generated by the generator is identical. The behavior of mock services depends only on RAML specification file loaded on startup. The mock service uses ramlification [19] to parse RAML specification into Python objects. Currently, mock services support RAML v0.8.

At the startup of the service, the Gateway parses the RAML specification file and bind all routers to one of four functions that will handle GET, POST, PUT, and DELETE requests. The part of this code is shown in Fig. 8.

```
def _init_url_rules(self):
  for resource in self.parser.resources:
    self.app.add_url_rule(
      rule=self._transrofm_path_raml_to_flask(resource.path),
      endpoint=resource.name,
      view_func=self.get,
      methods=['GET']
```

Fig. 8. Initialization of GET request handlers.

The request validator is implemented as a set of separated functions that validate parameters. The example of a validation function is shown in Fig. 9.

```
def validate_string(self, value, param):
  if param.min_length and param.min_len > len(value):
      raise Exception(message.ERR_STRING_MIN)
  if param.max_length and param.max_len < len(value):
      raise Exception(message.ERR_STRING_MAX)
```

Fig. 9. An example of validation function.

There are five main function implemented in the Gateway component: initialization function and four functions that handle HTTP requests. The implementation of the GET function is shown in Fig. 10.

```
def get(route, **kwargs):
    endpoint = Resolver.get_endpoint(route)
    Validator.validate_params(endpoint, request.args)
    body = BodyGenerator.generate_body(endpoint,'get')
    header = getHeader(endpoint,
                        resources.query_params, 'get')
    return header, body
```

Fig. 10. The implementation of GET requests handler.

To create a body of the response, the response generator parses the JSON Schema in the RAML file into a tree and performs a direct traversal of all the nodes in the tree, corresponding to the following procedure.

1. get_node function identifies the type of current node and call the special function for this JSON type (for instance, it can be a get_array, get_object or Pget_string function);
2. functions for generation objects and arrays calls get_node functions for the node children to fill inner data (see Fig. 11);

```
def get_array(self, node):
    n = utils.generate_int(self._array_min_count,
                            self._array_max_count)
    items = [self._get_node(node['items']) for _ in
                                            range(n)]
    return items
```

Fig. 11. The implementation of the get_array method.

3. the function generates the JSON data with the constraints imposed by the JSON schema to the current node.

6 Testing

Testing of mock service is conducted using the unit and end-to-end tests. Unit tests are developed using standard Python unit test framework. 40 Python unit tests have been developed to check the source code of the project.

To provide integration testing, we developed a series of tests that create a mock service by the RAML specification, and imitate mock service usage, sending a set of REST requests, and comparing received responses with expected ones (see Fig. 12). The example of the request and the response to it is shown in Fig. 13.

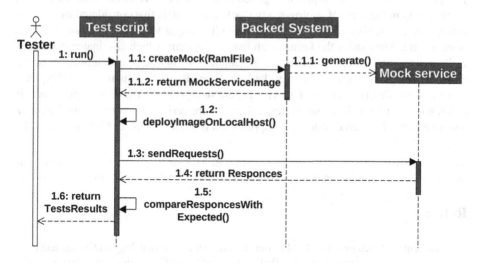

Fig. 12. Integration testing sequence diagram.

```
request:
GET http://127.0.0.1:5000/employees/1
response:
{
"fullName":"John Smith",
"department": "wood",
"email": "smith-toronto-1765@gmail.com"
}
```

Fig. 13. An example of test request and response.

Finally, this system was integrated into continuous integration system of the Naumen Service Desk project [20]. This project includes about 200 Selenium tests for the Android application. During the process of testing, the mobile client sends a series of REST requests to the server. All tests without mock service are completed within 60–80 min in one node. After integrating the mock service, the time required for testing has decreased by about 35%.

7 Conclusion

This article presents the design, architecture and implementation of the automatic mock service generation system. It provides generation of mock services based on the RAML specification in the form of deploy-ready Docker containers that considered as an alternative to complete virtualization approach providing a lightweight application execution environment. They share the kernel with the host system, which eliminates overheads while maintaining isolation between applications. We define four components in the mock service architecture: Gateway, Path resolver, Request validator, and Response generator. We describe the mock service generator algorithm. The developed system is tested with the unit and end-to-end tests. Services are verified for correct functioning in a real project. The source code of our application is available on our GitHub page [21].

Acknowledgment. The work was supported by the RFBR research project No. 15-29-07959 and by Act 211 Government of the Russian Federation, contract No. 02.A03.21.0011.

References

1. Savchenko, D., Radchenko, G.: Microservices validation: methodology and implementation. In: 1st Ural Workshop on Parallel, Distributed, and Cloud Computing for Young Scientists (Ural-PDC 2015). CEUR Workshop Proceedings, vol. 1531, Yekaterinburg, Russia, pp. 21–28 (2015)
2. Pahl, C.: Containerization and the PaaS cloud. IEEE Cloud Comput. **2**, 24–31 (2015)
3. Newman, S.: Learning Building Microservices. O'Reilly Media, Inc., Sebastopol (2015)
4. Kim, T., Park, C., Wu, C.: Mock object models for test driven development. In: Fourth International Conference on Software Engineering Research, Management and Applications, pp. 221–228. IEEE (2006)

5. Soltesz, S., Potzl, H., Fiuczynski, M., Bavier, A., Peterson, L.: Container-based operating system virtualization: a scalable, high-performance alternative to hypervisors. ACM SIGOPS Oper. Syst. Rev. **41**(3), 275–287 (2007)

6. Fielding, R.: Representational state transfer. In: Architectural Styles and the Design of Network-based Software Architecture, pp. 76–85 (2000)

7. Haupt, F.: A framework for the structural analysis of REST APIs. In: IEEE International Conference on Software Architecture (ICSA), pp. 55–58 (2017)

8. Cloves, C., Schmelmer, T.: Defining APIs. In: Microservices from Day One, pp. 59–74. Apress, New York (2016)

9. Surwase, V.: REST API modeling languages-a developer's perspective. IJSTE Int. J. Sci. Technol. Eng. **2**(10), 634–637 (2016)

10. API Blueprint. http://apiblueprint.com. Last accessed 15 Apr 2017

11. Voegler, J., Bornschein, J., Weber, G.: Markdown – a simple syntax for transcription of accessible study materials. In: Miesenberger, K., Fels, D., Archambault, D., Peňáz, P., Zagler, W. (eds.) ICCHP 2014, Part I. LNCS, vol. 8547, pp. 545–548. Springer, Cham (2014). https://doi.org/10.1007/978-3-319-08596-8_85

12. Mountebank - over the wire test doubles. http://www.mbtest.org. Last accessed 15 Apr 2017

13. Azzam, S., Al-Kabi, M.N., Alsmadi, I.: Web services testing challenges and approaches. In: Proceedings of the 1st Taibah University International Conference on Computing and Information Technology, pp. 291–296 (2012)

14. Tsouroplis, R., Petychakis, M., Alvertis, I., Biliri, E., Askounis, D.: Community-based API builder to manage APIs and their connections with cloud-based services. In: CAiSE Forum, pp. 17–23 (2015)

15. Merkel, D.: Docker: lightweight Linux containers for consistent development and deployment. Linux J. **2014**(239), 2 (2014)

16. JSON Schema: syntax and semantics. http://cswr.github.io/JsonSchema/. Last accessed 15 Apr 2017

17. Grinberg, M.: Flask Web Development: Developing Web Applications with Python. O'Reilly Media, Inc., Sebastopol (2014)

18. Elizabeth. http://elizabeth.readthedocs.io/en/latest/. Last accessed 15 Apr 2017

19. Ramlification, Python parser for RAML. https://github.com/spotify/ramlfications. Last accessed 15 Apr 2017

20. Naumen Service Desk. http://www.naumen.ru/products/service_desk/. Last accessed 15 Apr 2017

21. Raml-mock-service. https://github.com/veor12/raml-mock-service. Last accessed 15 Apr 2017

Architecture of Middleware to Provide the Multiscale Modelling Using Coupling Templates

Alexey Liniov[1]([✉]), Valentina Kustikova[1], Alexander Sysoyev[1],
Maxim Zhiltsov[1], Igor Polyakov[1], Denis Nasonov[2], and Nikolay Butakov[2]

[1] Lobachevsky State University of Nizhni Novgorod, Nizhny Novgorod, Russia
`alin@unn.ru`, `valentina.kustikova@gmail.com`, `sysoyev@vmk.unn.ru`,
`zhiltsov.max35@gmail.com`, `polykovio@mail.ru`
[2] ITMO University, St. Petersburg, Russia
`denis.nasonov@gmail.com`, `alipoov.nb@gmail.com`

Abstract. The Multiscale Modelling and Simulation approach is a powerful methodological way to identify sub-models and classify their interaction. The execution order and interaction of computational modules are described in the form of workflow. This workflow can be executed as a single HPC cluster job if there is a middleware which schedule modules execution on allocated resources. We present an architecture of such middleware called Wrapper which provides internal module execution scheduling, interconnection functionality, module migration between allocated resources and storing intermediate state of computations. This middleware is compatible with CLAVIRE (CLoud Applications VIRtual Environment) platform and acts as its execution mechanism.

Keywords: Supercomputing technologies · Parallel computing middleware · High-performance computing · Multiscale modelling · CLAVIRE

1 Introduction

By the moment a great number of high-performance purpose-oriented software has been developed to solve problems in different application fields. Most computational modules are developed by applied specialists using various numerical models, programming languages and parallel programming technologies. In most cases only source code and binaries are available. Joint usage of such modules requires integration of data formats, used technologies and platforms. First of all it is necessary to determine principles of combined use of different models implemented in such modules. However, the employment of such modules is rather difficult even in cases when interaction with the code developer is possible.

One of the approaches progressing in the field of composite applications description is Multiscale Modelling [1,2], presenting templates to combine computational modules. Multiscale Modelling offers several standard methods applying some models of different time and spatial scales: Extreme Scale Computing

© Springer International Publishing AG 2017
V. Voevodin and S. Sobolev (Eds.): RuSCDays 2017, CCIS 793, pp. 468–481, 2017.
https://doi.org/10.1007/978-3-319-71255-0_38

(ES), Hierarchical Multiscale Method of Computing (HMM), Replica Computing (RC) [3]. It defines the way to identify sub-models, classify the sub-model interactions as full or partial overlap of scales and specify the relation between the sub-models that could be represented as a task graph or workflow.

The development of the first multiscale modeling environments is carried out in specific applied fields. Among such environments, one can single out the Computational Materials Design Facility (CMDF) [4] that allows multi-scale multi-paradigm simulations of complex materials phenomena. This framework is based on a generic scripting environment, with the objective to enable simple setup of complex multi-scale simulation tasks. Interfaces between different modules, along with a central data structure allow straightforward communication between different simulation engines. CMDF uses the Python programming language to control the computational flow between disparate processing cores written in compiled languages (C/C++/Fortran) that carry out physicochemical calculations for multiscale/multiparadigm under a unified data model.

Morpheus [5] is another example of multiscale modeling environment. It allows the simulation and integration of cell-based models with ordinary differential equations and reaction-diffusion systems. It allows rapid development of multiscale models in biological terms and mathematical expressions rather than programming code. Morpheus separates modeling from numerical implementation by using a declarative domain-specific markup language.

Also, note Multiphysics Software Environment (MUSE) [6] for multiscale modeling in astrophysics. MUSE facilitates the coupling of existing codes written in different languages by providing inter-language tools and by specifying an interface between each module and the framework that represents a balance between generality and computational efficiency. MUSE has layered architecture. The top layer (flow control) is connected to the middle (interface layer) which controls the command structure for the individual applications. These parts and the underlying interfaces are written in Python, whereas the applications can be written in any language. The only constraint that code must meet to be wrapped as a module is that it is written in a programming language with a foreign function (C/C++, Fortran, C#, Java, Haskel etc.).

Later, ideas are formulated about the need to develop a universal environment that provides the possibility of carrying out a multiscale experiment, regardless of the specifics of the applied field. In this connection, the concept of a multiscale model is formalized and their classification is introduced [7]. Based on this classification the Multiscale Coupling Library and Environment (MUSCLE) [7] and its improved version MUSCLE 2 [8] are implemented. MUSCLE 2 is a component-based modeling tool inspired by the multiscale modeling and simulation framework, with an easy-to-use API which supports Java, C++, C, and Fortran. It assumes that a multiscale model is split into multiple coupled single scale submodels [7]. As a result, each submodel has inputs and outputs that can be coupled in a general way. Within one simulation, one submodel could for instance use hundreds of cores on a supercomputer, whereas another may have to make use of GPU-computing, and yet another needs high I/O performance.

Each submodel is managed by its own instance controller. The controller is an intermediary for any messages that a submodel sends or receives [8].

The proposed environment Wrapper is based on the same theoretical foundations of multiscale modeling as MUSCLE 2 [8]. In contrast to MUSCLE 2 we attempt to organize centralized submodels scheduling in accordance with statistics on its resource usage. To utilize computational resources efficiently one need to analyze parameters and statistical data related to utilization of hardware resources, execution of individual computational modules and composite application taken as a whole. In case of specific modules the relatively simple algorithms can be used [9,10], but provided huge computational facilities are implemented and complex applications are executed we shall use the more comprehensive approaches, such as Knowledge-Based Resource Management [11]. The middleware which we develop, is oriented to coupling with CLAVIRE (CLoud Applications VIRtual Environment) [12] which allows building composite applications using domain specific software available within distributed environment. CLAVIRE builds the workflow, reserve resources of high-performance computing system and launches Wrapper middleware in allocated resources.

2   Purpose of Wrapper Middleware

Wrapper middleware is a MPI program. Wrapper is launched in computational cluster by CLAVIRE scheduler, assumed as being executed in the node external towards the computational cluster. CLAVIRE scheduler ensures the delivery of input data to the cluster, the analysis of cluster resources and features of workflow to be executed, job formation for the cluster management system, launching Wrapper, as well as download the output data from the cluster.

Wrapper provides the following functionality.

- Collecting information about allocated resources of computational cluster.
- Dynamically assigning of computational modules to cluster nodes (with possibility of migration).
- Launching workflow execution.
- Data transmission between computational modules.
- Completing the workflow execution and release of computational cluster resources.

Wrapper architecture makes possible to schedule the execution of computational modules within the allocated resources, however the scheduler is not its subsystem. Scheduling algorithm implements by default static allocation of computational modules on cluster nodes, but we are going to provide integration with adapted version of CLAVIRE scheduler.

Figure 1 illustrates the structural diagram of subsystems interacting with Wrapper middleware.

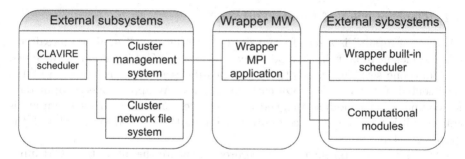

Fig. 1. Interacting subsystems: structural diagram

3   The Workflow Model

3.1   Introduction to the Workflow Model

Wrapper is oriented to use execution patterns presented as a workflow. The workflow contains information about composition of computational modules, possible sets of input as well as output data. The input data of computational module can be built based on output data of other modules or received from the scheduler (such option is required to launch the workflow execution). Figure 2 shows the example of workflow, consisting of the scheduler (S), 3 computational modules (CM-1, CM-2, CM-3) and description of relations between them. In this example the computational module can be executed if all input data has been received.

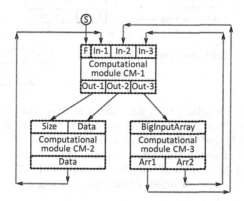

Fig. 2. Workflow: example. "S" is a scheduler. "CM-1", "CM-2", "CM-3" are computational modules. "F" and "In-1, In-2, In-3" are the sets of inputs of the module "CM-1"; "Out-1, Out-2, Out-3" is a set of outputs of the module "CM-1". "Size, Data" is a set of inputs of the "CM-2"; "Data" is a set of outputs of the "CM-2". "BigInputArray" is a set of inputs of the "CM-3"; "Arr1, Arr2" is a set of outputs of the "CM-3"

3.2   Concept "Set of Inputs/Outputs"

"The set of inputs" is a set of input data of computational module (hereafter –
CM), sufficient to launch the module execution. Each CM input can be included
only into one set. The description of all possible sets of inputs for each CM is
represented in the workflow. On each cluster node Wrapper collects input data
for computational module assigned to this node; once any full set of inputs is
collected it is transmitted to computational module (or CM is launched with the
prepared set).

"The set of inputs/outputs" is required to ensure the integrity of CM input
data structure as well as to support the concept of "launching output" (see
below). Here we shall point out that empty output and absence of output
are completely different situations because an empty output can be used (and
required) to form the set of inputs for another module.

The sets of inputs make it possible, using workflow, to describe the Multiscale Modelling templates including a number of integrated models (in time
and space); for example templates shown in Figs. 3 and 4 can be presented as
workflows on that figures.

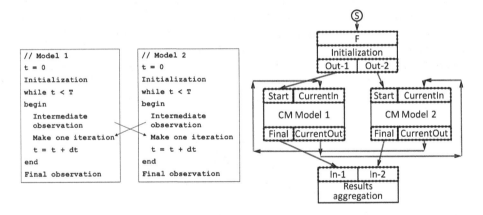

Fig. 3. The Multiscale Modelling with 2 models and integration in space: calculation
scheme and workflow. "CM Model 1", "CM Model 2" are computational modules
implementing "Model 1" and "Model 2" respectively. "F" and "Out-1, Out-2" are sets
of inputs and outputs of the Initialization module. "Start" and "CurrentIn" are sets
of inputs, "Final" and "CurrentOut" are sets of outputs of "CM Model 1" and "CM
Model 2". "In-1, In-2" is a set of inputs of the "Results aggregation" module

In both cases the execution starts after the scheduler sends the set of data
including one "S" output to the input of initialization module "F". For this
module one set of inputs is assigned containing only "F" input; that is why once
this input received the execution of the initialization module will start and the
outputs will be formed sufficient to launch CM models (two in the first case
and one in the second case). For example in the first case two outputs will be

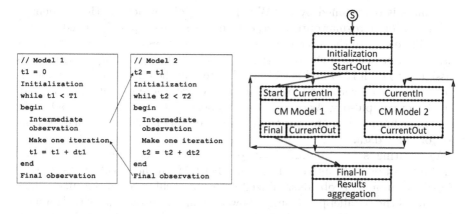

Fig. 4. The Multiscale Modelling with 2 models and integration in time: calculation scheme and workflow. "CM Model 1", "CM Model 2" are computational modules implementing "Model 1" and "Model 2" respectively. "F" and "Start-Out" are sets of inputs and outputs of the "Initialization" module. "Start" and "CurrentIn" are sets of inputs, "Final" and "CurrentOut" are sets of outputs of "CM Model 1". "CurrentIn" and "CurrentOut" are sets of input and outputs of "CM Model 2". "Final-In" is a set of inputs of the "Results aggregation" module

formed launching CMs for models 1 and 2. Thereafter CM for models 1 and 2 will be executed in a parallel way, in each iteration sending each other input sets sufficient for their next iteration. After the modeling is completed CMs form resulting data and transmit them to the module of result aggregation, which performs the final processing and completes the workflow execution.

Thus, workflow with sets of inputs can be implemented to organize the Multiscale computations based on Extreme Scale Computing (ES).

3.3   Workflow Modification. The "Module Instance" Concept

Usage of Multiscale computation template Hierarchical Multiscale Computing (HMM) implies the execution of many launches for CM model of less scale in one computation step of the model of larger scale. The number of launches for CM model of less scale can be unknown in advance and vary from iteration to iteration; and for efficient computations its necessary to parallel launch several copies of less-scaled CM model on different cluster nodes.

To support the HMM pattern the concepts "launching set of outputs", "module instance", "aggregating set of inputs" shall be introduced into workflow.

– "Module instance" is a computational module launched for processing one or several sets of inputs. Each module is identified by the pair (module identifier, instance identifier). Wrapper uses these pairs as CM addresses.
– "Launching set of outputs" (LSO) means that one or several module instances which receive the input from this set can be launched. The number of

instances is determined by the Wrapper scheduler which, at the moment of transmission of regular output set, determines if the outputs included into the set will be sent to the already launched CM instances or additional instances will be launched and the data shall be sent to them. A unique identifier is assigned to each LSO and it will be inherited by the output data of following modules. LSO are sent in blocks of arbitrary size, thereby simultaneous sending is not compulsory. As soon as the numbers of the first and last LSO in the block are known they are sent to the corresponding modules with aggregating set of input data.

– "Aggregating set of inputs" is used to collect several outputs of several instances of one CM into one input of other CM. In the workflow "Aggregating set of inputs" is obviously linked with "Launching set of outputs" which generates the parallel processing followed by aggregation. CM with "Launching set of outputs" for each block of outputs sent for processing shall obviously transmit numbers of the first and the last sets (corresponding outputs and inputs are automatically created in the sets of inputs/outputs) to all modules with aggregating input. The block of sets of inputs is transmitted to the module only after receiving all sets according to the first and last numbers.

Figure 5 shows the workflow example using HMM template.

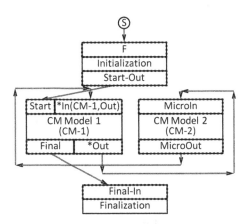

Fig. 5. Workflow for the Multiscale Modelling using Hierarchical Multiscale Computing (HMM). "CM Model 1 (CM-1)", "CM Model 2 (CM-2)" are computational modules. "F" and "Start-Out" are sets of inputs and outputs of the "Initialization" module. "Start" and "*In(CM-1, Out)" are sets of inputs of the "CM-1". "Final" and "*Out" are sets of outputs of "CM-1". "MicroIn" and "MicroOut" are a sets of inputs and outputs of the "CM-2". "Final-In" is a set of inputs of "Finalization" module

"CM Model 1" has one launching set of outputs which includes "*Out" output. Correspondingly, the arbitrary number of instances of "CM Model 2" can be launched. "CM Model 1" has also one aggregating set of inputs "*In(CM-1,Out)", where input data "In" are aggregated according to the blocks of outputs generated by the "Out" output of the same "CM Model 1".

4   Wrapper Architecture

First of all lets introduce the interconnection diagram of CM (Fig. 6).

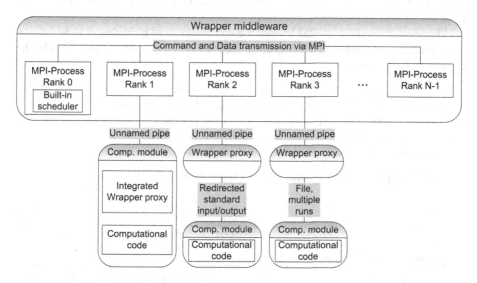

Fig. 6. Wrapper and CMs interconnection diagram. Wrapper middleware is a MPI program. Built-in scheduler is integrated into the Process with Rank 0. Another processes provide launching of computational modules and their communication

Adaptation of computational modules to execution with Wrapper can be performed by the following methods.

1. Integration between computational module and Wrapper at source code level: CM compiled and linked with a set of Wrapper functions providing the transmission of commands and data using mechanism of unnamed pipes (integrated or built-in Wrapper proxy). The computational module is launched one-time when the workflow execution starts, thereafter it is executed constantly, receiving and transmitting messages through the unnamed pipes.
2. The computational module is developed using the set of Wrapper functions which provide reading and parsing of input data and building the set of output data. The computational module is launched every time when Wrapper builds the full set of its input data (external Wrapper proxy). While launching the module 3 parameters are transmitted to it via environment variables: input file name, output file name, file name of the module state (the last parameter is used if the module shall save some data between iterations).

3. The computational module is compiled and operates independently receiving and transmitting input/output data using the standard input/output (external Wrapper proxy). The computational module is also launched one-time when the workflow execution starts and executed constantly, receiving and transmitting messages through the redirected standard input/output.

In the first mode Wrapper proxy requires to know the command lines in order to launch and stop the computational module. The computational module uses a provided specific interface for reading the input data and saving the output data.

The second mode is available only for computational modules in C and C++ programming languages. To integrate proxy into the computational module the following shall be done:

- design the computational module in a specific way;
- include the Wrapper proxy header files into CM source code;
- create a proxy object in the computational module;
- define the computation function to callback from proxy;
- launch proxy from the computational module.

The Wrapper structural diagram is shown in Fig. 7.

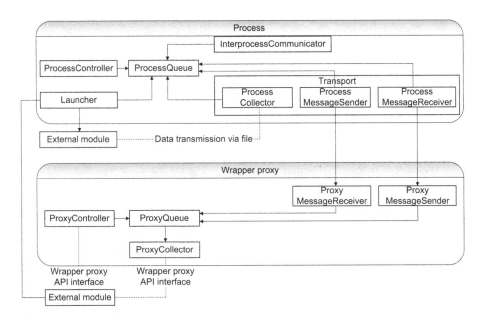

Fig. 7. The Wrapper middleware structure

Process contains components executed in MPI-process, Proxy contains components executed in the built-in Proxy. Structural components has the following purpose.

- ProcessQueue, ProxyQueue are message queues through which the interaction of other components is performed.
- InterprocessCommunicator provides the data transmissions between Wrapper processes using MPI technology.
- ProcessCollector, ProxyCollector receive input data for the computational modules and form the sets of inputs, distribute the sets of outputs and send them to the receivers. Besides collectors receive information about module migration. If collector does not prepare full set of inputs for module execution then it transfers current set of inputs back to the message queue. After that, InterprocessCommunicator sends data to the target computational module which was migrated.
- ProcessMessageSender, ProcessMessageReceiver, ProxyMessageSender, ProxyMessageReceiver provide data transmission between Wrapper and the computational modules with integrated Proxy.
- Launcher implements the launching of the computational modules.
- ProcessController, ProxyController manages allocation/release of other components.

5   The Results of Experiments

5.1   Computational Infrastructure

We used UNN Lobachevsky supercomputer. Nodes of the Linux segment we used have 2x Intel Sandy Bridge E5-2660 2.2 GHz processors (8 cores), 64 GB RAM, QDR InfiniBand network. We employed the Intel MPI and Intel C++ Compiler from the Intel Parallel Studio XE Cluster Edition 2017.

5.2   The Test Workflow

To perform the tests the following computational diagram has been used (see Fig. 8, hereafter "test workflow").

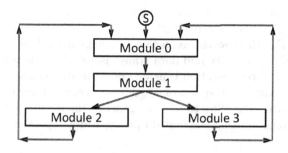

Fig. 8. The test computational diagram

Modules 0–4 do not perform any computations, they only receive the input data and send the output data. Module 0 sends the data block of the fixed size to

Module 1, Module 1 sends the copies of this block to Modules 2 and 3, Modules 2 and 3 send an empty message to Module 0.

5.3   Results

The performance of the test workflow has been estimated in terms of data transfer rate for two mechanism of data transmission to computational modules:

- using of external wrapper proxy and data transmission through the file;
- using proxy, integrated with the computational module.

During tests several iterations of workflow has been performed and average time for one iteration has been calculated. Data blocks from 10 bytes to 1 billion bytes has been used. Table 1 and Fig. 9 show the average execution time values for one iteration using data transmission through the file.

Table 1. Test workflow iteration times for external Wrapper proxy

#	Block size (B)	Average iteration time (s)
1	10	0.13
2	100	0.14
3	1 000	0.13
4	10 000	0.16
5	100 000	0.14
6	1 000 000	0.29
7	10 000 000	0.76
8	100 000 000	17.98
9	1 000 000 000	196.02

Test results show that the overhead for single iteration is approximately constant and makes up about 0.15 s. The transfer time starts to have a value only when the block size exceeds 1MB.

Table 2 and Fig. 10 show the average execution time values for one iteration using built-in Wrapper proxy and data transmission via unnamed pipes.

The overhead for one iteration is again constant and makes up about 0.05 s. Switching from using an external proxy to a built-in one reduces the transfer time of 1 GB data block from 196 s to 62 s. In general, the results of the experiment show the advantage of using the built-in proxy, and acceptable performance.

6   Application of Wrapper Middleware for "Restenosis" Modeling

Within our study we have adopted "Restenosis" application (computation of barrier reconstruction in blood vessels) in order to use the Wrapper middleware.

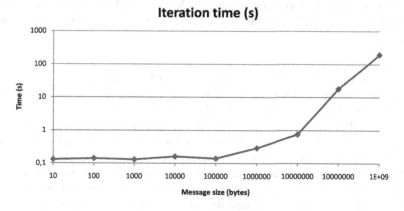

Fig. 9. The test workflow iteration times for external Wrapper proxy

Table 2. Test workflow iteration times for built-in Wrapper proxy

#	Block size (B)	Average iteration time (s)
1	10	0.054
2	100	0.053
3	1 000	0.05
4	10 000	0.06
5	100 000	0.05
6	1 000 000	0.06
7	10 000 000	0.12
8	100 000 000	6.39
9	1 000 000 000	61.92

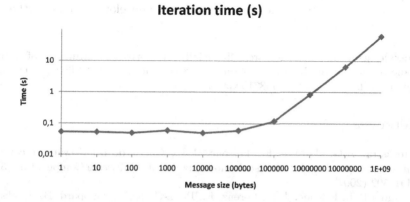

Fig. 10. The test workflow iteration times for built-in Wrapper proxy

Original version uses 8 computational modules and MUSCLE library, providing modules launch and data exchange between them. Adaptation includes the following stages to do.

1. Analyze launching methods of the computational modules.
2. Analyze data transmission workflow between the modules.
3. Analyze principles and mechanisms for implementation of interactions between modules using the MUSCLE library interface.
4. Develop "Restenosis" source code modifications which enable collecting and saving the sets of inputs and outputs. Make test launches and save the data sets (for further testing of adapted version).
5. Develop the set of Wrapper functions which implement reading and parsing of inputs as well as building sets of Wrapper outputs in programming languages used in the computational modules of "Restenosis" application (C, C++, Java).
6. Exclude the MUSCLE library from source code of the computational modules. Add Wrapper code.
7. Test the adapted computational modules.
8. Launch the adapted version of "Restenosis" using Wrapper middleware.

The adapted version of "Restenosis" on the test problem generates output that coincides with the original version, and shows comparable performance.

7   Conclusion

The paper describes extensions to the workflow model which enables the execution of composite tasks based on the Multiscale Modelling templates. The developed architecture of Wrapper middleware provides combined usage of the computational modules developed with different programming languages and technologies. Migration of the computational modules between cluster nodes becomes possible as well. Performance tests show acceptable results. The use of Wrapper in the modeling of "Restenosis" shows the possibility of using it for solving applied problems. The authors continue to develop and plan further use of Wrapper middleware.

Acknowledgments. This research financially supported by Ministry of Education and Science of the Russian Federation, Agreement #14.587.21.0024(18.11.2015). Unique Identification RFMEFI58715X0024.

References

1. Hoekstra, A.G., Lorenz, E., Falcone, J.-L., Chopard, B.: Toward a complex automata formalism for multiscale modeling. Int. J. Multiscale Comput. Eng. **5**(6), 491–502 (2007)
2. Borgdorff, J., Falcone, J.-L., Lorenz, E., Bona-Casas, C., Chopard, B., Hoekstra, A.G.: Foundations of distributed multiscale computing: formalization, specification, and analysis. J. Parallel Distrib. Comput. **73**, 465–483 (2013)

3. Alowayyed, S., Groen, D., Coveney, P., Hoekstra, A.: Multiscale Computing in the Exascale Era. arXiv preprint arXiv:1612.02467 (2016)
4. Computational Materials Design Facility (CMDF), http://web.mit.edu/mbuehler/www/research/CMDF/CMDF.htm
5. Starruby, J., Backy, W., Brusch, L., Deutsch, A.: Morpheus: a user-friendly modeling environment for multiscale and multicellular systems biology. Bioinformatics **30**(9), 1331–1332 (2014)
6. Portegies Zwart, S., McMillan, S., Nualláin, B.Ó., Heggie, D., Lombardi, J., Hut, P., Banerjee, S., Belkus, H., Fragos, T., Fregeau, J., Fuji, M., Gaburov, E., Glebbeek, E., Groen, D., Harfst, S., Izzard, R., Jurić, M., Justham, S., Teuben, P., van Bever, J., Yaron, O., Zemp, M.: A multiphysics and multiscale software environment for modeling astrophysical systems. In: Bubak, M., van Albada, G.D., Dongarra, J., Sloot, P.M.A. (eds.) ICCS 2008. LNCS, vol. 5102, pp. 207–216. Springer, Heidelberg (2008). https://doi.org/10.1007/978-3-540-69387-1_23
7. Borgdorff, J., et al.: Foundations of distributed multiscale computing: formalization, specification, and analysis. J. Parallel Distrib. Comput **73**, 465–483 (2013)
8. Borgdorff, J., Mamonski, M., Bosak, B., Groen, D., Belgacem, M.B., Kurowski, K., Hoekstra, A.G.: Multiscale computing with the multiscale modeling library and runtime environment. Procedia Comput. Sci. **18**, 1097–1105 (2013)
9. Kishimoto, Y., Ichikawa, S.: Optimizing the configuration of a heterogeneous cluster with multiprocessing and execution-time estimation. Parallel Comput. **31**(7), 691–710 (2005)
10. Dolan, E.D., Mor, J.J.: Benchmarking optimization software with performance profiles. Math. Program. **91**(2), 201–213 (2002)
11. Kovalchuk, S., Larchenko, A., Boukhanovsky, A.: Knowledge-based resource management for distributed problem solving. In: Wang, Y., Li, T. (eds.) Knowledge Engineering and Management. AISC, vol. 123, pp. 121–128. Springer, Heidelberg (2011). https://doi.org/10.1007/978-3-642-25661-5_16
12. Knyazkov, K.V., Kovalchuk, S.V., Tchurov, T.N., Maryin, S.V., Boukhanovsky, A.V.: CLAVIRE: e-Science infrastructure for data-driven computing. J. Comput. Sci. **3**(6), 504–510 (2012)

Anticipation Scheduling in Grid
with Stakeholders Preferences

Victor Toporkov[1], Dmitry Yemelyanov[1(✉)], and Anna Toporkova[2]

[1] National Research University "MPEI", Moscow, Russia
{ToporkovVV, YemelyanovDM}@mpei.ru
[2] National Research University Higher School of Economics, Moscow, Russia
atoporkova@hse.ru

Abstract. In this work, a job-flow scheduling approach for grid virtual organizations (VOs) is proposed and studied. Users' and resource providers' preferences, VOs internal policies, resources geographical distribution along with
local private utilization impose specific requirements for efficient scheduling
according to different, usually contradictive, criteria. With increasing level of
resources utilization, the set of available resources and corresponding decision
space are reduced. This further complicates the problem of efficient scheduling.
In order to improve overall scheduling efficiency, we propose an anticipation
scheduling approach based on a cyclic scheduling scheme. It generates a near
optimal but infeasible scheduling solution and includes a special replication
procedure for efficient and feasible resources allocation. Anticipation scheduling
is compared with the general cycle scheduling scheme and conservative backfilling using such criteria as average jobs' start and finish times as well as users'
and VO economic criteria: total execution time and cost.

Keywords: Scheduling · Grid · Resources · Utilization · Heuristic · Job
batch · Virtual organization · Cycle scheduling scheme · Anticipation ·
Replication

1 Introduction and Related Works

In grids with non-dedicated resources the computational nodes are usually partly utilized by local high-priority jobs coming from resource owners. Thus, the resources
available for use are represented with a set time intervals (slots) during which the
individual computational nodes are capable to execute parts of independent users'
parallel jobs. These slots generally have different start and finish times and a performance difference. The presence of a set of slots impedes the problem of resources
allocation necessary to execute the job flow from VOs users. Resource fragmentation
also results in a decrease of the total level of computing environment utilization [1, 2].

Application-level scheduling [3], as a rule, does not imply any global resource
sharing or allocation policy. Applications try to control grid resources independently.
Job flow scheduling in VOs [4, 5] supposes uniform rules of resource sharing and
consumption, in particular based on economic models [2, 4–6]. Usually there are three
parties in these models: users, resource owners, and VO administrators. General

© Springer International Publishing AG 2017
V. Voevodin and S. Sobolev (Eds.): RuSCDays 2017, CCIS 793, pp. 482–493, 2017.
https://doi.org/10.1007/978-3-319-71255-0_39

interaction and resources or services provisioning between these parties is performed by means of a certain currency. VO scheduling policy may offer optimization rules to satisfy both users' and VO common preferences (owners' and administrators' combined). The VO scheduling problems may be formulated as follows: to optimize users' criteria or utility function for selected jobs [6, 7], to keep resource overall load balance [8, 9], to have job run in strict order or maintain job priorities [10], to optimize overall scheduling performance by some custom criteria [11, 12], etc.

Users' preferences and VO common preferences may conflict with each other. Users are likely to be interested in the fastest possible running time for their jobs with least possible costs whereas VO preferences are usually directed to balancing of available resources load or node owners' profit boosting. In fact, an economical model of resource distribution per se reduces tendencies to cooperate [13]. Thus, VO economic policies in general should respect all members to function properly and the most important aspect of rules suggested by VO is their fairness. A number of works understand fairness as it is defined in the theory of cooperative games [7], such as fair job flow distribution [9], fair quotas [14, 15], fair user jobs prioritization [10], and non-monetary distribution [16]. In many studies VO stakeholders' preferences are usually ensured only partially: either owners are competing for jobs optimizing only users' criteria [6, 17], or the main purpose is the efficient resources utilization not considering users' preferences [18].

The goal of the current study is to design a general job-flow scheduling approach which will be able to find a tradeoff between VO stakeholders' contradictory preferences based on the cyclic scheduling scheme (CSS). CSS [19, 20] has fair resource share in a sense that every VO stakeholder has mechanisms to influence scheduling results providing own preferences. Thus, we elaborate a problem of parallel jobs scheduling in heterogeneous computing environment with non-dedicated resources considering users' individual preferences and goals.

The downside of a majority of centralized metascheduling approaches is that they lose their efficiency and optimization features in distributed environments with a significant workload. In such conditions of a *limited resources supply* overall job-flow execution makespan and individual jobs' finish time minimization become essential scheduling criteria. For example in [2], a traditional backfilling algorithm provided better scheduling outcome when compared to different optimization approaches in resource domain with a minimal performance configuration.

Main contribution of this paper is a CSS-based heuristic *anticipation* approach which retains scheduling efficiency and at the same time minimizes job-flow processing time. Initially this heuristic generates a near optimal but infeasible (anticipated) schedule. A special *replication* procedure is proposed and studied to ensure and provide a feasible scheduling solution.

The rest of the paper is organized as follows. Section 2 presents a general CSS fair scheduling concept. The proposed heuristic-based scheduling technique is presented in Sect. 3. Section 4 contains experiment setup and results for the proposed scheduling approach and its comparison with backfilling. Finally, Sect. 5 summarizes the paper.

2   Cyclic Alternative-Based Scheduling

Scheduling of a job flow using CSS is performed in time cycles known as scheduling intervals, by job batches [19, 20]. The actual scheduling procedure consists of two main steps. The first step involves a search for alternative scenarios of each job execution, or simply alternatives [21]. During the second step the dynamic programming methods [19, 20] are used to choose an optimal alternatives' combination. One alternative is selected for each job with respect to the given VO and user criteria. An example for a user scheduling criterion may be a job runtime, finish time, an overall running cost, etc. This criterion describes user's preferences for that specific job execution and expresses a type of an additional optimization to perform when searching for alternatives. Alongside with time (T) and cost (C) properties each job execution alternative has a user utility (U) value: user evaluation against the scheduling criterion. A common VO optimization problem may be stated as either minimization or maximization of one of the properties, having other fixed or limited, or involve Pareto-optimal strategy search involving both kinds of properties [3, 20, 22].

We consider the following relative approach to represent the user utility U. A job alternative with the minimum (best) user-defined criterion value Z_{min} corresponds to the left interval boundary ($U = 0\%$) of all possible job scheduling outcomes. An alternative with the worst possible criterion value Z_{max} corresponds to the right interval boundary ($U = 100\%$). In the general case, for each alternative with value Z, U is set depending on its position in [Z_{min}; Z_{max}] interval as follows: $U = \frac{Z - Z_{min}}{Z_{max} - Z_{min}} * 100\%$. Thus, each alternative gets its utility in relation to the "best" and the "worst" optimization criterion values user could expect according to the job's priority. The more some alternative corresponds to user's preferences the smaller is the U value.

For a fair scheduling model the second step of the VO optimization problem could be in form of: $C \rightarrow \max$, lim U (maximize total job flow execution cost, while respecting user's preferences to some extent: $U \leq U_{max}$); $U \rightarrow \min$, lim T (meet user's best interests, while ensuring some acceptable job flow execution time: $T \leq T_{max}$) and so on [19].

The launch of any job requires a co-allocation of a specified number of slots, as well as in the classic backfilling variation. A single slot is a time span that can be assigned to run a part of a parallel job. The target is to scan a list of available slots and to select a window of parallel slots with a "length" of the required resource reservation time. The user job requirements are arranged into a resource request containing a resource reservation time, characteristics of computational nodes (clock speed, RAM volume, disk space, operating system etc.), limitation on the selected window maximum cost.

ALP, AMP and AEP window search algorithms were discussed in [21]. The job batch scheduling performs consecutive allocation of a multiple nonintersecting in terms of slots alternatives for each job. Otherwise irresolvable collisions for resources may occur if different jobs will share the same time-slots. Sequential alternatives search and resources reservation procedures help to prevent such scenario. However in an extreme case when resources are limited or over utilized only at most one alternative execution could be reserved for each job. In this case alternatives-based scheduling result will be no different from First Fit resources allocation procedure [2]. First Fit resource selection algorithms [23] assign any job to the first set of slots matching the resource request conditions without any optimization.

3 Cyclic Anticipation Scheduling

In order to address the scheduling optimization problem the following anticipation heuristic for job batch scheduling is proposed. It consists of three main steps.

First, a set of all possible execution alternatives is found for each job not considering time slots intersections and without any resources reservation. The resulting intersecting alternatives found for each job reflect a full range of different job execution possibilities which user may expect on the current scheduling interval.

Second, CSS procedure [19, 20] is performed to select alternatives combination (one alternative for each job of the batch) optimal according to VO policy. The resulting alternatives combination most likely corresponds to an infeasible scheduling solution as possible time slots intersection will cause collisions on resources allocation stage. The main idea of this step is that obtained infeasible and *anticipated* solution will provide some *heuristic insights* on how each job should be handled during the scheduling. For example, if time-biased or cost-biased execution is preferred, how it should correspond to user criterion and VO administration policy and so on.

Third, a feasible resources allocation is performed. The resulting solution is both feasible and efficient as it reflects scheduling pattern obtained from a near-optimal reference solution – a *replication* step. The base for this replication is an Algorithm searching for Extreme Performance (AEP) described in details in [21]. AEP helps to find and reserve feasible execution alternatives most similar to those selected in the near-optimal infeasible solution.

We used AEP modification to allocate a diverse set of execution alternatives for each job. Originally AEP scans through a whole list of available time slots and retrieves one alternative execution satisfying user resource request and optimal according to the user custom criterion. During this scan, we saved all intermediate AEP search results to a dedicated list of possible alternatives.

For the replication purpose a new Execution Similarity criterion was introduced which helps AEP to find a window with a minimum distance to a reference alternative. Generally, we define a distance between two different alternatives (windows) as a relative difference or error between their significant criteria values. For example if reference alternative has C_{ref} total cost, and some candidate alternative cost is C_{can}, then the relative cost error E_C is calculated as $E_C = \frac{|C_{ref} - C_{can}|}{C_{ref}}$. If one needs to consider several criteria the distance D between two alternatives may be calculated as a linear sum of criteria errors: $D_l = E_C + E_T + .. + E_U$, or as a geometric distance in a parameters space: $D_g = \sqrt{E_C^2 + E_T^2 + .. E_U^2}$.

AEP modification with the Execution Similarity criterion is represented below.

Input Data:

    slotList – a list of available slots ordered non-decreasingly by their start time;

    job - a job for which the search is performed;

    refAlternative – reference alternative used to find similar job execution window.

Result:

    closestWindow – execution window similar to *refAlternative*

begin

```
  minDistance = MAX_VALUE;

  for each slot in slotList do
    if not(properHardwareAndSoftware(job, slot.node))
      continue;
    end if;

    windowSlotList.add(slot);
    windowStartTime = slot.startTime;

    for each wSlot in windowSlotList do
      minLength = wSlot.node.getWorkingTimeEstimate();
      if ((wSlot.endTime - windowStartTime) < minLength)
        windowSlotList.remove(wSlot);
      end if;
    end for;

    if (windowSlotList.size() ≥ job.nodesNeed)
      distance = calculateDistance(windowSlotList, refAlterna-
    tive);
      if (distance < minDistance)
        minDistance = distance;
        closestWindow = windowSlotList;
      end if;
    end if;
  end for;
end
```

In this algorithm an expanded window *windowSlotList* moves through a whole list of all available slots *slotList* sorted by their start time in ascending order. At each step any combination of *job.nodesNeed* slots inside *windowSlotList* can form a window that meets all the requirements to run the job. The main difference from the original AEP is that instead of searching for a window with a maximum single criterion value, we retrieve window with a minimum distance D_g or D_l to a reference execution alternative. Generally, this distance can reflect job execution preferences in terms of multiple criteria such as job execution cost, runtime, start time, finish time, etc.

4 Simulation Study

An experiment was prepared as follows using a custom distributed environment simulator [2, 19–21]. For our purpose, it implements a heterogeneous resource domain model: nodes have different usage costs and performance levels. A space-shared resources allocation policy simulates a local queuing system (like in GridSim or CloudSim [24]) and, thus, each node can process only one task at any given simulation time. The execution cost of each task depends on its execution time which is proportional to the dedicated node's performance level. The execution of a single job requires parallel execution of all its tasks.

The simulation environment was configured with the following features. The resource pool includes 80 heterogeneous computational nodes grouped in a single resource domain. A specific cost of a node is an exponential function of its performance value (base cost) with an added variable margin distributed normally as ±0.6 of a base cost. The scheduling interval length is 800 time quanta. The initial resource load with owner jobs is distributed hyper-geometrically resulting in 5% to 10% time quanta excluded in total.

Jobs number in a batch is 75. Nodes quantity needed for a job is a whole number distributed evenly on [2; 6]. Node reservation time is a whole number distributed evenly on [100; 500]. Job budget varies in the way that some of jobs can pay as much as 160% of base cost whereas some may require a discount. Every request contains a specification of a custom user criterion which is one of the following: job execution runtime or overall execution cost.

4.1 Replication Scheduling Accuracy

The first experiment is dedicated to a replication scheduling accuracy study. For this matter we conducted and collected data from more than 1000 independent job batch scheduling simulations. First, the general CSS was performed in each experiment for the following job-flow execution cost maximization problem $C \rightarrow \max$, $\lim U_a = 10\%$. U_a stands for the average user utility for one job, i.e. $\lim U_a = 10\%$ means that at average resulting deviation from the best possible outcome for each user did not exceed 10%. Next, linear and geometric replication algorithms were executed to replicate CSS solution using linear D_l and geometric D_g distance criteria. In the current experiment we used job execution cost error and processor time usage error to calculate distances.

In order to evaluate the resulting difference in scheduling outcomes, we additionally performed CSS algorithm ensuring users' individual preferences only ($\lim Ua = 0\%$) and ensuring VO preference by maximizing overall cost without taking into account users' criteria ($\lim Ua = 100\%$). These additional problems reflect extreme boundaries for scheduling results, which can be used to evaluate a relative replication error. Table 1 contains scheduling results for all these three problems and two replication algorithms.

Table 1. CSS replication average scheduling results

Job execution characteristic	C -> max, lim $Ua = 0\%$	C -> max, lim $Ua = 10\%$	Linear replication	Geometric replication	C -> max, lim $Ua = 100\%$
Cost	1283	*1349*	1353	1353	1475
Processor time	191.6	*191.2*	190.6	190.5	202.3
Finish time	367.1	*353.8*	356.2	356.4	358.5
U_a, %	0	*9.9*	17.6	17.8	65

The results indicate that both linear and geometric replication algorithms provided average scheduling parameters very close to the reference solution (indicated as bold in Table 1), and especially close against job execution cost and processor time usage, i.e. characteristics which were used for a replication distance calculation. For example, borderline problems provided average job execution cost (main job-flow optimization criterion) values 1283 and 1475 correspondingly. Reference intermediate solution provided 1349. And both replication algorithms ensured average job execution cost 1353 with only 2% deviation from reference solution against [1283; 1475] interval of possible scheduling outcomes. Although replication algorithms showed their efficiency with respect to integral job flow processing parameters (such as average job execution cost, runtime, finish time), individual user's preferences were considered to a lesser extent. It can be observed in the Table 1 that both replication algorithms provided average user utility U_a almost twice as much as the reference problem.

4.2   Anticipation and Backfilling Scheduling Comparison

The second experiment setup reiterates work [2] and is intended to compare anticipation scheduling procedure with a traditional backfilling algorithm. Backfilling is able to minimize the whole job-flow execution makespan as well as to generally follow the initial jobs relative queue order. These features make backfilling scheduling solution a good reference target for the anticipation scheduling scheme. The main criteria for comparison include average jobs' start and finish times as well as users' and VO economic criteria (such as execution time and cost). We used the following three algorithms for the comparison:

- CSS – the original cycle scheduling scheme;
- ANT – the anticipation scheduling procedure;
- BF – the conservative backfilling algorithm.

In a single experiment CSS and ANT solved $C \rightarrow$ max, lim $U_a = 10\%$ problem. Execution cost ($C \rightarrow$ min) and processor time ($T \rightarrow$ min) criteria were uniformly distributed between 75 user jobs generated in each experiment.

Important addition was introduced for ANT scheduling. In contrast with experiment series in Subsect. 4.1, job replication geometric distance D_g was calculated as $D_g = \sqrt{E_C^2 + E_T^2 + E_S^2}$, where additional element E_s stands for job start time error. As a reference start time value for each job we used start time obtained for a particular job by a prior backfilling scheduling. Thus, when searching for a job execution window we used infeasible solution for time and cost reference values, and a feasible backfilling solution as a reference for an attainable start time values complying with a queue priority.

To observe the behavior of the main scheduling parameters we conducted experiments with a different number N of computing nodes available during the scheduling: $N \in \{20, 25, 30, 40\}$.

Average job's start and finish times are presented in Figs. 1 and 2.

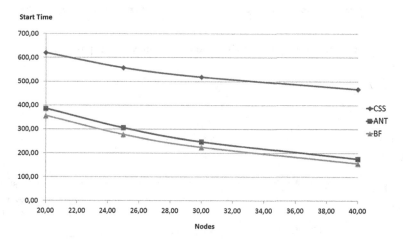

Fig. 1. Average jobs' start time in $C \to$ max, lim U problem

As can be seen in Figs. 1 and 2, backfilling provided better start and finish times for a job-flow execution compared to CSS and this result is consistent with [2]. In the current problem setup backfilling was able to finish the job flow execution almost twice earlier then CSS. It can be explained by $C \to$ max, lim U scheduling problem which required CSS to allocate resources for job-flow execution cost maximization considering contradictory user preferences, not minimizing jobs' completion times.

At the same time anticipation algorithm during each experiment solved the same $C \to$ max, lim U problem and provided jobs' start and finish times only 10% behind the backfilling scheduling outcome.

The details of anticipation scheduling can be examined in Figs. 3 and 4.

Figure 3 shows average job execution time provided by backfilling and anticipation algorithm. Additionally ANT T and ANT C represent average execution times obtained by anticipation scheduling for jobs with time minimization and cost minimization criteria correspondingly. As it can be observed, ANT and BF generally provided comparable execution times, which is not a direct optimization criterion for either of

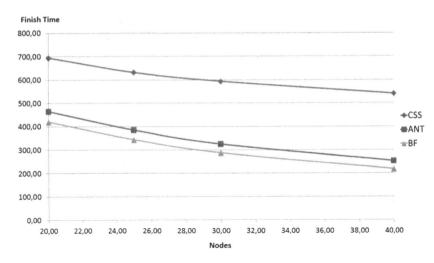

Fig. 2. Average jobs' finish time in $C \rightarrow$ max, lim U problem

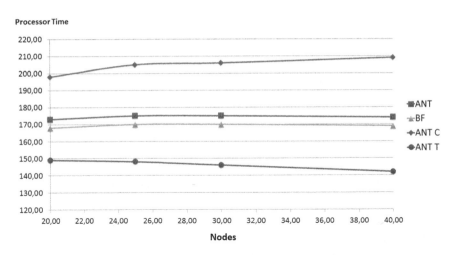

Fig. 3. Average jobs' execution time in $C \rightarrow$ max, lim U problem

them. At the same time ANT applied completely different scheduling policies for jobs with different private scheduling criteria. So that ANT T jobs used 25%–33% less processor time then ANT C jobs and 15% less compared to BF solution.

A similar pattern can be observed in Fig. 4, where average jobs' execution cost is presented. ANT and BF provided comparable general job-flow execution cost value. However ANT was able to consider user preferences and shared resources so that ANT C jobs execution cost was 10–15% less then ANT T jobs and 6–9% less compared to backfilling.

Summarizing the results, ANT is able to provide a general scheduling outcome similar to backfilling (with at most 10% error on job's start and finish times), and at the

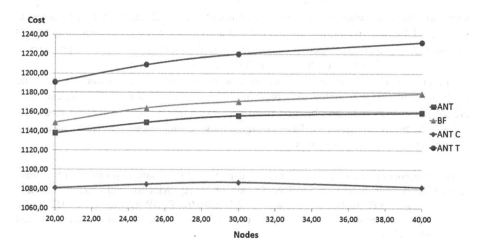

Fig. 4. Average jobs' execution cost in $C \rightarrow$ max, lim U problem

same time considers users' and VO preferences by efficiently solving $C \rightarrow$ max, lim U problem. Thereby the available resources are distributed between user jobs according to the predefined scheduling requirements (see Figs. 3 and 4). In our experiment set they include individual jobs execution preferences (for example, certain job's execution cost minimization) and a common job-flow scheduling policy (total job-flow execution cost maximization in our example).

Speaking of a whole job-flow scheduling policy it is worth noting that despite the cost maximization performed by ANT, backfilling still provided higher total job-flow execution cost (Fig. 4). This result may be explained by the need of ANT to additionally consider user preferences (lim $U_a = 10\%$), including user jobs with a cost *minimization* criterion. For example, in $C \rightarrow$ max, lim $U_a = 100\%$ problem, which performs cost maximization without taking into account user preferences, ANT provides 1–2% higher job-flow execution cost compared to backfilling, but does not reach original CSS by 10%. In this case ANT was limited by a start time reference (obtained from backfilling solution) and, thus, had fewer opportunities to use available resources for a total cost maximization as opposed to CSS.

5 Conclusions and Future Work

In this paper, we study the problem of fair job batch scheduling with a relatively limited resources supply. The main problem that arises is a scarce set of job execution alternatives which eliminates scheduling optimization efficiency.

We propose a heuristic anticipation scheduling which generates a near-optimal but infeasible reference solution and then replicates it to allocate a feasible accessible solution. The special replication procedure is proposed which provides 2–5% error from the reference scheduling solution. The obtained results show that the new heuristic approach provides flexible and efficient solutions for different fair scheduling scenarios. In case when computing environment with a limited set of resources is considered the anticipation algorithm is still able to allocate resources according to VO stakeholders' preferences, generally complies with queue priorities and provides a job-flow completion time up to 10% behind backfilling solution.

Future work will be focused on replication algorithm studies and its possible application to fulfill complex user preferences expressed in a resource request. Reference parameters may be obtained from user expectations or transformed from different scheduling solutions. Different weights may be introduced for errors calculation on different reference parameters.

Acknowledgments. This work was partially supported by the Council on Grants of the President of the Russian Federation for State Support of Young Scientists and Leading Scientific Schools (grants YPhD-2297.2017.9 and SS-6577.2016.9), RFBR (grants 15-07-02259 and 15-07-03401), and by the Ministry on Education and Science of the Russian Federation (project no. 2.9606.2017/8.9).

References

1. Dimitriadou, S.K., Karatza, H.D.: Job scheduling in a distributed system using backfilling with inaccurate runtime computations. In: Proceedings of the 2010 International Conference on Complex, Intelligent and Software Intensive Systems, pp. 329–336 (2010)
2. Toporkov, V., Toporkova, A., Tselishchev, A., Yemelyanov, D., Potekhin, P.: Heuristic strategies for preference-based scheduling in virtual organizations of utility grids. J. Ambient Intell. Humanized Comput. 6(6), 733–740 (2015)
3. Kurowski, K., Nabrzyski, J., Oleksiak, A., Weglarz, J.: Multicriteria aspects of grid resource management. In: Nabrzyski, J., Schopf, J.M., Weglarz, J. (eds.) Grid Resource Management. State of the Art and Future Trends, pp. 271–293. Kluwer Academic Publishers (2003)
4. Buyya, R., Abramson, D., Giddy, J.: Economic models for resource management and scheduling in grid computing. J. Concurrency Comput. 14(5), 1507–1542 (2002)
5. Rodero, I., Villegas, D., Bobroff, N., Liu, Y., Fong, L., Sadjadi, S.M.: Enabling interoperability among grid meta-schedulers. J. Grid Comput. 11(2), 311–336 (2013)
6. Ernemann, C., Hamscher, V., Yahyapour, R.: Economic scheduling in grid computing. In: Feitelson, D.G., Rudolph, L., Schwiegelshohn, U. (eds.) JSSPP 2002. LNCS, vol. 2537, pp. 128–152. Springer, Heidelberg (2002). https://doi.org/10.1007/3-540-36180-4_8
7. Rzadca, K., Trystram, D., Wierzbicki, A.: Fair game-theoretic resource management in dedicated grids. In: IEEE International Symposium on Cluster Computing and the Grid (CCGRID 2007), Rio De Janeiro, Brazil, pp. 343–350. IEEE Computer Society (2007)
8. Vasile, M., Pop, F., Tutueanu, R., Cristea, V., Kolodziej, J.: Resource-aware hybrid scheduling algorithm in heterogeneous distributed computing. J. Future Gener. Comput. Syst. 51, 61–71 (2015)
9. Penmatsa, S., Chronopoulos, A.T.: Cost minimization in utility computing systems. Concurrency Comput. Pract. Exp. 16(1), 287–307 (2014). Wiley
10. Mutz, A., Wolski, R., Brevik, J.: Eliciting honest value information in a batch-queue environment. In: 8th IEEE/ACM International Conference on Grid Computing, New York, USA, pp. 291–297 (2007)

11. Blanco, H., Guirado, F., Lérida, J.L., Albornoz, V.M.: MIP model scheduling for multi-clusters. In: Caragiannis, I., et al. (eds.) Euro-Par 2012. LNCS, vol. 7640, pp. 196–206. Springer, Heidelberg (2013). https://doi.org/10.1007/978-3-642-36949-0_22

12. Takefusa, A., Nakada, H., Kudoh, T., Tanaka, Y.: An advance reservation-based co-allocation algorithm for distributed computers and network bandwidth on QoS-guaranteed grids. In: Frachtenberg, E., Schwiegelshohn, U. (eds.) JSSPP 2010. LNCS, vol. 6253, pp. 16–34. Springer, Heidelberg (2010). https://doi.org/10.1007/978-3-642-16505-4_2

13. Vohs, K., Mead, N., Goode, M.: The psychological consequences of money. Science **314** (5802), 1154–1156 (2006)

14. Carroll, T., Grosu, D.: Divisible load scheduling: an approach using coalitional games. In: Proceedings of the Sixth International Symposium on Parallel and Distributed Computing, ISPDC 2007, p. 36 (2007)

15. Kim, K., Buyya, R.: Fair resource sharing in hierarchical virtual organizations for global grids. In: Proceedings of the 8th IEEE/ACM International Conference on Grid Computing, Austin, USA, pp. 50–57. IEEE Computer Society (2007)

16. Skowron, P., Rzadca, K.: Non-monetary fair scheduling cooperative game theory approach. In: Proceeding of SPAA 2013 Proceedings of the Twenty-Fifth Annual ACM Symposium on Parallelism in Algorithms and Architectures, pp. 288–297. ACM, New York (2013)

17. Dalheimer, M., Pfreundt, F.-J., Merz, P.: Agent-based grid scheduling with Calana. In: Wyrzykowski, R., Dongarra, J., Meyer, N., Waśniewski, J. (eds.) PPAM 2005. LNCS, vol. 3911, pp. 741–750. Springer, Heidelberg (2006). https://doi.org/10.1007/11752578_89

18. Jackson, D., Snell, Q., Clement, M.: Core algorithms of the Maui scheduler. In: Feitelson, D. G., Rudolph, L. (eds.) JSSPP 2001. LNCS, vol. 2221, pp. 87–102. Springer, Heidelberg (2001). https://doi.org/10.1007/3-540-45540-X_6

19. Toporkov, V., Yemelyanov, D., Bobchenkov, A., Tselishchev, A.: Scheduling in grid based on VO stakeholders preferences and criteria. In: Zamojski, W., Mazurkiewicz, J., Sugier, J., Walkowiak, T., Kacprzyk, J. (eds.) Dependability Engineering and Complex Systems. AISC, vol. 470, pp. 505–515. Springer, Cham (2016). https://doi.org/10.1007/978-3-319-39639-2_44

20. Toporkov, V., Toporkova, A., Tselishchev, A., Yemelyanov, D., Potekhin, P.: Metascheduling and heuristic co-allocation strategies in distributed computing. Comput. Inf. **34**(1), 45–76 (2015)

21. Toporkov, V., Toporkova, A., Tselishchev, A., Yemelyanov, D.: Slot selection algorithms in distributed computing. J. Supercomput. **69**(1), 53–60 (2014)

22. Farahabady, M.H., Lee, Y.C., Zomaya, A.Y.: Pareto-optimal cloud bursting. IEEE Trans. Parallel Distrib. Syst. **25**, 2670–2682 (2014)

23. Cafaro, M., Mirto, M., Aloisio, G.: Preference-based matchmaking of grid resources with CP-Nets. J. Grid Comput. **11**(2), 211–237 (2013)

24. Calheiros, R.N., Ranjan, R., Beloglazov, A., De Rose, C.A.F., Buyya, R.: CloudSim: a toolkit for modeling and simulation of cloud computing environments and evaluation of resource provisioning algorithms. J. Softw. Pract. Exp. **41**(1), 23–50 (2011)

The State-of-the-Art Trends in Education Strategy for Sustainable Development of the High Performance Computing Ecosystem

Sergey Mosin[(✉)] [iD]

Kazan Federal University, Kazan, Russia
smosin@ieee.org

Abstract. High-performance computing (HPC) plays very important role in the sphere of information technology as well as defines the strategic direction for inter- and trans-disciplinary breakthroughs ensuring the essential influence on local and global markets. The current status of HPC systems development in different countries is analyzed in the paper. The constraints for an active involvement the HPC in many business processes of different industrial, academic and research partners deal with low competence of the regular users and lack of the HPC proficient personnel. Both the technical infrastructure development and training the competent staff with wide range of the HPC related knowledge and skills are the strategic tasks of the national level. The second task is principal for stable development of the HPC ecosystems especially forwarding to the exascale era. The features of curricula focused on education in the HPC area are considered. The experience of implementation the education strategy of Kazan Federal University in the HPC field based on skills-driven model and partnership with IT-companies is discussed.

Keywords: High performance computing · Education · Trends · Sustainable development

1 Introduction

The comprehensive informatization of the state-of-the-art society and the active introduction of information technology into the business processes of all sectors of the economy determine the intensive development of hardware and software platforms and the IT sphere in a whole.

Nowadays there is an expansion of the range and complexity of the tasks demanded by the business community. Against this background, the demand of the labor market in IT specialists of different levels and qualifications increases: from the project managers of high level (leaders, architects, project managers, etc.) to rank-and-file executors (programmer, tester, technician, etc.). The efficiency of IT companies in many ways relays with the easiness of integration the university graduates into the processes of hardware-software co-design and implementation the systems of automated data processing in accordance with the requirements and specification of the customer.

© Springer International Publishing AG 2017
V. Voevodin and S. Sobolev (Eds.): RuSCDays 2017, CCIS 793, pp. 494–504, 2017.
https://doi.org/10.1007/978-3-319-71255-0_40

The reform of higher education in the Russian Federation is a consequence of the economic transformation of society associated with the transition from industrial orientation to the market. In this regard, the training of a specialist for a specific sector of industry gives way to a competence model of education, when theoretical knowledge is backed by the technological skills and the ability to use it in practice according to employers' requests. The application of the concept of practice-oriented education into the implementation of higher vocational study programs opens the possibility for training IT professionals with a combination of competencies that are in demand on the labor market and specified in the state and professional standards. In this case, universities and the corresponding educational programs acquire competitive advantages, providing greater appeal for entrants.

High-performance computing (HPC) plays very important role in the sphere of information technology (IT) as well as defines the strategic direction for inter- and trans-discipline breakthroughs ensuring the essential influence on local and global markets. Many countries which pretend on the global leadership have developed and implement national strategy for progress in the HPC and corresponding growing of industry, science and economy [1]. The HPC is very specific and key area in the IT sphere with the following distinctive features:

1. Complex infrastructure, which needs high quality specialists for use and support.
2. High direct and indirect cost on high-performance computer design, implementation, use and maintenance.
3. The unique architecture depending on the class of tasks or even specific task which should be solved. Each HPC system is designed for specific task.
4. Essential gap between hardware performance and available software possibilities.
5. Non-equal involvement of the HPC into interdisciplinary R&D.
6. A lot of skills and knowledge in different areas such as computer science, telecommunications, program engineering, power supply and consumption, applied and computational mathematics, management, etc., are required for efficient HPC user and computational scientists.

The constraints for active involvement the HPC in many business processes of different industrial, academic and research partners deal with low competence of the regular users, the problem originators and the general IT specialists in the HPC topics. The current situation can be changed by the complex modernization of the vocational study programs emphasizing the wide range of HPC applications. The common trends and experience of the Institute of Computational Mathematics and Information Technologies at Kazan Federal University in education specialists for sustainable development of the high-performance computing ecosystem are considered in the paper.

The rest of the paper is organized as the following: Sect. 2 provides an introduction into the HPC strategies of the main world players. The principal features of ACM Curricula for the HPC and NSF/IEEE-TCPP Curriculum Initiative on Parallel and Distributed Computing are considered in Sect. 3. Section 4 highlights the education strategy of Kazan Federal University for sustainable development of the HPC ecosystem. Final section ensures the concluding remarks.

2 The HPC Strategies of Different Countries

The advantage in both the development and application of high performance computing is a vital for countries' economic competitiveness and innovation potential [2]. Accordingly, many countries have made significant investments and fulfilled holistic strategies to position themselves at the forefront of the rivalry for the global HPC leadership. Many countries have created national programs that are investing large sums of money to wide use the existent HPC systems and develop exascale supercomputers. Mastering and active use the HPC systems, tools and technologies open the ways forward to generation of new significant technologies for overtaken the global grand challenges and improve both the innovative character of national economy and supremacy on the global market.

The state-of-the-art high performance computers are very complex systems, efficient exploiting of which requires a huge amount of financial, the power supply and human resources. There are a limited set of countries that can be able to design, manufacture and use the HPC systems. The influence and contribution of the high performance computers into scientific progress, industrial competitiveness, national security, and quality of life are significant. The open results of the world competitions for the HPC leadership are provided twice per year in the Top 500 ranking. The dynamics of changing the number of the most powerful supercomputers for some leading countries during the last three editions in Top 500 is shown in Fig. 1 [1].

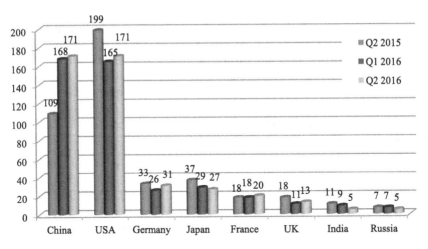

Fig. 1. The dynamic of changes the number of HPC systems in the leading countries at last three TOP500 editions

According to the last 48th Top 500 edition two countries, China and USA, dominate nowadays in both number and total performance of HPC systems (Table 1). The total performance of all 500 computers on the list is now 672 petaflops. The USA holds the narrowest of leads in the aggregate Linpack performance with 33.9% of the total and China is the second with 33.3%. The retention and especially augmentation the number of HPC systems in the list are very complex efforts for each country because all participants of Top 500 not stand still, but reinforce the processes to develop new systems with orientation onto the exascale architectures. Such stable development in many ways deals with strong progress in the integrated technologies [3], as well as the official state policy in the area of HPC technologies, granted the state programs and investment by government, industrial and business partners. The national strategies and financial resources for realization are represented in Table 2.

Table 1. HPC specification for top 10 countries in the 48th edition of Top 500

Country	Count	System share (%)	Rmax (GFlops)	Rpeak (GFlops)	Cores
USA	171	34,2	228,032,809	327,303,955	11,660,816
China	171	34,2	223,571,136	394,013,392	21,546,512
Germany	31	6,2	36,501,435	45,628,388	1,600,240
Japan	27	5,4	54,486,820	77,371,577	3,946,560
France	20	4	25,398,803	31,727,765	1,158,428
United Kingdom	13	2,6	27,602,596	31,682,369	1,148,968
Poland	7	1,4	6,162,214	8,157,370	208,284
Italy	6	1,2	14,062,113	21,140,514	606,312
India	5	1	3,092,368	4,456,051	133,172
Russia	5	1	4,411,812	6,515,928	181,070

Table 2. Short specification of the national HPC strategies

Country	HPC strategy/Program	Investment, $
USA	National Strategic Computing Initiative (NSCI)	320 million/year
China	13th Five-Year Development Plan (Develop Multiple Exascale Systems)	200 million/year (for next five years)
European Union	ETP4HPC; PRACE; ExaNeSt	1.1 in billion total allocated through 2020
Japan	Flagship2020 Program	@$200 million/year (for next five years)
India	National Supercomputing Mission	140 million/year (for 2016-2020)
South Korea	National Supercomputing Act	20 million/year (for 2016-2020)

China has made the HPC leadership a national priority. The leadership in high-performance computing for China is central to the country's goal of transitioning away from reliance on foreign technology to using home-made technology. For in-stance, the Sunway TaihuLight system placed the first place in the Top 500 list was developed by Chinese National Research Center of Parallel Computer Engineering & Technology using the state-of-the-art 260-core manycore processors ShenWei SW26010 designed by the National High Performance Integrated Circuit Design Center in Shanghai.

2.1   National Strategic Computing Initiative in USA

The National Strategic Computing Initiative (NSCI) was launched to advance the USA leadership in the HPC [4]. The NSCI is a whole-of-nation effort designed to create a cohesive, multi-agency strategic vision and Federal investment strategy, executed in collaboration with industry and academia, to maximize the benefits of HPC for the United States.

The NSCI seeks to accomplish five strategic objectives in the government collaboration with industry and academia [5]:

1. Accelerating delivery of a capable exascale computing system that integrates hardware and software capability to deliver approximately 100 times the performance of current 10 petaflop systems across a range of applications representing government needs.
2. Increasing coherence between the technology base used for modeling and simulation and that used for data analytic computing.
3. Establishing, over the next 15 years, a viable path forward for future HPC systems even after the limits of current semiconductor technology are reached (the "post-Moore's Law era").
4. Increasing the capacity and capability of an enduring national HPC ecosystem by employing a holistic approach that addresses relevant factors such as networking technology, workflow, downward scaling, foundational algorithms and software, accessibility, and workforce development.
5. Developing an enduring public-private collaboration to ensure that the benefits of the research and development advances are, to the greatest extent, shared between the United States Government and industrial and academic sectors.

The NSCI is supported and realized by many national agencies, which also have to develop an ambitious workforce development plan to educate the current generation and train the next generation of scientists and engineers to adopt HPC as an effective approach to solving problems of societal importance.

National Science Foundation (NSF) plays a central role in scientific discovery advances, the broader HPC ecosystem for scientific discovery, and workforce development. According to NSCI the NSF should [5]:

- Provide leadership in learning and workforce development to encompass support of basic HPC training for a broad user community as well as support for career path development for computational and data scientists;
- Increase engagement with industry and academia through existing programs;
- Support the broad deployment of NSCI technologies to increase the capacity and capability of the HPC ecosystem, enabling fundamental understanding across frontiers consistent with NSF scientific and engineering priorities;
- Lead the development of domestic and international collaborations that will advance transformative computational science and engineering with an integrated approach to high-end computing, data, networking, facilities, software, and multidisciplinary expertise, consistent with NSCI strategic objectives.

2.2 European HPC Strategy

HPC is considered as a high strategic importance for European society, competitiveness and innovation. The use of HPC has contributed significantly and increasingly to scientific progress, industrial competitiveness, national and regional security, and the quality of human life. HPC-enabled simulation is widely recognized as the third branch of the scientific method, complementing traditional theory and experimentation.

The European HPC strategy has three pillars [6]:

1. Developing the next generation of HPC technologies, applications and systems towards exascale;
2. Providing access to the best supercomputing facilities and services for the industry (including SMEs) and academia (Partnership for Advanced Computing in Europe – PRACE);
3. Achieving excellence in HPC application delivery and use through establishment of Centers of Excellence in HPC applications.

These pillars are complemented with awareness raising, training, education and skills development in HPC.

The European Technology Platform for High Performance Computing (ETP4HPC) is an industry-led think tank and advisory group of companies and research centers involved in the HPC technology research in Europe, which was formed in 2011 with the aim to build a world-class HPC Technology Supply Chain in Europe, increase the global share of European HPC and HPC technology vendors as well as maximize the benefit of HPC technology for the European HPC user community [7].

The PRACE ensures the wide availability of HPC resources on equal access terms, in order to strengthen the position of European industry and academia in the use, development and manufacturing of advanced computing products, services and technologies. The training an adequate number of professional personnel, including computational scientists, programmers, system administrators, technologists, etc. is considered as one of the key factor for successful development HPC ecosystem in the EU [8, 9].

The PRACE has an extensive education and training effort for effective use of the research infrastructure through seasonal schools, workshops and scientific and industrial seminars throughout Europe. Seasonal schools target broad HPC audiences, whereas workshops are focused on particular technologies, tools or disciplines or research areas.

All EU state programs in the field of HPC are oriented onto strengthening the position of European industry and academia in the use, development and manufacturing of advanced computing products, services and technologies [8].

3 Features of the Curricula Focused on Education in HPC

The major part of national strategies oriented onto development HPC as the key tasks defines generation and development of: (1) de-facto HPC systems emphasing in the short- and middle-terms onto exascale architectures; (2) infrastructure accumulating the public structures, private industry and business as well as academia involved into complex processes of development, support, maintenance and use the HPC systems; (3)

educational platform ensuring the training a huge number of required high qualified personnel responsible for effective use of existent hardware and software tools as well as generating a new knowledges and technologies in the HPC area, training the next generation of scientists, designers, engineers, users and task managers.

The implementation of the third task requires development of new or adaptation of existent curricula taking into accounts both the global tendencies of evolution the HPC tools and technologies and the local demands of the state sector, industry and business.

According to exhaustive analysis the efficient training of HPC professionals should be realized in the framework of computer science/computer engineering curricula by including the new specialized courses [10]. End users represent academic, research and industrial organizations and communities. Their applications in engineering, human, social and natural sciences are typically compute and/or data intensive. In some of the areas there are long traditions in using HPC, but in some areas computational science is just entering the domain. Therefore, delivering the introductory basic courses on HPC for master students especially in technical and technological fields in order to increase general awareness and knowledges about HPC possibilities and prepare qualified task managers is very important.

Many universities develop curricula on HPC based on ACM CS/CE Curricula [11, 12] and/or NSF/IEEE-TCPP Curriculum Initiative on Parallel and Distributed Computing [13]. The experience of different universities in adaptation and implementation the curricula focused on HPC and parallel and distributed computing is actively publicized and discussed [14–17].

3.1 ACM/IEEE-CS Joint Task Force: Computer Engineering and Computer Science Curricula

Main focus on the HPC is localized through the parallel and distributed computing techniques. The ACM/IEEE Computer Engineering curriculum [12] considers the following main aspects of HPC: (1) Computer architecture and organization with instruction-level and processor-level parallelism (multicore processor and multiprocessor system); (2) Distributed system architectures, high performance computing and networks, memory hierarchy architecture for single core and multicore systems; (3) Parallel algorithms and multi-threading; (4) Introduction to High Performance Computing, which covers the organization of high performance computer, design methods of parallel programming, performance model of programs, performance evaluation and optimization techniques, programming in MPI and OpenMP and algorithms in high performance computing.

The latest ACM/IEEE-CS Joint Task Force: Computer Science Curricula [11] proposal vastly upgraded the coverage of parallel thinking proposing topics such as: (1) Parallel and Distributed Computing; (2) Parallelism Fundamentals; (3) Parallel Decomposition; (4) Parallel Algorithms, Analysis, and Programming; (5) Parallel Architecture; (6) Parallel Performance; (7) Distributed Systems; and (8) Cloud Computing.

Parallel and distributed computing builds on foundations in many areas, including an understanding of fundamental systems concepts such as concurrency and parallel execution, consistency in state/memory manipulation, and latency. Communication and coordination among processes is rooted in the message-passing and shared-memory models of computing and such algorithmic concepts as atomicity, consensus, and conditional waiting.

Special attention is paid to software engineering, which considers different technologies, techniques and tools for software development with orientation on wide range of systems, such as real time systems; client-server systems; distributed systems; parallel systems; web-based systems; high integrity systems, etc. and specifics of parallel programming vs. concurrent programming.

3.2 NSF/IEEE-TCPP Curriculum Initiative on Parallel and Distributed Computing

The draft of parallel and distributed computing (PDC) curricula was designed by IEEE Computer Society Technical Committee on Parallel Processing (TCPP) with support of National Science Foundation (NSF) [13]. This document provides guidance and support for departments looking to expand the coverage of parallel and distributed topics in their undergraduate programs. According to the recommendations the problems of parallel and distributed computing fall into the following four knowledge areas:

(1) Architecture.
(2) Programming.
(3) Algorithms.
(4) Cross Cutting and Advanced Topics.

A primary goal of proposed curriculum is the definition for the computer science (CS)/computer engineering (CE) students and their instructors to receive periodic guidelines that identify aspects of PDC that are important to be covered, and suggest specific core courses in which their coverage might find an appropriate context. The proposed curriculum enables students to be fully prepared for their future careers in light of the technological shifts and mass marketing of parallelism through multicores, GPUs, and corresponding software environments, and to make a real impact with respect to all of the stakeholders for PDC, including employers, authors, and educators.

4 Education Strategy of Kazan Federal University in the HPC Field

Kazan Federal University (KFU) founded in 1804 nowadays is the biggest research and educational center in the Volga region federal district of Russia. The main priorities in the R&D area as well as innovations are organized and developed in the form of the following Strategic Academic Units (SAU): (1) 7P Translational Medicine; (2) Ecooil – global energy and resources for the materials of the future; (3) Astrochallenge: cosmology, monitoring, navigation, applications and (4) The quadrature of transforming teacher education – 4T. There are more than 150 OpenLabs and research centers involved in the state-of-the-art R&D projects as a part of SAU. The major part of research works uses numerical simulation, intellectual data analysis based on data mining and machine learning algorithms. The effective solution of many tasks may be obtained only using the up-to-date hardware and software tools oriented on parallel and distributed computations. Each SAU implements several tens trans- and interdisciplinary research works combining specialists from different scientific fields. IT professionals especially with strong experience in the HPC area play vital important role in the research groups.

The Institute of Computational Mathematics and Information Technologies at KFU (ICMIT) trains the IT specialists competent in the HPC technologies and tools on three levels of study: bachelor programs, master programs and Ph.D. programs. All curricula in the ICMIT are based on the ACM/IEEE CS Curriculum and NSF/IEEE-TCPP Curriculum Initiative on Parallel and Distributed Computing.

The educational process uses the practical skills-driven model. The professional courses combine theoretical knowledge and practical skills. The laboratory works are constructed in such way to master different technologies and tools of parallel and distributed programming for SMP, NUMA, MPP and Cluster architectures, using CUDA programming, OpenMP, OpenCL, MPI, threads programming, etc. The KFU HPC cluster system is used in education process as well. The cluster has the hybrid architecture and combines HPC subsystem, GPU-based cluster subsystem and Big Data processing subsystem. The total peak performance of KFU cluster consists of 39 TFLOPS.

Access of students to real HPC systems plays important role at training specialists adapted to real conditions and studying not only theory of parallel programming but also rules and processes specific in the HPC and data centers.

Additional workshops and short courses delivered by well experienced professionals from the partner's IT companies are important part of training process. The ICMIT regularly organizes such courses in partnership with Intel and NVidia, as well as some academic organization in the framework of Computer Science Club initiative. The KFU has close cooperation with the Supercomputing Consortium of Russian Universities [18].

Bachelor degree students receive basic competences in parallel and distributed computing. The master and Ph.D. students study advanced courses and combine training with R&D. Such multilayer education system allows generating different specialists for the local and global HPC ecosystems.

5   Conclusions

The global problems and tendencies in development of the HPC ecosystems were discussed. The necessity for continued collaboration and innovative initiatives is obvious and permanent grows. The number of required personnel competent in the HPC system design, implementation and maintenance; parallel programming and application development; numerical and computational modelling as well as task management is increased regularly due to developing the HPC centers and wide using the HPC systems by public organizations, private industry and business. The increased use of computational and information technologies brings innovation and efficiency in many production and business processes, generates products and services favoring the growing of industry, science and economy. The preparation and implementation the professional courses to train the new generation of specialists with knowledge and skills in mathematical simulation and modelling, intellectual data analysis and HPC using, administrating and management are very important tasks for future development the HPC ecosystems. The experience of Kazan Federal University in training IT specialists on three layers of study based on skills-driven model was described as well as concept of trans- and interdisciplinary collaboration in the project of Strategic Academic Units. The realization of introductory basic courses on the HPC for non-IT specialties can provide conditions for active use the HPC systems and technologies at interdisciplinary R&D in the short- and middle-terms.

Acknowledgement. The work is performed according to the Russian Government Program of Competitive Growth of Kazan Federal University.

References

1. Top 500 Homepage. https://www.top500.org. Last accessed 30 Apr 2017
2. Ezell, S.J., Atkinson, R.D.: The Vital Importance of High-Performance Computing to U.S. Competitiveness. ITIF, April 2016
3. Mosin, S.G.: The features of integrated technologies development in area of ASIC design. In: Proceedings of 9th Conference the Experience of Designing and Application of CAD System in Microelectronics (CADSM 2007), Lviv, Polyana, Ukraine, pp. 292–295 (2007)
4. The National Strategic Computing Initiative (NSCI) Homepage. https://www.nitrd.gov/nsci/index.aspx. Last accessed 30 Apr 2017
5. National Strategic Computing Initiative Strategic Plan (2016)
6. HORIZON 2020. High-Performance Computing Homepage. https://ec.europa.eu/programmes/horizon2020/en/h2020-section/high-performance-computing-hpc. Last accessed 30 Apr 2017
7. ETP4HPC strategic research agenda: 2015 update (2015)
8. High Performance Computing in the EU: Progress on the Implementation of the European HPC Strategy. IDC (2015). https://doi.org/10.2759/034719
9. Margetts, L.: Survey of Computing Platforms for Engineering Simulation, Supplementary Report, European Exascale Software Initiative (2015)
10. Monismith, D.R.: Incorporating parallelism and high performance computing into computer science courses. In: IEEE Frontiers in Education Conference (FIE), El Paso, TX, USA, 9 p. IEEE (2015)
11. ACM/IEEE-CS Joint Task Force, Computer Science Curricula 2013: Curriculum Guidelines for Undergraduate Degree Programs in Computer Science. http://www.acm.org/education/CS2013-finalreport.pdf. Last accessed 30 Apr 2017
12. ACM/IEEE Computer Engineering Curricula (2016). http://www.acm.org/binaries/content/assets/education/ce2016-final-report.pdf. Last accessed 30 Apr 2017
13. Prasad, S.K., Chtchelkanova, A., Dehne, F., Gouda, M., Gupta, A., Jaja, J., Kant, K., La Salle, A., LeBlanc, R., Lumsdaine, A., Padua, D., Parashar, M., Prasanna, V., Robert, Y., Rosenberg, A., Sahni, S., Shirazi, B., Sussman, A., Weems, C., Wu, J.: NSF/IEEE-TCPP Curriculum Initiative on Parallel and Distributed Computing - Core Topics for Undergraduates, Version I (2012). http://www.cs.gsu.edu/~tcpp/curriculum/. Last accessed 30 Apr 2017

14. Connor, C., Bonnie, A., Grider, G., Jacobson, A.: Next generation HPC workforce development: the computer system, cluster, and networking summer institute. In: Proceedings of the Workshop on Education for High Performance Computing (EduHPC 2016), Salt Lake City, Utah, USA, pp. 32–39. IEEE Press (2016)
15. Gergel, V., Liniov, A., Meyerov, I., Sysoyev, A.: NSF/IEEE-TCPP curriculum implementation at the State University of Nizhni Novgorod. In: IEEE 28th International Parallel & Distributed Processing Symposium Workshops, Phoenix, AZ, USA, pp. 1079–1084. IEEE (2014)
16. Škrinárová, J., Vesel, E.: Model of education and training strategy for the high performance computing. In: International Conference on Emerging eLearning Technologies and Applications (ICETA), Vysoke Tatry, Slovakia, 6 p. IEEE (2016)
17. Wilson, L.A., Dey, S.C.: Computational science education focused on future domain scientists. In: Proceedings of the Workshop on Education for High Performance Computing (EduHPC 2016), Salt Lake City, Utah, USA, pp. 19–24. IEEE Press (2016)
18. Supercomputing Consortium of Russian Universities Homepage. http://hpc.msu.ru/?q=node/136. Last accessed 30 Apr 2017

A Service-Oriented Infrastructure for Teaching Big Data Technologies

Oleg Sukhoroslov[1,2,3](✉)

[1] Institute for Information Transmission Problems of the Russian Academy
of Sciences (Kharkevich Institute),Moscow, Russia
`sukhoroslov@iitp.ru`
[2] Yandex School of Data Analysis, Moscow, Russia
[3] Higher School of Economics, Moscow, Russia

Abstract. The paper presents an experience in incorporating Big Data
technologies into introductory parallel and distributed computing courses
and building a service-oriented infrastructure to support practical exercises involving these technologies. The presented approach helped to provide a smooth practical experience for students with different technical
background by enabling them to run and test their MapReduce and
Spark programs on a provided Hadoop cluster via convenient web interfaces. This approach also enabled automation of routine actions related
to submission of programs to a cluster and evaluation of programming
assignments.

Keywords: Big Data · Parallel programming · Distributed computing ·
Hadoop · MapReduce · Spark · Web-based interfaces · Web services

1 Introduction

The explosive growth of data observed in a variety of areas from research to
commerce, commonly referred to as the Big Data phenomenon, requires the use
of high-performance resources and efficient means for storing and processing
large amounts of data. During the last decade, the distributed data processing
models such as MapReduce [1] and technologies like Hadoop [2] and Spark [3] are
emerged. Modern HPC systems such as clusters are being increasingly used for
running data-intensive applications in science and technology. Therefore there is
a growing demand to incorporate relevant programming models and technologies
into a parallel and distributed computing (PDC) teaching curriculum.

The introduction of Big Data technologies in a PDC course brings a number
of challenges. First, these technologies are noticeably different from traditional
parallel programming technologies (e.g., MPI), by using other programming languages (e.g., Java, Scala or Python) and computing models (e.g., MapReduce or
Spark RDD). The Big Data applications are also quite different from the traditional HPC applications, which often motivates the development of specialized
courses. Second, currently it is not possible to easily collocate Big Data and

V. Voevodin and S. Sobolev (Eds.): RuSCDays 2017, CCIS 793, pp. 505–515, 2017.
https://doi.org/10.1007/978-3-319-71255-0_41

HPC applications on a single computing cluster due to incompatible resource managers and resource allocation policies. This necessitates the provision of dedicated computing infrastructure for such applications, e.g., Hadoop cluster. Third, the implementation of practical exercises is challenging due to the inherent complexity of involved systems and user interfaces. This is particularly true for undergraduate or non-technical students without prior Linux background. While the similar problem exists for traditional HPC systems, Big Data systems have specific interfaces that should be taken into account.

This paper reports an experience on solving the mentioned challenges while teaching two introductory PDC courses at the Yandex School of Data Analysis (YSDA) and the Higher School of Economics (HSE). The Parallel and Distributed Computing course at YSDA is an introductory PDC course for MSc students that features the following topics: concurrency, parallel programming and distributed data processing. The similar course in HSE is for BSc students from the Faculty of Computer Science. Both courses consider distributed computing models and platforms for processing of large data sets.

In particular, the paper describes the software infrastructure and high-level web services implemented in order to support practical exercises involving Big Data technologies. The presented service-based approach helped to provide a smooth practical experience for students with different technical background by enabling them to run and test their programs on a Hadoop cluster via convenient web interfaces. This approach also enabled automation of routine actions related to submission of programs to a cluster and evaluation of homework solutions.

The paper is structured as follows. Section 2 discusses related work. Section 3 provides an overview of the developed infrastructure. Section 4 describes the computing infrastructure and how it was adapted to accommodate both HPC and Big Data applications. Section 5 provides an overview of Everest, a web-based distributed computing platform used for building the presented services. Section 6 describes the generic services for running MapReduce and Spark programs and the problem-specific services for evaluating solutions of related programming assignments. Section 7 concludes and discusses future work.

2   Related Work

The use of web technologies for building convenient interfaces to HPC systems has been exploited since the emergence of the World Wide Web. For example, in [4] authors describe several prototypes of web-based parallel programming environments, including the Virtual Programming Laboratory (VPL) used for teaching parallel programming. The emergence of grid computing and the web portal technology enabled development of grid portals facilitating access to distributed computing facilities. For example, [5] describes an experience of building a grid portal to support an undergraduate parallel programming course.

The web-based interfaces have also been exploited to support submission and automated evaluation of programming assignments in PDC courses. For example, in [6] authors describe a framework enabling implementation of web portals for automated testing of student programming assignments in distributed programming courses. Among the recent works, [7] describes a web-based application for automated assessment and evaluation of source code in the field of parallel programming. In [8] authors present a similar web-based system for running and validating parallel programs written in different programming paradigms.

The web technologies are also being actively used nowadays for supporting Massive Open Online Courses (MOOC) with a large number of attendees. For example, WebGPU is a web-based system developed to support GPU programming assignments in the Heterogeneous Parallel Programming course [9]. In [10] authors describe the "Introduction to Parallel Computing" course that is developed on the base of Moodle learning management system and supports automatic evaluation of parallel programs.

While the previously mentioned systems support teaching traditional PDC topics, currently there exists only a few web-based environments focused on teaching Big Data technologies. The only similar project is the WebMapReduce (WMR) [11], which provides a simplified web interface to Hadoop designed for teaching the MapReduce computing model. The WMR portal allows students to write mappers and reducers in a variety of languages. The programs are executed on a Hadoop cluster or in a testing environment that mimics the behavior of Hadoop while running within a single thread. In contrast to WMR, the presented infrastructure is more generic by supporting other technologies and computing models beyond MapReduce, e.g., Spark, and addressing additional challenges such as automated evaluation of homework assignments.

In addition, a variety of open source and commercial systems are currently emerging that provide convenient web interfaces for working with Big Data technologies [12–14], including interactive notebooks and dashboards. While not specifically designed for teaching, these systems can also be used in educational activities. The presented infrastructure relies on one of such interfaces, namely Hue [12], for browsing the data stored on a Hadoop cluster.

3   Infrastructure Overview

A high-level overview of the infrastructure used to support practical exercises in the mentioned courses is presented on Fig. 1.

The computing infrastructure consists of a dedicated cluster with 20 nodes which is split into two partitions for running HPC and Big Data workloads. The students can optionally request a direct access to the cluster command line via SSH. However, the default way to access the cluster is via a set of provided web services that automate submission and execution of parallel programs on the cluster. There are two main types of such services. The so called generic services can be used to run arbitrary programs for some technology, e.g., MPI or MapReduce. There are also problem-specific services that can be used for submission and evaluation of solutions for homework assignments. The services are developed and deployed on Everest, a web-based distributed computing platform [15, 16] which supports integration with computing resources via special software

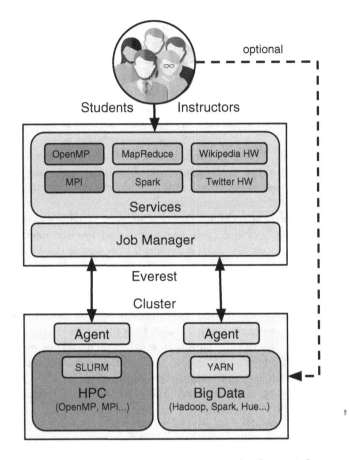

Fig. 1. Architecture of supporting computing and software infrastructure

agents. These agents are deployed on the cluster and are used by the Everest job manager for execution of programs submitted via the services.

The main advantage of the service-based approach is the ease of use and ubiquity in comparison to the command line environment. Such environment and queuing systems used on the cluster are unfamiliar and too low-level for many students. The execution of programs of the cluster also implies manual copying of required files that can be automated by the services, which is very convenient for quick demonstrations in class. Another advantage of the presented approach is the reduced administration overhead, since it does not require creation of cluster accounts for each student. The management of students in Everest can be automated by creating a dedicated user group and configuring a secret code for self-registration by the students. Finally, the use of special services for evaluation of homework assignments can provide an instant feedback for the students which enhances the learning experience.

4   Computing Infrastructure

As was previously mentioned, Big Data and HPC technologies use different resource managers and resource allocation policies. For example, while the execution of MPI applications on a cluster is usually managed by the batch system such as SLURM or PBS, the execution of MapReduce programs is managed by the YARN service, which is a part of Apache Hadoop platform. Also, while MPI programs allocate and use a fixed subset of cluster resources, MapReduce programs can dynamically allocate and release resources during their execution. Therefore it is very hard to use a single resource manager for both types of workloads.

To accommodate both HPC and Big Data applications the cluster was split into two separate partitions. The first partition, managed by the SLURM batch system and using the NFS file system, is dedicated for running HPC applications such as MPI and OpenMP programs. The second partition, managed by the YARN service and using the HDFS file system (also a part of Hadoop platform), is dedicated for running Big Data applications such as Hadoop MapReduce and Spark programs. The second partition also has a number of other Big Data technologies installed such as Hive, HBase and Kafka.

Having two separate cluster partitions brought an issue of efficient cluster utilization when one of the partitions is underutilized, for example when the students study MPI programming and use only the HPC partition. Currently the size of each partition can be changed by the administrator by manually stopping and starting the SLURM and YARN daemons on the cluster nodes. Given the known schedule of practical exercises by different courses using the cluster, the manual tuning of partition sizes proved to be sufficient. However, a more sophisticated automated tuning based on a current load can also be implemented in the future.

Both partitions have configured limits of resource usage per program which is essential in order to avoid the excessive use of cluster resources by inefficient or misbehaving programs. The HPC partition imposes a limit on the wall clock time used by a program. However, it is not possible to use a similar metric for Big Data applications since their run time can depend on the current cluster load. Therefore an alternative metric of consumed core-seconds was used to limit the resource consumption for the second partition. Since Hadoop YARN doesn't support enforcement of resource usage limits, a special script was developed that periodically checks the current resource consumption of running programs and kills those that exceeded the configured limits. The preemption in YARN scheduler is turned off in order to ensure stable execution and measurements, especially for Spark programs.

5   Everest Overview

Everest [15] is a web-based distributed computing platform used for building the services of the described infrastructure. In this section we provide a brief overview of this platform.

Everest provides users with tools to quickly publish and share computing applications as web services. The platform also manages execution of applications on external computing resources attached by users. In contrast to traditional distributed computing platforms, Everest implements the PaaS model by providing its functionality via remote web and programming interfaces. A single instance of the platform can be accessed by many users in order to create, run and share applications with each other. The platform is available online to all interested users [16].

Everest supports development and execution of computing applications following a common model. An application has a number of inputs that constitute a valid request to the application and a number of outputs that constitute a result of computation corresponding to some request. Upon each request Everest creates a new job consisting of one or more computational tasks generated by the application according to the job inputs. The tasks are executed by the platform on computing resources specified by a user.

To simplify creation of applications Everest provides a generic skeleton for command-line applications that makes it possible to avoid programming while adding an application. In addition to description of application inputs and outputs, the user should specify the command pattern parametrized by input values and describe the mappings between inputs/outputs and files read/produced by the application.

An application is automatically published as a RESTful web service with a unified interface. This enables programmatic access to applications, integration with third-party tools and composition of applications into workflows. The platform's web user interface also generates a web form for running the application via web browser. The application owner can manage the list of users that are allowed to run the application.

Instead of using a dedicated computing infrastructure, Everest performs execution of application tasks on external resources attached by users. The platform implements integration with standalone machines and clusters through a developed program called *agent*. The agent runs on the resource and acts as a mediator between it and Everest enabling the platform to submit and manage computations on the resource. Everest manages execution of tasks on remote resources and performs routine actions related to staging of input files, submitting a task, monitoring a task state and downloading task results.

6   Services

A number of web services have been developed using the Everest platform in order to simplify and automate execution of various types of parallel programs by the students on the cluster.

In order to create an application an instructor should specify via Everest Web UI application's metadata, input and output parameters, mapping of parameters to the executed command and files, etc. The core part of the application is a wrapper that takes input parameters and manages execution of a parallel program on the cluster. The wrapper can be written in any programming language since Everest runs it via command line. It usually performs program compilation, preparing of execution environment, submitting the program via queuing system, etc. The development of such wrapper is currently the most difficult part of the process, however once implemented its parts can be reused for other applications.

6.1   Generic Execution Services

The following generic services have been developed for execution of different types of programs using Big Data technologies on the cluster. These services can be used to run an arbitrary program of some specific type.

Two generic services were implemented for running Hadoop MapReduce programs. The first service supports programs written in Python using the Hadoop Streaming interface, targeting students without Java skills. The submit form of this service is presented on Fig. 2. The second service supports Java programs using the Hadoop Java API. Both services allow specifying program files, command line arguments, input and output paths in HDFS file system, number of reduce tasks and additional Hadoop options. The wrapper script performs submission of MapReduce job, monitors the job's state and updates status information displayed in Everest. When the job is running, a student is provided with a link to the job status page in the Hadoop web interface. After the job is completed the total resource usage in core-seconds is displayed along with a link to the job history interface with task logs. This provides enough information to troubleshoot failed programs or evaluate the program's efficiency.

Two similar services were implemented for running Apache Spark programs written in Python or Scala/Java on the cluster. In comparison to the MapReduce services, the Spark services have more sophisticated runtime parameters such as the number of executors, cores and memory per executor. It is also possible to specify the minimum ratio of registered executors to wait for before starting computations. This enables students to examine various trade-offs related to using different values of runtime parameters. The corresponding wrapper script is also more sophisticated. It allows to limit the maximum amount of physical resources requested by the program and the number of concurrent jobs per user. The wrapper script also computes the effective resource usage for a Spark program by excluding core-seconds spent while waiting for the executors.

Upon the program submission the student is redirected to the job page that displays dynamically updated information about the job state. The job page also includes sections containing general information about the job, inputs specified by the student and outputs produced by the job. For teaching purposes the services were configured to automatically share all jobs submitted by the students with the instructors group, so that in case of a problem a student can just send a link to a failed job to the instructor.

Due to the large size of input data and produced results, in addition to running programs on Hadoop cluster it was essential to provide a way to easily browse files stored in the HDFS file system without fully downloading them. This was achieved by using Hue [12], a web interface for Hadoop which includes

Hadoop MapReduce (Streaming)

About	Parameters	Submit Job	Discussion

Job Name

Hadoop MapReduce (Streaming)

Input path

/data/vm

Path to the file or directory in HDFS you want to use as the input data for the job.

Output path

/user/pdc/VASECHKIN/wordcount

Path to the directory in HDFS where you want to save the output of the job.

Mapper command

mapper.py

Command to use as mapper, e.g. "mapper.py"

Reducer command

reducer.py

Command to use as reducer, e.g. "reducer.py"

Combiner command

Optional command to use as combiner, e.g. "combiner.py"

Number of reducers

1

Specify the number of reduce tasks you want to use (maximum is 50). Specify zero if you do not want

Required files

[+ Add file...]

[+ Add item]

Specify files implementing mapper, reducer, or combiner.

Options

Additional options to pass to Hadoop.

Email Notification

☐ Send me email when the job completes

Your job will be automatically shared with: @pdc-instructors

Request JSON

► Submit

Fig. 2. The submit form of the service for running Hadoop MapReduce programs

a convenient HDFS file browser. Hue also provides a web interface for running jobs, however it is more complicated and low level in comparison to the developed services.

6.2 Services for Programming Assignments

The evaluation of programming assignments requires a significant effort and is one of the key scalability bottlenecks in terms of a number of students. The generic services described above can be used for quick demonstrations, practical exercises and projects. However, they usually do not provide a feedback needed to validate solutions to programming assignments. For example, whether the program produced a correct result or has a good performance. Such immediate feedback is crucial for students since it helps to avoid manual validation and to focus on the solution. This feedback can also help instructors to reduce the time and effort needed to grade the solution.

A set of problem-specific services have been implemented for automated evaluation of homework assignments related to Big Data technologies. These services are implemented on Everest using the same approach as the previously discussed generic execution services. However, in this case the wrapper is replaced by a test suite for the given assignment.

The first assignment is dedicated to the MapReduce programming model and its implementation in Hadoop. The students should use MapReduce to build inverted index of the contents of Wikipedia pages. The solution of this assignment requires multiple MapReduce steps such as computing a list of frequent words excluded from the index and building an index itself for English and Russian versions of Wikipedia. Since it is difficult to implement an interface for specifying and running all these steps, the provided service doesn't perform the execution of solutions and only checks the provided results. The student should pass to the service the HDFS paths to the produced indexes. The service runs a script that checks that the index conforms to all requirements specified in the assignment. The students should include the link to test results in the homework report along with the links to all program runs via generic MapReduce services used to build the indexes. Instructors can view all programs created by a student by following these links in Everest. The generic services and job history web interfaces provide enough information to evaluate the efficiency of each program.

The second assignment is dedicated to using Apache Spark and its Resilient Distributed Datasets (RDD) programming model. The students should compute a number of results given a graph of follower relationships between Twitter users, such as the average count of followers, the most popular users and the number of users that can be reached by a tweet from popular users. The solution of this assignment also requires multiple steps, however, in contrast to MapReduce, these steps can be run as a single job in Spark. Nevertheless, to provide the maximum flexibility, the similar approach was used as in the previous assignment by implementing a service that only checks the produced results. This enabled students to incrementally compute and check different results. Again the students were asked to provide links to all submissions via generic services used to produce all results.

7   Conclusion

The paper presented an experience in incorporating Big Data technologies into introductory PDC courses and building a service-oriented infrastructure to support practical exercises involving these technologies. The presented approach helped to provide a smooth practical experience for students with different technical background by enabling them to run and test their MapReduce and Spark programs on a provided Hadoop cluster via convenient web interfaces. This approach also enabled automation of routine actions related to submission of programs to a cluster and evaluation of programming assignments.

Future work will focus on improving the presented infrastructure and publishing the service implementations to enable other educators to reproduce the presented approach using the Everest platform.

Acknowledgments. This work is supported by the Russian Science Foundation (project No. 16-11-10352).

References

1. Dean, J., Ghemawat, S.: MapReduce: simplified data processing on large clusters. Commun. ACM **51**(1), 107–113 (2008)
2. White, T.: Hadoop: The Definitive Guide. O'Reilly Media Inc, Sebastopol (2012)
3. Zaharia, M., Chowdhury, M., Franklin, M.J., Shenker, S., Stoica, I.: Spark: cluster computing with working sets. HotCloud **10**(10–10), 95 (2010)
4. Dincer, K., Fox. G.C.: Design issues in building web-based parallel programming environments. In: 1997 Proceedings of the Sixth IEEE International Symposium on High Performance Distributed Computing, pp. 283–292. IEEE (1997)
5. Tourino, J., Martin, M.J., Tarrio, J., Arenaz, M.: A grid portal for an undergraduate parallel programming course. IEEE Trans. Educ. **48**(3), 391–399 (2005)
6. Maggi, P., Sisto, R.: A grid-powered framework to support courses on distributed programming. IEEE Trans. Educ. **50**(1), 27–33 (2007)
7. Schlarb, M., Hundt, C., Schmidt, B.: SAUCE: a web-based automated assessment tool for teaching parallel programming. In: Hunold, S., Costan, A., Giménez, D., Iosup, A., Ricci, L., Gómez Requena, M.E., Scarano, V., Varbanescu, A.L., Scott, S.L., Lankes, S., Weidendorfer, J., Alexander, M. (eds.) Euro-Par 2015. LNCS, vol. 9523, pp. 54–65. Springer, Cham (2015). https://doi.org/10.1007/978-3-319-27308-2_5
8. Nowicki, M., Marchwiany, M., Szpindler, M., Bała, P.: On-line service for teaching parallel programming. In: Hunold, S., Costan, A., Giménez, D., Iosup, A., Ricci, L., Gómez Requena, M.E., Scarano, V., Varbanescu, A.L., Scott, S.L., Lankes, S., Weidendorfer, J., Alexander, M. (eds.) Euro-Par 2015. LNCS, vol. 9523, pp. 78–89. Springer, Cham (2015). https://doi.org/10.1007/978-3-319-27308-2_7
9. Heterogeneous Parallel Programming. https://www.coursera.org/course/hetero
10. Gergel, V., Kustikova, V.: Internet-oriented educational course "Introduction to Parallel Computing": a simple way to start. In: Voevodin, V., Sobolev, S. (eds.) RuSCDays 2016. CCIS, vol. 687, pp. 291–303. Springer, Cham (2016). https://doi.org/10.1007/978-3-319-55669-7_23
11. Garrity, P., Yates, T., Brown, R., Shoop, E.: Webmapreduce: an accessible and adaptable tool for teaching map-reduce computing. In: Proceedings of the 42nd ACM Technical Symposium On Computer Science Education, pp. 183–188. ACM (2011)
12. Hue. http://gethue.com/
13. Databricks Platform. https://databricks.com/product/databricks
14. Cloudera Data Science Workbench. https://www.cloudera.com/products/data-science-and-engineering/data-science-workbench.html

15. Sukhoroslov, O., Volkov, S., Afanasiev, A.: A web-based platform for publication and distributed execution of computing applications. In: 2015 14th International Symposium on Parallel and Distributed Computing (ISPDC), pp. 175–184, June 2015

16. Everest. http://everest.distcomp.org/

JobDigest – Detailed System Monitoring-Based Supercomputer Application Behavior Analysis

Dmitry Nikitenko[(✉)] ⓘ, Alexander Antonov ⓘ, Pavel Shvets ⓘ, Sergey Sobolev ⓘ,
Konstantin Stefanov ⓘ, Vadim Voevodin ⓘ, Vladimir Voevodin ⓘ,
and Sergey Zhumatiy ⓘ

Research Computing Center, Lomonosov Moscow State University,
Moscow 119234, Russian Federation
{dan,asa,shpavel,sergeys,cstef,vadim,voevodin,serg}@parallel.ru

Abstract. The efficiency of computing resources utilization by user applications can be analyzed in various ways. The JobDigest approach based on system monitoring was developed in Moscow State University and is currently used in everyday practice of the largest Russian supercomputing center of Moscow State University. The approach features application behavior analysis for every job run on HPC system providing: the set of dynamic application characteristics - time series of values representing utilization of CPU, memory, network, storage, etc. with diagrams and heat maps; the integral characteristics representing average utilization rates; job tagging and categorization with means of informing system administrators and managers on suspicious or abnormal applications. The paper describes the approach principles and workflow, it also demonstrates JobDigest use cases and positioning of the proposed techniques in the set of tools and methods that are used in the MSU HPC Center to ensure its 24/7 efficient and productive functioning.

Keywords: HPC · Supercomputing · Efficient computing · Resource utilization · Job dynamics · Application efficiency · Parallel programming

1 Introduction

The JobDigest[1] approach follows monitoring of HPC application performance principles – one of the possible ways of resource utilization efficiency analysis for both applications and supercomputers. Studying of the resource utilization type and rate, determining bottlenecks in programs, hardware and/or their interplay are the typical usage scenarios for such methods. The characteristics of the HPC system components state serve the basis for these approaches. For example, various CPU load types, incoming and outgoing network traffic on the node, number of floating point or integer operations, accelerator usage, memory-related operations like number of load and store operations, misses in the cache memory of different levels and so on, input/output activity, and many other characteristics.

---

[1] The JobDigest® is a registered trademark in Russian Federation. The application for an invention of the JobDigest approach was filed.

© Springer International Publishing AG 2017
V. Voevodin and S. Sobolev (Eds.): RuSCDays 2017, CCIS 793, pp. 516–529, 2017.
https://doi.org/10.1007/978-3-319-71255-0_42

Most of available monitoring systems do not process the data at computing nodes to reduce the impact on job execution [1–7]. At the same time there are uprising monitoring systems that reasonably distribute the processing of data and secure low influence on job execution [8, 9]. The key advantage of monitoring-based approaches is general independence on the code, absence of necessity to make any changes to the program source code or binary that is being analyzed. This allows reducing the influence on program execution and at the same time expanding the range of jobs that can be analyzed.

Different approaches to job performance monitoring are known [10–17]. In this paper the details of the JobDigest approach to supercomputer job analysis based on system monitoring data is described. According to the approach principles, the special report that gives all-round view over the job behavior built of diverse dynamic and integral characteristics is generated for every job, even if it has not successfully finished.

2 Approach Principles

The JobDigest report represents the detailed information of job behavior starting from basic job information to the precise info on computing, storage and network resource utilization.

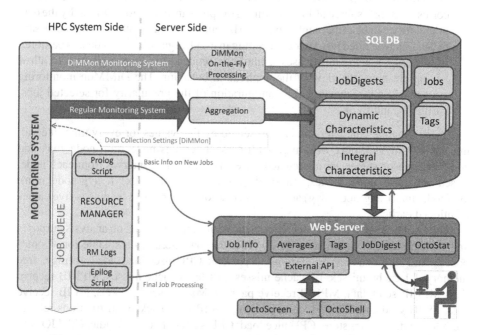

Fig. 1. General proposed workflow based on system monitoring data analysis.

The key feature of the proposed approach is the possibility of job behavior analysis for any and every run with no code preparations by the user [18, 19].

The JobDigest is built on the basis of system monitoring data and job details from the resource manager [20]. As soon as the job is assigned to the set of nodes, it becomes possible to bind the system monitoring data collected from the corresponding nodes to form the profile of application execution (Fig. 1).

It is supposed that only one job at a time is allocated to the node. If a node can be assigned to several jobs, such a situation can also be handled to some extent but only if process to core pinning is enabled.

3 The JobDigest Report

The report consists of several blocks that focus on various scopes of analysis. Altogether the blocks shape all-round basis for application behavior study (Fig. 2).

A. **General job information.** The table includes the general information on job and its node allocation from the resource manager: job ID, user account, run command, working directory, output file, partition, time limit, submit/start/end timestamps, status, duration, number of allocated cores and nodes, core hours, list of allocated nodes.

B. **Dynamical job characteristics.** Dynamic characteristics represent the rate of corresponding resource utilization during the program execution. Originally the data granularity can be rather high, up to 10 Hz, but for the most jobs it is not necessary to have such detailed information, keeping in mind large volumes of data to be stored in this case. For the most cases, granularity reduced to 5 min is enough to allow having clear overall view of the application behavior. The DiMMon monitoring system will allow dynamical reconfiguration of the granularity for selected jobs, partitions, and so on.

Every dynamic characteristic is represented by five values for every time interval: min, max, min_avg, max_avg, avg. With present settings node_min, node_max and node_avg sensor values are aggregated for every 5 min time interval from each node. Based on these three values, min(node_min), max(node_max) и avg(node_min), avg(node_max), avg(node_avg) are calculated across all job nodes leading to five values mentioned earlier.

The available set of dynamic characteristics can vary depending on analysis purposes and system settings. There are over 20 different characteristics available for «Lomonosov» and «Lomonosov-2» systems at present: CPU user load, load average, free memory, L1 cache misses, L2 cache misses, last level cache misses, MPI IB receive data, MPI IB send data, MPI IB receive packets, MPI IB send packets, FS IB receive data, FS IB send data, FS IB receive packets, FS IB send packets, instructions retired, memory load, memory store, CPU nice load, CPU system load, CPU idle, CPU IO wait load, CPU IRQ load, CPU soft IRQ load, etc.

There are three ways of studying dynamic characteristics supported by JobDigest at present: diagrams, CSV tables for using with external analysis and visualization tools, and heat maps. Heat maps are 2D charts. Horizontal axis corresponds to time, and vertical axis corresponds to used nodes. The dot color represents dynamic characteristic

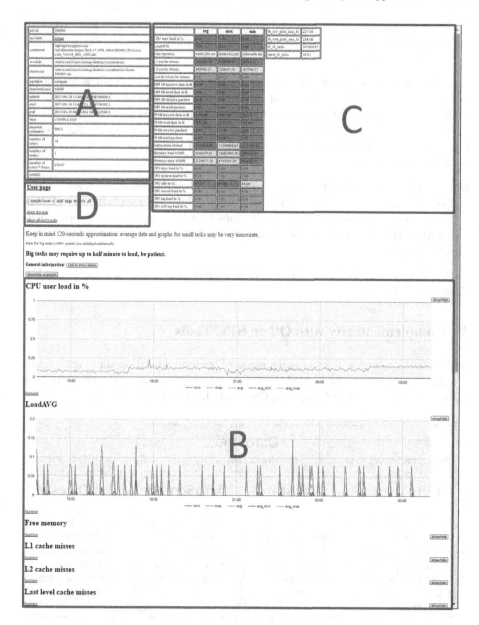

Fig. 2. JobDigest report blocks: A – general job information, B – dynamic job characteristics, C – integral job characteristics, D – tags and job categories.

value from minimum (red color) to maximum (green color) among all the values during the job execution. An example of such heat map is shown in Fig. 8.

C. **Integral job characteristics.** The integral job characteristics represent average resource utilization and are built on the base of dynamic job characteristics. Every integral characteristic is provided as a set of minimum, average, and maximum levels during the whole job run. Every of these three values can be highlighted according to preset and calculated thresholds.

In the same block there are also given some other characteristics built as derivatives of averages. For example, the following average characteristics can be useful and are available in the report: IB receive packet size for FS, IB receive packet size for MPI, IB send packet size for FS, IB send packet size for MPI, and some memory-related characteristics like level 1 to level 3 cache miss ratio and memory load plus memory store to level 1 cache miss ratio.

D. **Job categories and tags.** Based on calculated integral characteristics and general job information every job is tagged after it is finished. The tags help to divide jobs into categories, helping to find specific jobs later by category or categories intersection. The tags mark the scale of the job, partition, duration, resource utilization specifics, etc.

The examples of the report and its blocks are provided in the "Evaluation and Use Cases" section of the paper.

4   Complementarity with Other HPC Tools

JobDigest interaction with other components of the toolkit mentioned in Figs. 1 and 3 is widely used every day in the MSU HPC Center [21]. All these components have been developed in the Research Computing Center of Moscow State University.

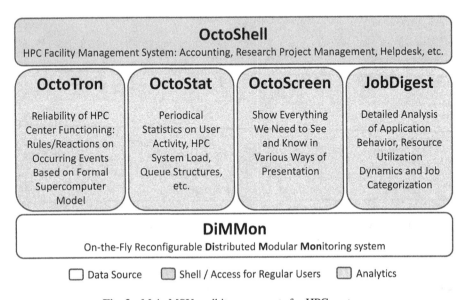

Fig. 3. Main MSU toolkit components for HPC centers.

DiMMon. This monitoring system is a promising scalable reconfigurable data collector with elements of on-the-fly analysis. In comparison to other monitoring systems, it allows performing much of data processing before saving source data to the database. For example, it allows creating JobDigest reports and calculating averages for the running jobs.

OctoScreen. Administrators and system managers can proceed to the more detailed job JobDigest reports from specifically generated job lists provided by OctoScreen. Such lists can be formed by various criteria, for example, by job owner, responsible organization for the research project that the jobs belong to, geographical criteria, research areas and so on [22].

OctoStat. Provided by Octostat analysis of daily statistics on queue structure and top resource consuming projects, users and accounts can be successfully amplified having the possibility to look at the details of any job owned by a specific account, user or research project. This is especially valuable for suspicious jobs that are found according to extremely low activity or suspicious node allocation.

OctoTron. The resilience system logs all problems with storage, network, compute nodes and the interfaces [23]. If something goes wrong, it is possible to track the jobs that could have been harmed by the issues with the known node set in the specified time period and create the list of potentially affected job runs. One can see more details to find out if there was really something wrong with the job proceeding to the JobDigest reports of any of these jobs.

OctoShell. If the HPC system is not a dedicated one and there is a number of users and research projects that should not be allowed to see the working results and activity details of each other, it is essential to have a special system that would serve as a single entry point for all or, at least, most of services. The OctoShell system is used in the MSU HPC Center for such purposes [24, 25]. All the information on job runs is provided to users through this system according to user access permissions and settings in the account area on the website [26].

As shown above, JobDigest is highly integrated in the HPC center administration and management workflow as a valuable analytical part of the used toolkit.

5 Implementation Details

The workflow of the current version of JobDigest is presented on Fig. 4. The numbers in circles present the sequence of stages.

1. System monitoring data from agents of monitoring system on the nodes of HPC system is sent to the aggregation service.
2. The aggregation service filters data and reduces granularity. Data is further saved into the database. It is not pinned to jobs yet.

3. As soon as job starts, prolog script of resource manager informs server-side application on the web server on job run. Server-side script writes the initial info into the database.

4. When the job finishes, epilog script informs server-side application on the web server on job end. Server-side script updates the job record in the database and initiates job data processing in background mode: integral characteristics are being calculated and written into the database. Job tagging is performed afterwards based on integral characteristics and general job information, the tags assignment is written into the database.

5. Special script gathers information on all run jobs and checks if all the jobs are represented in the database. This is done once per day to tackle possible problems with network during the job run and other issues that could have prevented the job info getting processed.

6. External services and applications can access job information and monitoring data, as well as daily statistics data in JSON format via HTTP protocol using special API. For example, regular users access the JobDigest reports and get job run statistics via OctoShell.

7. Administrators can access job info and corresponding monitoring data, job lists, extended JobDigest reports, statistics and special visualization templates as HTML pages by HTTP with authorization.

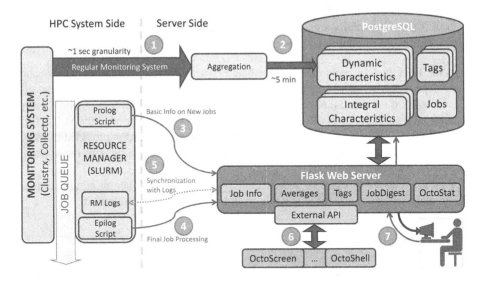

Fig. 4. JobDigest workflow stages.

At present the following software components are used:

– PostgreSQL, version 8.4.20;
– Flask web server, version 0.11.1;
– SLURM resource manager [27], «Lomonosov» HPC system – v.2.5.6, «Lomonosov – 2» system - v.15.0.8.1;

- Modified ClustrXWatch agents are used now as the data source [28];
- Google Charts are used for diagrams in the JobDigest reports;
- Highcharts are used for heat map generation.

The latest versions of the JobDigest components are available at GihHub [29]. As for now the custom installation is a bit tricky. As soon as the development of DiMMon monitoring system allows, the out-of-the-box package will be available.

6 Evaluation and Use Cases

The first approach evaluation was achieved as a result of MSU team efforts under the joint RU-EU HOPSA project [30].

The described approach is now widely used in the everyday practice of the Moscow State University Supercomputer Center, the largest HPC collaborative facility in Russia having «Lomonosov» and «Lomonosov-2» systems with a total of over 4 PFlops peak performance at present, over 500 collaborative research projects and thousands of scientists using the computing facility in 24/7 mode. This results in processing of thousands of supercomputer jobs per day [31], and the special JobDigest report generation is provided for any and every of those. In this section we illustrate some examples of using JobDigest report and its data.

6.1 Integral Job Characteristics

Every JobDigest report contains a block with integral job characteristics that represent general resource utilization rate for the whole run. It is typically MIN, MAX and AVG values for every dynamic characteristic available, see the description in subsection C of Sect. 3 of the paper.

These integral characteristics are available not only from inside the report. Most of them are provided together with the job list according to access permissions of the user (own jobs for user, selected jobs for the experts, or all jobs for administrators). In many cases the general average resource utilization rate is already can be sufficient for preliminary analysis and/or job selection for detailed study.

Figure 5 shows the example of such a table with integral characteristics that can help to find the hanged jobs among the currently run jobs according to extremely low CPU_user.

Lomonosov running tasks

id	account	partition	t_start	num_cores	avg cpu_user	avg loadavg	comment
1109414	9_jwcjwnku	regular4	2017-05-18 13:04:46	128	0.08	0.01	HANGED?!
1409111	dhannerv	gpu	2017-05-15 16:30:45	128	60.97	7.78	
1440012	scuznz_1309	regular6	2017-05-15 22:27:30	48	46.87	12.00	
1440013	nakrezhwaod_1719	regular6	2017-05-16 07:24:30	48	40.11	10.00	
1409030	seiabeznok_5749	regular6	2017-05-15 20:53:04	48	47.52	12.00	
1440005	glui3affaya	regular6	2017-05-16 01:41:40	192	35.89	3.01	

Fig. 5. JobDigest list with suspicious jobs.

The layout of such a job list can vary and can contain many integral characteristics at a time.

One of the possible promising ways of using integral characteristics is application scalability analysis that can be performed analyzing changing integral job characteristics of sequence of runs [32].

6.2 Job Categories and Tags

Integral characteristics values can serve as the basis for job categorization and tagging according to belonging to the specific job group or type [33]. For example (Fig. 6), if no GPU usage is observed in GPU partition job, one can suspect a cheating - the regular CPU job could have been intentionally put to GPU partition just because of the shorter queue, blocking GPU resources for GPU oriented jobs.

Lomonosov task table

id	account	t_start	t_end	state	cores_hour	num_core	duration	partition	cpu_use	loadavg	gpu_load	ib_rcv_data
		2017-04-26 08:51:19	2017-04-26 16:02:47	CANCELLED	1840.92	256	431.47	gpu	0.08	0.04	0.00	1543.59
		2017-04-23 15:25:26	2017-04-26 15:25:49	TIMEOUT	9216.82	128	4320.38	gpu	44.33	8.04	0.00	4257495.73
		2017-04-23 08:51:03	2017-04-26 08:51:16	TIMEOUT	18432.92	256	4320.22	gpu	44.79	8.01	0.00	4178453.95
		2017-04-23 00:08:06	2017-04-25 22:08:20	TIMEOUT	8960.50	128	4200.23	gpu	43.65	8.01	0.00	272222397.13
		2017-04-24 13:29:34	2017-04-25 16:26:55	COMPLETED	862.59	32	1617.35	gpu	43.29	8.00	0.00	37734.45
		2017-04-24 12:58:47	2017-04-25 15:21:39	COMPLETED	2110.49	80	1582.87	hdd4,regular6,gpu	34.41	7.99	0.00	128912186.16
		2017-04-24 07:40:16	2017-04-25 05:26:57	COMPLETED	1742.24	80	1306.68	hdd6,regular6,gpu	35.91	7.97	0.00	139101891.21
		2017-04-24 09:20:05	2017-04-24 12:58:47	COMPLETED	291.60	80	218.70	hdd6,regular6,gpu	34.88	7.98	0.00	131083555.43
		2017-04-21 09:39:58	2017-04-24 09:20:05	COMPLETED	3440.09	48	4300.12	gpu	43.14	8.04	0.00	40936.26
		2017-04-21 09:39:49	2017-04-24 07:40:14	TIMEOUT	8960.89	128	4200.42	gpu,regular4	43.67	8.00	0.00	220447106.14
		2017-04-21 16:36:08	2017-04-24 01:36:08	TIMEOUT	4560.00	80	3420.00	regular4,regular6,gpu	34.26	7.99	0.00	126557870.06

<< Prev Next >>

Fig. 6. JobDigest list of "low_gpu_load" tagged categories.

Another promising way of job categorization and finding job anomalies bases on data mining principles and processing of historical dynamic job characteristics, the first results are described in [34].

As shown above, the job lists can be equipped with basic filtration tools by time, account name, categories (tags), and can contain adjustable number of general and integral job characteristics. In JobDigest, the job categories and tags are shown as described in subsection D of Sect. 3.

6.3 Dynamic Job Characteristics

The dynamics of execution and behavior of any job can be best observed by analyzing the behavior of dynamic characteristics and their interplay. There are three general modes of analysis available: CSV export for external tools, diagrams, and heat maps.

Figure 7 illustrates changing of dynamic job characteristics during the job execution with over 200 processes used. The blue color lines (upper) correspond to maximum values of all the processes, green lines correspond to averages. In the second part of job execution it can be seen that the average values of all characteristics become very low - almost drop to zero level. At the same time the maximum value still represents normal activity. Knowing that the number of processes of the job was about several hundred, such a behavior can be possibly explained by the activity of just a few processes while the rest hundreds of processes are in a wait state. In any case, such a behavior found using the JobDigest report is suspicious.

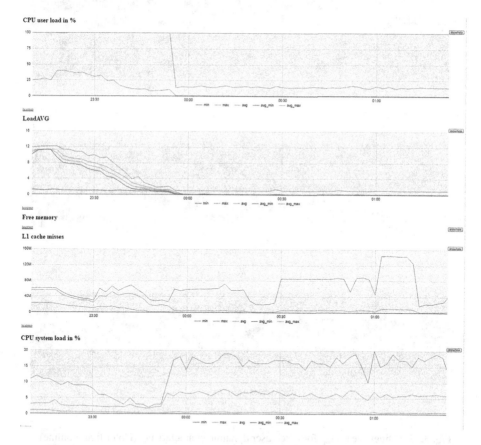

Fig. 7. JobDigest dynamic characteristic diagrams example. (Color figure online)

The main goal of the JobDigest report is to detect bottlenecks and application performance degradation for any job in whole job flow, and special tools are required for further deep analysis to locate exact reasons in the program.

These diagrams are available in JobDigests as described in subsection B of Sect. 3 of the paper.

Heat maps. Another useful way of analyzing dynamic characteristics in the JobDigest report is studying heat maps. Figure 8 represents the heat map for CPU_user of the previous example. Only one node activity is clearly seen on the second stage of program run by the max values heat map, and looking at average values heat map it can be supposed that there was just one core used on that single node.

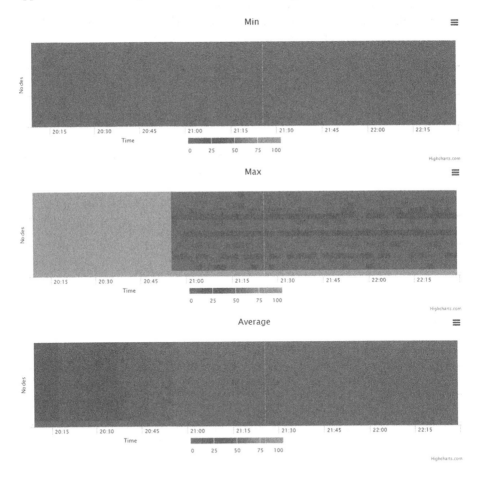

Fig. 8. JobDigest heat map for CPU_user dynamic characteristic. (Color figure online)

Interesting examples of JobDigest usage are found everyday by support team of MSU supercomputer center. Some of those have been described in publications as a result of performed research on the reasons of performance degradation and specifics of resource utilization by various program types and program models [35].

7 Conclusions

The practiced approach based on system monitoring data analysis features possibility to have an in-depth view into every job launch without any changes to the program source codes or binaries to detect application performance issues.

The developed JobDigest reports provide valuable all-round view over the application execution behavior and resource utilization profile and rate.

All the described tools are developed in Moscow State University and are currently being used in the MSU HPC Center in 24/7 mode. The proposed system monitoring-based approach has been also evaluated in Uppsala University (Sweden) and showed promising results, Uppsala University team and SNIC support are hereby highly acknowledged.

The developed tools are available as an open source software, contributions are welcome.

Acknowledgements. The results were obtained in the Research Computing Center of M.V. Lomonosov Moscow State University. The work is funded in part by the Russian Found for Basic Research, grants №17-07-00719, №16-07-01121, and Russian Presidential study grant (SP-1981.2016.5).

References

1. Zenoss. http://www.zenoss.org. Last accessed 10 May 2017
2. Zabbix. http://www.zabbix.com. Last accessed 10 May 2017
3. Cacti®. http://www.cacti.net. Last accessed 10 May 2017
4. Massie, M.L., et al.: The ganglia distributed monitoring system: design, implementation, and experience. Parallel Comput. **30**(7), 817–840 (2004)
5. The OpenNMS project. http://www.opennms.org. Last accessed 10 May 2017
6. Nagios - the industry standard in IT infrastructure monitoring. http://www.nagios.org. Last accessed 10 May 2017
7. Collectd – The system statistics collection daemon. https://collectd.org. Last accessed 10 May 2017
8. Stefanov, K.S., Voevodin, Vl.V.: Distributed modular monitoring (DiMMon) approach to supercomputer monitoring. In: Proceedings of the 2015 IEEE International Conference on Cluster Computing, pp. 502–503. IEEE (2015). https://doi.org/10.1109/CLUSTER.2015.83
9. Stefanov, K.S., Voevodin, Vl.V., Zhumatiy, S.A., Voevodin, Vad.V.: Dynamically reconfigurable distributed modular monitoring system for supercomputers (DiMMon). Procedia Comput. Sci. **66**, 625–634 (2015). Elsevier B.V. https://doi.org/10.1016/j.procs.2015.11.071
10. Gunter, D., Tierney, B., Jackson, K., Lee, J., Stoufer, M.: Dynamic monitoring of high-performance distributed applications. In: Proceedings of the 11th IEEE International Symposium on High Performance Distributed Computing, pp. 163–170 (2002)

11. Mellor-Crummey, J., Fowler, R.J., Marin, G., Tallent, N.: HPCVIEW: a tool for top-down analysis of node performance. J. Supercomput. **23**(1), 81–104 (2002)

12. Jagode, H., Dongarra, J., Alam, S., Vetter, J., Spear, W., Malony, A.D.: A holistic approach for performance measurement and analysis for petascale applications. In: Allen, G., Nabrzyski, J., Seidel, E., van Albada, G.D., Dongarra, J., Sloot, P.M.A. (eds.) ICCS 2009, Part II. LNCS, vol. 5545, pp. 686–695. Springer, Heidelberg (2009). https://doi.org/10.1007/978-3-642-01973-9_77

13. Adhianto, L., Banerjee, S., Fagan, M., Krentel, M., Marin, G., Mellor-Crummey, J., Tallent, N.R.: HPCTOOLKIT: tools for performance analysis of optimized parallel programs. Concurrency Comput. Pract. Exp. **22**(6), 685–701 (2010)

14. Eisenhauer, G., Kraemer, E., Schwan, K., Stasko, J., Vetter, J., Mallavarupu, N.: Falcon: on-line monitoring and steering of large-scale parallel programs. In: Proceedings of the Fifth Symposium on the Frontiers of Massively Parallel Computation, pp. 422–429 (1995)

15. Kluge, M., Hackenberg, D., Nagel, W.E.: Collecting distributed performance data with dataheap: generating and exploiting a holistic system view. Procedia Comput. Sci. **9**, 1969–1978 (2012)

16. Mooney, R., Schmidt, K.P., Studham, R.S.: NWPerf: a system wide performance monitoring tool for large Linux clusters. In: 2004 IEEE International Conference on Cluster Computing (IEEE Cat. No.04EX935), pp. 379–389 (2004)

17. Ries, B., et al.: The paragon performance monitoring environment. In: Proceedings of Supercomputing 1993, pp. 850–859 (1993)

18. Nikitenko, D.A.: Complex approach to performance analysis of supercomputer systems based on system monitoring data. Numer. Meth. Programm. New Comput. Technol. **15**, 85–97 (2014)

19. Antonov, A.S., Zhumatiy, S.A., Nikitenko, D.A., Stefanov, K.S., Teplov, A.M., Shvets, P.A.: Analysis of dynamic characteristics of job stream on supercomputer system. Numer. Meth. Program. New Comput. Technol. **14**(2), 104–108 (2013)

20. Nikitenko, D.A., Adinets, A.V., Bryzgalov, P.A., Stefanov, K.S., Voevodin, Vad.V., Zhumatiy, S.A.: Job Digest - approach to analysis of application dynamic characteristics on supercomputer systems. Numer. Meth. Program. New Comput. Technol. **13**, 160–166 (2012)

21. Voevodin, V., Voevodin, V.: Efficiency of exascale supercomputer centers and supercomputing education. In: Gitler, I., Klapp, J. (eds.) ISUM 2015. CCIS, vol. 595, pp. 14–23. Springer, Cham (2016). https://doi.org/10.1007/978-3-319-32243-8_2

22. Nikitenko, D.A., Zhumatiy, S.A., Shvets, P.A.: Making large-scale systems observable — another inescapable step towards exascale. Supercomput. Front. Innovations **3**(2), 72–79 (2016). https://doi.org/10.14529/jsfi160205

23. Antonov, A., Nikitenko, D., Shvets, P., Sobolev, S., Stefanov, K., Voevodin, V., Voevodin, V., Zhumatiy, S.: An approach for ensuring reliable functioning of a supercomputer based on a formal model. In: Wyrzykowski, R., Deelman, E., Dongarra, J., Karczewski, K., Kitowski, J., Wiatr, K. (eds.) PPAM 2015, Part I. LNCS, vol. 9573, pp. 12–22. Springer, Cham (2016). https://doi.org/10.1007/978-3-319-32149-3_2

24. Nikitenko, D.A., Voevodin, Vl.V., Zhumatiy, S.A.: Resolving frontier problems of mastering large-scale supercomputer complexes. In: ACM International Conference on Computing Frontiers (CF 2016), Como, Italy, pp. 349–352. ACM, New York, 16–18 May 2016. https://doi.org/10.1145/2903150.2903481

25. Nikitenko, D.A., Voevodin, Vl.V., Zhumatiy, S.A.: Octoshell: large supercomputer complex administration system. In: Russian Supercomputing Days International Conference, Moscow, Russia. CEUR Workshop Proceedings, vol. 1482, pp. 69–83, 28–29 September 2015

26. Nikitenko, D., Stefanov, K., Zhumatiy, S., Voevodin, V., Teplov, A., Shvets, P.: System monitoring-based holistic resource utilization analysis for every user of a large HPC center. In: Carretero, J., et al. (eds.) ICA3PP 2016. LNCS, vol. 10049, pp. 305–318. Springer, Cham (2016). https://doi.org/10.1007/978-3-319-49956-7_24
27. Slurm Workload Manager. http://slurm.schedmd.com. Last accessed 10 May 2017
28. Clustrx. http://www.hpcc.unical.it/hpc2010/ctrbs/tkachev.pdf. Last accessed 10 May 2017
29. JobDigest components. https://github.com/srcc-msu/job_statistics. Last accessed 10 May 2017
30. Mohr, B., Hagersten, E., Giménez, J., Knüpfer, A., Nikitenko, D., Nilsson, M., Servat, H., Shah, A., Voevodin, Vl., Winkler, F., Wolf, F., Zhukov, I.: The HOPSA workflow and tools. In: Proceedings of the 6th International Parallel Tools Workshop, Stuttgart (2012)
31. Voevodin, Vl.V., Zhumatiy, S.A., Sobolev, S.I., Antonov, A.S., Bryzgalov, P.A., Nikitenko, D.A., Stefanov, K.S., Voevodin, Vad.V.: Practice of "Lomonosov" Supercomputer. Open Systems J. **7**, 36–39 (2012). Open Systems Publ., Moscow
32. Antonov, A., Teplov, A.: Generalized approach to scalability analysis of parallel applications. In: Carretero, J., et al. (eds.) ICA3PP 2016. LNCS, vol. 10049, pp. 291–304. Springer, Cham (2016). https://doi.org/10.1007/978-3-319-49956-7_23
33. Nikitenko, D.A., Voevodin, Vl.V., Voevodin, Vad.V., Zhumatiy, S.A., Stefanov, K.S., Teplov, A.M., Shvets, P.A.: Supercomputer application integral characteristics analysis for the whole queued job collection of large-scale HPC systems. In: 10th Annual International Scientific Conference on Parallel Computing Technologies, PCT 2016. CEUR Workshop Proceedings, vol. 1576, Arkhangelsk, Russian Federation, pp. 20–30, 29–31 March 2016
34. Voevodin, Vl.V., Voevodin, Vad.V., Shaikhislamov, D.I., Nikitenko, D.A.: Data mining method for anomaly detection in the supercomputer task flow: numerical computations: theory and algorithms. In: The 2nd International Conference and Summer School, Pizzo calabro, Italy. AIP Conference Proceedings, vol. 1776, pp. 090015-1–090015-4, 20–24 June 2016. https://doi.org/10.1063/1.4965379
35. Andreev, D.Yu., Antonov, A.S., Voevodin, Vad.V., Zhumatiy, S.A., Nikitenko, D.A., Stefanov, K.S., Shvets, P.A.: A system for the automated finding of inefficiencies and errors in parallel programs. Numer. Meth. Program. New Comput. Technol. **14**(2), 48–53 (2013)

Author Index

Printed in the United States
by Bookmasters

Printed in the United States
By Bookmasters